CATALYST PREPARATION
Science and Engineering

Edited by
John Regalbuto

CRC Press
Taylor & Francis Group
Boca Raton London New York

CRC Press is an imprint of the
Taylor & Francis Group, an informa business

CRC Press
Taylor & Francis Group
6000 Broken Sound Parkway NW, Suite 300
Boca Raton, FL 33487-2742

First issued in paperback 2020

ISBN-13: 978-0-367-57772-8 (pbk)
ISBN-13: 978-0-8493-7088-5 (hbk)

Library of Congress Card Number 2006016073

Library of Congress Cataloging-in-Publication Data

Catalyst preparation: Science and Engineering / edited by John Regalbuto.
    p. cm.
  Includes bibliographical references and index.
  ISBN-13: 978-0-8493-7088-5 (alk. paper)
  ISBN-10: 0-8493-7088-4
  1. Catalysts--Handbooks, manuals, etc. 2. Catalysts--Synthesis--Handbooks, manuals, etc. I. Regalbuto, John R. (John Robert), 1959-

QD501.H216 2007
541'.395--dc22
    2006016073

Visit the Taylor & Francis Web site at
http://www.taylorandfrancis.com

and the CRC Press Web site at
http://www.crcpress.com

Dedication

This book is dedicated to the memory of Professor James A. Schwarz. At his November 8, 1999 lecture to the Catalysis Club of Chicago ("The Nature of Protons on the Surfaces of Catalytic Materials") Professor Schwarz was introduced by Professor Regalbuto with the following poem.

My Hero

Hold it right there, folks,
We have with us tonight
A pioneer of catalyst
Preparation science.

Of his impact on my research
I will gladly tell the tale.
I may have laid some concrete but
Jim Schwarz has blazed the trail.

He's a man of the likes
There are very few others.
This professor is my hero,
Mentor and soul brother.

Preface

Catalysis is vital to the world's economy and standard of living, yet relatively little attention is paid to optimizing catalytic materials through rational methods of preparation. More and more we hear the call to "transform the art of catalyst preparation into a science." A robust answer to that call is the first chapter-themed book devoted to catalyst synthesis, *Preparation of Solid Catalysts* (Ertl, Knözinger, and Weitkamp, Eds., Wiley-VCH, 1999). This excellent compendium of materials and preparation methods contains material now almost a decade old. I hope that *Catalyst Preparation: Science and Engineering* can contribute to worldwide efforts in catalyst preparation science.

I suggest that these efforts require a substantially different and complementary focus from traditional catalysis. Along traditional paths there is much excellent progress with state of the art analytical and computational methods to obtain structure-function relationships: correlations of the atomic-scale physical and chemical makeup of catalytic materials to their reactivity. This research tells us why a particular catalyst works well or poorly, and in the best computational cases, precisely which catalytic material should be prepared and with what atomic arrangement.

Getting there is another story. We might eliminate a great deal of empiricism in catalyst development with a fundamental understanding of the genesis of structure; from an antecedent preparation-structure relationship. How can a particular catalyst composition and morphology be effectively synthesized? When must attention be paid to the initial distribution of catalyst precursors on a support surface, and when does migration during pretreatment render this inconsequential? What methods are readily scalable and industrially feasible?

A focus on the synthesis of heterogeneous solids is inherently interdisciplinary: material science, colloid chemistry, geophysics. However, the same characterization tools and computational methods that are used to elucidate structure-function correlations can be brought to bear on preparation-structure relationships. While laborers are perhaps few, the field of catalyst preparation science is ripe! I hope this message comes across clearly in the content to follow.

The book was inspired by several sources. At the beginning of my career I recall muttering excitedly about the elegant simplicity of J. P. Brunelle's landmark paper (*Pure and Applied Chemistry* 1978, 50, 1211) on the electrostatic nature of metal adsorption onto oxides, and seeing it confirmed in a comprehensive study of Pd anion and cation adsorption onto alumina (C. Contescu and M. I. Vass, *Applied Catalysis* 33, 1987, 259). I have been greatly influenced by the research, and moreover the friendship, of the late James A. Schwarz, who laid much groundwork with prolonged, systematic studies of the chemical fundamentals of many catalyst impregnation systems.

Jim's work in preparation fundamentals culminated in his (along with coauthors Cristian Contescu and Adriana Contescu) 1995 review, "Methods for Preparation of Catalytic Materials" (*Chem. Rev.* 1995, 95, 477). The synthesis of bulk materials was described in an initial section on three-dimensional chemistry followed by a section on two-dimensional chemistry which reviewed the various aqueous, organic, vapor, and solid phase methods to apply catalyst precursors to support surfaces.

It was around this outline that a two day, four-session symposium, "The Science and Engineering of Catalyst Preparation," was organized for the 227th ACS meeting in Anaheim, California. Much of the material in this book stems from that symposium, and the order has been largely retained. I had the pleasure of cochairing one of those sessions with Jim, and have deeply felt his absence in the subsequent editing of this book.

Thus the first section to follow pertains to the synthesis of bulk materials including amorphous and mesoporous oxide supports (chapters 1–4), heteropoly-acids (chapter 5), and colloidal metals (chapter 6). The second section covers the synthesis of heterogeneous materials, and has been divided into syntheses in nanoscale domains (chapters 7–10) and those based on two-dimensional metal complex-substrate interactions (chapters 11–14), or a clever way around non-interacting precursors via viscous drying (chapter 15). Effects of drying (chapter 16) and pretreatment (chapter 17) comprise the third section of the book.

A final source of inspiration has been the quadrennial conference founded by Profs. Bernard Delmon, Pierre Jacobs, and George Poncelet in 1975: The International Symposium on the Scientific Bases for the Preparation of Heterogeneous Catalysts. With its ninth meeting in 2006, this symposium continues to be the hallmark worldwide conference on catalyst preparation fundamentals. The proceedings of all these symposia are published by Elsevier, all but the first in *Studies in Surface Science and Catalysis*. These tomes contain vast deposits of information in far-ranging areas. I'm honored to have had Professor Delmon contribute a prognostication on the future of catalyst preparation (chapter 18) to this work.

I wish to thank the Degussa Corporation of Hanau-Wolfgang, Germany, the Petroleum Research Fund of the American Chemical Society, and the ACS' Division of Colloid and Surface Science for a matching grant. Financial support from these institutions facilitated the participation of a large number of international researchers in the The Science and Engineering of Catalyst Preparation symposium.

John R. Regalbuto
Department of Chemical Engineering
University of Illinois at Chicago

About The Editor

John R. Regalbuto is a professor of chemical engineering at the University of Illinois at Chicago. He has twice served as president of the Chicago Catalysis Club and was on the organizing committee of the 15th North American Meeting of the Catalysis Society held in Chicago. He is active in organizing catalysis sessions for meetings of the American Chemical Society and the American Institute of Chemical Engineers. He has lectured around the world on catalyst preparation and characterization.

Contributors

Michael D. Amiridis
University of South Carolina
Columbia, South Carolina

Lekhal Azzeddine
Rutgers University
Piscataway, New Jersey

Justin Bender
University of Illinois at Chicago
Chicago, Illinois

Ken Brezinsky
University of Illinois at Chicago
Chicago, Illinois

Laura E. Briand
Universidad Nacional de La Plata
Buenos Aires, Argentina

Hung-Ting Chen
Iowa State University
Ames, Iowa

Bernard Delmon
Université Catholique de Louvain
Louvain-la-Neuve, Belgium

D. Samuel Deutsch
University of South Carolina
Columbia, South Carolina

Lawrence D'Souza
International University
Bremen, Germany

Jennifer Dunn
University of Illinois at Chicago
Chicago, Illinois

Luis A. Gambaro
Universidad Nacional de La Plata
Buenos Aires, Argentina

Bruce C. Gates
University of California
Davis, California

John W. Geus
Utrecht University
Utrecht, Netherlands

Benjamin J. Glasser
Rutgers University
Piscataway, New Jersey

Benoît Heinrichs
Université de Liège
Liège, Belgium

Seong Huh
Iowa State University
Ames, Iowa

Nathalie Job
Université de Liège
Liège, Belgium

Dieter Kerner
Degussa AG, Aerosil & Silanes
Hanau-Wolfgang, Germany

Johannes G. Khinast
Rutgers University
Piscataway, New Jersey

William V. Knowles
Rice University
Houston, Texas

Alexander I. Kozlov
Northwestern University
Evanston, Illinois

Harold H. Kung
Northwestern University
Evanston, Illinois

Mayfair C. Kung
Northwestern University
Evanston, Illinois

Stephanie Lambert
Université de Liège
Liège, Belgium

Victor X.-Y. Lin
Iowa State University
Ames, Iowa

Catherine Louis
Université Pierre et Marie Curie
Paris, France

Silvana R. Matkovic
Universidad Nacional de La Plata
Buenos Aires, Argentina

Michael O. Nutt
Rice University
Houston, Texas

Gerard M. Pajonk
Université Claude Bernard-Lyon 1
Villeurbanne, France

Jean-Paul Pirard
Université de Liège
Liège, Belgium

Geoffrey L. Price
University of Tulsa
Tulsa, Oklahoma

Ryan Richards
International University
Bremen, Germany

Graciela M. Valle
Universidad Nacional de La Plata
Buenos Aires, Argentina

Christopher T. Williams
University of South Carolina
Columbia, South Carolina

Eduardo E. Wolf
University of Notre Dame
South Bend, Indiana

Michael S. Wong
Rice University
Houston, Texas

Wen-Mei Xue
Northwestern University
Evanston, Illinois

Shandong Yuan
Northwestern University
Evanston, Illinois

Table of Contents

Part I

Synthesis of Bulk Materials

1 Flame Hydrolysis for Oxide Supports

Dieter Kerner

CONTENTS

1.1 MANUFACTURE

The first product produced by flame hydrolysis was pyrogenic (or fumed) silica. It was developed during World War II by the German chemist Harry Kloepfer, driven by the need to find an alternative to carbon black based on local resources. As sand is available everywhere and Kloepfer had knowledge about the gas black process, he developed the idea of the flame hydrolysis of silicontetrachloride (SiCl$_4$, made by carbochlorination of sand) to pyrogenic silica. Today, this process is known as the AEROSIL® process [1], invented in 1942 by Kloepfer [2]. The details of this process were published in 1959 and later [3, 4]. A simplified flow sheet of the flame hydrolysis process is shown in Figure 1.1.

Silicontetrachloride is vaporized, mixed with dry air and hydrogen, and then fed into the burner. During the combustion, hydrogen and oxygen form water, which quantitatively hydrolyzes the SiCl$_4$, forming nanoscaled primary particles of SiO$_2$. The particle size or specific surface area can be adjusted by the flame parameters. The reaction products silica and hydrochloric acid are cooled down by heat exchangers before entering the solid/gas separation system. Filters or cyclones separate the silica from the off-gas (hydrochloric acid, combustion gases). The residual hydrochloric acid adsorbed on the large surface of the silica particles is removed in a fluidized bed or rotary kiln reactor. The chemical reaction can be described as a high-temperature hydrolysis of SiCl$_4$:

$$2H_2 + O_2 \rightarrow 2H_2O \tag{1.1}$$

$$SiCl_4 + 2H_2O \rightarrow SiO_2 + 4HCl \tag{1.2}$$

3

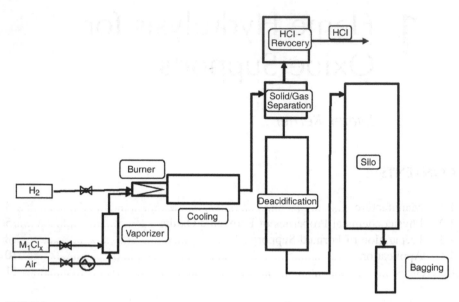

FIGURE 1.1 Schematic drawing of the flame hydrolysis process.

$$SiCl_4 + 2H_2 + O_2 \rightarrow SiO_2 + 4HCl \qquad (1.3)$$

Silicontetrachloride vapors are hydrolyzed in a hydrogen-oxygen flame by water vapors (Equation 1.2), which have been formed by the combustion of hydrogen in oxygen or air (Equation 1.1). The overall reaction is given in Equation 1.3.

The general mechanism of formation and growth of the silica particle occurring in the flame reactor can be described as follows [5–7]: At first, spherical primary particles are formed by the flame hydrolysis through nucleation. Further growth of these primary particles is accomplished by the reaction of additional $SiCl_4$ on the surface of the already generated particles. By means of coagulation (collision of primary particles) and coalescence (sintering), primary particles form aggregated structures [8–12]. Accumulations of primary particles can be described by the expressions *aggregate* and *agglomerate*. An aggregate is a cluster of particles held together by strong chemical bonds. Agglomerates, however, are defined as loose accumulations of particles (aggregates) sticking together by, for example, hydrogen bonds and van der Waals forces. For definitions and illustrations, see Reference 13. Figure 1.2 shows the growth of particles to aggregates and agglomerates in relation to the residence time in the flame reactor. The formation of primary particles is already completed within a short time, whereas the extent of aggregation and agglomeration increases with the residence time in the reactor [14]. Theoretical investigations are supporting the experimental results [15, 16].

The size of the primary particles and the degree of aggregation and agglomeration can be influenced by the flame temperature, the hydrogen-to-oxygen ratio, the silicontetrachloride concentration, and the residence time in the nucleation and aggregation zone of the flame reactor. Besides silicontetrachloride, methyltrichlorosilane (MTCS) or trichlorosilane (TCS) are used as raw materials in the AEROSIL process. Also, non-chloride-containing raw materials like D4 (octamethyltetrasiloxane) and

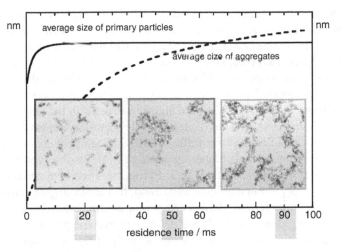

FIGURE 1.2 Average primary particle and aggregate size.

tetramethoxysilane (TMOS) have been reported [17, 18]. The worldwide production capacity for pyrogenic silica was estimated to be above 120,000 tons per annum in 2002 [19]. It is by far the most widely produced pyrogenic oxide [20]. The main producers are Degussa (AEROSIL®), Cabot (CAB-O-SIL®), Wacker (HDK®), and Tokuyama (REOLOSIL®). The process is not limited to SiO_2. Aluminum oxide and titanium dioxide have been made based on the respective metal chlorides and are today available on a commercial scale, as well as silica-alumina mixed oxides [8]. Various other pyrogenic oxides have been reported by Kleinschmit et al. [21, 22].

1.2 PHYSICOCHEMICAL PROPERTIES OF PYROGENIC OXIDES

One key parameter is the surface area, which is mainly determined by the size of the primary particles formed during the nucleation process in the flame reactor. For commercial pyrogenic silica products, surface areas lie between 50 and 400 m^2/g^{-1}, which corresponds to primary particles of approximately 40 to 7 nm. The silica consists of amorphous, spherical particles without an inner surface area, which are aggregated and agglomerated, as described in Section 1.1. Materials with a specific surface area above 300 m^2/g^{-1} have a certain amount of micro-porosity instead of a further reduced primary particle size. In pyrogenic silica, the SiO_4 tetrahedrons are arranged randomly. Therefore, a distinct and well-defined x-ray pattern—as obtained with any crystalline silica modification occurring in nature (like quartz or christobalite)—cannot be observed [8]. The x-ray photograph of pyrogenic silica is characterized by the absence of a diffraction pattern, indicating an entirely amorphous material. In contrast to the crystalline forms, this material does not cause silicosis.

Although made in an oxygen-hydrogen flame, pyrogenic alumina (AEROXIDE® Alu C, Degussa) has a crystalline structure consisting of the thermodynamically metastable γ- and δ-forms instead of the stable α-form [23]. The primary particle size is in the range of 13 nm; the specific surface area is ~100 m^2/g^{-1}. At temperatures above 1200°C, pyrogenic alumina can be transformed to α-Al_2O_3, which is associated with a decrease in the specific surface area and enlargement of the primary particles. Pyrogenic titania (AEROXIDE TiO$_2$ P 25, Degussa) has an average primary particle size of about 20 nm and a specific surface area of about 50 m^2/g^{-1} [23]. This product differs significantly from pigmentary TiO_2 produced through precipitation processes, with average particle diameters of several hundred nm and surface areas in the range of 10 m^2/g^{-1}.

In the pyrogenic TiO_2, the thermodynamically metastable anatase phase (~80%) is the dominant modification over rutile (~20%). At temperatures above 700°C, a lattice transformation towards higher amounts of rutile is observed and associated with a decrease of specific surface area. Pyrogenic zirconium dioxide (VP ZrO_2 from Degussa) is predominantly monoclinic and, to a lesser degree, tetragonal. Here also, despite the flame genesis, the low-temperature form is found to a greater degree than the high-temperature form. This ZrO_2 has an average primary particle size of 30 nm and a specific surface area of about 40 m^2/g^{-1} [23]. At temperatures around 1100°C, the conversion from a monoclinic to a tetragonal structure is observed, followed by another transformation at about 2300°C from tetragonal to cubic. The transformation from monoclinic to tetragonal is associated with an increase of density (monoclinic: 5.68 g/cm^{-1}; tetragonal: 6.10 g/cm^{-1}) and has a negative effect on the thermal resistance of, for example, molded ceramic parts, up to the point where the formation of microcracks is observed [24]. By addition of dopants, high-temperature modifications can be stabilized far below their original transformation point [24–26]. Y_2O_3 is the most common stabilization aid. TEM photographs of four different pyrogenic oxide grades are shown in Figure 1.3.

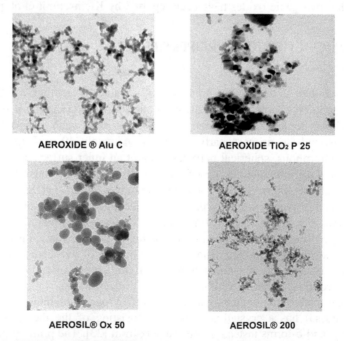

FIGURE 1.3 TEM photograph of different pyrogenic oxides.

Table 1.1 provides an overview of the physicochemical properties of pyrogenic oxides.

Besides their comparatively large specific surface area, their well-defined spherical primary particles, and their high chemical purity, the surface chemistry of pyrogenic oxides is another reason for their use in catalytic applications. Siloxane (Si-O-Si) and silanol groups (Si-OH) are the main functional groups (Figure 1.4) on the surface of pyrogenic silica.

The hydrophilic character of the silanol groups dominates the surface chemistry and makes pyrogenic silica wettable. The density of the silanol groups can be determined, for example, by the reaction with lithium aluminum hydride and varies between 1.8 and 2.7 Si-OH nm^{-2} [8]. By exposure to humid conditions (water), siloxane groups react to form additional silanol groups on the surface of pyrogenic silica. This reaction can be reversed (dehydroxylation) by heating pyrogenic silica up to temperatures above 150°C. Through infrared spectroscopy, changes in the moisture balance at different temperatures of pyrogenic silica can easily be detected and followed [27–29]. The hydroxyl groups are acidic, resulting in an isoelectric point at a pH value of 2. Pyrogenic Al_2O_3 has basic hydroxyl groups at its surface that react weakly alkaline in water, corresponding

TABLE 1.1
Physicochemical Properties of Pyrogenic Oxides

		AEROSIL OX50	AEROSIL 200	AEROXIDE AluC	AEROXIDE TiO₂ P 25	VP ZrO²
Specific surface area by BET (1)	m^2g^{-1}	50±15	200±25	100±15	50±15	40±10
Average size of primary particles	nm	40	12	13	21	30
Tapped density (2)	g/l	~130	~50	~50	~130	~80
Density (3)	g/ml	~2.2	~2.2	~3.2	~3.7	~5.4
Purity (4)	Wt. %	$SiO_2 \geq 99.8$	$SiO_2 \geq 99.8$	$Al_2O_3 \geq 99.6$	$TiO_2 \geq 99.5$	$ZrO_2 \geq 96$
Loss on drying (5)	Wt. %	1.5	1.5	5	1.5	2
Loss on ignition (6, 7)	Wt. %	1	1	3	2	3
Isoelectrical point at pH		2	2	9	6.5	8.2
Phase composition		Amorphous	Amorphous	γ and δ modification	80% anatase, 20% rutile	Monoclinic and tetragonal

1. According to DIN 66131.
2. According to DIN ISO 787/XI, JIS K 5101/18 (not screened).
3. Determined with a pynknometer.
4. Based on material ignited for 2 hours at 1000°C.
5. According to DIN ISO 787/II, ASTM D 280, JIS K 5101/21.
6. According to DIN 55921, ASTM D 1208, JIS K 5101/23.
7. Based on material dried for 2 hours at 105°C.

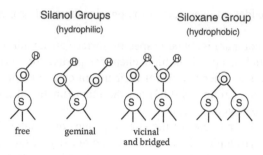

Silanol Groups
(hydrophilic)

Siloxane Group
(hydrophobic)

free   geminal   vicinal
and bridged

FIGURE 1.4 Surface groups of pyrogenic silica.

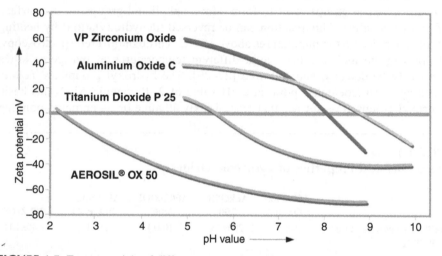

FIGURE 1.5 Zeta potentials of different pyrogenic oxides.

to an isoelectric point at a pH value of 9. The total dehydroxylation of alumina results in the presence of aluminum ions located at the surface, coordinated by only five rather than six oxygen atoms, thus representing Lewis acid centers. These centers can either add pyridine or be rehydroxylated through the adsorption of water [30]. Depending on the coordination of the hydroxyl groups at the surface of titania, an acidic as well as a basic character of these groups can be observed. Although the acidic sites react with ammonia and diazomethane, the basic sites can be detected by exchange reactions with anions. The acidic and basic hydroxyl groups are also reflected in an isoelectric point at a pH value of 6.5 [31]. Zirconia shows a similar surface chemistry. However, relative to titania, zirconia has more basic rather than acidic sites, resulting in an isoelectrical point at a pH value of 8.2. Figure 1.5 illustrates the zeta potential of the pyrogenic oxides mentioned in this chapter.

1.3 PREPARATION OF FORMED SUPPORTS

The fine and fluffy pyrogenic oxides are not very convenient for use in catalysis unless the powder shall not be part of the final product. Shaping processes can help make more convenient supports, especially for the use in fixed-bed reactors. The requirements for an excellent support are summarized in Reference 32:

- Well-defined chemical composition
- High purity
- Well-defined surface chemistry
- No sintering at high temperatures
- Good abrasion resistance and crushing strength
- Well-defined porosity, pore size distribution, and pore volume
- Easy separation from the reactants

Pyrogenic oxides in powder form only fulfill part of the requirements. Catalysts made by impregnation of oxide powders were of only academic interest for a long time because, in technical scale, problems arose when separating the powder from the products. These disadvantages can be overcome by size enlargement, meaning compaction of the powder. Options for size enlargement to make formed supports are [33] agglomeration, spray drying or spray granulation, pressure compaction, or extrusion. Agglomeration processes are not convenient for pyrogenic oxides because of their fluffiness. In the spray drying process, a suspension of pyrogenic oxides in water is fed via a spraying device into the chamber of a spray dryer. Microgranulates of pyrogenic silica [34, 35], alumina [36], and titania [37] have been reported using the spray drying method. The properties of the resulting spheres can be controlled by the solid content of the pyrogenic oxide in the suspension, the type of spraying devices, and the residence time and temperature in the spraying chamber. Typically, spheres in the range of 10–150 μm can be achieved. The pore size and pore size distribution can be adjusted by selecting oxides with different particle size distributions and surface areas, respectively.

In lab scale, pressure compacting (making tablets) is used for basic investigations on the processability and performance of newly developed recipes, whereas in pilot and production scale, extrusion is the preferred method. Basically, both processes follow the same schematic procedure. The desired properties can be achieved using auxiliaries like plasticizers or pore building substances. A simple way of making tablets from pyrogenic silica, alumina, and titania using silica sol as a binder and polyfunctional alcohols as plasticizer is described in Reference 38. The corresponding pyrogenic oxide is mixed with water, silica sol, and glycerol; pelletized; dried at ambient temperature; and calcined at 550°C. The tablets consist of 50–60% void volume, and the initial surface area of the powder is reduced by less than 20%. Deller et al. describe the use of various auxiliaries like kaolin, graphite, sugar, starch, urea, and wax as binders and pore building agents for making pellets of pyrogenic silica [39, 40], alumina [41], zirconia [42], and silica-alumina mixed oxides [43]. A detailed description of the extrusion process of pyrogenic titania is given in References 44 and 45. It consists of four crucial process steps: kneading the raw materials, extrusion, drying of the green bodies, and calcination. Figure 1.6 shows a picture of an extruder and the corresponding extrudates made.

FIGURE 1.6 Extruder and extrudates.

1.4 APPLICATIONS

Because of their unique properties like purity and well-defined and accessible surface, pyrogenic oxides are used as model substances. A review up to the year 1980 is given in Reference 46 about the investigation of silica (26 references), alumina (11 references), and titania (4 references). As can be seen, the major focus of mainly basic investigations was on pyrogenic silica rather than the other oxides. In commercial applications, silica supports play an important role in the synthesis of vinylacetate monomers: Wunder in Reference 47 describes the use of Pd- or Au-impregnated cylindrical supports made of either pyrogenic silica or a mixture of silica and alumina. Formed silica or alumina is also used as a support in catalysts where the support is impregnated with Pd/K/Cd, Pd/K/Ba, or Pd/K/Au, giving a selectivity of above 90% [48, 49]. Bankmann et al. [50] describe improved vinylacetate catalysts based on formed silica and silica/alumina with various shapes. Silica supports in the form of pellets, beads, or globular shape impregnated with phosphoric acid are used in the hydration of olefins in a fixed bed reactor [51].

In a rather basic study of the gas phase polymerization of ethane, it could be shown that the activity of catalysts using pyrogenic silica as support for metallocenes is 10 times higher than with other silica supports [52]. Pyrogenic alumina is used in automotive catalysts. Modern three-way catalysts consist of an alumina washcoat containing one or more of the elements Pt, Rh, Pd, and so-called storage components for NOx and oxygen, respectively. Liu and Anderson investigated the stability of stored NOx [53, 54]. NOx is stored under lean-burn conditions on an alumina-supported alkaline earth oxide component (10% BaO on Al_2O_3), released during intermittent rich/stoichiometric periods, and reduced by hydrogen, CO, or hydrocarbons over the noble metal component. Similarly, oxygen can be stored under oxygen-rich conditions and released under oxygen-lean conditions using CeO/ZrO_2 on alumina. In Reference 55, oxygen storage in fresh, thermally aged, and oxychlorinated treated catalysts has been studied.

In recent years, titania in general gained a lot of interest in the field of photocatalysis. Here, the titania is not only support but also catalyst. Various review articles with hundreds of references have been published in the last 10 years [56–59]. Mills and Lee [60] reported about a Web-based overview of current commercial applications. One big field of application is the treatment of (waste) water and air by photodegradation of inorganic compounds (like ammonia and nitrates) and organic substances (like chlorinated aliphatic and aromatic compounds) as well as volatile organic compounds (VOCs) in the air. Even 2,4,6,-trinitrotoluene (TNT) can be completely destroyed under aerobic conditions by the use of AEROXIDE TiO_2 P 25 [61]. Another field is the use of titania as sensitizer in the photodissociation of water. First investigations took place in the early 1970s by Fujishima and Honda [62], followed by Graetzel et al. in the early 1980s [63–67]. Graetzel further improved the titania catalyst by depositing RuO_2 and Pt on the surface or doping the titania with Nb_2O_5. They also used sensitizers like $Ru(bpy)_3^{2+}$, Ru(bpy)2(4,4'-tridecyl-2,2'- bpy)2+, and 8-hydroxy-orthoquinoline.

Surprisingly, the photocatalytic activity of AEROXIDE TiO_2 P 25 is higher than expected, most likely due to the specific mixed crystal structure with approximately 80% by weight of anatase and 20% of rutile [68]. TiO_2 P 25 is often regarded as the reference in photocatalytic investigations [58, 69]. Both crystal forms are tetragonal but with different dimensions of the elementary cell. According to Hurum et al. [70], the rutile acts as an antenna to extend the photoactivity into visible wavelengths, and by the special structural arrangement, catalytic "hot spots" are created at the rutile–anatase interface.

The extraordinary photocatalytic performance of AEROXIDE TiO_2 P 25 in comparison to other nanoscaled titania particles has been published in several papers: It is, for example, useful in the degradation of humic acid [71], of phenol and salicylic acid [72], of 1,4dichlorobenzene [73], and in the photocatalytic reduction of Hg(II) [74]. It is also used in the oxidation of primary alcohols to aldehydes [75] or in the photopolymerization of methyl methacrylate [76]. Its use in cement can help reduce environmental pollution [77, 78]. A detailed study is reported by Bolte [79]. The results show that crystal size and filling ratio in mass are more important than the modification of the titania. Pyrogenic titania is not only useful in photocatalysis but also in other catalytic applications.

It is the base material for DeNOx catalysts [80–82] and where selective hydrogenations are required [50]. Also, a broad field is the use in Fischer-Tropsch catalysts [83–87].

REFERENCES

1. AEROSIL® is a registered trademark of Degussa AG.
2. H. Kloepfer, *DE 762723,* Degussa, 1942.
3. L. J. White, G. J. Duffy, *J. Ind. Eng. Chem.* 51, 232, 1959.

4. E. Wagner, H. Brünner, *Angew. Chem.* 72, 744, 1960.
5. D. Schaefer, A. Hurd, *Aerosol Science and Technology* 12, 876–890, 1990.
6. S. Pratsinis, *J. Colloid Interface Sci.* 124, 416–427, 1988.
7. G. Ulrich, *Combustion Science and Technology* 4, 47–57, 1971.
8. Degussa AG, Düsseldorf, *Technical Bulletin Fine Particles Nr. 11*, 2003.
9. S. Friedlander, *Smoke, Dust and Haze—Fundamentals of Aerosol Behaviour*, John Wiley & Sons, New York, 1977.
10. T. Johannessen, S. Pratsinis, H. Livbjerg, *Chemical Engineering Science* 55, 177–191, 2000.
11. W. Koch, S. Friedlander, *J. Colloid Interface Sci.* 140, 419–427, 1990.
12. J. Landgrebe, S. Pratsinis, *J. Colloid Interface Sci.* 139, 63–86, 1990.
13. Degussa AG, Düsseldorf, *Technical Bulletin Pigments Nr. 60*, 1986.
14. O. Arabi-Katbi, S. Pratsinis, P. Morrison, C. Megaridis, *Combustion and Flame* 124, 560–572, 2001.
15. F. Kruis, K. Kusters, S. Pratsinis, B. Scarlett, *Aerosol Science and Technology* 19, 514–526, 1993.
16. A. Schild, A. Gutsch, H. Mühlenweg, S. E. Pratsinis, *J. Nanoparticle Research* 1, 305–315, 1999.
17. R. Schwarz, P. Kleinschmit, *DE 3016010*, Degussa, 1981.
18. D. Wright, S. Jesseph, H. Cilly, *Silicon for the Chemical Industry VII*, MS Trollfjord, Tromsoe-Bergen, Norway, 201–218, September 21–24, 2004.
19. W. Küchler, *Chemische Technik, 5. Auflage, Band 3*, 853, 2005.
20. M. Ettlinger, *Ullmann's Encycl. Ind. Chem. 5th ed.* A23, 635–642, 1993.
21. A. Liu, P. Kleinschmit, *DE 3 611 449* Degussa, 1986.
22. G. Kriechbaum, P. Kleinschmit, *Angew. Chem. Adv. Mater.* 101, 1446–1453, 1989.
23. Degussa AG, Düsseldorf, *Technical Bulletin Pigments Nr. 56*, 2001.
24. *Grundzüge der Keramik*, Skript zur Vorlesung Ingenieurkeramik I, Professur für nichtmetallische Werkstoffe, ETH Zürich, 37–43, 2001.
25. *Ullmann's Encyklopädie der technischen Chemie*, Band 24, 4. Auflage, Verlag Chemie GmbH, Weinheim, 694–696, 1983.
26. H.-J. Bargel, G. Schulze, *Werkstoffkunde*, 8. überarbeitete Auflage, Springer Verlag Berlin Heidelberg, 312–314, 2004.
27. L. Zhuravlev, *Colloid and Surfaces*, A 173, 1–38, 2000.
28. R. Iler, *The Chemistry of Silica*, Wiley-Interscience, New York, 1979.
29. A. Legrand, *The Surface Properties of Silicas*, Wiley & Sons, New York, 1998.
30. J. Peri, *J. Physic. Chem.* 69, 220–230, 1965.
31. H.-P. Boehm, *Angew. Chemie* 78, 617–652, 1966.
32. B. Despeyroux, K. Deller, H. Krause, *Chemische Industrie*, 10, 48, 1993.
33. Kirk-Othmer (3), Vol. 21, 77, 1978.
34. H. Biegler, G. Kallrath, DE
35. K. Deller et al., EP 0 725 037, Degussa, 1996.
36. J. Meyer et al., US 6 743 269, Degussa, 2004.
37. H. Gilges et al., EP 1 078 883, Degussa, 2001.
38. M. Ettlinger et al., DE 3 132 674, Degussa, 1983.
39. K. Deller et al., DE 3 803 895, Degussa, 1989.
40. K. Deller et al., DE 3 912 504, Degussa, 1990.
41. K. Deller et al., DE 3 803 897, Degussa, 1989.
42. K. Deller et al., DE 3 803 898, Degussa, 1989.
43. K. Deller et al., DE 3 809 899, Degussa, 1989.
44. M. Bankmann, R. Brand, B. Engler, J. Ohmer, *Catalysis Today*, 14, 225–242, 1992.

45. R. Brand et al., DE 4 012 479, Degussa, 1991.
46. D. Koth, H. Ferch, *Chem.-Ing.-Tech.* 52 (8), 628–634, 1980.
47. F. Wunder et al., DE 3 803 900, Hoechst, 1989.
48. R. Abel et al., EP 0 634 214, Hoechst, 1995.
49. H. Krause et al., EP 0 916 402, Degussa, 1998.
50. M. Bankmann, B. Despeyroux, H. Krause, J. Ohmer, R. Brand, *Stud. Surf. Sci. Catal.* 75, 1781–1784, 1993.
51. R. Cockman et al., EP 0 578 441, BP Chemicals, 1993.
52. M. Walden, Dissertation, Universität Essen, October 2002.
53. Z. Liu, J. Anderson, *J.Catalysis* 224, 18–27, 2004.
54. Z. Liu, J. Anderson, *J.Catalysis* 228, 243–253, 2004.
55. R. Daley, S. Christou, A. Efstathiou, J. Anderson, *Applied Catalysis B: Environmental* 60, 119–129, 2005.
56. M. Hoffmann, S. Martin, W. Choi, D. Bahnemann, *Chem. Rev.* 95, 69–96, 1995.
57. A. Linsebigler, G. Lu, J. Yates, *Chem. Rev.* 95, 735–758, 1995.
58. A. Mills, S. Le Hunte, *J. Photochem. and Photobiol. A: Chemistry* 108, 1–35, 1997.
59. A. Fujishima, T. Rao, D. Tryk, *J. Photochem. and Photobiol. C: Photochemistry Reviews* 1, 1–21, 2000.
60. A. Mills, S.-K. Lee, *J. Photochem. and Photobiol. A: Chemistry* 152, 233–247, 2002.
61. Zh. Wang, Ch. Kutal, *Chemosphere* 30, 1125–1136, 1995.
62. A. Fujishima, K. Honda, *Nature* 238, 37–38, 1972.
63. E. Borgarello, J. Kiwi, E. Pelizzetti, M. Visca, M. Graetzel, *Nature* 289, 158–160, 1981.
64. E. Borgarello, J. Kiwi, E. Pelizzetti, M. Visca, M. Graetzel, *J. Am. Chem. Soc.* 103, 6324–6329, 1981.
65. M. Graetzel, *Acc. Chem. Res.* 14, 376–384, 1981.
66. V. Houlding, M. Graetzel, *J. Am. Chem. Soc* 105, 5695 5696, 1983.
67. M. Graetzel, *Dechema Monographien* 106, 189–204, 1987.
68. Degussa AG, Düsseldorf, *Schriftenreihe Fine Particles Nr.* 80, 2003.
69. A. Mills, A. Lepre, N. Elliott, Sh. Bhopal, I. Parkin, S. O'Neill, *J. Photochem. and Photobiol. A: Chemistry* 160, 213–224, 2003.
70. D. Hurum, A. Agrios, K. Gray, T. Rajh, M. Thurnauer, *J. Phys. Chem. B* 107, 4545–4549, 2003.
71. C. Uyguner, M. Bekbolet, *Int. J. Photoenergy* 6, 73–80, 2004.
72. K. Chhor, J. Bocquet, C. Colbeau-Justin, *Mater. Chem. Phys.* 86, 123–131, 2004.
73. J. Papp, S. Soled, K. Dwight, A. Wold, *Chem. Mater.* 6, 496–500, 1994.
74. X. Wang, S. Pehkonen, A. Ray, *Electrochimica Acta* 49, 1435–1444, 2004.
75. M. Malati, N. Seger, *J. Oil Col. Chem. Assoc.,* 64, 231–233, 1981.
76. C. Dong, X. Ni, *J. Macromolecular Sci. A* A 41, 5, 547–563, 2004.
77. L. Cassar et al. WO 98/05601, Italcementi, 1998.
78. M. Lackhoff, X. Prieteo, N. Nestle, F. Dehn, R. Niessner, *Applied Catalysis B: Environmental* 43, 205–216, 2003.
79. G. Bolte, *Cement International,* 3, 92–97, 2005.
80. H. Hellebrand et al. DE 3938 155, Siemens, 1990.
81. R. Brand et al. EP 0 385 164, Degussa, 1990.
82. R. Brand et al. DE 3 740 289, Degussa, 1989.
83. M. Vannice et al. US 4 042 614, Exxon, 1976.
84. I. Wachs et al. US 4 559 365, Exxon, 1984.
85. K. Thampi, J. Kiwi, M. Graetzel, *Nature* 327, 506–508, 1987.
86. R. Fiato et al. US 4 749 677, Exxon, 1988.
87. S. Plecha et al. US 6 117 814, Exxon, 2000.

2 Amine-Assisted Synthesis of Aluminum Oxide

Alexander I. Kozlov, Mayfair C. Kung, Wen-Mei Xue, Harold H. Kung, and Shandong Yuan

CONTENTS

ABSTRACT

Alumina prepared by conventional hydrolysis of aluminum compound precursors is covered by surface hydroxyl groups. High-temperature calcination is needed in order to expose the surface coordination unsaturation of Al ions. A new method, which involves stoichiometric hydrolysis of an amine-Al alkoxide monomeric complex, can generate alumina with a surface that is covered with far fewer hydroxyls without high-temperature postsynthesis treatment. In this method, the coordination unsaturation site of Al is protected with an amine throughout the preparation process. The bound amine on the alumina surface can be exchanged with other bases, and the final solid is a Lewis-acid catalyst and catalyzes reactions such as aminolysis of epoxide. The chemistry in the preparation of such an alumina is described.

2.1   INTRODUCTION

Aluminum compounds are often used as Lewis acids for synthesis and catalysis [e.g., 1, 2, 3, 4, 5, 6, 7]. The Lewis acidity is a consequence of the coordination unsaturation of the Al atom and its ability to accept an electron pair. However, many of these compounds are in either the gaseous or liquid state under ordinary conditions. For ease of separation and processing, it is often desirable to use solid Lewis acids. Aluminum oxide is a solid and is widely used as an adsorbent, ion-exchange material, catalyst, catalyst support, and membrane, as well as a component in electronic materials. However, some suitable forms of pretreatment are necessary to generate a surface of the desired properties. Typically, alumina is prepared by hydrolysis of aluminum ions to form aluminum oxyhydroxide, which is then calcined to form alumina. For example, hydrolysis of mineral aluminum salts generates boehmite (AlO(OH)), which forms alumina upon heating. Hydrolysis of Al alkoxide using a limited amount of water, on the other hand, produces a sol gel that can be converted to alumina also by heating. The surface of these as-prepared alumina, after exposure to ambient, is covered with hydroxyl groups. Dehydroxylation at elevated temperatures is the usual method to remove surface hydroxyl groups with concomitant generation of surface Lewis acidity, and it has been studied extensively [8].

Recently, a new method to synthesize aluminum oxide that contains high surface Lewis acidity relative to surface hydroxyls without dehydroxylation at elevated temperatures was reported [3]. It was shown that this oxide, compared with one prepared by a conventional method, contains a much higher density of surface Lewis acid sites relative to hydroxyl groups. Furthermore, these surface Lewis acid sites are catalytically active. The solid catalyzes ring opening of cyclopentene oxide with piperidine to form 2-piperidylcyclopentanol selectively [4]. In this chapter, a description of the preparation method and characterization of the steps in the synthesis will be presented.

The concept behind the synthesis strategy is protection of the Lewis acid site (i.e., the coordination unsaturation) of aluminum in the alkoxide precursor with an amine base throughout the hydrolysis process (Scheme 2.1). The amine protection is coupled with controlled hydrolysis using a stoichiometric amount of water. The approach differs from other methods where amine or amine-inorganic complexes are used primarily as structure directing agents [9, 10, 11] and for pore structure control [12, 13, 14], while placing little emphasis on controlling the surface properties.

Sufficient details will be presented for the readers to evaluate the potential and limitations of such an approach. Thus, the relevant interaction of amines with Al alkoxide and the reactions of the amine-Al alkoxide adduct will be described. This includes the formation of the amine-Al alkoxide precursor complexes for various

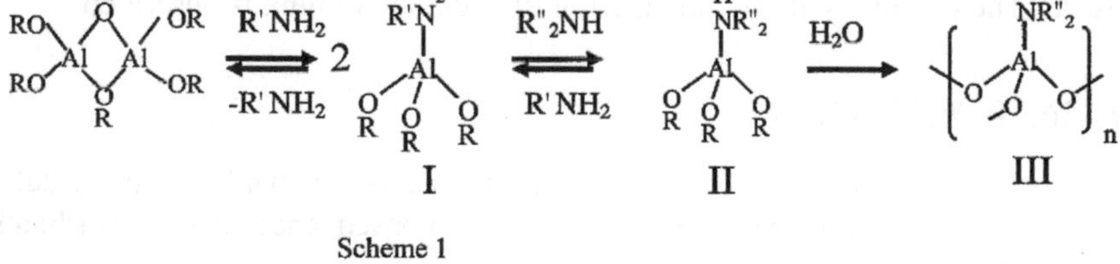

Scheme 1

SCHEME 2.1

amines, the stability of these complexes, and reactions at different stages of the synthesis. As will become obvious later, ^1H, ^{13}C, and ^{27}Al NMR spectroscopy are very informative tools. Characterization of the hydrolyzed products includes use of ^{13}C CP MAS NMR, ^{27}Al MAS NMR, and FTIR of adsorbed amines.

2.2 PREPARATION AND PROPERTIES OF AMINE-Al ALKOXIDE COMPLEXES

Because of the strong tendency of the Al atoms in the common precursor compounds (e.g., Al alkoxide, Al chloride) to form donor-acceptor complexes with Lewis bases, they accept the electron pairs of the oxygen of the alkoxy ligand or chlorine atom of another molecule to form dimers and oligomers when no other bases are present. The presence of other Lewis bases would compete with these ligands, and if their binding with Al is stronger and kinetically feasible, they could displace the chloride or alkoxide, thereby dissociating the dimer or oligomer into a monomeric Al alkoxide-base adduct (first step in Scheme 2.1). For example, monomeric amine-Al t-butoxide complex can be formed quantitatively by the addition of a stoichiometric excess of amine [3] or THF [15] to the Al t-butoxide. Various amines, phosphines, ethers, and phosphine oxides have been shown to form adduct with Al halides and alkylaluminum compounds [16].

The reaction of Al alkoxides with amine can be followed readily with ^1H NMR. For example, mixing different amounts of n-octylamine with aluminum t-butoxide in d_8-toluene converted the two singlets at 1.39 and 1.51 in the ^1H NMR spectra of the butoxide (with a peak area ratio of 2:1 due to the terminal and bridging t-butoxy groups, respectively [17]) into a singlet at 1.4 to 1.46 due to the formation of a monomeric amine-aluminum t-butoxide adduct (I in Scheme 2.1, R' = n C_8H_{17}) [3]. The equilibrium conversion of the dimer to monomer was about 16% for an n-octylamine/aluminum ratio of 1.0. Complete conversion to the monomeric species was observed only at an N/Al ratio of ca. 10. The equilibration occurred slowly (hours) at room temperature. Increased temperature favored faster transformation but the equilibrium was shifted toward the $(Al(OBu^t)_3)_2$ dimer due to entropic effect. Removal of n-octylamine from the reaction mixture by evacuation regenerated the starting $(Al(OBu^t)_3)_2$ dimer. Similarly, ^{13}C NMR confirmed the existence of dimeric $(Al(OBu^t)_3)_2$ in the starting compound with four singlets at 76.9 and 69.5 for bridging and terminal $(O-C-(CH_3)_3)$ and 32.3 and 34.2 for bridging and terminal $(O-C-(CH_3)_3)$. Interaction of $(Al(OBu^t)_3)_2$ and n-octylamine resulted in new monomeric alkoxide species with δ 67.5 $(O-C-(CH_3)_3)$ and δ 34.2 $(O-C-(CH_3)_3)$.

At room temperature, the Al-bound n-octylamine exchanged rapidly with free amine, as indicated by broadening of the α-, β-, and γ-carbon peaks in the ^{13}C NMR spectra [3]. In fact, in the presence of a large excess of octylamine, there were no peaks that could be attributed to the bound amine. When the temperature was lowered to $-75°C$, a new set of peaks at δ 40.7, 31.9, and 27.3 was detected, which were upfield from the set of peaks assigned to free amine. The upfield shifts are consistent with amine complexed with a Lewis acid [18]. The intensity ratio of these two sets of peaks was close to the ratio expected for free and bound amines. The peak area ratio of amine α-C at δ 40.7 to the methyl carbon in the alkoxide group was close to the expected ratio for an $Al(OBu^t)_3$(n-octylamine) complex.

The reactions of $[Al(OBu^t)_3]_2$ dimer with other amines follows a similar trend. Al t-butoxide reacts readily with other primary amines, such as n-propylamine, n-butylamine, ethylene diamine, propargylamine, and tris-(2-aminoethyl)amine, to form aluminum alkoxide monomers. All the monomers show a singlet in the region δ 1.39–1.46 in the ^{1}H NMR spectra due to the methyl protons of the t-butoxy groups (Table 2.1). Similarly, ^{13}C NMR resonances of the monomer fall in the range of δ 68.1–67.5 and 34.6–34.1 for tertiary and primary carbons of the t-BuO group, respectively. The equilibrium towards monomer formation is more favorable for multidentate amines than amines containing only one amino group. Thus, ethylene diamine and tris-(2-aminoethyl)amine converted 70 and 98%, respectively, of the dimer to monomeric species at an amine-to-alkoxide ratio of 1.1. Under similar conditions, the conversion was only 16% for n-octylamine.

Table 2.1 summarizes the NMR peak positions of the t-butoxy group in various amine-Al alkoxide adducts in toluene. Due to rapid exchange with free amines in many cases, no attempts were made to obtain peaks due to the bound amine ligands. Therefore, their values are not shown.

The reactions of $[Al(OBu^t)_3]_2$ with piperidine (Pipy), 4-butylpyridine, and other sterically hindered amines are very slow at room temperature. For example, less than 1% of the Al t-butoxide was transformed to the $Al(OBu^t)_3$(Pipy) complex after

---

TABLE 2.1
NMR Data of Aluminum t-Butoxide-amine Adducts[a]

		t-Bu group	
Amine	Amine:Al ratio	^{1}H NMR, ppm	^{13}C NMR, ppm
None	0	1.51, 1.39	76.9, 69.5, 34.5, 32.3
n-Octylamine	1	**1.46**	n/m
		1.51, 1.39	
	10	**1.40**	**68.1, 34.6**
n-Propylamine	1	**1.46**	n/m
		1.51, 1.38	
	10	**1.42**	**67.8, 34.2**
n-Butylamine	1	**1.46**	n/m
		1.51, 1.38	
Piperidine	1.5	**1.43**	**67.8, 34.2**
	10	**1.40**	**68.0, 34.4**
Ethylene diamine	1.1	**1.39**	**68.0, 34.1**
		1.50, 1.36	76.2, 69.0, 33.9, 31.6
Tris-(2-aminoethyl)-amine	1.1	**1.40**	**67.5, 34.2**
		1.50, 1.36	76.1, 68.6, 33.9, 31.8

[a] Solvent is d_8-toluene, room temperature, C_{Al} = 0.24 M. Peaks in bold are assigned to amine-Al t-butoxide adducts. Other values are for the Al t-butoxide dimer for comparison.

6 days in the presence of 10-fold excess of piperidine. These observations, together with the absence of any detectable, dissociated aluminum alkoxide, suggest that the Reaction R2.1 does not proceed by first dissociation of the alkoxide dimer followed by binding of amine to the monomer. One would not expect a strong difference between primary and secondary amines, then. Instead, the results suggest an association-dissociation process, in which the reaction rate depends on how easily an amine can approach the Al atom in the alkoxide dimer. Thus, the rate would be slower for an amine with a more sterically hindered nitrogen. An association-dissociation process also implies that the reaction proceeds via a penta-coordinated Al transition state. Such a penta-coordinated species has been postulated to be involved in catalytic reactions of Al Lewis acids [1].

$$Al_2(O^tBu)_6 + 2N = 2Al(O^tBu)_3N \text{ equilibrium const } K \qquad (R2.1)$$

Because the binding of amine is essential for protection of the coordination unsaturation site of Al during synthesis, it is useful to know the binding constants K_B of various amines to Al alkoxide. For Al t-butoxide, K_B can be defined by Equation 2.1 for Reaction R2.2:

$$Al(O^tBu)_3 + N = Al(O^tBu)_3N \qquad (R2.2)$$

$$K_B = \frac{[AN]}{[A][N]} \qquad (2.1)$$

where AN is the amine-Al alkoxide complex, A is the Al alkoxide monomer, and N is the free amine. K_B is related through Equation 2.3 (A_2 is the alkoxide dimer) to the equilibrium constant K of Reaction R2.1 by the dissociation constant K_D of the Al t-butoxide dimer (Reaction R2.3):

$$Al_2(O^tBu)_6 = 2 Al(O^tBu)_3 \qquad (R2.3)$$

$$K_D = \frac{[A]^2}{[A_2]} \qquad (2.2)$$

$$K_B = \sqrt{\frac{K}{K_D}} \qquad (2.3)$$

Whereas the equilibrium constant K can be determined relatively readily for many primary amines, it is much more difficult for secondary or sterically hindered amines because of the slow reaction. This kinetic limitation can be overcome by associative substitution of $Al(OBu^t)_3(Pr^nNH_2)$ with the hindered amine (Reaction R2.4), which proceeds much more rapidly:

$$Al(O^nPr)_3N + N' = Al(O^nPr)_3N' + N \qquad\qquad (R2.4)$$

Using either Reaction R2.1 or R2.4, the binding equilibria K of a number of amines were determined. Because K_D is a constant for a given Al alkoxide, the relative values of K obtained experimentally can be used to provide the relative values of the binding constants K_B. Binding of amine to Al alkoxide is a Lewis acid-base interaction. Thus, one expects that the binding would be stronger for stronger bases. There are at least two different scales of basicity of amine, depending on the medium: the gas phase basicity and the pK_a (of the protonated amine) in an aqueous solution. They measure the interaction of an amine with a proton. Figure 2.1 shows a plot of $\log(K)$ versus pK_a (2.1a) or gas phase basicity (2.1b) for Al t-butoxide. It can be seen that pK_a gives a better linear correlation than gas phase basicity for

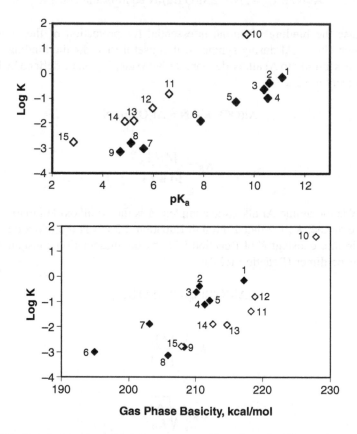

FIGURE 2.1 Correlation of equilibrium constants with amine pK_a (2.1a, top) and gas phase basicity (2.1b, bottom). The filled data points are for aliphatic amines: 1. piperidine, 2. n-butylamine, 3. n-propylamine, 4. s-butylamine, 5. $C_6H_5CH_2NH_2$, 6. propargylamine, 7. p-toluidine, 8. m-toluidine, and the open data points are for aromatic amines: 10. 4-dimethylaminopyridine, 11. 4-methoypyridine, 12. 4-t-butylpyridine, 13. pyridine, 14. 3-methoxypyridine, and 15. 3-chloropyridine.

the aliphatic amines, and different correlations apply to the aliphatic amines and the aromatic amines.

It is somewhat unexpected that aqueous pK_a could correlate with the equilibrium constant determined in toluene. One would expect that the hydrocarbon solvent used is much less effective in screening ionic charge than water, in which the pK_a values are determined. Consequently, correlation with gas basicity would be better. This is clearly not the case. One possible interpretation is that, being polar molecules, both the Al t-butoxide dimer and the amine-Al butoxide monomer could form aggregates in toluene. In these aggregates, the close proximity of the butoxy groups may function as polar solvent molecules to offer non-negligible solvation stabilization. Effects due to such aggregation have been used to explain the higher enthalpy and equilibrium constants of phenol-dimethylacetamide adducts in cyclohexane than in carbon tetrachloride [19].

The linear correlation between pK_a and $log(K)$ is consistent with the computational results that bonding in an amine-Al adduct is mostly electrostatic interaction with very little covalent contribution [20]. It also suggests that the inductive effect of the substituents plays a similar role in both the protonation and the donor-acceptor interaction, although it has a larger effect in protonation.

The binding equilibrium information is useful when choosing an appropriate amine as protection agent of the coordination unsaturation site of Al for this synthesis. Piperidine is one of the strongly bound amines that had been used. In the specific example reported in Reference 3, the piperidine-Al t-butoxide adduct was formed by associative substitution with the n-PrNH$_2$ adduct. The ^{1}H NMR spectrum of a mixture containing $[Al(OBu^t)_3]_2$ and n-PrNH$_2$ indicated the formation of monomer by the presence of one type of $tert$-butoxy group with δ 1.42 at an N:Al ratio of 20. The monomeric species remained intact after addition of piperidine followed by n-PrNH$_2$ removal with dry N$_2$ purge or evacuation. Prolonged evacuation at room temperature produced a white solid which, when redispersed in d$_8$-toluene, contained only monomeric Al species. The ^{1}H NMR spectrum and the corresponding ^{13}C and ^{27}Al NMR spectra were consistent with the Al(OBu^t)$_3$(Pipy) structure with tetrahedral coordination around aluminum (Scheme 2.1, **II**, R"$_2$ = Pipy). The ^{1}H NMR in d$_8$-toluene at room temperature was assigned as follows: δ 3.10 (d, br, 2H, NCH_{eq}H$_{ax}$CH$_2$), 2.49 (m, 2H, NCH$_{eq}$$H_{ax}CH_2$), 1.43 (s, 27H, AlOC(C$H_3$)$_3$ and 1H, CH$_2$CH$_2$CH_{eq}H$_{ax}$), 1.28 (d, br, 2H, NCH$_2$CH_{eq}H$_{ax}$), 0.97 (m, 1H, CH$_2$CH$_2$CH$_{eq}$$H_{ax}$), 0.79 (m, 2H, NCH$_2CH_{eq}$$H_{ax}$). The ^{13}C NMR was assigned as follows: 67.8 (AlOC(CH$_3$)$_3$), 47.0 (br, NCH$_2$, free piperidine), 34.2 (AlOC(CH$_3$)$_3$), 46.0 (NCH$_2$, bound piperidine), 26.0 (NCH$_2$CH$_2$, free piperidine), 25.7 (NCH$_2$CH$_2$, bound piperidine), 24.8 (NCH$_2$CH$_2$CH$_2$, free piperidine), 23.7 (NCH$_2$CH$_2$CH$_2$, bound piperidine). The ^{27}Al NMR was assigned as 58.6 (s, br). Partial decomposition of Al(OBu^t)$_3$(Pipy) was observed when the complex was heated to 70°C *in vacuo*.

It is interesting that the bound piperidine exhibited separate resonance peaks for the axial and equatorial protons. This probably reflects the fact that the inversion-rotation processes of the stronger bound piperidine is slow. The equatorial protons are located downfield by 0.5–0.6 ppm because of the deshielding effect of σ-electrons of the C-C bonds [21, 22].

2.3 HYDROLYSIS OF Al(OBu^t)$_3$(C$_8$H$_{17}$NH$_2$) AND Al(OBU^t)$_3$(Pipy) TO FORM Al$_2$O$_3$-AMINE

Hydrolysis of amine-Al alkoxide complexes can be carried out by adding water slowly to the solution. We have attempted two different procedures. For Al(OBu^t)$_3$(C$_8$H$_{17}$NH$_2$) adduct, hydrolysis was carried out by dropwise addition of water dispersed in a mixture of anhydrous toluene and n-octylamine. Excess amine was used in order to minimize dissociation of the adduct. The hydrolysis process was followed by ^{1}H and ^{27}Al NMR. As the hydrolysis proceeded, the $tert$-butoxy ligand of aluminum disappeared, while t-butanol appeared quantitatively in solution (Table 2.2). Due to the large excess of amine present in solution and the rapid exchange between bound and free amines, little useful information could be derived from the amine peaks. The concentration of Al detectable with NMR decreased steadily with the degree of hydrolysis, although the solution remained clear, suggesting that oligomers of Al species were formed. The detectable amount of butoxy groups remained constant, suggesting that the undetected Al species did not contain butoxy groups. Hydrolysis was terminated when there was no detectable amine-Al t-butoxide monomer. After hydrolysis and solvent removal at 70°C, there was still

---

TABLE 2.2
Hydrolysis of Monomeric Al(OBu^t)$_3$-amine Adducts

Amine-Al adduct	Degree of hydrolysis[a] %	Detectable Al[b], %	Amine balance[c], %	t-BuO balance[c], %
Al(OBu^t)$_3$(C$_8$H$_{17}$NH$_2$)	0	100.0	100.0	100.0
	9.0	90.0	110.0	98.5
	27.0	75.0	107.0	98.5
	45.0	58.0	105.0	104.0
	71.0	41.0	98.2	101.0
	100.0	2.0	115.0	98.3
Al(OBu^t)$_3$(Pipy)	0	100.0	100.0	100.0
	19.7	80.6	105.0	100.0
	46.0	60.3	100	86.4
	66.4	41.7	85.7	75.3
	87.5	26.1	71.4	61.7
	100.0	0	67.0	63.2[d]

[a] Defined as (decrease in amine-Al(OBu^t)/initial amine-Al(OBu^t)) × 100 as determined by ^{1}H NMR.
[b] By ^{27}Al NMR.
[c] By ^{13}C NMR; sum of peak intensities of bound and free amine (for amine balance) or of all butyl groups (for butoxy balance) relative to those at beginning of reaction.
[d] Including 9.3 and 1.3% of t-BuOH and piperidine recovered from LN2 trap, respectively.

---

significant amount of octylamine occluded in the solid (~5.4 Al/N as determined by titration).

For the $Al(OBu^t)_3(Pipy)$ adduct, because piperidine is much more tightly bound, hydrolysis could be carried out without excess amine. Water was introduced via the gas phase in an N_2 stream passed over a stirred toluene solution of the Al-amine complex. As shown by the data in Table 2.2, the solution NMR signal intensities of the amine, the butoxy group, as well as Al decreased as the hydrolysis progressed. The ^{13}C MAS NMR of the dried solid showed a t-BuO/piperidine ratio of 2.9, which was consistent with the solution data. Analysis of carbon, hydrogen, and nitrogen content of the solid suggested a composition of $AlO_{1.24}(C_4H_9O)_{0.52}(C_5H_{11}N)_{0.15}$. The solid is labeled Al_2O_3-Pipy (**III**, R"$_2$ = Pipy). Hydrolysis of $Al(OBu^t)_3(Pipy)$ in the presence of excess propylamine was also attempted to see whether additional protection of Lewis acid site could be achieved. However, the resultant dried solid appeared to be similar (with the exception that a small amount of adsorbed propylamine was also present in addition to piperidine).

Stoichiometric amounts of water were used in these methods, because excess water would increase the tendency for water to displace the protective amine. Likewise, the mode of water addition is also important, as local high concentration of water can also have deleterious effect. Of the two methods used, gas-phase introduction of water is a more controlled method. As shown in Table 2.2, with the gas-phase method, condensed aluminum alkoxide oligomers are formed in solution during the hydrolysis, as indicated by the poor balance of detectable butoxy groups beyond 40% hydrolysis. That is, some of the butoxy groups have become undetectable by solution NMR. In contrast, using the liquid-phase hydrolysis method that involves mixing two liquids together, despite using a high stirring rate and a very dilute concentration of water in a toluene-amine solvent, practically all the butoxy groups could be detected in solution any time, indicating that all three butoxy groups of one Al alkoxide molecule are hydrolyzed rapidly. We believe this to be a consequence of the high local concentration of water at the mixing point. Admission of water via the vapor phase can avoid high local concentration of water.

2.4   CHARACTERIZATION OF Al$_2$O$_3$-AMINE

2.4.1   NMR SPECTROSCOPY OF THE DRIED HYDROLYZED SOLID

^{27}Al MAS NMR spectra of the dried hydrolysis products Al_2O_3-$C_8H_{17}NH_2$ (**IV**) and Al_2O_3-Pipy (**III**) are shown in Figure 2.2. The spectrum of γ-Al_2O_3 prepared by calcination of a boehmite gel, Al_2O_3(MPD), is also shown for comparison (Spectrum 2.2e). It exhibits two distinctive resonances at 60 and 3 ppm due to tetrahedral (Al^{IV}) and octahedral Al (Al^{VI}), respectively. The NMR spectra of **IV** and **III** (Figure 2.2a and Figure 2.2b, respectively) show an additional strong signal at δ 30–35, which can be ascribed to Al^V or distorted Al^{IV}. Deconvolution of the Spectrum 2.2b shows that this peak accounts for 40–50% of the Al, whereas about 15% of Al species is Al^{IV}. If a mixture of $(Al(OBu^t)_3)_2$ and piperidine is hydrolyzed instead of the $Al(OBu^t)_3(Pipy)$ complex, the resulting solid Al_2O_3(Pipy) possesses only about 10% Al^{IV} and 20% Al^V (Spectrum 2.2c). On the other hand, if excess water is used to

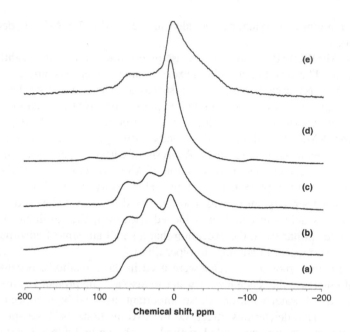

FIGURE 2.2 ^{27}Al MAS NMR of (a) Al_2O_3-$C_8H_{17}NH_2$ **IV**, (b) Al_2O_3-Pipy **III**, (c) Al_2O_3(Pipy), (d) Al_2O_3-Pipy hydrolyzed using excess H_2O, and (e) Al_2O_3(MPD).

hydrolyze $Al(OBu^t)_3$(Pipy) (3.5 times stoichiometry), the concentrations of Al^{IV} + Al^V relatively to Al^{VI} decrease drastically, and only about 2% of the aluminum atoms are Al^{IV} (Spectrum 2.2d).

^{13}C CP MAS NMR spectrum of Al_2O_3-$C_8H_{17}NH_2$ exhibits peaks at δ 41.6, 32.5, 30.2, 28.0, 23.2, and 14.4 due to n-octylamine adsorbed on the oxide surface. These compare well with peaks at δ 41.2, 32.2, 29.8, 26.5, 22.8, and 12.5 for octylamine adsorbed on Al_2O_3(MPD) dehydroxylated at 900°C. The resonance of α-C of the amine at δ 41.6 is very broad compared to other peaks, but its position agrees with octylamine bound to a Lewis acid. For Al_2O_3-Pipy dried at 60°C under vacuum, features due to nonhydrolyzed t-butoxy groups are clearly observed at δ 34.5 and 68.3. Broad lines due to adsorbed piperidine appear at δ 45.1 and 25.5. Thermal treatment at 170°C results in significant shifts for both groups of signals. t-Butoxy resonances are observed at δ 65.5 and 29.5, whereas piperidine peaks shift to δ 23.0 and 43.3. Exchange of piperidine in **III** with n-propylamine produced the Al_2O_3-$PrNH_2$ **V** sample for which n-$PrNH_2$ peaks dominate the spectrum, appearing at δ 42.2, 24.1, and 9.2 due to α-, β-, and γ-carbons of adsorbed n-$PrNH_2$, respectively.

2.4.2 IR MEASUREMENTS

FTIR and DRIFT spectroscopy can be used to monitor exchange of the bound amines on the dried, hydrolyzed solids. Although FTIR offers better signal-to-noise, handling the powder sample is easier with DRIFT, especially if the sample tends to pick up moisture, as is the case of these samples. The DRIFT results on these samples have been published [3], and similar conclusions can be drawn with the FTIR results shown here. However, it is clear that there was significant uptake of water by the FTIR sample during sample preparation for measurement, as evidenced by the much higher intensity of the broad envelope from 2500 to 3800 cm^{-1}.

Exchange of bound amine from synthesis with other amines can be examined. Spectrum 2.3a shows the FTIR spectrum for Al_2O_3-Pr^nNH_2 **V** prepared by treating Al_2O_3-Pipy **III** in refluxing n-propylamine followed by drying at 60°C *in vacuo*. The band at 1592 is due to the $-NH_2$ group, and bands at 1470, 1460, 1384, and 1360 are the -CH deformation bands [23]. There are unresolved bands near 2900 cm^{-1} due to C-H stretches. Upon exposure to ammonia gas at 60°C, the intensity of the propylamine bands decreases very rapidly for the initial 15 min (Spectrum 2.3b), and new features appear at around 1621 and 1258 cm^{-1}. Upon further exchange of the surface propylamine with ammonia, these features become more intense and are clearly seen as broad bands at 1621 and 1258 cm^{-1} (Figure 2.3c), which are characteristic of the asymmetric and symmetric vibrations of NH_3 coordinated to Al^{3+}[24]. No further changes in the spectrum can be observed after 95 min (Spectrum 2.3c), although n-propylamine bands of low intensities are still detectable that can be due to occluded propylamine. Thus, adsorbed propylamine can be replaced by NH_3. On the other hand, the broad envelope from 2500 to 3800 cm^{-1} is little affected by the amine exchange. It is possible that it is due to internal OH groups that are not accessible to the amines.

Figure 2.4 shows the effect of exposing **III** to pyridine vapor (~1.3 kPa) at room temperature and at 60°C followed by evacuation at the temperature of exposure. Spectrum 2.4a shows that, at room temperature, intense pyridine bands appeared. The 1441 and 1492 cm^{-1} bands are the characteristic 19b and 19a ring vibration [25], and the band around 1576 cm^{-1} is the 8b ring vibration of all pyridine present

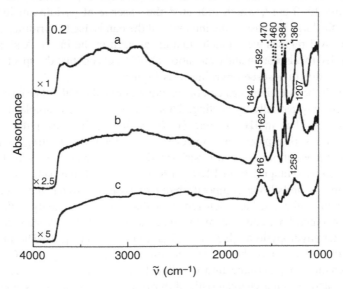

FIGURE 2.3 FTIR spectra of Al_2O_3-Pr^nNH_2: (a) after heating *in vacuo* at 60°C; (b) subsequent exposure to NH_3 for 15 min at 60°C; and (c) exposure to NH_3 for 95 min.

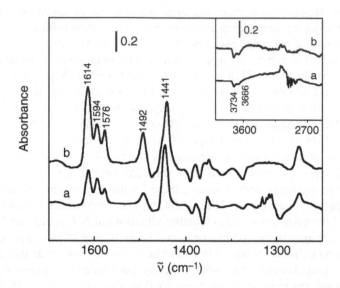

FIGURE 2.4 Difference spectra of Al$_2$O$_3$-Pipy **III** before and after room-temperature exposure to pyridine followed by evacuation (a), and (b) the same treatment at 60°C. The spectra were collected at the treatment temperatures.

[26]. The low-intensity band at 1594 cm^{-1} may be a composite band of pyridine species hydrogen bonded to surface hydroxyl and pyridine adsorbed on Al^{VI}, according to Morterra [26]. The band around 1614 cm^{-1} is the 8a ring vibrational mode of pyridine coordinated to the Lewis acid sites [27, 28, 29]. There is no absorption band around 1540 cm^{-1}, which indicated the absence of pyridinium ions [27]. In contrast to the large increase in the intensity of the bands due to pyridine, the changes in the hydroxyl regions were much smaller, as seen in the insert. A similar observation applied to adsorption and evacuation at 60°C, except that the spectral intensity of the bands due to adsorbed pyridine was higher.

Although the bound amines can be exchanged readily with other bases, they can be desorbed only partially by heating. This is shown by the example in Figure 2.5 for bound piperidine in **III**. Heating **III** *in vacuo* significantly reduced the band intensities of adsorbed piperidine. However, even after 120°C heating, there was still a substantial amount of adsorbed piperidine detected as bands at 1470, 1457, 1360, and the broad band ranging from 1270 to 1200 cm^{-1} [30]. The broad asymmetric band at 1646 cm^{-1} is probably a composite band of adsorbed water and decomposition product of piperidine. It has been reported that Al$_2$O$_3$ can catalyze the dehydrogenation of piperidine, and one of the intermediates of dehydrogenation, 1,2,5,6-isomer of tetrahydropyridine, has an IR band at 1655 cm^{-1} [30]. The intensities of these bands decreased further at higher temperatures, and by 200°C, significant dehydrogenation of piperidine into pyridine had taken place, as evidenced by the 1612 cm^{-1} band, which is characteristic of pyridine adsorbed on Lewis acid site. After 300°C heating, there were still residual organic species on the surface.

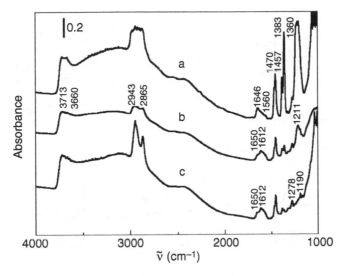

FIGURE 2.5 FTIR spectra of Al_2O_3-Pipy **III** after heating at: (a) 120°C; (b) 200°C; and (c) 250°C.

Adsorption of pyridine to one of these samples heated to 250°C resulted in the spectra shown in Figure 2.6. Spectrum 2.6a is the difference spectrum due to pyridine adsorption at room temperature. Bands are observed at about 1615, 1610, 1592, 1574, 1489, 1444, 1220, and 1155 cm⁻¹, which are typical bands of adsorbed pyridine. Desorption of pyridine at 150°C resulted in the difference spectrum 5b. The thermal treatment removes the more weakly bound pyridine (1592 and 1574 cm⁻¹ bands) without affecting those more strongly bound to Lewis acid sites. It is interesting to note that adsorption of pyridine causes a concomitant change in the hydroxyl region. Unlike adsorption on conventionally prepared alumina, where adsorption of pyridine caused reduction in the intensities of several OH bands [28], pyridine adsorption on **III** that had been heated to 250°C caused reduction in the 3747 cm⁻¹ band exclusively (Figure 2.6). This might indicate that the hydroxyls on **III** are much more uniform than on a conventional alumina.

2.5 GENERAL DISCUSSION

The common aluminum oxide precursors such as $AlCl_3$ and $Al(OR)_3$ are Lewis acids that exist as dimers (e.g., Al_2Cl_6) and oligomers (e.g., Al ethoxide). For the preparation described here, the alkoxides are more suitable precursors. In the presence of a strong Lewis base such as an amine, Al_2Cl_6 dimers transform into monomeric Lewis acid-base adducts [31, 32]. However, exposure to water would hydrolyze the Al-Cl bond to form HCl, which would subsequently react with the amine to form amminium chloride and destroy the amine-aluminum adduct. On the other hand, aluminum alkoxide produces alcohol on hydrolysis, thus avoiding this problem.

Not every amine is equally effective in protecting the Al coordinatively unsaturated site. Amines that exchange rapidly with free amine in solution are unlikely

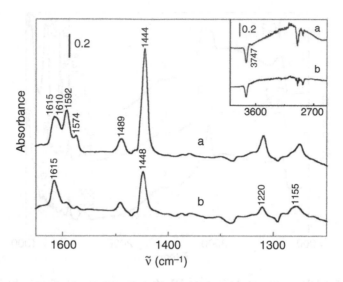

FIGURE 2.6 FTIR difference spectra due to pyridine adsorption on Al_2O_3-Pipy **III** heated *in vacuo* to 250°C, collected at room temperature before and after pyridine adsorption followed by evacuation. (a) Pyridine adsorption at room temperature, and (b) followed by heating *in vacuo* at 150°C.

to be effective because they would be readily displaced by water during hydrolysis. Primary amines, including multidentate ones, belong to this group. For these amines, the α- and β-C peaks in the ^1H NMR spectra are broad, indicating rapid exchange between free and bound amine species.

On the other hand, using a stronger base should lead to a more stable Al-amine adduct. Secondary amines are stronger, but due to steric hindrance, the reaction of secondary amines with the $(Al(OBu^t)_3)_2$ dimer is very slow. This problem can be overcome by the associative ligand substitution approach shown by Reaction R2.4. Addition of piperidine to a solution containing monomeric $Al(OBu^t)_3(Pr^nNH_2)$ results in the formation of $Al(OBu^t)_3(Pipy)$, even at room temperature. The complex is stable at room temperature and provides the desired protective function to the Lewis acid site of alumina during hydrolysis.

The consequence of losing the protective amine is the formation of alumina similar to conventional preparation. This is illustrated in the preparation when H_2O in excess of stoichiometric amount was used in the hydrolysis. The resulting alumina possesses much less Al of coordination IV and V (Figure 2.2), and no piperidine was detected on its surface as determined by ^{13}C (CP) MAS NMR. A similar observation of effect of low H_2O:alkoxide ratio has been reported by Coster and Fripiat [33]. Simply having excess piperidine present in the solution does not offer as effective protection as bound ones. The ratio of $Al^{IV} + Al^V$ to Al^{VI} peaks was much lower in $Al_2O_3(Pipy)$ than that in Al_2O_3-Pipy **III**.

Preservation of Lewis acidity through hydrolysis and formation of the final solid was confirmed by both ^{13}C MAS NMR and FTIR of adsorbed amines. In ^{13}C NMR, the chemical shifts of the α- and β-carbon peaks of the bound piperidine in

Al_2O_3-Pipy **III** center at δ 25.5, which is distinct from the shift of δ 22.5 in piperidine hydrochloride. Likewise, the IR band frequencies of the bound amine agree with those bound to Lewis acid sites. The NH_2 bending frequency of propylamine confirms that the amine is not protonated [23]. The surface is very dynamic, and the piperidine and propylamine can easily be replaced by NH_3 and pyridine. Exposure of the sample to a low vapor pressure of pyridine followed by evacuation at 60°C results in intense bands of pyridine characteristic of pyridine bound to a Lewis site. The appearance of intense pyridine adsorption bands is accompanied by only relatively very small changes in the isolated hydroxyl bands (Figure 2.4). This is unlike that reported in the literature, where the positive bands of adsorbed pyridine are accompanied by large and complex changes in the hydroxyl region [28]. After **III** was heated to 250°C, followed by pyridine adsorption, the surface has changed as the intensity of the pyridine adsorption relative to the intensity of the hydroxyl band is reduced.

ACKNOWLEDGMENTS

Support was provided by the U.S. Department of Energy, Office of Science, Basic Energy Sciences.

REFERENCES

1. S.G. Nelson, B.-K. Kim, T.J. Peelen, *J. Amer. Chem. Soc.* 122, 9318, 2000.
2. D. Chakraborty, A. Rodriguez, E.Y.-X. Chen, *Macromolec.* 36(15), 5470–5481, 2003.
3. A.I. Kozlov, M.C. Kung, W. Xue, H.H. Kung, *Angewandte Chem. Int. Ed.,* 42, 2415, 2003.
4. W. Xue, M.C. Kung, A.I. Kozlov, K.E. Popp, H.H. Kung, *Catal. Today,* 85(24), 219–224, 2003.
5. W. Braune, J. Okuda, *Angewandte Chem. Int. Ed.* 42(1), 64–68, 2003.
6. X.S. Zhao, M.G.Q. Lu, C. Song, *J. Molec. Catal. A: Chem.* 191(1), 67–74, 2003.
7. M. Tabuchi, T. Kawauchi, T. Kitayama, K. Hatada, *Polymer* 43(25), 7185–7190, 2002.
8. H. Knözinger, P. Ratnasamy, *Catal. Rev. Sci. Eng.* 17, 31, 1978.
9. A. Corma, *Chem Rev.* 97, 2373, 1997.
10. M.E. Davis, *Nature* 417, 813, 2002.
11. J.Y. Ying, C.P. Mehnert, M.S. Wong, *Angew. Chem. Int. Edn. Eng.* 38, 56, 1999.
12. S.A. Bagshaw, T.J. Pinnavaia, *Angew. Chem. Int. Edn. Eng.* 35, 1102, 1996.
13. Z.R. Zhang, T.J. Pinnavaia, *J. Am. Chem. Soc.* 124, 12294, 2002.
14. V. Gonzalez-Pena, I. Diaz, C. Marquez-Alvarez, E. Sastre, J. Perez-Pariente, *Mesopor. Micropor. Mat.* 44, 203, 2001.
15. C.G. Lugmair, K.L. Fujdala, T.D. Tilley *Chem. Mater.* 14, 888, 2002.
16. A. Haaland, in *Coordination Chemistry of Aluminum*, G.H. Robinson (Ed.), VCH, New York, p. 1, 1993.
17. V.J. Shiner, Jr., D. Whitaker, V.P. Fernandez, *J. Amer. Chem. Soc.* 85, 2318, 1963.
18. G. Ofori-Okai, S. Bank, *Heteroatom Chem.* 3, 235, 1992.
19. W. Partenheimer, T.D. Epley, R.S. Drago, *J. Amer. Chem. Soc.* 90, 3886, 1968.
20. V. Jonas, G. Frenking, M.T. Reetz, *J. Amer. Chem. Soc.* 116, 8741, 1994.

21. R.M. Silverstein, G.C. Bassler, T.C. Morrill, *Spectroscopic Identification of Organic Compounds*, 5th ed.; John Wiley & Sons, New York, p. 175, 1991.
22. S.J. Schauer, C.H. Lake, C.L. Watkins, L.K. Krannich, D.H. Powell, *J. Organomet. Chem.* 549, 31, 1997.
23. S.D. Williams, K.W. Hipps, *J. Catal.* 78, 96, 1982.
24. J. Shen, R.D. Cortright, Y. Chen, J.A. Dumesic *J. Phys. Chem.* 98, 8067, 1994.
25. C.H. Kline, J. Turkevich, *J. Chem. Phys.,* 12, 300, 1944.
26. C. Morterra, A. Chiorino, G. Ghiottoa, E. Garrone, *J. Chem. Soc. Farady I* 74, 271, 1978.
27. C. Morterra, G. Magnacca, *Catal. Today* 27, 497, 1996.
28. X. Liu, R.E. Truitt, *J. Am. Chem. Soc.* 119, 9856, 1997.
29. H. Stolz, H. Knözinger, *Z. Polym.,* 75, 271, 1979.
30. D. Ouafi, D. Maugé, J.C. Duchet, J.C. Lavalley, *React. Kinet. Catal. Lett.* 38, 95, 1989.
31. D.F. Grant, R.C.G. Killean, J.L. Lawrence, *Acta Crystall. B-Struct.* B25, 377, 1969.
32. A. Ahmed, W. Schwarz, J. Weidlein, H. Hess, *Z. Anorg. Allg. Chem.* 434, 207, 1977.
33. D. Coster, J.J. Fripiat, *Chem. Mater.* 5, 1204, 1993.

3 Aerogel Synthesis

Gerard M. Pajonk

CONTENTS

ABSTRACT

Aerogels are ultra-porous solid materials exhibiting large pore volumes (porosities $\geq 90\%$), high surface areas (hundreds of m^2/g), and low bulk densities. They can be directly prepared as powders, chunks, thin films, or monoliths of many tens of cm^2, opaque, translucent, or transparent to light. Despite their low bulk densities, aerogel powders can be fluidized. They are obtained through the combination of sol-gel chemistry and subsequent drying under supercritical conditions with respect to the liquid phase filling their porosities. When the high-temperature supercritical process is applied (with alcohols, acetone, etc.), the catalysts are hydrophobic, and with the low-temperature supercritical method (with CO_2) or when they are dried under the form of ambigels, they are hydrophilic. But in every case, they can be made hydrophilic or hydrophobic at will, in a reversible process, through cogelation or derivatization methods with the intervention of a thermal treatment in the first case and with a hydrophobic reactant in the second case. Aerogels are typically nanomaterials. For all these properties, aerogel solids are very interesting candidates for heterogeneous applications.

3.1 INTRODUCTION

Scientific experts in heterogeneous catalysis are always willing to improve their capacities to propose more efficient catalytic solids by modifying those already existing or by developing new synthesis using already existing reactants (including those that are transformed in the catalytic reactor by changing the nature of the feeds). At the present time, it is clear that catalysis can and must contribute to the protection of our environment (abatement of pollution at all stages, purification of the reactants and products of the catalytic process, preparation of the catalysts themselves) and also to the development of new forms of energy. For the first aim, old catalysts or reactor feeds must be substituted by new, cleaner ones that reach greater selectivity in particular (no side products) in lower energetic conditions (chiefly reaction temperatures). For the second one, entirely new processes must be proposed such as fuel cells, catalytic gas combustors, catalytic gas turbines, and better natural gas reformers for the attainment of dihydrogen.

Aerogels have been successfully tested for a myriad of such applications: hydrogenation, dehydrogenation, isomerization, reforming, three-way catalyst converters, selective oxidations, NOx abatement, CO or CO_2 hydrogenation, VOC degradation, enzyme aerogel-encapsulated reactions like long chain alcohol esterification, and large olefin molecules epoxidation, at the lab scale, at least (1, 2).

It is recalled that the sol-gel method involves very common conditions of temperature not in excess of 80°C in most cases, ambient pressure, and low energy. Typically, it pertains to what is called *soft chemistry* or *chimie douce* (3–6). Once the gel is made, one can choose many conditions of drying in order to obtain the solid: simple evaporation of the liquid phase leading to the formation of xerogels (3), supercritical evacuation of the liquid phase at low temperature and pressure (at around the ambient) or at high temperature and pressure (at around 250°C and 100 bars), both giving aerogels. A new method giving aerogel-like solids has been in progress since the 1990s. It consists of drying the wet gel at ambient pressure by evaporation after exchange of the pore fluid by liquids, developing low surface tension coefficients (cyclohexane, for example), in order to minimize the pore collapse due to the capillary gradient of pressures inside the pores.

Another method is the capping of the surface hydroxyl groups of the solid pore walls by nonpolar molecules like trimethylchlorosilane (silylation) in the case of silica, for example, and recovering of the porosity through the springback effect after the evaporation process. In both cases, the resulting solid is named *ambigel* (3, 7). It is generally accepted that ambigels are not so highly developed solids as aerogels from the textural point of view, but nonetheless they are much closer to them than xerogels are. Finally, one can also dry a wet gel using freeze-drying, which gives cryogels (5), which exhibit textural properties similar to those of aerogels but cannot give films or monoliths, to the author's knowledge. The main advantage of the synthesis of cryogels over aerogels is that no high-pressure vessel is needed to get them.

In this chapter, the aerogel synthesis method is detailed and recent catalytic examples, from 1999 up, will be presented and discussed. Because of their potentialities, ambigels, which do not require the use of an autoclave, are also briefly described.

3.2 THE PRINCIPLES OF THE AEROGEL PROCESS

Silica and, to a less extent, alumina aerogels are detailed below because most of the catalytic aerogels are silica- or alumina-based composites. As mentioned in the Abstract section, the synthesis of an aerogel results from a two-step process: First, a solution-to-sol-to-gel is prepared, and second, this gel is supercritically dried in an autoclave.

3.2.1 THE SOLUTION-TO-SOL-TO-GEL STEP

The main reaction is the condensation of a sol of the gel precursor that is a dispersion of colloidal particles, the dimensions of which range between 1 and 1000 nm (not a true solution), into a gel. A gel can be considered as a macromolecule (or polymer) of a volume equal to that of the liquid mixture that has given birth to it, as shown in Figure 3.1. The gel is apparently a one-phase piece of a solid semielastic material. In fact, it is a two-phase solid made out of a liquid entrapped in the porosity of a solid one. The pores are filled by the liquid phase, which is always characterized by its never-zero-surface tension coefficient γ and which is at the origin of the formation of a concave meniscus in the pore between the liquid and the vapor phases. The higher the surface tension coefficient, the greater the capillary forces. The corresponding capillary forces are given by the well-known Laplace equation, and they exert forces attaining 100 tons/cm on the walls of the pore during their emptying steps. Thus, simple evaporation cannot preserve the porosity developed in the gel when it is still wet. Some surface tension coefficients are given in Table 3.1.

Commonly, gels are obtained by a hydrolysis step followed by a condensation reaction, like in the case of silica and most of the aerogels of catalytic interest. The destabilization of a sol into a gel, which mainly occurs by modifying the pH of the dispersion (see the DLVO theory)[8], can also be performed. The first situation, which is by far the most encountered, is now described. Generally, an alkoxide in solution

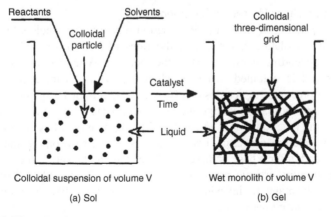

FIGURE 3.1 The sol-gel process.

TABLE 3.1
Some Surface Tension Coefficients at 20°C

Liquid	γ in erg cm^{-2} at 20°C	Liquid	γ in erg cm^{-2} at 20°C
Water	72.8	Benzene	29.0
Monoethanolamine	48.4	Acetic acid	27.4
Diethylene glycol	≈48.5	Chloroform	25.3
Ethylene glycol	46.5	Acetone	23.3
Aniline	≈44.5	Methanol	22.6
Dimethylsulfoxide	43.0	Ethanol	22.3
Phenol	40.9	Heptane	20.3
Dimethylformamide	36.8	Di-isopropylamine	20.0
Formamide	36.0	Diethylether	17.1
Dioxane	≈34.5	CO$_2$ (liquid)	≈1.0–2.0
Dichloroethane	32.2		

in an organic solvent is hydrolyzed in the presence (or not) of a catalyst according to Equation 3.1:

3.1°   -M-OR (precursor molecule) + H$_2$O → -M-OH + ROH,

where the alkoxide can be Si(OCH$_3$)$_4$ (tetramethoxysilane).

Equation 3.2 and Equation 3.3 describe the condensation reaction:

3.2°   -M-OH + -M-OH → -M-O-M- + H$_2$O

3.3°   -M-OH + -M-OX → -M-O-M + X-OH,

where X can be either H or an alkyl group.

Roughly speaking, hydrolysis/condensation reactions are nucleophilic substitutions (olations, oxolations). Generally, the rates of hydrolysis and condensation are different, depending upon the pH of the medium. When the medium is acid, a polymer-like gel is obtained, whereas when the pH is basic, a colloidal gel is prepared. The precursor molecules can be either organic or inorganic ones.

According to the partial charge model, silicon organic precursors are not very reactive, and the reactions must be catalyzed by an acid or a base or even by both in a two-step gelation process: First, an acid is added to promote the hydrolysis phenomenon and, after a base is mixed, to favor the condensation reactions. On the contrary for alumina organic precursors (high partial charge) like Al-sec-butylate, one has the advantage to introduce chelating molecules like ß-diketones metal complexes (acetylacetonate or acac, ethylacetoacetate or etac) in order to control the rate of gelation. Thus, when two or more precursors of different chemical hydrolysis or condensation reactivity are used in a cogelation step or stepwise, one can play with the relative rates of reaction to tailor the final solid composite catalysts. Water is used to perform the hydrolysis reaction; this defines a water/precursor molar ratio h, also called the hydrolysis parameter, so that with current metal alkoxides $M(OR)_m$ three cases are possible:

1. h < 1: No gelation nor precipitation reactions occur.
2. 1 < h < m: A polymeric gel may be formed.
3. h > m: Particulate gels, cross-linked polymer gels, or precipitate may occur.

The rate of addition of water also exerts an influence on the qualities of the gels. When both rates of hydrolysis and condensation are slow, a sol is formed; when they are both fast, a colloidal gel or a gelatinous precipitate is obtained. A polymeric gel is made when the rate of hydrolysis is faster than the rate of condensation, and in the reverse situation, controlled precipitation is the result. The gel can be aged *in situ* and washed before being dried, as shown in Figure 3.2.

The principal advantages to make a gel are numerous. It provides a homogeneous means of mixing the precursors at the molecular level, it involves low-cost energy and environmentally friendly chemistry (soft chemistry), it allows molecular mixing in all proportions, and of course it develops nanochemical solids.

It is worth mentioning that other parameters are also important in the synthesis of gels: (a) the chemical nature of the precursors, (b) the nature of the solvent or

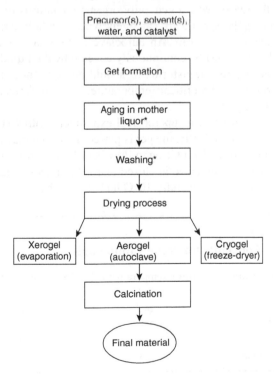

FIGURE 3.2 Sol-gel and drying flowchart. *The aging and washing steps are optional.

dispersing medium, polar or not, aprotic or not, (c) the chemical nature of the catalysts used for the synthesis, (d) the application of an aging step or not, (d) the temperature of the sol-to-gel reaction, (e) the mixing process: mechanical or ultrasonical (solventless method is possible for sonogels, for example), (f) the use of a drying control chemical additive (DCCA), and (g) the pH of the medium.

The next but not necessarily last step is the drying one, which is now described below.

3.2.2 THE SUPERCRITICAL DRYING STEP

In order to avoid the development of the capillary gradient stresses responsible for the collapse of the porosity (and subsequently the surface area) of the dry solid, one must operate a process where the surface tension of the liquid phase is zeroed. Such a situation is easily realized by drying the wet gels above the critical temperature T_c and pressure P_c of the liquid phase (i.e., at supercritical temperature and pressure) because when one liquid reaches (or exceeds) its critical temperature, its surface tension vanishes. The critical parameters are functions of the chemical nature of the liquid; for instance, Table 3.2 gives some values for common liquids.

Hydrogels can never be supercritically dried with respect to water because their critical constants are so high that all gels are peptized and recrystallized, leading to very poor textural properties. Consequently, before supercritical extraction of the wet aqua (or hydro)-gel, water must be exchanged with a more "gentle" liquid medium. This is the reason why a generally organic medium is used for the sol-to-gel step giving orgagels, specifically alcogels when alcohols are the medium or acetogels when acetone is chosen. An autoclave is necessary because during the heating of the gels, they must be continuously covered by the liquid until the critical conditions are met, where it transforms in a fluid phase without the building of a liquid-vapor interface. At the critical temperature, a gas is formed, exhibiting the same density as the liquid.

Figure 3.3 is a diagram of a supercritical experiment with CH_3OH, and Figure 3.4 shows an autoclave. Practically, the gel is placed in the autoclave with a sufficient amount of liquid and flushed by a nitrogen flow in order to extract all traces of water contained in air, then the autoclave is closed and subsequently heated at the supercritical temperature selected, generally 10°C higher than the corresponding T_c. After

TABLE 3.2
Critical Constants for some Compounds

Compound	Critical temperature in °C	Critical pressure in bars
Water	374	218
Ethanol	240	63
Methanol	240	79
Acetone	235	43
Carbon dioxide	31	73

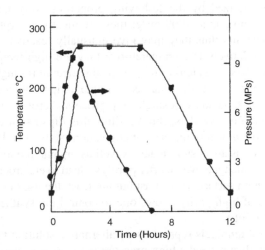

FIGURE 3.3 Temperature against pressure diagram during a high-temperature supercritical drying run with methanol.

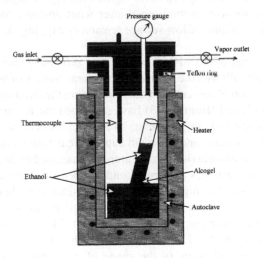

FIGURE 3.4 Autoclave schematic.

equilibration, the outlet valve is opened to reach the ambient pressure under isothermal conditions (T_c + 10°C, in general). When the pressure is the atmospheric one, the autoclave is again flushed and filled by technical nitrogen; only after this step, the cooling of the autoclave is performed (under nitrogen) and the lid is opened when the temperature is the ambient one.

It is always possible to introduce a reducing reactant like dihydrogen in the autoclave when a supported metal catalyst is synthesized.

3.2.3 SOME ADVANTAGES OF THE AERO(GEL) SYNTHESIS

Aerogels are characterized by the following properties: very large porosity and surface areas up to more than 1000 m^2/g; they are mostly amorphous XRD solids in a metastable state and thus they may give unusually reactive solid phases after heat treatments at relatively low temperatures; they are high-temperature-resisting solids that can develop several tens of m^2/g even at temperatures higher than 1000°C; and all compositions are easily obtained (simple oxides, mixed oxides, metal-supported oxides, etc.), mostly in a single-step process. Aerogels can develop many morphologies: monoliths, powders, chunks, films after dip or spin-coating. They can be made hydrophobic *in situ* in the case of cogelation (9, 10, 11). Hybrid composite organic/inorganic solids are easy to make as well as organic monolithic copolymers like resorcinol/formaldehyde, which after pyrolysis lead to electroconducting carbon monoliths of very high surface areas. Last but not least, the use of commercial metal organic complexes of high purity allows one to control the synthesis steps with an excellent reproducibility.

The synthesis of ambigels represents an alternative solution to the preparation of aerogels, which always need a high-pressure vessel (autoclave). However, up to now, there is no catalytic application described in the literature. The main drawback is probably the time-consuming step of the liquid exchange in highly porous textures compared to the autoclave method, no matter what ambigel method is chosen: simple evacuation of a liquid of low surface tension or capping (3, 7, and references therein).

Recently, in the end of the 90s, an exciting and promising chemistry, performed in supercritical fluids, allowed the fabrication of a large variety of new aerogel solids not needing the addition of solvents and thus shortening the duration of the synthesis. Shakesheff et al. (12) and Brunner (13) have published the features of such preparations in supercritical CO_2 in particular. The direct formation—in one step—of silica microparticles (spheres and fibers) in supercritical acetone (high temperature) or supercritical carbon dioxide (low temperature) is described by Moner-Girona et al. (14). Here, the sol-gel step was performed under supercritical conditions; when the aerogels were synthesized at high temperature, the silica particle dimensions were between 1.2 and 2.2 μm, depending on the hydrolysis h values (see Section 3.1.1, the higher the h, the larger the particle dimensions). The structure of these silica particles was examined by TEM, SEM, and AFM techniques, showing nanoparticles of 5–30 nm diameter and pores in the range of 5–50 nm, characteristic of the materials that are similar to conventional aerogels. Smirnova and Arlt (15) observed that supercritical CO_2 itself is a very efficient catalyst for the sol-gel step and plays two roles, being a catalyst and an extraction supercritical fluid at the same time. Supercritical CO_2 was even found to be a better gelation catalyst than ammonia or hydrochloric acid.

It seems that new opportunities exist in aerogel synthesis, taking advantage of both the ambigel method and the direct sol-gel chemistry under supercritical conditions. It is likely that these new methods could be of a great potential interest for scientists involved in the preparation of catalytic materials.

3.3 SOME EXAMPLES OF THE SYNTHESIS OF CATALYTIC AEROGELS

Liquid-phase hydrogenation nickel-alumina-based catalysts were prepared by Suh et al. (16) in order to convert benzophenone into benzhydrol and, further, diphenyl methane at 130°C according to the following reactions:

$$C_6H_5\text{-}CO\text{-}C_6H_5 + H_2 \rightarrow C_6H_5\text{-}CHOH\text{-}C_6H_5 + H_2 \rightarrow C_6H_5\text{-}CH_2\text{-}C_6H_5 + H_2O,$$

and to hydrogenate a fatty acid such as soybean oil, which is a mixture of C_{18} fatty acids containing one or more double bonds and tri-esters of glycerol. The aerogel catalyst was made from Al-sec-butylate and nickel acetate dissolved in ethanol and dried with respect to CO_2 and contained an amount of nickel of 20 wt%. They tested an Ni-Al cogel, a nickel-impregnated alumina aerogel, and an impregnated commercial γ-alumina with the same contents in nickel. All catalysts were reduced by dihydrogen at 450–700°C, and thereafter they were passivated at room temperature by a flux of 2% air in helium mixture. The catalytic results demonstrated that the best catalyst for the conversion of benzophenone in terms of conversion percentage and selectivity—both near 100%—was the cogel one, then the alumina aerogel-impregnated one, and finally the nickel on γ-alumina.

Concerning the soybean oil conversion, the best one was again the cogel, followed by the Ni on γ-alumina catalyst, whereas the worst one was the Ni on alumina aerogel. Another study of the catalytic behavior of nickel-alumina aerogels containing 15 up to 50 wt% of metal was published by Krompiec et al. (17). The catalysts were prepared under the form of cogels from the same reactants as in the previous study (15), but etac was reacted with the Al-sec-butylate in order to complex the aluminum precursor. The gels were aged for 5 days prior to their drying in a stream of supercritical carbon dioxide. The resulting solids were reduced in a stream of dihydrogen at 550°C. Five reactions test were performed, and the corresponding experimental conditions are listed in Table 3.3; for all the tests, the conversion and selectivity were 100%.

In both studies (16, 17), the catalysts exhibited pore size distributions characteristic of alumina in general; i.e., they were mesoporous solids. They attempted to selectively reduce NO by propene in the presence of dioxygen at temperatures

---

TABLE 3.3
Catalytic Reaction Tests on a 50 wt% Nickel on Alumina Aerogel

Reaction test	Reaction temperatures range in °C	Reaction product
Propene hydrogenation	65–210	Propane
1-Butene hydrogenation	65–210	Butane (no isobutene at all)
Conversion of CO by NO	200–300	Carbon dioxide and dinitrogen
Carbon dioxide methanation	320–360	Methane
Cyclohexene dehydrogenation	300–350	Benzene

between 200 and 600°C on pure alumina aerogel and galia containing alumina composite aerogel (18). Alumina was made from Al-sec-butylate in the presence of etac in ethanol; galium nitrate was added in the former mixture, with different loadings varying from 38 up to 66 wt%. The gels were obtained under the form of monoliths and dried with respect to supercritical ethanol. The addition of galia Ga_2O_3 had two consequences on the texture and morphology of the pure alumina aerogel, compared to the composite one. The pore volume and the specific surface area, as well as the average pore size, increased with the addition of galia versus pure alumina. Particles of pure alumina aerogel either were microfibrous or adopted the shape of needles, whereas in the composite aerogel particles, they shift to cubic ones or granules when the content of galia was >50 wt%. Between 350 and 500°C, the NO selective conversion was significantly higher than with pure alumina, and it was recorded that the oxidation of propene followed the same trend.

Wang and Willey (19) synthesized fine iron oxide particles (Fe_2O_3) made out of a solution of iron (III)acetylacetonate in methanol and water; this solution (no gel was formed at room temperature) was poured into an autoclave and evacuated with respect to the conditions of supercritical methanol. The iron oxide aerogel developed a specific surface area of 10 m^2/g. The primary particle dimensions were found to be 8–30 nm, as shown by XRD technique. The catalytic test run was the partial oxidation of methanol in the autoclave in the presence of supercritical CO_2 at temperatures varying from 225 up to 325°C and the pressure was 91 bars. The main reaction product formed was dimethyl-ether, small amounts of formaldehyde, and methyl formate with a selectivity below 10% for both minor products. A 20% iron oxide-molybdenum oxide aerogel tested in the same supercritical conditions showed a very good selectivity of 94% for formaldehyde, the other product being only dimethyl-ether.

It is well known that Au is catalytically inert in its bulk state, but nanoparticles (particles of about 5 nm) of gold are very active catalytic sites for oxidation, epoxidation of propene, and the water-gas-shift reactions, even at low temperatures. Tai et al. (20) fabricated titania-TiO_2-and titania-coated silica aerogels for the oxidation of CO at temperatures as low as –40°C. In order to preserve the high division state of gold, a wet gel of titania or titania-coated silica was impregnated by a solution of toluene containing Au nanoparticles passivated by dodecanethiol. The titania precursor was titanium-tetraisopropoxide in ethanol; silica was prepared with tetramethoxysilane in methanol. The coating was obtained by contacting a solution of titanium precursor in toluene. All gels were supercritically dried with respect to CO_2. A calcination treatment at 400°C prevented the formation of sulfate, a poison produced during the combustion of thiols. This precaution did not vary the Au nanoparticle sizes, which were 2.2 and 4 nm for the titania-coated silica and the titania, respectively. The Au loading was of a few wt% (less than 5%). CO was converted into carbon dioxide with a conversion of 100% at all reaction temperatures (–40 to 0°C). Titania-coated silica catalysts were more active than the Au-titania one.

In order to run photocatalytic oxidative remediation of airborne VOC on titania, Cao et al. (21) made low-density silica-titania aerogels by first preparing a silica aerogel and then impregnating the monolithic silica by a solution containing the Ti precursor. Finally, the impregnated silica aerogels were one more time supercritically dried but with supercritical CO_2. The catalysts were revealed to be active for the UV-photo-oxidation of both CO and acetone.

Kalies et al. (22) studied the formate species hydrogenation on a Pt-zirconia aerogel catalyst. Zirconia aerogel was synthesized by the hydrolysis of a solution of Zr(IV) isopropylate in isopropanol and the solvent evacuation in supercritical conditions with respect to the alcohol. The zirconia aerogel was further contacted with a solution of hexachloroplatinic acid in isopropanol to give a Pt wt% of 0.5; the Pt-impregnated zirconia was dried with supercritical isopropanol. All solids were XRD amorphous. The reaction was monitored by FTIR and showed that the formate species originating from the adsorption of CO/He was hydrogenated by H_2 into methoxy moieties only when Pt was present on zirconia. These species were further converted into methane through a reverse spillover mechanism (23). On pure zirconia, the formate species were not hydrogenated because of the inability of zirconia to dissociate dihydrogen.

The total combustion of ethylacetate—a good VOC model molecule—was achieved on bulk chromia-based aerogels by Rotter et al. (23). Aqueous chromium nitrate nanohydrate was gelled in the presence of urea and dried either with CO_2 (C_5 sample) at 45°C or methanol at 295°C (C_8 sample). The low-temperature drying led to a narrow pore size distribution (pores of 5 nm), whereas the high-temperature drying exhibited a wide pore distribution from 5 to 50 nm. Another series of composite aerogels was based upon Co-Ce-Mn and Cu oxides, which were deposited on C_8 chromia aerogels by impregnation, followed by methanol supercritical drying. The promoted chromia aerogels showed significantly lower light off temperatures, total combustion of the VOC with a selectivity of 95% in CO_2, than the pure chromia aerogels, and the ceria chromia exhibited the highest catalytic activity, which was explained by the ceria ability to store oxygen species measured by its oxygen storage capacity (OSC).

New bio-organic aerogels like chitosan can be very useful catalysts for fine chemistry purposes. Chitosan is a natural polysaccharide derived from chitin and constitutes the largest biomass polysacharide. Valentin et al. (25) imagined drying with supercritical carbon dioxide gels of beads of chitosan. The so-obtained chitosan aerogel was able to catalyze the esterification reaction between fatty acid—lauric acid—and glycidol, yielding the corresponding monoglyceride-α-monolaurin- at 70°C. Freezed chitosan (cryogel) was totally inactive, whereas the chitosan aerogel exhibited a high glycidol conversion (98%) and a selectivity of 71% calculated on the basis of the glycidol reactant after 24 h of reaction. The same team extended this previous work in making an inorganic/bio-organic aerogel, chitosan-silica, using again CO_2 to extract the alcoholic solvent (ethanol). The chitosan beads were mixed with tetraethoxysilane and NaF as catalyst. The hybrid aerogel showed a fibrous shape analogous to chitosan as seen by SEM technique. When the core-shell morphology was observed for the microspheres, the core was found to be the hybrid aerogel and the shell was made of pure silica aerogel. The same catalytic test reaction with lauric acid and glycidol was run and showed a conversion of 82%, a selectivity into α-monolaurin near 65% (26).

An example of forming a mesoporous alumina aerogel without reacting an alkoxide as an alumina precursor is described in a paper by Nguyen et al. (27).

To make the alumina gel, they used aluminum chloride and propylene oxide in ethanol at room temperature, propylene epoxide was used to scavenge protons, and the wet gel was dried with supercritical carbon dioxide. These aerogels had a specific surface area of 1147 m²/g after a thermal treatment in vacuum at 210°C and kept a very unusual high surface area of 247 m²/g after 2 h at 1000°C in the presence of dioxygen. The aerogel sample remained XRD amorphous up to 800°C, and at higher temperatures, the formation of γ-alumina was detected. According to the results claimed here, a new form of simple alumina aerogel, which resists very well towards high-temperature treatments, is a promising candidate for catalytic composite systems in total catalytic combustion reactors and three-way exhaust catalysts.

3.4 CONCLUSION

Though most of the papers published in the open literature deal with aerogels as materials only pertaining to materials science, their properties can be usefully applied to heterogeneous catalytic systems. A look at References 1, 2, 4, 5, and 6 and the references therein clearly demonstrates myriad aerogel catalyst synthesis and applications. The flexibility of the sol-gel method is at the origin of this situation, and the many ways to dry the gels into xerogels, ambigels, cryogels, and aerogels give birth to a vast number of possible combinations. Tailor-made aerogels as well as ambigels are easy to synthesize from commercial-grade reactants. Up to now and despite the fact that ambigels are studied only by materials scientists, it is obvious that they deserve to be systematically studied for catalytic applications because of their simplicity and safety of fabrication. As previously mentioned for aerogels, it seems always possible to finely tune their properties, a quality highly appreciated in heterogeneous catalysis.

ACKNOWLEDGMENT

The author is grateful to ACS, PRF, and Keller Companies for their support.

REFERENCES

1. G.M. Pajonk, *Catal. Tod.* 52, 3, 1999.
2. G.M. Pajonk, *Recent Res. Devel. Catalysis* 2, 1, 2003.
3. C. Alie, A.J. Lecloux and J.P. Pirard, *Recent Res. Devel. Non-Crystalline Solids* 2, 335, 2002.
4. G.M. Pajonk and A.V. Rao, *Recent Res. Devel. Non-Crystalline Solids* 1, 1 2001.
5. A.C. Pierre and G.M. Pajonk, *Chem. Reviews* 102, 4243, 2002.
6. G.M. Pajonk, *Colloid Polym. Sci.* 281, 637, 2003.
7. D.R. Rolison and B. Dunn, *J. Mater. Chem.* 11, 963, 2001.
8. E.J.W. Verwey and J.Th.G. Overbeek, *Theory of Stability of biophilic colloids*, Elsevier, Amsterdam, 1948.
9. A.V. Rao, S.D. Baghat, G.M. Pajonk and P. Barboux, accepted paper for *J. Non-Cryst. Solids, Proceedings of the 7th Inter. Symposium on Aerogels,* Alexandria, VA, 2–5 November 2003, 2004.

10. A.V. Rao, M.M. Kulkarni, G.M. Pajonk, D.P. Amalnerkar and T. Seth, *J. Sol-Gel Sci. and Technol.* 27, 103, 2003.
11. A.V. Rao, R.R. Kalesh and G.M. Pajonk, *J. Mater. Sci.* 38, 4407, 2003.
12. H. Woods, M.C.G. Silva and C. Nowel, *J. Mater. Chem.*, 11, 1663, 2004.
13. G. Brunner, Supercritical Fluids as Solvents and Reactant Media, Elsevier, 2004.
14. M. Moner-Girona, A. Roig, E. Molins and J. Llibre, *J. Sol-Gel Sci. and Technol.* 26, 645, 2003.
15. I. Smirnova and W. Arlt, *J. Sol-Gel Sci. and Technol.* 28, 175, 2003.
16. D.J. Suh, T.J. Park, S.H. Lee and K.L. Kim, *J. Non-Cryst. Solids* 285, 309, 2001.
17. S. Krompiec, J. Mrovec-Bialon, K. Skutil, A. Dokowicz, L. Pajak and A.B Jarzebski, *J. Non-Cryst. Solids* 315, 297, 2003.
18. T. Chono, H. Hamada, M. Haneda, H. Imai and H. Hirashima, *J. Non-Cryst. Solids,* 285, 333, 2001.
19. C.T. Wang and R.J. Willey, *Catal. Tod.* 52, 83, 1999.
20. Y. Tai, J. Murakami, K. Tairi, F. Ohashi, M. Date and S. Tsubota, *Appl. Catal. A: Gen.* 268, 183, 2004.
21. S. Cao, K.L. Yeung and P.L. Yue, *Proceedings 13th International Congress on Catalysis*, Paris, 11–16 July 2004.
22. H. Kalies, N. Pinto, G.M. Pajonk and D. Bianchi, *Appl. Catal. A: Gen.* 202, 197, 2000.
23. G.M. Pajonk, *Appl. Catal. A: Gen.* 202, 157, 2000.
24. H. Rotter, M.V. Landau, M. Carrera, D. Goldfarb and M. Herskowitz, *Appl. Catal. B: Environ.* 47, 111, 2004.
25. R. Valentin, K. Molvinger, F. Quignard and D. Brunel, *Proceedings 13th International Congress on Catalysis*, Paris, 11–16 July 2004.
26. K. Molvinger, F. Quignard and D. Brunel, *Proceedings 13th International Congress on Catalysis,* Paris, 11–16 July 2004.
27. M.T. Nguyen, T.J Park and D.J. Suh, *Proceedings 13th International Congress on Catalysis*, Paris, 11–16 July 2004.

4 Fine-Tuning the Functionalization of Mesoporous Silica

Hung-Ting Chen, Seong Huh, and Victor S.-Y. Lin

CONTENTS

4.1 INTRODUCTION

Design and synthesis of materials with well-defined particle size, morphology, and ordered mesoporosity (2 to 10 nm in pore diameter) is a burgeoning area of current research in materials chemistry. These materials have two different surfaces, the *interior pore surface* and the *exterior particle surface*, which offer many advantages over solid particle materials. The mesoporous structure provides a size and functional group selective microenvironment that allows encapsulation of the desired molecular moieties and shelters these molecules from exposure to the external environment. The unique structural features of these mesoporous materials are important prerequisites for utilization in diverse areas, such as catalysis, chromatographic supports, controlled release of drugs or agrochemicals, development of medical implants, miniaturization of electronic devices, sensor design, and formation of semiconductor nanostructures.[1-12] To realize these applications, the desired mesoporous material should also have the following features:

1. Chemically, thermally, and mechanically stable structure
2. Ordered particle and pore morphology
3. Large surface area and tunable pore size
4. Selectively functionalizable interior and exterior surfaces

A major breakthrough in fabrication of mesoporous material was the development at Mobil Corporation of the MCM family of mesostructured silicas by utilizing surfactants as structure-directing templates to generate a range of MCM-type mesoporous silica structures with tunable pore size and pore morphology, such as MCM-41 and MCM-48 silicas consisting of hexagonal channels and cubic pores, respectively.[13, 14] Over the past decade, several other mesoporous silica materials with ordered porous structures, such as SBA-,[15, 16] MSU-,[17, 18] and FSM-type[19] of mesoporous silicas, have also been developed. The typical synthesis of these structurally well-defined mesoporous silicas is based on a surfactant micelle templating approach. In an acidic or basic aqueous solution, organic surfactants, such as Pluronic P123 triblock copolymer[15, 16] and cetyltrimethylammonium bromide (CTAB),[13, 14] first form self-assembled micelles. These micelles serve as a structure-directing template that can interact with oligomeric silicate anions via hydrogen bonding or electrostatic interaction during the condensation reaction of tetraethoxysilane (TEOS). By either calcinations or acid extraction, the organic surfactants are removed, leaving an inorganic mesoporous silica framework. Depending on the specific synthetic condition, a disordered, hexagonal, or cubic pore structure of mesoporous silica can be obtained.

4.2 CONVENTIONAL METHODS FOR ORGANIC FUNCTIONALIZATION OF MESOPOROUS SILICA MATERIALS

To utilize these mesoporous silica materials for the aforementioned applications, it is important to develop methods for controlling the degree of organic functionalization, so that the physical and chemical properties of the mesoporous silicas can be fine-tuned. Generally, organic functional groups can be immobilized to the silica surface through covalent or noncovalent interactions between the organic moiety and the surface silanol group. For most applications, functionalization through covalent bonding is preferred in order to circumvent the undesired leaching problem.

4.2.1 Postsynthesis Grafting Method

Among various surface functionalization methods, the postsynthesis grafting method is the most popular approach for covalently incorporating organic functionalities to mesoporous silica materials. As shown in Figure 4.1, this method is based on a condensation reaction between a given organosilane, such as $RSi(OR')_3$, $RSiCl_3$, or $HN(SiR_3)_2$, and the surface free silanol ($\equiv Si\text{-}OH$) and geminal silanol ($=Si(OH)_2$) groups of the surfactant-removed mesoporous silica in dry nonpolar solvent, such as toluene and benzene. Even though a wide variety of organic functionalities have been introduced to the mesoporous silica surface via this straightforward approach, it has been found that most materials functionalized via the grafting method contain an inhomogeneous surface coverage of organic functional groups.[20] This result has been attributed to the diffusion-dependent mass transport issue associated with these 3D mesoporous materials. Given that the silanols located on the external surface and the pore opening are kinetically more accessible than those of the interior pore surface, most organic functional groups introduced by grafting have been shown to be located on the external surface or congregated at the mesopore opening.

4.2.2 Organosiloxane/Siloxane Co-condensation Method

Another common approach for preparing organically functionalized mesoporous silica materials is the co-condensation method, which is a direct synthesis method where a given organoalkoxysilane is introduced to the aqueous solution of CTAB and TEOS during the condensation reaction (Figure 4.2). In order to efficiently incorporate organic functional groups to the mesoporous silica surface, the organo-silane precursors need to compete with silicate anions to interact favorably with the surfactant micelles by either electrostatic or other noncovalent interactions

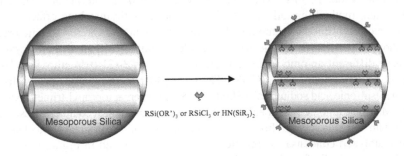

RSi(OR')₃ or RSiCl₃ or HN(SiR₃)₂

Mesoporous Silica   Mesoporous Silica

FIGURE 4.1 Schematic representation of functionalization of mesoporous silicas by the postsynthesis grafting method.

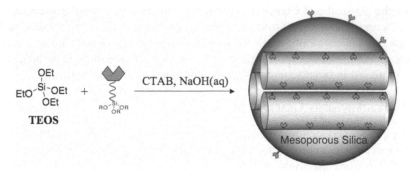

FIGURE 4.2 Schematic representation of functionalization of mesoporous silicas by the co-condensation method.

during the acid- or base-catalyzed condensation of silicate. Therefore, the choice of the organosilane precursors for the co-condensation reaction is limited to those with organic functional groups that would be soluble in water and could tolerate the extreme pH conditions that are required for the synthesis of mesoporous silicas and the subsequent removal of surfactants. Furthermore, the amount of functional groups introduced by the co-condensation method often cannot exceed 25% surface coverage without destroying the structural integrity and the long-range periodicity of the synthesized materials.[7] Despite these limitations, it has been found that the spatial distribution of the pore surface-immobilized organic groups in the mesoporous silica materials functionalized by the co-condensation method are more homogeneous than those of the postsynthesis grafting method as recently reviewed by Stein and coworkers.[7, 20]

4.3 INTERFACIAL DESIGNED CO-CONDENSATION METHOD

Because of the aforementioned disadvantages of current synthetic methods, developing new techniques for controlling the loading as well as the spatial distribution of organic functional groups on the mesoporous surface is of keen interest of many research groups worldwide. To this goal, we recently developed an interfacial designed co-condensation method that utilizes an effective electrostatic matching effect between the cationic cetyltrimethylammonium bromide surfactant and various anionic organotrimethoxysilanes.[21] We synthesized disulfide-containing organotrimethoxysilanes that have different anionic functional groups, such as 3-(3'-(trimethoxysilyl)-propyldisulfanyl)propionic acid (CDSP-TMS), 2-[3-(trimethoxysilyl)-propyldisulfanyl]ethanesulfonic acid sodium salt (SDSP-TMS), and mercaptopropyl-trimethoxysilane (MP-TMS). The anionic functional groups of these organosilanes could electrostatically match with the cationic cetyltrimethylammonium bromide surfactant micelles in an NaOH-catalyzed condensation reaction of TEOS as depicted in Figure 4.3.

Three organically functionalized mesoporous silica nanoparticles, MSN-COOH, MSN-SO$_3$H, and MSN-SH, were synthesized. TEOS and a corresponding organotrimethoxysilane (10 mol% to the TEOS) were added dropwise to an aqueous solution

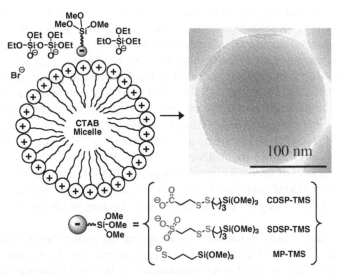

FIGURE 4.3 Schematic representation of the utilization of anionic organoalkoxysilanes for controlling the functionalization of the mesoporous silica nanoparticle (MSN) materials. The MCM-41 type mesoporous channels are illustrated by the parallel stripes shown in the TEM micrograph of the MSN-SH material. (Figure modified from Reference 21.)

of CTAB and NaOH mixture at 80°C. After stirring for 2 h, the white solid products were isolated by filtration and washed with methanol. The surfactant-removed mesoporous silica nanoparticle (MSN) materials were obtained by an acid extraction. All three organically functionalized MSNs exhibited spherical particle shapes, with an average particle diameter of 200 nm, as shown in the TEM micrograph (Figure 4.3). The powder x-ray diffraction (XRD) of these materials showed a typical MCM-41 diffraction pattern, including (100), (110), and (200) peaks.[21] Type IV BET isotherms without any hysteresis were observed in all materials.[21] The pore volumes, BET surface areas, and Barret-Joyner-Halenda (BJH) average pore diameters of these MSNs are very similar, as summarized in Table 4.1.

By treating MSN-COOH and MSN-SO$_3$H with a reducing agent, dithiothreitol (DTT), the disulfides were chemically converted to free thiols in these materials. UV/Vis absorbance measurements and elemental analyses of the resulting thiol-functionalized MSNs were performed to quantify the amount of thiols in each

TABLE 4.1
N$_2$ Sorption Analysis of the Organically Functionalized MSNs

Materials	BET surface area (m²/g)	Pore volume (cm³/g)	BJH average pore diameter (Å)
MSN-SH	999	0.793	25.8
MSN-COOH	920	0.657	27.3
MSN-SO$_3$H	863	0.755	28.1

material. Given that only the chemically accessible thiols could react with 2-ald-rithiols and yield 2-pyridothione in the solution, as shown in Scheme 4.1, the amount of surface-immobilized thiols in each material was determined by measuring the supernatant concentration of 2-pyridothione. The results indicated that the surface concentration of the chemically accessible thiol functional group increased from MSN-SH (0.56 ± 0.01 mmol/g), MSN-COOH (0.97 ± 0.01 mmol/g), to MSN-SO$_3$H (1.56 ± 0.01 mmol/g). However, the amounts of thiols in these three MSN materials, measured by element analysis, exhibited a different trend in the order of MSN-COOH < MSN-SO$_3$H < MSN-SH, as summarized in Table 4.2. Although total loading of thiol groups in the MSN-SO$_3$H material was lower than MSN-SH, the amount of chemically accessible thiol groups in MSN-SO$_3$H material was 3 times higher than that of MSN-SH. These results indicated that the least hydrated sulfonate-containing SDSP-TMS gave rise to the highest loading of chemically accessible organic groups.

Furthermore, relative to the carboxylate group, the thiolate functionality offered less stabilization of the CTAB micelle aggregates and, therefore, gave rise to the lowest loading. This observation was consistent with the results discovered by Larsen and Magid.[22, 23] They reported on the anionic lyotropic series (citrate < CO_3^{2-} < SO_4^{2-} < $CH_3CO_2^-$ < F^- < OH^- < HCO_3^- < Cl^- < NO_3^- < Br^- < $CH_3C_6H_4SO_3^-$) for interaction with the CTAB surfactant micelle based on the enthalpy of transfer of the salt from the water to solution of 0.1 M CTAB. They concluded that anions less hydrated than Br^-, such as sulfonate, would be able to replace Br^- and bind tightly to the cetyltrim-ethylammonium head group of the CTAB molecule, thereby effectively mitigating the repulsion between these cationic head groups and stabilizing the micelle structure. In short, our results have shown that the amount of chemically accessible organic functional groups of MCM-41 silicas can be fine-tuned by carefully designing the interfacial electrostatic matching between the surfactant head groups and the desired organic functional group precursors in co-condensation reaction.

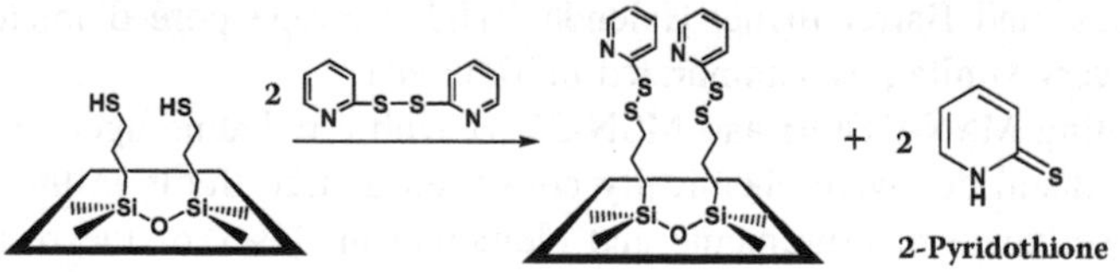

2-Pyridothione

SCHEME 4.1

TABLE 4.2
Elemental Analysis of the Organically Functionalized MSNs

Materials	C%	H%	S%
MSN-SH	13.23 ± 0.01	2.74 ± 0.01	10.09 ± 0.01
MSN-COOH	8.60 ± 0.01	2.33 ± 0.01	5.98 ± 0.01
MSN-SO$_3$H	9.79 ± 0.01	2.60 ± 0.01	7.95 ± 0.01

4.4 CONTROL OF PARTICLE MORPHOLOGY OF ORGANICALLY FUNCTIONALIZED MSNS BY INTERFACIAL DESIGNED CO-CONDENSATION METHOD

The ability to fine-tune the concentration and surface distribution of the organic functional groups alone, without considering the surface morphology, is not sufficient to design efficient single-site heterogeneous catalysts. Only through combined control of both surface functionalization and pore/particle morphology can one achieve the ability to deconvolute and study all the interconnected properties of the catalytic system, such as reactivity, selectivity, and mass transport.

To develop a synthetic strategy of mesoporous silicas that can provide such control, we have synthesized monofunctionalized mesoporous silica nanoparticles (MSNs) with several organosilanes (Scheme 4.2), including 3-aminopropyltrimethoxysilane (APTMS), *N*-(2-aminoethyl)-3-aminopropyltrimethoxysilane (AAPTMS), 3-[2-(2-aminoethylamino)ethylamino]propyltrimethoxysilane (AEPTMS), ureidopropyltrimethoxysilane (UDPTMS), 3-isocyanatopropyltriethoxysilane (ICPTES), 3-cyanopropyltriethoxysilane (CPTES), and allyltrimethoxysilane (ALTMS), by the aforementioned co-condensation method with a minor modification.[24] For simplification, we refer to these materials as X-MSN, where X represents the corresponding organotrialkoxysilane, and MSN stands for mesoporous silica nanoparticles.

We have discovered that the particle morphology of the organically functionalized MSN materials strongly depends on the organic precursors, as showed in the FE-SEM micrographs (Figure 4.4). The MSNs functionalized with hydrophilic organic precursors, such as APTMS, AAPTMS, AEPTMS, and UDPTMS, resulted in larger particle sizes, whereas the MSNs functionalized with hydrophobic organosilanes, such as ICPTES, CPTES, and ALTMS, formed smaller particles. The ^{13}C and ^{29}Si solid-state NMR spectra of these MSN materials confirmed the covalently

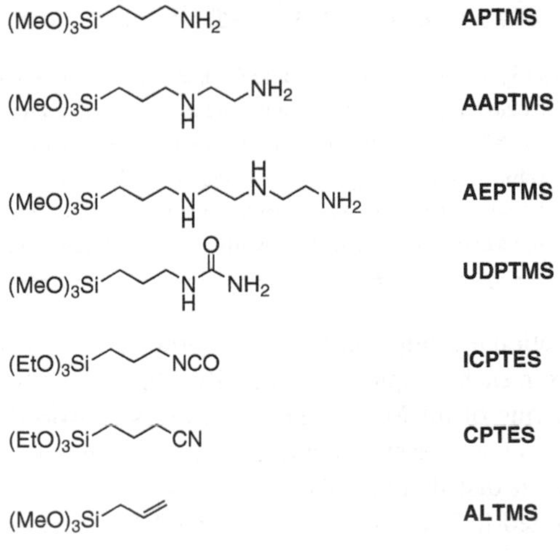

SCHEME 4.2

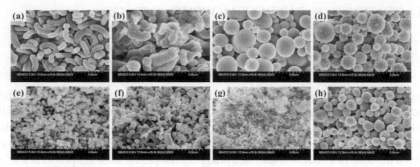

FIGURE 4.4 FE-SEM images of (a) AP-MSN, (b) AAP-MSN, (c) AEP-MSN, (d) UDP-MSN, (e) ICP-MSN, (f) CP-MSN, (g) AL-MSN, and (h) pure MCM-41 silica synthesized by our condensation reaction condition without adding any organic functional group. The image magnification is the same for all of the images (scale bar = 3 μm). (Figure modified from Reference 24.)

TABLE 4.3
Structural Properties of the Organically Functionalized Mesoporous Silica Materials

Sample	d_{100} (Å)[a]	a_0 (Å)[a]	S_{BET} (m²/g)[a]	V_p (cm³/g)[a]	W_{BJH} (Å)[a]	$d_{pore\,wall}$ (Å)[a]	Amount of Organic Group (%)[b]
AP-MSN	39.8	46.0	721.7	0.45	23.7	22.3	12
AAP-MSN	41.3	47.7	664.6	0.48	25.9	21.8	5
AEP-MSN	38.4	44.4	805.8	0.57	26.0	18.4	7
UDP-MSN	43.7	50.5	1022.4	0.78	28.6	21.9	6
ICP-MSN	39.8	46.0	840.1	0.66	25.8	20.2	14
CP-MSN	39.4	45.5	1012.5	0.68	23.5	22.0	10
AL-MSN	33.7	38.9	1080.5	0.65	19.7	19.2	11
MCM-41	38.1	44.0	767.1	0.55	25.5	18.5	–

[a] The BET surface area (S_{BET}), the mesopore volume (V_p), and the mean mesopore width (W_{BJH}) were obtained from the nitrogen adsorption/desorption data. The d_{100} numbers represent the d-spacing corresponding to the main (100) XRD peak. The unit cell size (a_0) is calculated from the d_{100} data using $a_0 = 2d_{100} / 3^{1/2}$. The pore wall thickness ($d_{pore\,wall} = a_0 - W_{BJH}$). [b] The amounts of organic functional groups incorporated to the silica materials were estimated from the ^{29}Si DPMAS. [c] Pure MCM-41 silica synthesized under the same reaction condition without any addition of organoalkoxysilane.

bonded organic functional groups on the mesoporous silica surface.[24] The amounts of functional groups in each sample were quantified based on ^{29}Si DPMAS spectra.[24]

The BET isotherms of all MSNs, analyzed by N_2 adsorption/desorption techniques, were Type IV, characteristic of cylindrical pore structures.[24] The BET surface areas and BJH pore size distributions of these MSNs with different organic functional groups are summarized in Table 4.3. The powder XRD spectra of these materials are shown in Figure 4.5.

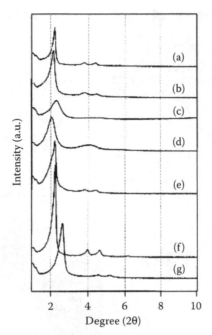

FIGURE 4.5 XRD spectra of the surfactant-removed (a) AP-MSN, (b) AAP-MSN, (c) AEP-MSN, (d) UDP-MSN, (e) AL-MSN, (f) CP-MSN, and (g) ICP-MSN. (Figure modified from Reference 24.)

A characteristic of hexagonal MCM-41 diffraction pattern, including (100), (110), (200) peaks with the spacing ratio of $1:\sqrt{3}:\sqrt{4}$ was observed in AP-MSN, AAP-MSN, CP-MSN, ICP-MSN, and AL-MSN. For AEP-MSN and UDP-MSN samples, broad peaks at 4.52° and 4.10°, respectively, were observed, indicating a disordered porous structure. As depicted in Figure 4.6, the TEM micrographs corroborate the different porous structures between these materials.

As depicted in Scheme 4.3, organotrialkoxysilanes with nonpolar groups (R_2) tend to stabilize the formation of long individual cylindrical micelles by intercalating their hydrophobic groups to the micelles. The uniform organization of the trialkoxy-silyl group at the Gouy-Chapman region of the surface of micelles would help the rapid cross-linking/condensation between the micelle-oriented trialkoxysilyl groups in the basic NaOH aqueous solution, as depicted in Scheme 4.3. The resulting side-on packing of the silicate-coated cylindrical micelles would give rise to small rod-like nanoparticles. Similar phenomenon was also observed by Cai and coworkers, while utilizing NaOH and NH_4OH as catalysts to manipulate the particle morphology of the MCM-41 silicas without organic functional groups.[25] For more hydrophilic organoalkoxysilane precursors, such as AAPTMS, AEPTMS, and UDPTMS, the polar groups (R_1) are not favored by the interaction with the surfactants. This would inhibit the formation of long micelles and reduce their tendency toward side-on condensation. The difference in the rate of condensation between organosilicate-coated micelles versus that of the free silicate (TEOS) molecules would likely be

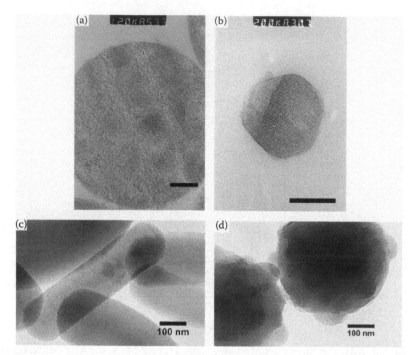

FIGURE 4.6 TEM micrographs of (a) AEP-MSN, (b, c) CP-MSN, and (d) UDP-MSN materials. Image (a) and (b) represent ultramicrotomed samples (all scale bar = 100 nm). (Figure modified from Reference 24.)

small. Because of the lack of thermodynamic incentives for the silicate-coated micelles to pack in an ordered fashion, such co-condensation reactions should yield particles with randomly oriented pore structures. A similar phenomenon was recently reported by Sadasivan and coworkers, who observed the growth of the mesoporous silica particle in the direction that is perpendicular to the pore-alignment upon the introduction of amine-containing organoalkoxysilanes.[26]

4.5 SELECTIVE CATALYSIS BY MSN MATERIALS SYNTHESIZED VIA INTERFACIAL DESIGNED CO-CONDENSATION METHOD

4.5.1 ORGANICALLY FUNCTIONALIZED MSN FOR HETEROGENEOUS NUCLEOPHILIC CATALYSIS

By introducing the interfacial charge-matching concept to the synthesis of functionalized mesoporous silica materials, we were able to control the spatial distribution of the organic functional groups in MSNs with defined particle and pore morphology. Such control is particularly important for designing selective catalysts that can mimic enzymes in taking advantage of the spatially well-organized catalytic

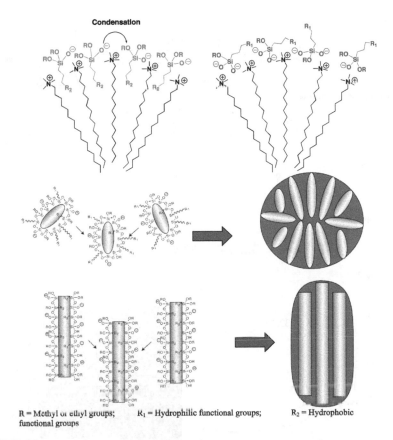

R = Methyl or ethyl groups; R₁ = Hydrophilic functional groups; R₂ = Hydrophobic functional groups

SCHEME 4.3

groups situated in a three-dimensional cavity (enzyme active site). To demonstrate that these structurally well-defined MSN materials can indeed serve as selective catalysts with high recyclability and stability, we have synthesized a 4-(dimethylamino)pyridine functionalized mesoporous silica nanoparticle (DMAP-MSN) material, which exhibited a superior reactivity and product selectivity for several industrially important reactions, such as Baylis-Hillman, acylation, and silylation.[27]

The DMAP-MSN material was synthesized by adding 4-[N-[3-(triethoxysilyl)propyl]-N-methyl-amino]pyridine (DMAP-TES) along with TEOS to an aqueous solution of sodium hydroxide with a low concentration of CTAB under the aforementioned co-condensation condition (Scheme 4.4). The resulting DMAP-MSN material showed a disordered mesoporous structure. The disordered pore structure of DMAP-MSN was confirmed by XRD and TEM (Figure 4.7). The DMAP-MSN material exhibited a spherical particle shape with an average diameter 400 nm, as showed in the scanning electron micrograph (SEM) (Figure 4.7). The measured BET surface area and BJH pore size distribution of this DMAP-functionalized material were 835 m²/g and 2.0 nm, respectively. A Type IV BET isotherm without any significant hysteresis was observed in the N₂ sorption analysis of this organically functionalized material.[27]

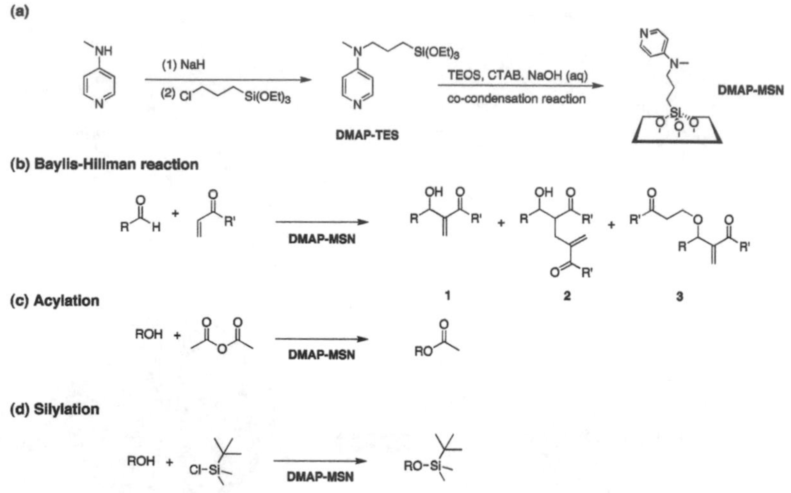

(a)

(1) NaH

(2) Cl————Si(OEt)₃

DMAP-TES

TEOS, CTAB, NaOH (aq)
co-condensation reaction

DMAP-MSN

(b) Baylis-Hillman reaction

DMAP-MSN

1   2   3

(c) Acylation

ROH +

DMAP-MSN

(d) Silylation

ROH + Cl-Si

DMAP-MSN

SCHEME 4.4.

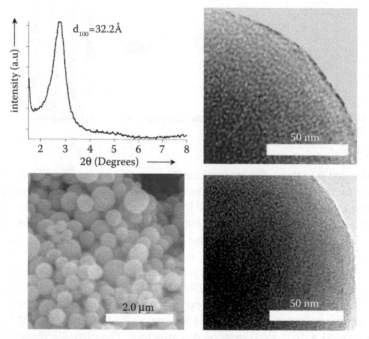

FIGURE 4.7 (a) XRD spectrum of DMAP-MSN. (b) TEM micrograph of DMAP-MSN, scale bar = 50 nm, (c) SEM micrograph of DMAP-MSN, scale bar = 2.0 μm, (d) TEM micrograph of DMAP-MSN after 10 runs of acylations, scale bar = 50 nm. (Figure modified from Reference 27.)

The chemical shifts obtained from the ^{13}C solid-state NMR spectra of DMAP-MSN matched well with the solution data of the organic precursors, as shown in Figure 4.8. The results confirmed the presence of the DMAP functionality on the

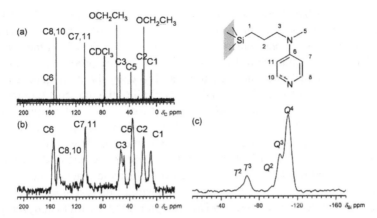

FIGURE 4.8 (a) ^{13}C NMR spectrum of DMAP-TES in CDCl$_3$ solution. (b) ^{13}C CPMAS spectrum of DMAP-MSN resulting from 12,000 scans acquired with a delay of 1 s in a 5 mm probe (v_R = 10 kHz). During each CP period of 1.5 ms, v_{RF}^H was ramped between 16 and 40 kHz (in 11 steps), while v_{RF}^H was set to 36 kHz. The v_{RF}^H fields of 83 kHz and 65 kHz were applied to protons during initial excitation and high power decoupling, respectively. (c) ^{29}Si DPMAS spectrum of DMAP-MSN obtained with the same probe using CPMG acquisition (10 echoes). A total of 600 scans were collected with a delay of 300 s to allow the complete relaxation of ^{29}Si nuclei. (Figure modified from Reference 27.)

mesoporous silica surface. The loading of DMAP on the silica estimated from the ^{29}Si-DPMAS spectra was 1.2 ($\pm$0.1) molecules per nm^2 (1.6 $\pm$ 0.15 mmol/g), and the number of silanol groups was 2.6 ($\pm$0.2) per nm^2. Direct comparison between the chemical shifts of ^{13}C NMR spectra of the mono- or di-protonated species of 4-(dimethylamino)pyridine and those of the DMAP-MSNs indicated that there was no protonated forms present in the DMAP-MSN sample (see Table 4.4).[28]

The catalytic performances of the spherical DMAP-MSN catalyst for three different nucleophilic reactions (Baylis-Hillman, acylation, and silylation reactions) were examined (Scheme 4.4). For Baylis-Hillman reaction, only a catalytic amount

TABLE 4.4
^{13}C Chemical Shifts Observed in DMAP-MSN and in the Reference Compounds (DMAP-TES and 4NMe$_2$-Py).

Sample	Solvent	C1	C2	C3	C5	C6	C7, C11	C8, C10
DMAP–TES	CDCl$_3$	7.8	20.1	54.0	37.5	153.5	106.6	149.8
DMAP–MSN	Solid-state	8.8	19.3	53.0	35.8	153.8	106.1	147.8
4NMe$_2$-Py[a]	CDCl$_3$				38.4	153.6	106.0	149.1
	CF$_3$COOH in CDCl$_3$				40.1	153.6	105.6	139.9
	CF$_3$SO$_3$H in CD$_3$NO$_2$				47.3	157.8	120.6	145.6

[a] Data taken from Reference 28.

(30 mol%) of DMAP-MSNs was needed. Among the examined different α,β-unsaturated ketone reactants, the reactivity follows this order: methyl vinyl ketone > cyclopentenone > cyclohexenone, as summarized in Table 4.5. In the case of the activated aldehyde, 4-nitrobenzaldehyde, only the desired product was obtained in high yield. In contrast, the same reaction catalyzed by the homogeneous DMAP molecules resulted in a mixture of products, including the diadduct (compound 2), Michael addition product (compound 3), and some oligomerized products (Scheme 4.5a). Given that the diadduct 2 and the Michael addition side product 3 could only be generated from the Baylis-Hillman product 1, the excellent product selectivity of DMAP-MSN could be attributed to the difference in the rate of diffusion to the active sites located inside the pores between the aldehyde reactant and compound 1, which would serve as the reactant for the undesired side reactions. To investigate this matrix effect, we carried out two DMAP- and DMAP-MSN-catalyzed Baylis-Hillman reactions by using compound 1 (1.0 e.q.) as reactant to interact with methyl vinyl ketone (2.0 e.q.) in THF/H_2O solution at 50°C for 24 h as depicted in Scheme 4.5b. Indeed, the reaction catalyzed by homogeneous DMAP catalyst gave rise to a mixture of products. On the contrary, no reaction was observed in the case of DMAP-MSN. The results support the hypothesis that the MSN matrix could regulate the reaction selectivity by preferentially allowing certain reactants to access the catalytic sites.

For acylation and silylation reactions, the steric effect significantly influenced product yield in the DMAP-MSN system. To further investigate the recyclability and stability of DMAP-MSN catalysts, acylation of alcohols was chosen as the model reaction.[27] Our results showed that the DMAP-MSN catalysts could be recycled for at least 10 times without losing any catalytic reactivity. The turnover number (TON) of DMAP-MSN catalysts was as high as 3340 for 24 days.

4.5.2 TRANSITION METAL-FUNCTIONALIZED MSN FOR CONFORMATION DIRECTING POLYMERIZATION

To utilize the hexagonally packed cylindrical channels of MSNs as a conformation directing scaffold to encapsulate conjugated polymers for directional energy- or electron-transfer, we prepared two Cu(II)-functionalized mesoporous silica (Cu-MSN) and alumina (Cu-MAL) materials by co-condensation and postgrafting/impregnation method, respectively.[29] By using the copper catalyzed oxidative coupling reaction, 1,4-diethynylbenzene was polymerized to poly(phenylene butadiynylene) (PPB) inside the channels of Cu-MSN, as shown in Figure 4.9. The rate and the extent of polymerization within the mesoporous channels depend on (1) the amount and the spatial distribution of catalytic sites, and (2) the diffusion rate of organic monomers and their local concentration near the catalytic sites. These factors could be manipulated by tuning the pore size and the degree of surface functionalization of the catalytic groups.

Incorporation of the Cu(II) complex to the mesopore surface was accomplished by the aforementioned co-condensation method using a methanolic solution of Cu(II) complex as a precursor. The solution was prepared by mixing copper(II) bromide and N-(2-aminoethyl)-aminopropyltrimethoxysilane (AAP-TMS). The measurements of

TABLE 4.5
DMAP-MSN Catalyzed Baylis-Hillman Reaction[a]

$$R \overset{O}{\underset{H}{\diagup}} + \diagup\diagdown^{EWG} \xrightarrow[\substack{R = Aryl \\ EWG = COR'}]{Catalyst} 1 + 2^* + 3^*$$

* Side products **2** and **3** are not observed in all reactions.

Entry	Aldehyde	Ketone	Catalyst	Product	Yield [%][b]
1			DMAP-MSN		86
2			DMAP-MSN		49
3			DMAP-MSN		99
4			DMAP-MSN		49

(continued)

TABLE 4.5 (Continued)
DMAP-MSN Catalyzed Baylis-Hillman Reaction[a]

Entry	Aldehyde	Ketone	Catalyst	Product	Yield [%][b]
5			DMAP-MSN		25
6[c]			DMAP-MSN		50
7[c]			DMAP-MSN		25
8			DMA-SiO$_2$[d]		20
9			MCM-41		NR[e]

[a] Reaction condition: p-nitrobenzaldehyde (0.25 mmol), α,β-unsaturated ketone (0.5 mmol), and catalyst (50 mg, 30 mol%) in THF/H$_2$O = 3:1 (2 ml) at 50°C for 24 h;
[b] isolated yield; [c] aldehyde/ketone = 1:4 at 50°C for 3 d; [d] 3-(dimethylamino)propyl-functionalized silica gel; [e] no reaction.

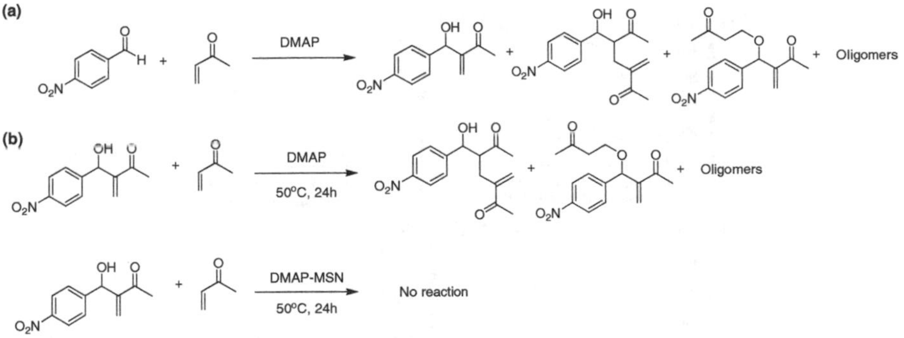

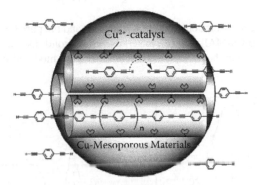

SCHEME 4.5

FIGURE 4.9 Schematic representation of Cu²⁺-functionalized mesoporous silica (Cu-MSN) and alumina (Cu-MAL) catalysts for oxidative polymerization of 1,4-diethynylbenzene into conjugated oligo(phenylene butadiynylene). (Figure modified from Reference 29.)

powder XRD,[29] BET isotherm, and BJH pore size distribution of the Cu(II)-immobilized mesoporous silica nanoparticles (Cu-MSN) indicated that the material is a typical MCM-41 type mesoporous silica with a high surface area of ca. 655.0 m^2/g and a BJH average pore diameter of 25.6 Å (Figure 4.10). The SEM micrograph of Cu-MSN showed exclusively spherical-shaped particles.[29] A Cu(II)-impregnated mesoporous alumina (Cu-MAL) was prepared by using a literature-reported procedure.[30] The Cu-MAL material posed a BET surface area of 448.0 m^2/g and a BJH average pore diameter of 68.0 Å. The amounts of Cu(II) in Cu-MSN and Cu-MAL materials were determined to be ca. 5.2×10^{-4} and 1.3×10^{-5} mol/g, respectively.

Polymerization of 1,4-ethylnylbenzene was performed with Cu-functionalized materials in the refluxed pyridine for 56 h. The resulting PPB/mesoporous silica or alumina composites exhibited very different photophysical properties. In the case of Cu-MSN, a 245 nm bathochromic shift (emission $\lambda_{max} = 643.5$ nm) of the fluorescence emission spectra of the PPB polymer was observed, as shown in Figure 4.11A. The result indicated that the extent of π-conjugation in the PPB polymer synthesized within the mesoporous channels of Cu-MSN is comparable to that of PPB synthesized in homogeneous solution. Conversely, the polymer catalyzed by Cu-MAL showed

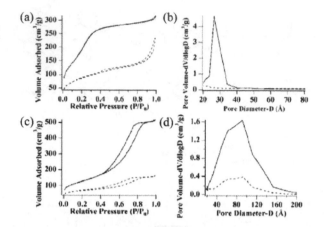

FIGURE 4.10 Nitrogen adsorption/desorption isotherms and the pore size distributions of the Cu-MSN, (a), (b), and Cu-MAL, (c), (d), materials before (solid line) and after (dashed line) PPB polymerization within the mesoporous channels. (Figure modified from Reference 29.)

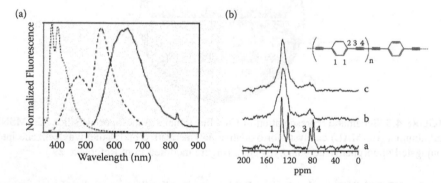

FIGURE 4.11 (A) Normalized fluorescence emission spectra of 1,4-diethynylbenzene (dotted line) and the PPB-containing composite materials of Cu-MAL and Cu-MSN catalysts (dashed and solid lines, respectively) after 56 h of polymerization, (B). ^{13}C CPMAS spectra of (a) structurally aligned PPB polymer catalyzed by Cu-MSN, (b) PPB polymer synthesized with Cu-MAL, and (c) bulk PPB. The spectra were measured at room temperature, using 4680 scans, 10 s intervals, sample rotation rate of 5 kHz, contact time of 2 ms, CW ^{1}H decoupling at 70 kHz, and sideband suppression (TOSS). (Figure modified from Reference 29.)

two bathochromic-shifted fluorescence peaks (λ_{max} = 472.5 and 551.5 nm) under the same reaction condition. Both shifts are significantly smaller than that of Cu-MSN, indicating a shorter π-conjugation of the PPB polymer in the case of Cu-MAL.

The information of spatial arrangement and conformation alignment of the polymer chains in both materials was obtained by solid-state ^{13}C CP-MAS NMR spectra. The ^{13}C NMR spectrum of PPB/Cu-MSN composite material exhibited four well-resolved, narrow resonance peaks, as depicted in Figure 4.11B. In contrast, the ^{13}C NMR spectra of homogeneously synthesized PPB polymer and the PPB/Cu-MAL composite showed two broad resonance peaks. The observed broadening reflects the conformational heterogeneity in the three-dimensional arrangements of polymeric chains. Additional significant differences are observed in spectra (b) and (c) with respect to (a): (1) both resonances are not only broader, but comprise more than two lines, and (2) the intensities of sp^2 and sp peaks do not match the 3:2 ratio expected for PPB. These spectral features are consistent with considerable polydiacetylene-type cross-linking that occurred in the bulk polymer and Cu-MAL. In the case of Cu-MSN, the sharpness of the observed resonances and lack of cross-linking demonstrate that isolated molecular wires are formed within the parallel channels of Cu-MSN.

4.6 INCORPORATION OF MULTIPLE ORGANIC FUNCTIONAL GROUPS WITH PRECISELY CONTROLLED RELATIVE CONCENTRATIONS AND PARTICLE MORPHOLOGY

For several successful heterogeneous single-site catalyst systems, recent studies have indicated that the support materials often play an active role as cofactors by participating and cooperatively facilitating reactions. The ability to harness important cofactors to the support of the heterogenized single-site catalysts through multifunctionalization would surely enhance the overall reactivity and selectivity. Therefore, we have synthesized multifunctionalized mesoporous silicas as new support materials, where one functionality can be dedicated as the linker for the surface immobilization of the homogeneous catalyst of interest and the other(s) will serve as cofactor to manipulate the reactivity or selectivity of the resulting catalytic system.

First, we have developed a synthetic method that offers the ability to tune the relative ratio of different functional groups without altering the pore and particle morphology of the mesoporous silica material. As previously described, the difference in hydration abilities of various organoalkoxysilane precursors in basic aqueous solutions of CTAB surfactant micelles can not only influence the loading of a particular functional group under our co-condensation condition, but also dictate the pore and particle morphology of the resulting mesoporous material. By introducing two organoalkoxysilanes with different structure-directing abilities as precursors of our co-condensation reaction, we can take advantage of such a difference between the two organoalkoxysilanes and utilize one precursor with stronger structure-directing ability to create the desired pore and particle morphology and employ the other for selective immobilization of catalysts. This strategy allows us to generate a series of multifunctionalized mesoporous silica materials with control of both morphology and functionalization.

As a proof of principle, we synthesized a series of bifunctionalized MSN materials with AEP and CP functional groups.[31] These MSN materials were prepared by varying the molar ratio of organosilane precursors, AEPTMS and CPTMS, while keeping the total amount of organosilanes the same (12.8 mol% to TEOS). The SEM micrographs of all AEP/CP-MSNs showed exclusively spherical particles (Figure 4.12). Given the fact that the monofunctionalized AEP-MSN

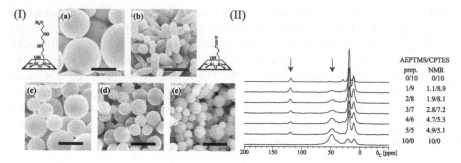

FIGURE 4.12 (I) FE-SEM images of (a) AEP-MSN, (b) CP-MSN, (c) 5/5 AEP/CP-MSN, (d) 3/7 AEP/CP-MSN, (e) 1/9 AEP/CP-MSN. Scale bar = 1 μm for all the micrographs. (II) ^{13}C CPMAS spectra of monofunctionalized (top and bottom traces) and bifunctionalized (middle traces) AEP/CP-MPs. Arrows highlight the resonances that are unique for each species and thus were used for quantitative analysis. The numbers represent the molar ratio between two components used for preparation (left column) and obtained from analysis of NMR spectra (right column). (Figure modified from Reference 31.)

and CP-MSN are spherical and rod-shaped particles, respectively, the shape of bifunctional AEP/CP-MSNs apparently is governed by the structure-directing ability of the AEPTMS precursor in the co-condensation reaction. The powder XRD spectra[31] and TEM micrographs (Figure 4.13) of AEP/CP-MSNs materials indicated that these bifunctionalized MSNs share the same wormhole-like porous structures as observed in the case of the monofunctionalized AEP-MSN. The relative concentrations between the two organic functional groups were measured by solid-state NMR. As shown in Figure 4.12II, the loading of these functional groups could indeed be tuned by introducing different molar ratios of organosilanes to the co-condensation reaction.

4.7 SYNERGISTIC CATALYSIS BY MULTIFUNCTIONALIZED MESOPOROUS SILICA NANOPARTICLES

4.7.1 GATE KEEPING EFFECT: TUNING REACTION SELECTIVITY BY BIFUNCTIONALIZED MSN CATALYSTS

Our ability to anchor two types of groups on mesopore walls allows us to tether not only the catalyst but also other functional moieties. If these auxiliary groups are chosen properly, they can select only certain molecules to enter the pores and get converted to products. In such a multifunctional catalyst, the selectivity depends on the selectivity of the gatekeeper. To investigate the influence of gate-keepers, three bifunctionalized MSN catalysts were synthesized by introducing equal amounts of AEP-TMS with UDP-TMS, MP-TMS, or AL-TMS to our co-condensation reaction.[32] The resulting MSNs are functionalized with a common AEP group and three different secondary groups, UDP, MP, and AL functionalities. As depicted in Figure 4.14, all three MSNs are spherical in shape and showed

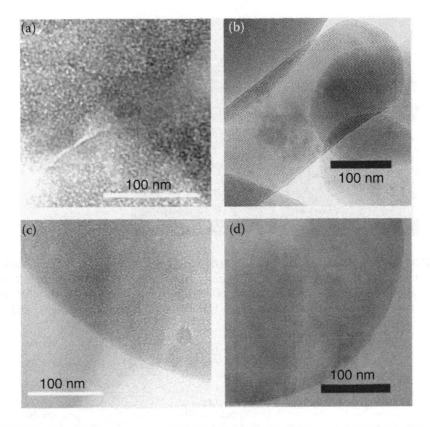

FIGURE 4.13 TEM micrographs of (a) AEP-MSN, (b) CP-MSN, (c) 5/5 AEP/CP-MSN, and (d) 1/9 AEP/CP-MSN. Images of (a), (c), and (d) represent ultramicrotomed samples. (All scale bar = 100 nm). (Figure modified from Reference 31.)

disordered mesoporous structures, as indicated by powder XRD and TEM analyses. The measured BET surface area of AEP/UDP-MSN, AEP/MP-MSN, and AEP/AL-MSN were 759.6, 778.7, and 703.5 m2g–1.32 The BJH pore distribution measurement of AEP/UDP-MSN and AEP/MP-MSN exhibited similar pore diameters as 26.0 and 22.9 Å. The AEP/AL-MSN catalyst, functionalized with more hydrophobic organosilanes, showed a smaller pore diameter, ca. 15 Å. The quantification of the organic groups in each MSN catalyst, by ^{13}C and ^{29}Si solid-state NMR spectroscopy, indicated that all MSNs contained approximately the same total amounts of organic functional groups (1.0 mmol/g in AEP/UDP-MSN, 1.4 mmol/g in AEP/MP-MSN, and 1.3 mmol/g in AEP/AL-MSN).[32] Furthermore, the relative molar concentration between AEP and the secondary groups were very similar (1.17, 1.04, and 1.13 for AEP/UDP-MSN, AEP/MP-MSN, and AEP/AL-MSN, respectively).

As depicted in Figure 4.15, a competitive nitroaldol (Henry) reaction was performed on these bifunctional MSNs in a solution of nitromethane with equal

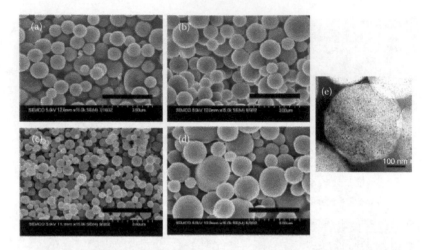

FIGURE 4.14 FE-SEM images of (a) AEP/UDP-MSN (0.7–1.3 μm), (b) AEP/MP-MSN (0.7–1.6 μm), (c) AEP/AL-MSN (0.2–0.7 μm), and (d) AEP-MSN (0.8–2.1 μm). TEM micrograph of the ultramicrotomed sample (60 to 90 nm thickness) of AEP/AL-MSN (e). Scale bar is 3 μm for the (a–d) micrographs and 100 nm for the (e) micrograph. (Figure modified from Reference 32.)

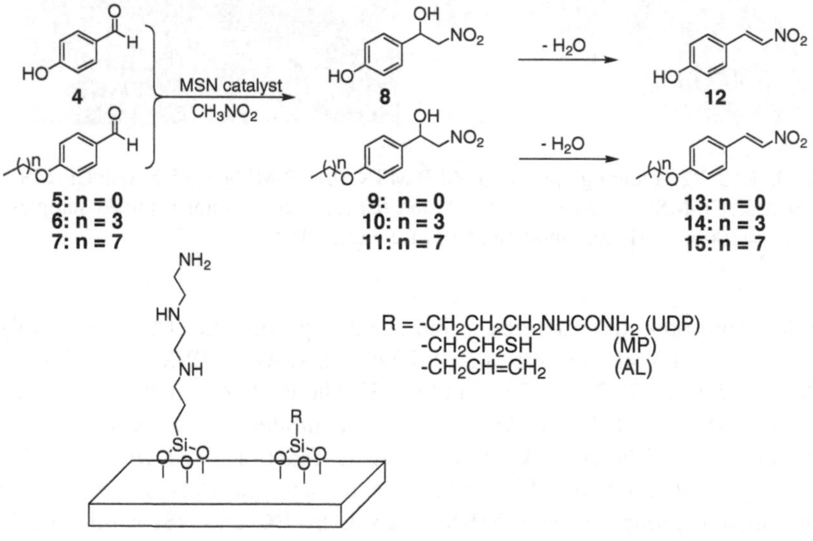

FIGURE 4.15 Competitive nitroaldol reaction. Reaction condition: 50 mg MSN catalyst, 10 ml CH_3NO_2, 5.0 mmol aldehydes, 263 K for 24 h (Figure modified from Reference 32.)

amounts of 4-hydroxybenzaldehyde and one of the three 4-alkoxybenzaldehydes (**5, 6,** and **7**). The ratios of the nitroalkene products (**13/12, 14/12,** and **15/12**) were used to determine the reaction selectivity of each MSN material. Interestingly, the mono-functionalized AEP-MSN and the hydrophilic bifunctionalized AEP/UDP-MSN catalysts did not show any reaction selectivity for all three combinations of the reactants, as illustrated in Figure 4.16. However, an increase of reaction selectivity

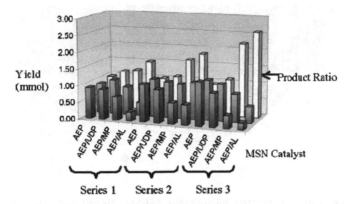

FIGURE 4.16 Histogram of the competitive nitroaldol reactions. Blue bars: yield of **12** (mmol); red bars: yield of **13, 14,** or **15** (mmol); white bars: molar ratio of products (**13/12, 14/12,** or **15/12**). Series 1, 2, and 3 are experiments conducted with reactants **4** and **5, 4** and **6,** and **4** and **7,** respectively. (Figure modified from Reference 32.)

towards the nonpolar and more hydrophobic alkoxybenzaldehyde reactants (**5, 6,** and **7**) was clearly observed in the cases of AEP/MP-MSN and AEP/AL-MSN catalysts, where the catalytic AEP groups are situated in mesopores decorated with hydrophobic groups. The results suggested that the secondary hydrophobic group (MP and AL) play a significant role in preferentially allowing more hydrophobic reactants to penetrate into the mesopores and react with the AEP catalytic group. To demonstrate this hydrophobic solvation effect, a solubility test for compound **4** and **12** in 1-propanethiol was conducted to simulate the penetration/dissolution of these benzaldehydes with different polarities into the mercaptopropyl-functionalized mesopores. Given the fact that compound **4** (4-hydroxybenzaldehyde) was insoluble in 1-propanethiol, whereas compound **12** was completely miscible with 1-propanethiol, the selectivity of our catalysts most likely originated from the variation of physicochemical properties of the bifunctionalized mesopores, such as polarities and hydrophobicity.

4.7.1 COOPERATIVE CATALYSIS BY GENERAL ACID AND BASE BIFUNCTIONALIZED MSN CATALYSTS

Enzymes engaged in carbonyl chemistry often employ both general acid and base catalytic residues in the active sites to cooperatively activate specific substrates.[33] Recently, several synthetic catalytic systems have utilized the double hydrogen bonding capability of urea or thiourea functionality as a general acid catalyst to activate carbonyl compounds in homogeneous reactions.[34, 35] An important prerequisite for the construction of a heterogeneous system with catalytic cooperativity would be to multifunctionalize a solid support with control of the relative concentrations and proper spatial arrangements between these functional groups.

Recently, we reported on a new cooperative catalytic system comprised of a series of bifunctionalized mesoporous silica nanosphere materials with various relative

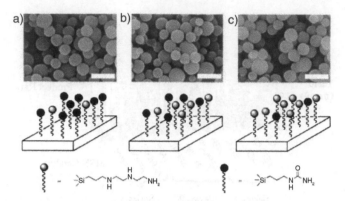

FIGURE 4.17 Scanning electron micrographs (SEMs, above) and schematic drawings (below) of the bifunctional MSNs: (a) 2/8 AEP/UDP-MSN, (b) 5/5 AEP/UDP-MSN, and (c) 8/2 AEP/UDP-MSN. Scale bar = 2.0 μm. (Figure modified from Reference 36.)

concentrations of a general acid, ureidopropyl group, and a base, 3-[2-(2-amino-ethylamino)ethylamino]propyl group.[36] These bifunctionalized MSN catalysts with the initial molar ratio of organosilane precursors (AEP/UDP = 2/8, 5/5, and 8/2) are monodisperse spherical particles with similar sizes, as shown in Figure 4.17. Disordered pore structures, same as AEP/CP-MSNs, were observed in all AEP/UDP-MSNs.[36] The BET surface areas of 2/8, 5/5, and 8/2 AEP/UDP-MSN catalysts were 938.7, 759.6, and 830.4 m^2/g, respectively. The corresponding BJH average pore diameters of these MSNs are 27.8, 22.9, and 25.9 Å. The total surface concentrations of the organic functional groups (AEP + UDP) in the 2/8, 5/5, and 8/2 AEP/UDP-MSNs were determined by solid-state ^{13}C and ^{29}Si NMR spectroscopy to be 1.3, 1.0, and 1.5 mmol/g.[36] The concentration ratios of AEP/UDP were 2.5/7.5, 5.4/4.6, and 6.7/3.3, respectively.

Three chemical reactions, aldol, Henry, and cyanosilylation, were examined by using these bifunctionalized silicas as catalysts, as depicted in Figure 4.18. A common electrophile, 4-nitrobenzaldehyde and three different nucleophiles (acetone, nitromethane, and trimethylsilyl cyanide) were used as reactants. We anticipate that this bifunctionalized MSN system with the UDP group serving as a general acid that can activate carbonyl groups, and the AEP group functioning as a general base that is capable of (1) generating the enamine with acetone in the aldol condensation,[37, 38] (2) de-protonating nitromethane in the Henry reaction,[32, 39, 40] and (3) facilitating the formation of a hypervalent silicate nucleophile with trimethylsilyl cyanide in cyanosilylation,[41] could cooperatively catalyze these reactions, as depicted in Scheme 4.6. Indeed, the turnover numbers of all AEP/UDP-MSNs were higher than that of AEP-MSN as shown in Table 4.7.

The reaction rates of all three reactions were significantly accelerated (up to 4 times) by the AEP/UDP-MSN catalysts. Interestingly, among the MSNs with different AEP/UDP ratios, 2/8 AEP/UDP-MSNs were the most effective bifunctional catalysts in all three reactions. Given that only UDP group is capable of activating carbonyl group, this observation suggests that the activation of carbonyl groups could be the rate-determining step in these heterogeneous reactions. In contrast, the

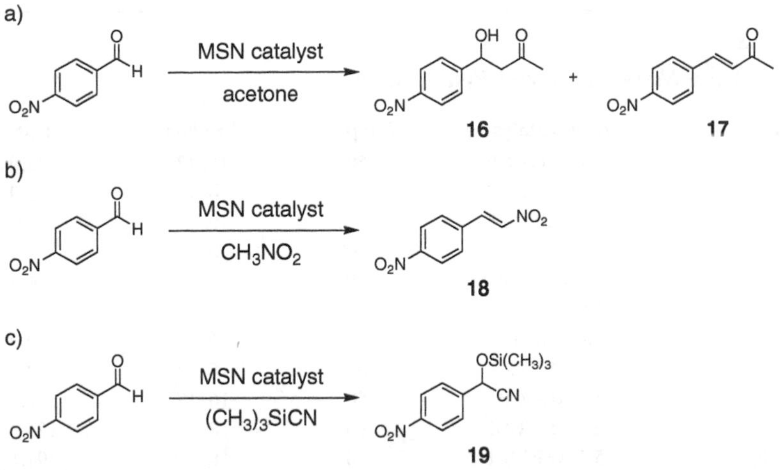

a)

b)

c)

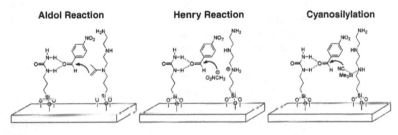

FIGURE 4.18 Three model reactions catalyzed by the MSN catalysts: aldol reaction (a), Henry reaction (b), and cyanosilylation (c). (Figure modified from Reference 35.)

| Aldol Reaction | Henry Reaction | Cyanosilylation |

SCHEME 4.6

TABLE 4.6
Textural Properties of the Surfactant-free Materials

Sample	d_{100} (Å)	a_0 (Å)	S_{BET} (m²/g)	V_p (cm³/g)	W_{BJH} (Å)	$d_{pore\ wall}$ (Å)
5/5 AEP/CP-MP	37.1	42.9	722.2	0.38	—	—
4/6 AEP/CP-MP	42.1	48.6	822.6	0.51	26.4	22.2
3/7 AEP/CP-MP	38.1	44.0	980.8	0.56	26.4	17.6
2/8 AEP/CP-MP	37.4	43.2	801.4	0.43	20.5	22.7
1/9 AEP/CP-MP	36.5	42.2	966.6	0.51	20.7	21.5

---

reactions catalyzed by a physical mixture of AEP-MSN and UDP-MSN showed significantly smaller TONs in comparison with those of the 5/5 AEP/UDP-MSN catalyst, confirming the synergistic effect between the AEP and UDP groups. To examine whether the activity enhancement was caused by the surface dilute effect of the AEP group, we studied the catalytic performance of two other bifunctionalized MSN catalysts (2/8 and 5/5 AEP/CP-MSNs). These two MSN catalysts are

TABLE 4.7
TONs of the MSN-Catalyzed Reactions[a]

Reaction	MSN catalyst	Temp °C	Product	TON
Aldol	2/8 AEP/UDP	50	16, 17	22.6
	5/5 AEP/UDP	50	16, 17	11.9
	8/2 AEP/UDP	50	16, 17	8.6
	AEP	50	16, 17	5.4
	Physical mixture[b]	50	16, 17	6.4
Henry	UDP	50	16, 17	0.0
	Pure MSN[c]	50	16, 17	0.0[d]
	2/8 AEP/CP	50	16, 17	12.4
	5/5 AEP/CP	50	16, 17	9.3
	2/8 AEP/UDP	90	18	125.0
Cyanosilylation	5/5 AEP/UDP	90	18	91.1
	8/2 AEP/UDP	90	18	65.8
	AEP	90	18	55.9
	Physical mixture[b]	90	18	79.2
	UDP	90	18	5.8
	Pure MSN[c]	90	18	0.0[d]
	2/8 AEP/CP	90	18	78.0
	5/5 AEP/CP	90	18	71.0
	2/8 AEP/UDP	50	19	276.1
	5/5 AEP/UDP	50	19	170.5
	8/2 AEP/UDP	50	19	109.4
	AEP	50	19	111.4
	Physical mixture[b]	50	19	126.9
	UDP	50	19	45.9
	Pure MSN[c]	50	19	43.0[d]

[a] TON = mmol product/mmol catalyst during 20 h of reaction time for aldol and Henry reactions and 24 h for cyanosilylation reaction using 20 mg of MSN. [b] Physical mixture = 20 mg of AEP-MSN + 20 mg of UDP-MSN. [c] Nonfunctionalized MCM-41. [d] % yield.

functionalized with the nucleophilic AEP group, along with different surface concentrations of a catalytically inert cyanopropyl group. If the surface dilution effect indeed played a role in our bifunctionalized MSN system, replacing the carbonyl-activator (UDP group) with an inert CP group should not lower the TONs in all three reactions. In fact, the TONs of AEP/CP-MSNs were significantly lower than those of the AEP/UDP-MSNs, indicating that the surface dilution effect could not account for the observed rate enhancements by the AEP/UDP-MSNs. These results support the hypothesis that the superior rate enhancements in these reactions catalyzed by the bifunctional acid-base MSN catalysts are most likely originated from a cooperative effect between the base (AEP) and the general acid (UDP) catalysts that are anchored on the mesopore surface.

4.8 CONCLUSION

In this review, we have outlined the approach of using an interfacial designed co-condensation method for the synthesis of a series of organically functionalized mesoporous silica nanoparticles with various particle morphologies. By introducing organoalkoxysilanes with different functional groups as structure-directing reagents to the NaOH-catalyzed condensation reaction of TEOS, we have constructed various sophisticated organic/inorganic hybrid nanomaterials. The surface concentrations of multiple organic functionalities in these materials could be tuned, while keeping the particle and pore morphology of the solid support constant. We have demonstrated that several important catalytic principles, such as gatekeeping and cooperative catalysis, can be realized by using these novel materials as heterogeneous catalysts with high selectivity and efficiency. We envision that the further development on this new synthetic method would lead to the control of the spatial location and distribution of organic functional groups in a variety of structurally well-defined mesoporous metal oxide materials that are important for many catalytic applications.

REFERENCES

1. Corma, A., From microporous to mesoporous molecular sieve materials and their use in catalysis. *Chemical Reviews* 97, (6), 2373–2419, 1997.
2. Ford, D. M.; Simanek, E. E.; Shantz, D. F., Engineering nanospaces: Ordered mesoporous silicas as model substrates for building complex hybrid materials. *Nanotechnology* 16, (7), 458–475, 2005.
3. Giri, S.; Trewyn, B. G.; Stellmaker, M. P.; Lin, V. S. Y., Stimuli-responsive controlled-release delivery system based on mesoporous silica nanorods capped with magnetic nanoparticles. *Angewandte Chemie, International Edition* 44, (32), 5038–5044, 2005.
4. Lai, C.-Y.; Trewyn, B. G.; Jeftinija, D. M.; Jeftinija, K.; Xu, S.; Jeftinija, S.; Lin, V. S. Y., A mesoporous silica nanosphere-based carrier system with chemically removable CdS nanoparticle caps for stimuli-responsive controlled release of neurotransmitters and drug molecules. *Journal of the American Chemical Society* 125, (15), 4451–4459, 2003.
5. Lin, V. S. Y.; Lai, C.-Y.; Huang, J.; Song, S.-A.; Xu, S., Molecular recognition inside of multifunctionalized mesoporous silicas: Toward selective fluorescence detection of dopamine and glucosamine. *Journal of the American Chemical Society* 123, (46), 11510–11511, 2001.
6. Radu, D. R.; Lai, C.-Y.; Wiench, J. W.; Pruski, M.; Lin, V. S. Y., Gatekeeping layer effect: A poly(lactic acid)-coated mesoporous silica nanosphere-based fluorescence probe for detection of amino-containing neurotransmitters. *Journal of the American Chemical Society* 126, (6), 1640–1641, 2004.
7. Stein, A.; Melde, B. J.; Schroden, R. C., Hybrid inorganic-organic mesoporous silicates—nanoscopic reactors coming of age. *Advanced Materials* 12, (19), 1403–1419, 2000.
8. Ciesla, U.; Schuth, F., Ordered mesoporous materials. *Microporous and Mesoporous Materials* 27, (2–3), 131–149, 1999.
9. Davis, M. E., Ordered porous materials for emerging applications. *Nature* 417, (6891), 813–821, 2002.

10. Schuth, F.; Schmidt, W., Microporous and mesoporous materials. *Advanced Materials* 14, (9), 629–638, 2002.

11. Stein, A., Advances in microporous and mesoporous solids—highlights of recent progress. *Advanced Materials* 15, (10), 763–775, 2003.

12. Ying, J. Y.; Mehnert, C. P.; Wong, M. S., Synthesis and applications of supramolecular-templated mesoporous materials. *Angewandte Chemie, International Edition* 38, (1/2), 56–77, 1999.

13. Kresge, C. T.; Leonowicz, M. E.; Roth, W. J.; Vartuli, J. C.; Beck, J. S., Ordered mesoporous molecular sieves synthesized by a liquid-crystal template mechanism. *Nature* 359, (6397), 710–712, 1992.

14. Beck, J. S.; Vartuli, J. C.; Roth, W. J.; Leonowicz, M. E.; Kresge, C. T.; Schmitt, K. D.; Chu, C. T. W.; Olson, D. H.; Sheppard, E. W.; et al., A new family of mesoporous molecular sieves prepared with liquid crystal templates. *Journal of the American Chemical Society* 114, (27), 10834–10843, 1992.

15. Zhao, D.; Feng, J.; Huo, Q.; Melosh, N.; Frederickson, G. H.; Chmelka, B. F.; Stucky, G. D., Triblock copolymer syntheses of mesoporous silica with periodic 50 to 300 angstrom pores. *Science* 279, (5350), 548–552, 1998.

16. Zhao, D.; Huo, Q.; Feng, J.; Chmelka, B. F.; Stucky, G. D., Nonionic triblock and star diblock copolymer and oligomeric surfactant syntheses of highly ordered, hydrothermally stable, mesoporous silica structures. *Journal of the American Chemical Society* 120, (24), 6024–6036, 1998.

17. Bagshaw, S. A.; Prouzet, E.; Pinnavaia, T. J., Templating of mesoporous molecular sieves by nonionic polyethylene oxide surfactants. *Science* 269, (5228), 1242–1244, 1995.

18. Tanev, P. T.; Pinnavaia, T. J., A neutral templating route to mesoporous molecular sieves. *Science* 267, (5199), 865–867, 1995.

19. Inagaki, S.; Fukushima, Y.; Kuroda, K., Synthesis of highly ordered mesoporous materials from a layered polysilicate. *Journal of the Chemical Society, Chemical Communications* (8), 680–682, 1993.

20. Lim, M. H.; Stein, A., Comparative studies of grafting and direct syntheses of inorganic-organic hybrid mesoporous materials. *Chemistry of Materials* 11, (11), 3285–3295, 1999.

21. Radu, D. R.; Lai, C.-Y.; Huang, J.; Shu, X.; Lin, V. S. Y., Fine-tuning the degree of organic functionalization of mesoporous silica nanosphere materials via an interfacially designed co-condensation method. *Chemical Communications* (10), 1264–1266, 2005.

22. Larsen, J. W.; Magid, L. J., Calorimetric and counterion binding studies of the interactions between micelles and ions. Observation of lyotropic series. *Journal of the American Chemical Society* 96, (18), 5774–5782, 1974.

23. Larsen, J. W.; Magid, L. J., Enthalpies of binding of organic molecules to cetyltrimethylammonium bromide micelles. *Journal of Physical Chemistry* 78, (8), 834–839, 1974.

24. Huh, S.; Wiench, J. W.; Yoo, J.-C.; Pruski, M.; Lin, V. S. Y., Organic functionalization and morphology control of mesoporous silicas via a co-condensation synthesis method. *Chemistry of Materials* 15, (22), 4247–4256, 2003.

25. Cai, Q.; Luo, Z.-S.; Pang, W.-Q.; Fan, Y.-W.; Chen, X.-H.; Cui, F.-Z., Dilute solution routes to various controllable morphologies of MCM-41 silica with a basic medium. *Chemistry of Materials* 13, (2), 258–263, 2001.

26. Sadasivan, S.; Khushalani, D.; Mann, S., Synthesis and shape modification of organo-functionalized silica nanoparticles with ordered mesostructured interiors. *Journal of Materials Chemistry* 13, (5), 1023–1029, 2003.

27. Chen, H.-T.; Huh, S.; Wiench, J. W.; Pruski, M.; Lin, V. S. Y., Dialkylaminopyridine-functionalized mesoporous silica nanosphere as an efficient and highly stable heterogeneous nucleophilic catalyst. *Journal of the American Chemical Society* 127, (38), 13305–13311, 2005.

28. Dega-Szafran, Z.; Kania, A.; Nowak-Wydra, B.; Szafran, M., UV, 1H and 13C NMR spectra, and AM1 studies of protonation of aminopyridines. *Journal of Molecular Structure* 322, (1–3), 223–232, 1994.

29. Lin, V. S. Y.; Radu, D. R.; Han, M.-K.; Deng, W.; Kuroki, S.; Shanks, B. H.; Pruski, M., Oxidative polymerization of 1,4-diethynylbenzene into highly conjugated poly(phenylene butadiynylene) within the channels of surface-functionalized mesoporous silica and alumina materials. *Journal of the American Chemical Society* 124, (31), 9040–9041, 2002.

30. Bagshaw, S. A.; Pinnavaia, T. J., Mesoporous alumina molecular sieves. *Angewandte Chemie, International Edition in English* 35, (10), 1102–1105, 1996.

31. Huh, S.; Wiench, J. W.; Trewyn, B. G.; Song, S.; Pruski, M.; Lin, V. S. Y., Tuning of particle morphology and pore properties in mesoporous silicas with multiple organic functional groups. *Chemical Communications* (18), 2364–2365, 2003.

32. Huh, S.; Chen, H.-T.; Wiench, J. W.; Pruski, M.; Lin, V. S. Y., Controlling the selectivity of competitive nitroaldol condensation by using a bifunctionalized mesoporous silica nanosphere-based catalytic system. *Journal of the American Chemical Society* 126, (4), 1010–1011, 2004.

33. Nakayama, T.; Suzuki, H.; Nishino, T., Anthocyanin acyltransferases: specificities, mechanism, phylogenetics, and applications. *Journal of Molecular Catalysis B: Enzymatic* 23, (2–6), 117–132, 2003.

34. Pihko, P. M., Activation of carbonyl compounds by double hydrogen bonding: An emerging tool in asymmetric catalysis. *Angewandte Chemie, International Edition* 43, (16), 2062–2064, 2004.

35. Schreiner, P. R., Metal-free organocatalysis through explicit hydrogen bonding interactions. *Chemical Society Reviews* 32, (5), 289–296, 2003.

36. Huh, S.; Chen, H.-T.; Wiench, J. W.; Pruski, M.; Lin, V. S. Y., Cooperative catalysis by general acid and base bifunctionalized mesoporous silica nanospheres. *Angewandte Chemie, International Edition* 44, (12), 1826–1830, 2005.

37. Kubota, Y.; Goto, K.; Miyata, S.; Goto, Y.; Fukushima, Y.; Sugi, Y., Enhanced effect of mesoporous silica on base-catalyzed aldol reaction. *Chemistry Letters* 32, (3), 234–235, 2003.

38. List, B., Enamine catalysis is a powerful strategy for the catalytic generation and use of carbanion equivalents. *Accounts of Chemical Research* 37, (8), 548–557, 2004.

39. Gelman, F.; Blum, J.; Avnir, D., Acids and bases in one pot while avoiding their mutual destruction. *Angewandte Chemie, International Edition* 2001, 40, (19), 3647–3649, 2001.

40. Liu, J.; Shin, Y.; Nie, Z.; Chang, J. H.; Wang, L.-Q.; Fryxell, G. E.; Samuels, W. D.; Exarhos, G. J., Molecular assembly in ordered mesoporosity: A new class of highly functional nanoscale materials. *Journal of Physical Chemistry A* 104, (36), 8328–8339, 2000.

41. Kantam, M. L.; Sreekanth, P.; Santhi, P. L., Cyanosilylation of carbonyl compounds catalyzed by a diamino-functionalized mesoporous catalyst. *Green Chemistry* 2, (2), 47–48, 2000.

5 The Environmentally Friendly Synthesis of Heteropolyacids

Graciela M. Valle, Silvana R. Matkovic,
Luis A. Gambaro, and Laura E. Briand

CONTENTS

ABSTRACT

Fosfotungstic and fosfomolybdic Wells-Dawson heteropolyacids, $H_6P_2W_{18}O_{62}.xH_2O$ and $H_6P_2Mo_{18}O_{62}.xH_2O$, respectively, were synthesized through ion exchange with a higher yield (~90%) than the conventional organic route (~70%).

Wells-Dawson heteropolyacids are obtained when the corresponding ammonium salt $[(NH_4)_6P_2W_{18}O_{62}.13H_2O$ and $(NH_4)_6P_2Mo_{18}O_{62}.12H_2O]$ is kept in contact with an acid resin [about (1:0.8) salt:resin weight ratio] up to 3 days. The use of an organic media instead of an aqueous media greatly favors the completeness of the exchange.

Isopropanol chemisorption and temperature programmed surface reaction (TPSR) towards propylene allowed the determination of the number of active acid sites and the activation energy in order to compare the acid properties of HPAs with various catalytic materials.

The HPAs possess a higher number of active acid sites than monolayer supported materials due to their pseudoliquid behavior. Moreover, a stressed electronic influence of the oxide support (W-O-Support) and the central phosphorous atom (W-O-P) on the acid strength was observed.

5.1   INTRODUCTION

Heteropolyanions $[X_yM_xO_m]^{n-}$ are composed of a close-packed framework of metal-oxygen octahedrons, M-O (M = Mo^{6+}, W^{6+}, V^{5+}) surrounding a central atom, X (Si^{4+}, P^{5+}, etc) [1]. Nowadays, many researchers focus on Keggin $[XM_{12}O_{40}]^{3-}$ and Wells-Dawson-type $[X_2M_{18}O_{62}]^{6-}$ anions due to their promising application as catalytic materials (2, 3). The free acid form of the heteropolyanions (i.e., Keggin $H_3XM_{12}O_{40}$ and Wells-Dawson $H_8X_2M_{18}O_{62}$-type compounds) are solid superacids because their acid strength is greater than 100% H_2SO_4. Therefore, many heteropolyacids are more active catalysts than conventional organic and inorganic acids in liquid-phase reaction. The insolubility of the heteropolyacids in many liquid organic substances allows an easy separation and reutilization of these catalysts [3, 4].

Undoubtedly, the catalytic processes based on heteropolyacids are an outstanding contribution in the development of environmental benign technologies. However, the synthesis of these compounds is rather complicated; it involves the use of dangerous chemicals and produces harmful liquid and gaseous wastes.

Fosfotungstic $H_6P_2W_{18}O_{62}.24H_2O$, fosfomolybdic $H_6P_2Mo_{18}O_{62}.nH_2O$ (n= 33–37), and arsenic-molybdic $H_6As_2Mo_{18}O_{62}.nH_2O$ (n= 25–35) Wells-Dawson acids have been synthesized. The heteropolyacids are synthesized through the etherate method, electrodialysis, ion exchange, and precipitation with sulfuric acid [5].

The etherate method requires a strongly acidified aqueous solution of the heteropolyanion (from an aqueous-soluble heteropoly-salt) that is shaken with diethyl ether in order to separate three phases: an upper ether layer, an aqueous layer, and a heavy oily layer. This layer contains an etherate of the heteropolyacid whose nature is unknown. The etherate is decomposed with water, and the solution is evaporated until the acid crystallizes. According to early studies by Wu and our own experience, this method does not yield pure $H_6P_2Mo_{18}O_{62}.nH_2O$ but a mixture with fosfomolybdic Keggin-type acid $H_3PMo_{12}O_{40}$ [6].

This technique requires expensive and dangerous chemicals, possesses a low yield towards the acid, produces an important amount of harmful wastes (concentrated HCl and ether mixture), and is time consuming.

Kozhevnikov et al. applied electrodialysis to synthesize iso and heteropolyacids of various structures in highly concentrated aqueous solutions [7]. In general, this method avoids the use of diethyl ether and inorganic acids because the solutions are electrochemically acidified during operation. However, the synthesis of the $H_6P_2W_{18}O_{62}$ is rather complicated because it requires the electrodialysis of a mixture of H_3PO_4 and Na_2WO_4 up to a certain pH value, then a thermal treatment (6 h at 423 K) and dialysis again to obtain the desired product.

More recently, Wijesekera et al. patented the synthesis of fosfomolybdic Wells-Dawson-type polyoxometallates through an ion exchange methodology in aqueous media [8, 9]. Although infrared spectroscopy and nuclear magnetic resonance were applied to establish the purity of the compounds, no investigation of the surface properties was performed. Moreover, the ion exchange in organic media should also be also explored because molybdenum-based heteropoly compounds are unstable in aqueous solution [10].

The present investigation presents a detailed examination of the conditions (solvents, substrate-resin ratio, time of the exchange) to optimize the synthesis of fosfotungstic and fosfomolybdic Wells-Dawson acids through the ion exchange method. Additionally, isopropanol chemisorption and temperature programmed surface reaction was applied to obtain information of the nature and number of the active acid sites.

5.2 EXPERIMENTAL

5.2.1 SYNTHESIS OF THE $(NH_4)_6P_2W_{18}O_{62}.13H_2O$ AND $(NH_4)_6P_2Mo_{18}O_{62}.12H_2O$ SALTS

The heteropolyacids were synthesized through the ion exchange of Wells-Dawson ammonium salts of the $(NH_4)_6P_2W_{18}O_{62}.13H_2O$ and $(NH_4)_6P_2Mo_{18}O_{62}.12H_2O$. The salts were synthesized according to methods reported in the literature [10, 11].

Additionally, the fosfotungstic Wells-Dawson acid $H_6P_2W_{18}O_{62}$ was synthesized according to the etherate method with a 71.4% yield. The details of the synthesis have been published previously [12].

5.2.2 ION EXCHANGE METHOD

A Dowex HCR-W2 (Sigma-Aldrich) acid resin was used in the ion exchange experiments. This material possesses a 1.9 meq/ml exchange capacity, according to the vendor. The density of the wet resin was determined as 1.18 g/ml, which results in a 2.4 meq/g exchange capacity. According to these data, the stoichiometric exchange ratios would be (1:0.6) and (1:0.8) salt:resin weight ratio for the

$(NH_4)_6P_2W_{18}O_{62}.13H_2O$ (MW. 4708 g/mol, 6×10^{-3} meq/mol) and $(NH_4)_6P_2Mo_{18}O_{62}.12H_2O$ (MW. 3106 g/mol, 6×10^{-3} meq/mol) salts, respectively.

The ion exchange experiments were performed both in a column and batch-wise. The flow rate of the effluent in the column experiments was calculated as the ratio between the volume of effluent collected per unit of time and volume of wet resin.

The starting ammonium salts were dissolved in absolute ethanol (Merck, 99.8%), distilled water, or a 50:50 ethanol:water mixture prior to the contact with the resin. The resultant solution after the exchange was allowed to dry at room temperature prior to the infrared analysis.

5.2.3 ADDITIONAL MATERIALS

The acidity of the HPA was compared with several tungsten- and molybdenum-based materials such as bulk tungsten trioxide, monolayer supported tungsten species, nanoparticulate mesoporous tungsten-zirconium oxide, and Keggin heteropolyacids.

Bulk WO_3 was obtained through the thermal decomposition of $(NH_4)_6H_2W_{12}O_{40}$ (Fluka A.G.) at 1073 K for 4 h. Monolayer supported tungsten species over oxide supports (8% WO_3/TiO_2 and 6% WO_3/ZrO_2) were synthesized through incipient-wetness impregnation and were kindly provided by Lehigh University (Bethlehem, PA) [13].

The mesoporous 30.5% WO_3/ZrO_2 nanoparticles (~5 nm) were provided by Rice University (Houston). The details of the synthesis procedure are reported in the literature [14].

Additionally, commercial fosfotungstic $H_3PW_{12}O_{40}.xH_2O$ (Fluka puriss.) and fosfomolybdic $H_3PMo_{12}O_{40}.xH_2O$ acids (Sigma-Aldrich) were used.

5.2.4 CHARACTERIZATION TECHNIQUES

5.2.4.1 Infrared Spectroscopy

The presence of the ammonium ion after the exchange and the stability of the Wells-Dawson structure were followed through infrared spectroscopy. The analysis was performed with FTIR Bruker IFS 66 equipment under ambient conditions. The solid samples were diluted with KBr and pressed into thin wafers.

5.2.4.2 Raman Spectroscopy

The analysis was performed with a new state-of-the-art UV-VIS Raman spectrometer (Horiba-Jobin Yvon Labram-HR) equipped with a notch filter to reject the Rayleigh scattering, a single monochromator stage, and a CCD detector cooled to 140 K. The excitation source is 325 nm and 545 nm from an He-Cd laser (Kimmon) and the scattered photons are directed and focused onto a single monochromator (LabRam HR Jobin Yvon). A UV-sensitive LN CCD detector (Spex) is used to collect the signal. The laser power employed on the sample is only 0.2 mW, with a collection time of 40 s.

5.2.4.3 Thermo-Gravimetric and Differential Thermal Analyses (TGA-DTA)

TGA-DTA analyses were performed in a Shimadzu TGA-50H and DTA-50, respectively. The samples were heated at 10 K/min in a stream of helium (40 sccm) from room temperature to 1073 K. Typically, 100 mg of the sample placed in a platinum cell was used for both analyses.

5.2.4.4 Nuclear Magnetic Resonance ^{31}P NMR

The purity of the Wells-Dawson structure was followed through nuclear magnetic resonance of phosphorous. Liquid NMR spectra were performed with a Bruker AM 500 spectrograph both in D_2O and deuterated methanol CD_3OD under ambient conditions. The equipment operates at a frequency of 202.458 MHz with 11.3 µs pulses. The analysis involved 8–800 pulse responses with a resolution of 0.25 Hz per point.

5.2.4.5 Specific Surface Area and Pore Volume Determination

The adsorption-desorption isotherm of N_2 at 77 K was determined with a Micromeritics Accusorb 2020 surface area and porosity analyzer. The samples were degassed at 373 K under vacuum until a pressure of 10^{-3} Torr was reached. The specific surface area, total pore volume and diameter, and micropores volume and area were determined with the BET, Barret-Joyner-Halenda (BJH), and t-plot methods, respectively.

5.2.4.6 Chemisorption and Temperature Programmed Surface Reaction Spectroscopy

The number and properties of the active sites were determined by adsorption of isopropanol at room temperature followed by temperature programmed surface reaction. Typically, 20–90 mg of the sample were used during the experiments. The samples were pretreated at 673 K (373 K for $H_6P_2Mo_{18}O_{62}.xH_2O$) for 1 h under a flow of pure oxygen prior to chemisorption and TPSR analysis. They were allowed to cool down to the adsorption temperature under a flow of helium (35 cm^3(NTP) min^{-1}). Successive pulses of 0.5 µl of isopropanol (Merck P.A., 100%) were dosed through a heated septum until the saturation of the heteropolyacid was reached. The adsorption process was monitored *in situ* through a mass spectrometer and a conductivity cell that detects the nonadsorbed alcohol or the species desorbed from the sample. After the saturation, the samples were heated up to 723 K at 10 K/min for the temperature programmed surface reaction experiment. The species resulting from the reaction of the surface species were detected in the mass spectrometer and recorded in a computer. The following *m/e* ratios were employed to identify the desorbed species: C_3H_7OH, *m/e* = 45; C_3H_6, *m/e* = 41; di-isopropyl ether, *m/e* = 31 and 87; acetone, *m/e* = 43 and 58; H_2O, *m/e* = 18; CO_2, *m/e* = 44; and CO, *m/e* = 28. The details of the calibration procedures of isopropanol and propylene were published before [15].

Temperature programmed desorption analysis was also performed over the heteropolyacids without previous thermal treatments in order to identify the temperature of desorption of water and ethanol (m/e =31 and 46) that remain adsorbed after the ion exchange.

5.3   RESULTS AND DISCUSSION

5.3.1   OPERATIVE CONDITIONS FOR THE SYNTHESIS OF HETEROPOLYACIDS THROUGH ION EXCHANGE

The influence of the operation conditions, such as amount of resin (salt:resin weight ratio), nature of the solvent, and temperature and time in the exchange of ammonium by protons ions, were investigated in batch experiments. A great variety of conditions were tested; however, only the most representative results are included in order to simplify the discussion.

Figure 5.1 shows the infrared spectra of the materials obtained through the ion exchange of the fosfotungstic salt with the acid resin performed in aqueous and ethanol media both in batch and column. Additionally, the spectrum of the fosfotungstic Wells-Dawson acid $H_6P_2W_{18}O_{62}.24H_2O$ synthesized through the etherate method and the starting ammonium fosfotungstic Wells-Dawson salt $(NH_4)_6P_2W_{18}O_{62}.13H_2O$ are presented for comparison.

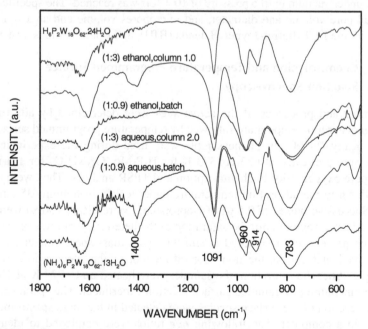

FIGURE 5.1 Infrared spectra of the $(NH_4)_6P_2W_{18}O_{62}.13H_2O$ salt, the $H_6P_2W_{18}O_{62}.24H_2O$ heteropolyacid obtained through the conventional organic method and the materials obtained through the ion exchange of the salt in a column containing (1:3) salt:resin weight ratio (1.0 min^{-1} and 2.0 min^{-1} indicates the flow rate of the effluent), and 3 days in batch with a (1:9) ratio, both in aqueous and organic media.

The spectrum of the fosfotungstic Wells-Dawson acid possesses an intense band at 1091 cm^{-1} assigned to the stretching mode of the P-O species that is considered a fingerprint of the Wells-Dawson heteropolyanion $P_2W_{18}O_{62}^{6-}$ structure. The bands at 960, 914, and 783 cm^{-1} correspond to the W-O species of the cage structure surrounding the central phosphorous species [12]. The spectra of the materials after being in contact with the acid resin in aqueous media possess a band at 1400 cm^{-1} that corresponds to NH_4^+ species, indicating that the exchange was not complete even after 3 days.

In contrast, the organic media favors the ion exchange based on the observation that the pure fosfotungstic heteropolyacid is obtained after 3 days in contact with (1:0.9) salt:resin weight ratio. Further experiments demonstrated that contact times and salt:resin ratios below 3 days and (1:0.9) do not yield the acid.

Additionally, the ion exchange was carried in a glass column with a (1:3) salt:resin weight ratio. The experiments were performed with varying flow rates (0.8 to 3.6 BV/min) due to the fact that this parameter was difficult to reproduce from one experiment to another. This is a typical drawback of the exchange columns that possess a regular two-way stopcock. In contrast with the results obtained batchwise, the exchange of the fosfotungstic salt was complete when the experiment was carried in aqueous medium but not in ethanol.

Surprisingly, the yield of the synthesis was 95.5%, which is higher than the etherate route for fosfotungstic acid (see Section 5.2.1).

^{31}P NMR analysis was not useful to determine the effectiveness of the exchange because both the starting ammonium fosfotungstic Wells-Dawson salt and the acid possess similar signals at −10.78, −11.55, and −12.28 ppm, which makes it difficult to distinguish both substances (spectra not shown).

The experiments with the ammonium fosfomolybdic salt with various salt:resin weight ratios demonstrated that a complete exchange is achieved with the stoichiometric (1:0.8) salt:resin exchange ratio. Again, the comparison of the infrared spectra of the materials obtained in aqueous (spectra not shown) and organic media indicate that the ion exchange is greatly favored in organic media.

Figure 5.2 shows the effect of time and temperature in the ion exchange of the ammonium fosfomolybdic Wells-Dawson salt in organic media (ethanol and a 50:50 ethanol:water mixture). The spectra of the material obtained after 6 and 24 h of ion exchange in ethanol and after 3 days in ethanol:water mixture possess the characteristic signals of the starting Wells-Dawson salt $(NH_4)_6P_2Mo_{18}O_{62}.12H_2O$. The fosfomolybdic Wells-Dawson structure presents the infrared signal of the central phosphorous species at 1078 and 904 cm^{-1}, and Mo-O species at ~936 and 769 cm^{-1} [10, 16]. Additionally, the ammonium salt possesses the infrared signal of the NH_4^+ species at 1403 cm^{-1}, similarly to the fosfotungstic salt.

The fosfomolybdic acid $H_6P_2Mo_{18}O_{62}.23H_2O$ is obtained after 3 days of ion exchange, with a 92.2% yield. Similar results were obtained either at room temperature or below, which clearly demonstrated that the temperature does not have any positive influence in the exchange.

Further studies through ^{31}P NMR were performed in order to establish the purity of the fosfomolybdic acid obtained through ion exchange. Figure 5.3 shows that the ammonium fosfomolybdic Wells-Dawson salt possesses a single signal at about

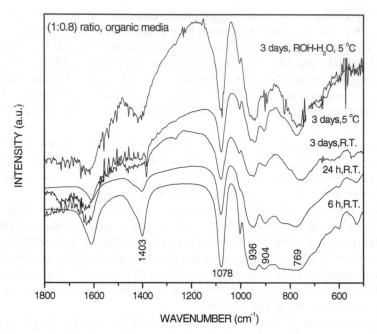

FIGURE 5.2 Infrared spectra of the $(NH_4)_6P_2Mo_{18}O_{62}.12H_2O$ salt and the materials obtained through the ion exchange of the salt in contact with (1:0.8) salt:resin weight ratio and variable time and temperature of exchange in ethanol and ethanol:water media.

−1.61 ppm in deutered methanol [16]. The spectrum of the salt partially exchanged shows additional signals at −1.67 ppm (−2.42 ppm in D_2O) assigned to the fosfomolybdic Wells-Dawson acid along with the characteristic signal of the Keggin $PMo_{12}O_{40}^{3-}$ species at −3.64 ppm (−3.99 ppm in D_2O) [17]. A third one, at −2.10 ppm, could not be identified. However, it might be attributed to an interaction with the organic solvent because it is not observed in deutered water.

Previous studies demonstrated that the fosfomolybdic Wells-Dawson salt $(NH_4)_6P_2Mo_{18}O_{62}$ decomposes in the lacunar Keggin-type anion $H_xPMo_{11}O_{39}^{(7-x)-}$ in aqueous media [10]. Nevertheless, the degradation of the structure is avoided in organic solvents such as alcohols [16]. The observation that traces of Keggin species appear in the analysis performed in CD_3OD, along with the acid obtained in organic media (without the presence of nonexchanged salt), unambiguously demonstrates that those species are produced due to the decomposition of the acid. Further degradation of the fosfomolybdic Keggin anion towards the lacunar species is avoided due to the highly acid medium [17].

Raman spectroscopy allowed determining the purity of the heteropolyacids in the solid state. Figure 5.4 shows the spectra of the starting salts and the corresponding acids, both fully hydrated. The absence of the signals of WO_3 (803, 713, 607, and 275 cm^{-1}) and MoO_3 (987, 815, 664, and 279 cm^{-1}) provides evidence of the purity of the fosfotungstic and fosfomolybdic heteropolyanions phases, respectively.

The intense bands of the tungsten-based heteropoly-compounds above 950 cm^{-1} (centered at 996 and 1001 cm^{-1} for $H_6P_2W_{18}O_{62}$ and $(NH_4)_6P_2W_{18}O_{62}$, respectively)

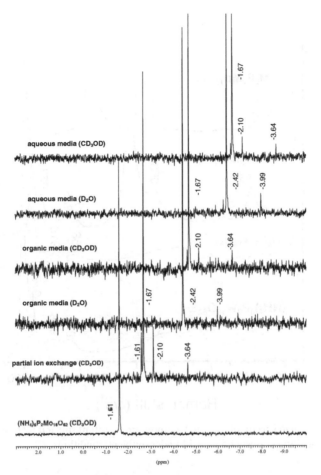

FIGURE 5.3 ^{31}P nuclear magnetic resonance spectra of pure ammonium fosfomolybdic Wells-Dawson salt and after being partially and completely exchanged with the acid resin in aqueous and organic media. The spectra were taken both in deuterated water (D_2O) and methanol (CD_3OD).

are attributed to W = O stretches within WO_6 octahedral species. Moreover, the bands between 800 and 950 cm^{-1} belong to symmetric O-W-O stretches of poly-tungstate species [13].

Similarly, the molybdenum-based HPCs possess Raman bands due to polymer-ized, octahedral $Mo_7O_{24}^{6-}$, or $Mo_8O_{26}^{4-}$ clusters. The signals in the 830–1000 cm^{-1} range are attributed to the stretching modes of the terminal Mo = O bonds, and the signals at ~600 cm^{-1} (653 and 715 cm^{-1} for $(NH_4)_6P_2Mo_{18}O_{62}$ and 668 cm^{-1} for $H_6P_2Mo_{18}O_{62}$) might be assigned to the Mo-O-Mo stretching [18].

5.3.2 THERMAL STABILITY OF WELLS-DAWSON HETEROPOLYACIDS

Temperature programmed desorption analysis gave insights on the species evolved upon heating of the heteropolyacids that were synthesized through ion exchange

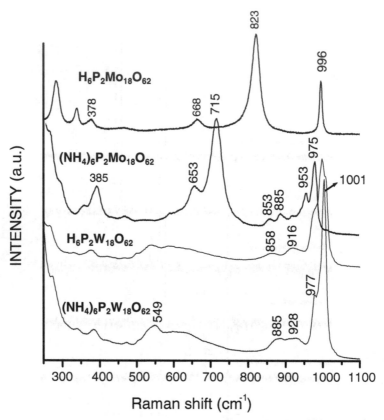

FIGURE 5.4 Raman spectra of hydrated ammonium fosfotungstic and fosfomolybdic Wells-Dawson salts and the corresponding acid under ambient conditions.

using ethanol as solvent. Figure 5.5 and Figure 5.6 show the distribution of substances produced during the temperature programmed desorption analysis of the acids (without previous thermal treatments).

The fosfomolybdic acid presents two intense signals due to water loss centered at 403 and 543 K. Complementary thermo-gravimetric analysis revealed that 24 and 3 water moles per mol of acid are released consecutively.

The TPD analysis demonstrated that ethanol strongly interacts with the heteropolyanion structure because it desorbs as CO/CO_2 above 673 K. No molecular desorption is observed during the analysis (see the $m/e = 31$ signal), which suggests that the alcohol is chemisorbed within the structure.

Further, infrared analyses of the heteropolyacid after calcination at various temperatures demonstrated that the Wells-Dawson structure maintains intact during the first dehydration process and begins a structural rearrangement towards a Keggin structure at 473 K (see Figure 5.7). Only the signals belonging to the $PMo_{12}O_{40}^{3-}$ anion at 1065 cm^{-1} (P-O species) and 960 cm^{-1} (Mo-O species) are observed at 523 K, indicating that the Wells-Dawson structure was completely degraded to a

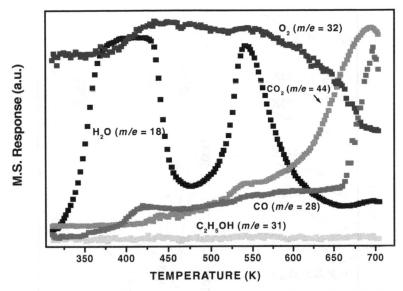

FIGURE 5.5 Temperature programmed desorption spectra of $H_6P_2Mo_{18}O_{62} \cdot 27H_2O$.

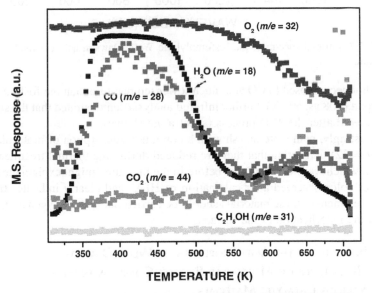

FIGURE 5.6 Temperature programmed desorption spectra of $H_6P_2W_{18}O_{62} \cdot xH_2O$.

Keggin-type structure [17]. The infrared signals at 569, 871, and 994 cm^{-1} indicate that the heteropoly anion collapses completely towards MoO_3 at 873 K.

The fosfotungstic acid also shows a broad signal due to water loss from room temperature toward 553 K and a smaller one centered at 523 K (see Figure 5.6). These observations are in agreement with the consecutive loss of water of crystallization associated with the protons as $H^+(H_2O)_n$, H_3O^+, and $H_5O_2^+$ species reported in the literature [12, 19]. Although no molecular desorption of ethanol was observed,

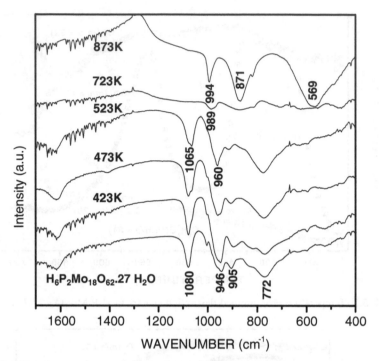

FIGURE 5.7 Infrared spectra of the fosfomolybdic Wells-Dawson acid calcined at various temperatures.

the alcohol decomposed to CO at a much lower temperature than the fosfomolybdic counterpart (408 vs. 694 K). Further infrared analysis demonstrated that the structure was unaltered after the TPD process (spectra not shown).

Interestingly, both materials showed a continuous desorption of molecular oxygen ($m/e = 32$), indicating that they are reduced during the thermal treatment. This behavior, which was never reported before in the literature, might explain the typical change of color observed after calcination of HPAs and clearly indicates that any thermal treatment of these materials should be carried out under oxygen atmosphere to avoid the modification of the surface.

5.3.3 INSIGHTS ON THE ACID PROPERTIES OF WELLS-DAWSON HPAs THROUGH MOLECULAR PROBES: COMPARISON WITH VARIOUS CATALYTIC MATERIALS

5.3.3.1 Number and Nature of the Accessible Active Acid Sites

Isopropanol was used as a probe molecule to characterize the acidity of heteropolyacid compounds because the product's distribution upon reaction depends on the nature of the surface active sites. Strong Brönsted (H^+) and Lewis acid sites catalyze the dehydration of isopropanol to propylene (di-isopropyl ether over weak Lewis acid sites), and redox/basic sites lead to the dehydrogenation of the alcohol to acetone.

Previous investigations demonstrated for the first time in the literature that *in situ* isopropanol chemisorption and quantitative temperature programmed surface reaction are suitable to be used to determine the nature, number, and acid strength of the surface/bulk active sites of tungsten oxide-based catalysts and particularly of the heteropoly compounds [15].

The chemisorption of isopropanol at 313 K leads to the coverage of the heteropolyacids with a stable monolayer of adsorbed isopropoxy species and avoids further surface reaction. These adsorbed intermediate-reactive alkoxy species further react and desorb as propylene (or other product depending on the nature of the site) upon controlled heating during the TPSR experiment. Therefore, the quantification of the desorbed product is proportional to the total number of acid sites active on isopropanol dehydration towards propylene.

Table 5.1 presents the atoms coordination, specific surface area, external and microporous area, total pore volume, micropore volume, and diameter (in the mesoporous range) of the Wells-Dawson and Keggin HPAs, along with monolayer supported and mesostructured tungsten-based materials and bulk WO_3. Table 5.2 compares the temperatures of isopropanol chemisorption, number of surface active acid sites (Ns), temperatures of propylene desorption, and activation energy of isopropoxy surface reaction towards propylene.

Isopropanol adsorption was performed at several temperatures in order to determine the more suitable conditions for the alcohol chemisorption on the active sites avoiding further reaction. The observation that no molecular isopropanol is detected in the TPSR spectra indicates that no physisorption (or weak chemisorption) of the alcohol is produced even at 313 K over heteropolyacids (spectra not shown). The maximum amount of surface isopropoxy species over tungsten oxide WO_3 and monolayer supported tungsten oxide species was obtained through isopropanol adsorption at 383 and 343 K, respectively.

The number of available surface sites for isopropanol adsorption of bulk tungsten trioxide and monolayer supported tungsten oxide catalysts, even mesoporous nanoparticles, is orders of magnitude (0.9–6 $\mu mol/m^2$) lower than the heteropolyacids (8–55 $\mu mol/m^2$). This observation cannot be attributed to the surface structure because all these materials possess polymerized WO_6 species with octahedral coordination (see Table 5.1). Moreover, no correlation is observed with the specific surface area and the pore diameter because high surface area, mesoporous monolayer supported catalysts possess lower Ns than HPAs. In fact, Table 5.1 shows that those materials do not possess microporosity; the specific surface area is only external surface area. This observation shows that the adsorption of alcohol occurs exclusively at the outermost surface layer of WO_3 and monolayer supported tungsten oxide catalysts.

In general, the structure of the heteropolyacids possesses a certain contribution of microporosity, and the pseudoliquid behavior that allows the adsorption of alcohol on the external surface and within the tertiary structure of heteropoly compounds (see Table 5.1). The abnormally high Ns of the fosfomolybdic Wells-Dawson acid is attributed to the presence of ethanol derived species, modifying the pore structure of the HPA. In fact, previous investigations demonstrated that the number of surface acid sites is directly influenced by the hydration water because fully

Table 5.1

Synthesis methodology, structure, specific surface area, BJH adsorption cumulative pore volume (between 17 and 3000 Å pore width), t-plot micropore volume and area, and BJH adsorption average pore diameter of tungsten- and molybdenum-based bulk and supported catalysts

Catalyst	Synthesis	Structure	S_{BET} (m²/g)	External surface area (m²/g)	Micropore area (m²/g)	V_p (cm³/g)	Vmicropore[c] (cm³/g)	Pore diameter (nm)
WO_3	Thermal decomposition	Octahedral	1.3	1.4	—	5.4×10^{-3}	—	19.0
8% WO_3/TiO_2	Incipient wetness[a]	Octahedral[a]	46.7	47.2	—	0.4	—	32.6
6% WO_3/ZrO_2	Incipient wetness[a]	Octahedral[a]	33.8	33.1	—	0.2	—	28.3
30.5% WO_3/ZrO_2	Precipitation w/template	Octahedral	45.8	46.8	—	3.4×10^{-2}	—	2.7
$H_3PW_{12}O_{40}$	Commercial	Octahedral[b]	5.3	4.1	1.2	8.7×10^{-3}	5.0×10^{-4} (6%)	6.6
$H_3PMo_{12}O_{40}$	Commercial	Octahedral[b]	15.5	8.9	6.7	1.4×10^{-2}	3.0×10^{-3} (22%)	6.7
$H_6P_2W_{18}O_{62}$	Organic	Octahedral	3.7	2.0	1.7	1.2×10^{-2}	7.7×10^{-4} (6%)	25.5
$H_6P_2W_{18}O_{62}$	Ion exchange/no calcination	Octahedral	1.1	0.8	0.3	3.5×10^{-3}	1.3×10^{-4} (4%)	19.5
$H_6P_2W_{18}O_{62}$	Ion exchange/calcined	Octahedral	1.4	—	2.0	1.5×10^{-3}	9.4×10^{-4} (64%)	55.2
$H_6P_2Mo_{18}O_{62}$	Ion exchange (hydrated)	Octahedral	2.5	1.7	0.8	8.0×10^{-3}	3.6×10^{-4} (5%)	21.4

[a] From Reference 13.
[b] From Reference 5.
[c] The value between brackets represents the contribution of microporous volume to the total pore volume in percentage.

Table 5.2
Temperature of isopropanol chemisorption, maximum number of active sites for isopropanol chemisorption, and activation energy of surface reaction of tungsten- and molybdenum-based bulk and supported catalysts

Catalyst	Temperature of adsorption (K)	Ns (μmol/m^2)	Tp (K)	Ea (kcal/mol)
WO_3	383	0.9	429	29.1
8% WO_3/TiO_2	343	1.0	429	29.1
6% WO_3/ZrO_2	343	0.2	467	31.7
30.5% WO_3/ZrO_2	343	6.4	439	29.8
$H_3PW_{12}O_{40}$	313	19.8	385	26.0
$H_3PMo_{12}O_{40}$	313	3.7	372	25.7
$H_6P_2W_{18}O_{62}$ org.	313	8.1	369	24.9
$H_6P_2W_{18}O_{62}$ exch.	313	10.1	372	25.1
$H_6P_2Mo_{18}O_{62}$	313	54.5	373	25.2

hydrated fosfotungstic Wells-Dawson acid produces twice the amount of propylene (52 μmol/g) than the anhydrous acid (25 μmol/g) [15]. Moreover, a continuous drop of the amount of propylene was observed with the decrease of the degree of hydration of the Wells-Dawson structure. This observation shows that the loss of water leads to the shortening of the distance between the Wells-Dawson units and the decrease of the available active sites for isopropanol chemisorption. On line with this observation, the TPD analysis of the fosfomolybdic Wells-Dawson acid showed that ethanol is strongly retained into the structure because it desorbs as COx above 673 K. Therefore, the HPA still possesses those species after calcination at 373 K, isopropanol chemisorption, and the TPSR processes. The effect of the organic media as a template of the porous structure is also observed on the fosfotungstic Wells-Dawson HPA synthesized through ion exchange in ethanol media and further calcined. In fact, the acid possesses 64% of total pore volume due to microporous volume after calcination at 673 K, compared with 4% before ethanol desorption. This observation might explain why the calcined acid possesses 20% more available active acid sites (10.1 μmol/m^2) than the acid synthesized through the conventional organic procedure (8.1 μmol/m^2) (see Table 5.2).

5.3.3.2 Strength of the Active Acid Sites: Activation Energy for Surface Species Reaction

Previous work by Wachs et al. demonstrated that the information obtained during the temperature programmed surface reaction of adsorbed intermediate species over powdered samples enables us to determine kinetic parameters, such as the activation energy Ea of a surface reaction [20]. The authors demonstrated that the Redhead equation allows determining the activation energy Ea,

$$Ea / (R \times Tp^2) = v / B \exp (-Ea / (R \times Tp)),$$

where R is the gas phase constant; B, the heating rate; Tp, the temperature of desorption (°K) of the surface species; and ν, the first-order Arrhenius rate constant pre-exponential factor ($\sim 10^{13}$ s^{-1}).

This equation was applied here to calculate the activation energy for the dehydration of isopropoxy species towards propylene as a measurement of the acid strength of the catalytic materials, discussed in the previous section.

The higher activation energy ($\sim$29 kcal/mol) of surface reaction over bulk tungsten trioxide and monolayer supported tungsten oxide catalysts than HPAs ($\sim$25 kcal/mol) is attributed to the differences in the nature of the active sites of those catalysts. Spectroscopic studies of dehydrated monolayer supported tungsten oxide catalysts demonstrated that the surface structure is composed of polymerized octahedrally coordinated WOx species possessing W = O (mono-oxo) terminal bond and bridging W-O-S (S = tungsten or support cation) bonds. Adsorption of NH$_3$ and pyridine bases also demonstrated that monolayer supported metal oxide catalysts (Re, Cr, Mo, W, V, Nb, etc.) possess mainly Lewis acid sites and Brönsted sites as a minor contribution [21].

The structure of Wells-Dawson and Keggin-type heteropolyacids is composed of a central phosphorous atom PO$_4$ surrounded by a cage of tungsten atoms. Similarly to monolayer supported tungsten oxide catalysts, each tungsten atom composes WO$_6$ octahedral units with one terminal double-bonded oxygen and is linked together through W-O-W bonds. However, many studies demonstrated that HPAs possess Brönsted acid sites (WOH) stronger than conventional solid acids such as SiO$_2$-Al$_2$O$_3$, sulfated zirconia, and titania [23].

Moreover, the acid strength is influenced by the electronegativity of the oxide support (ZrO$_2$ and TiO$_2$) and the central phosphorous atom of the HPAs. Figure 5.8

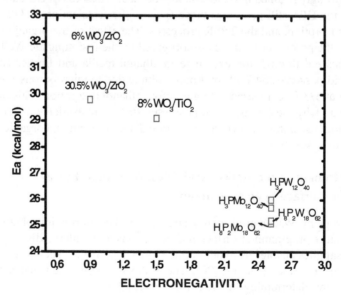

FIGURE 5.8 Activation energy Ea of the surface reaction of chemisorbed isopropoxy species toward propylene over monolayer supported tungsten species and bulk heteropolyacids.

shows that the higher the electronegativity (P > Ti > Zr), the lower the activation energy and the higher the acidity of the materials. This evidence suggests that an electronic deficiency of the surface active sites improves their acid strength.

5.4 CONCLUSIONS

The present investigation demonstrates that the ion exchange methodology is suitable to be used for the synthesis of fosfotungstic and fosfomolybdic Wells-Dawson acids in a higher yield than the conventional organic route.

The chemisorption of isopropanol at 313 K leads to the coverage of the catalysts with a monolayer of adsorbed isopropoxy species and avoids surface reaction. Heteropoly compounds are highly active towards isopropanol dehydration above that temperature. Temperature programmed reaction analysis shows that the decomposition of adsorbed isopropoxy species towards propylene at about 373 K, which shows the acid nature of the active sites.

The number of available acid sites of Keggin and Wells-Dawson heteropolyacids is significantly higher than bulk WO_3 and even monolayer supported tungsten oxide species. This observation shows that the alcohol is adsorbed on both the surface and the bulk, in agreement with the pseudoliquid phase behavior of the heteropolyanion structure. Moreover, the activation energy of surface decomposition of adsorbed isopropoxy species towards propylene is significantly lower on the heteropolyacids than the other tungsten oxide-based catalysts. The higher acid strength of the HPAs rather than WO_3 and monolayer supported tungsten oxide catalysts is attributed to the presence of Brönsted acid sites and the electronic deficiency caused by the highly electronegative central phosphorous atom.

ACKNOWLEDGMENTS

The authors acknowledge Prof. I. E. Wachs for the Raman analysis and monolayer supported catalysts; Will Knowls for mesoporous tungsten-zirconia sample; and the financial support provided by Consejo Nacional de Investigaciones Científicas y Técnicas (CONICET; PEI No. 6132 and NSF-CONICET collaboration program Res. No. 0060); Agencia Nacional de Promoción Científica y Tecnológica (PICT; 14-12161/02); and Universidad Nacional de La Plata (project X378).

REFERENCES

1. N. Mizuno and M. Misono, *Chem. Rev.* 98, 1999.
2. T. Okuhara, N. Mizuno, M. Misono, in *Catalytic Chemistry of Heteropoly Compounds*, D.D. Eley, W.O. Haag and B. Gates (Eds.), Academic Press Inc., Adv. Catal. 41, 113, 1996.
3. L.E. Briand, G.T. Baronetti, H.J. Thomas, *Appl. Catal.* 256, 37, 2003.
4. P. Vázquez, L. Pizzio, C. Cáceres, M. Blanco, H. Thomas, E. Alesso, L. Finkielsztein, B. Lantaño, G. Moltrasio, J. Aguirre, *J. Mol. Catal. A: Chemical* 161, 223, 2000.
5. M.T. Pope, in *Heteropoly and Isopoly Oxometalates*, Springer-Verlag, Berlin, 1983.

6. H. Wu, *J. Biol. Chem.* 43, 189, 1920.
7. G.M. Maksimov, R.I. Maksimovskaya, I.V. Kozhevnikov, Russ. *J. Inorg. Chem.* 39, 595, 1994.
8. T.P. Wijesekera, J.E. Lyons, P.E. Ellis, Jr., U.S. Patent 6,060,419, May 9, 2000.
9. T.P. Wijesekera, J.E. Lyons, P.E. Ellis, Jr., U.S. Patent 6,169,202, January 2, 2001.
10. L.E. Briand, G.M. Valle, H.J. Thomas, *J. Mater. Chem.* 12, 299, 2002.
11. M. Filowitz, R.K.C. Ho, W.G. Klemperer and W. Shum, *Inorg. Chem* 18, 93, 1979.
12. G. Baronetti, L. Briand, U. Sedran, H. Thomas, *Appl. Catal.* 172, 265, 1998.
13. D.S. Kim, M. Ostromecki, I.E. Wachs, J. Mol. *Catal. A. Gen.* 106, 93, 1996.
14. M.S. Wong, E.S. Jeng, J.Y. Ying, Nano Lett. 1, 637, 2001.
15. L.A. Gambaro, L.E. Briand, Appl. Catal. 264, 151, 2004.
16. G.M. Valle, L.E. Briand, Mat. Lett. 57, 3964, 2003.
17. A. Concellón, P. Vázquez, M. Blanco, C. Cáceres, *J. Coll. Int. Sc.* 204, 256, 1998.
18. H. Hu, I.E. Wachs, S. Bare, J. Phys. Chem. 99, 10897, 1995.
19. J.E. Sambeth, G.T. Baronetti, H.J. Thomas, J. Molec. *Catal. A: Chemical* 191, 35, 2003.
20. I.E. Wachs, J.-M. Jehng, W. Ueda, *J. Phys. Chem.* B 109, 2275, 2005.
21. I.E. Wachs, *Coll. Surf.* 105, 143, 1995.

6 Synthesis of Metal Colloids

Lawrence D'Souza and Ryan Richards

CONTENTS

6.1 INTRODUCTION

Colloidal metals—originally called sols—first generated interest because of their intense colors, which enabled them to be used as pigments in glass or ceramics. The most extensively studied metal as a colloid is gold because of its well-known optical properties. The research on colloidal gold dates back to the 18th century. Hans Heinrich Helcher, who was a philosopher as well as a doctor, published a formal report on colloidal gold in 1718.[1] During the 19th and 20th century, colloidal gold was used extensively in medicine, dying silk fabrics, and colored glasses and ceramics. In 1857, Faraday reported the formation of deep red solutions of colloidal gold by reduction of an aqueous solution of chloroaurate ($AuCl_4^-$) using phosphorus in CS_2 (a two-phase system) in a well-known work. He studied the effect on optical properties of thin films of colloidal gold under mechanical force.[2] In the 19th century, a great deal of theoretical and experimental research was conducted on many colloidal metals, especially from d-block elements.

In the 21st century, the field is booming with novel applications in science and technology. Nanoparticulate metal clusters/colloids are defined as isolable particles in the nanometer size range, which are prevented from agglomerating by protecting shells. They can be redispersed in water ("hydrosols") or organic solvents ("organosols"). The number of potential applications of these colloidal particles is growing rapidly because of the unique electronic structure of the nanosized metal particles and their extremely large surface areas.[3–8] Mono- and bimetallic colloids can be used as precursors for a new type of catalyst that is applicable in both the homogeneous and heterogeneous phases.[9, 10] Nanoparticles, comprised of one or two different metal elements, are of considerable interest from both the scientific and technological points of view.[11] The catalytic properties and the electronic structure of the nanomaterials can be tailored by changing the cluster size, composition, and structure.[12] The primary applications of metal nanosystems are in catalysis, whereas those of metal oxides are in cosmetics, sensors, ceramics, catalyst supports, and the petroleum industries. Nearly 80% of chemical industries employ heterogeneous catalysis to produce chemicals, which account for 35% of the world gross product; obviously, the cost and efficiency of the catalysts employed in these processes directly influence the world economy. Herein, we will provide an overview of colloidal methods and properties as well as the approaches being used to bridge the fields of nanotechnology and catalysis.

6.2 DEFINITIONS

6.2.1 Modern Classification of Nanoscale Materials

The nomenclature of nanomaterials in the literature has traditionally been poorly defined and confusing, until Schmid provided a clear-cut definition as described below.[13] Finke has also pointed out that the lack of precise compositional, size, or purity data for cluster or colloids also contributes to the nomenclature problem.[14] Any system can now be classified into discrete single-metal complexes; metal clusters including nanoclusters <100 Å; traditional metal colloids >100 Å; and bulk metal. This system of classification is followed in the present literature.[14]

6.2.2 Magic Number or Full Shell Clusters

Clusters with full shell atomic lattices are known as magic number clusters; they have an extra stability due to this packing nature.[15] The number of surface atoms S_n in "nth" shell for different crystal systems is as follows:

Polyhedron Formulas	
Truncated tetrahedron	$S_n = 14 \, n^2 + 2$
Cuboctahedron or twinned cuboctahedron	$S_n = 10 \, n^2 + 2$
Truncated octahedron	$S_n = 30 \, n^2 + 2$
Truncated cube	$S_n = 46 \, n^2 + 2$
Triangular prism	$S_n = 4 \, n^2 + 2$

The atomic arrangement in magic number crystals is more compact and symmetrical, which is favored over noncompact and unsymmetrical arrangements. This compactness gives more stability for a crystalline solid. For example, the atomic arrangement of three, four, six, or eight spheres of equal or nearly equal sizes is more compact and symmetrical than five, seven, or nine spheres.

6.3 METHODS OF COLLOID PREPARATIONS

Techniques for the preparation of metal cluster/nanoparticles can be classified into three primary categories: condensed phase, gas phase, and vacuum methods. In condensed phase synthesis, metal and semiconductor nanoparticles are prepared by means of chemical synthesis, which is also known as wet chemical preparation. In gas phase synthesis, metal is vaporized, and the vaporized atoms are condensed in the presence or absence of an inert gas. In vacuum methods, the metal of interest is vaporized with high-energy Ar, Kr ions, or laser beams in a vacuum, and thus generated metal vapor is deposited on a support.

6.3.1 Wet Chemical Preparations

Nanoparticles are obtained by two general approaches: top down and bottom up. In top down methods, bulk metals are mechanically ground to the nanosize and stabilized by using a suitable stabilizer.[16, 17] The problem with this method is difficulty in achieving the narrow size distribution and control over the shape of the particles. Moreover, bimetallic nanoparticles with core shell structures cannot be obtained by this method.

In bottom up methods, nanoparticles are obtained by starting with molecular precursors and building up. It is the bottom up approach that is really in the spirit of nanoscience, in that here, materials are built up from nature's smallest building blocks, atoms and molecules. For the preparation of metal nanoparticles, metal salts are usually reduced by using suitable reducing agents like hydrides, citrate, hydrazine, and hydrogen, or by electrochemical reduction or irradiating with high energetic radiations.

Metal salts are usually reduced in the presence of suitable capping agents such as thiols, amines, surfactants, or polymers that stabilize the nanoparticles from agglomeration. The chemical reduction of transition metal salts in the presence of stabilizing agents to generate zerovalent metal colloids in aqueous or organic media was first published by Faraday in 1857.[2] Turkevich later further established the standard protocol for the preparation of metal colloids.[3, 4, 18] He studied the detailed mechanism of the formation of gold colloid by citrate reduction and found it to act as both reducing agent and stabilizer in this system. Additionally, he found that sodium citrate reduction is an autocatalytic process, and reduction curves consist of four regions: an induction period, a rapid rise at the beginning of the nucleation, a linear portion, and finally a decay portion. Turkevich further stated that the growth process follows an exponential law of growth.[3] Data from modern analytical techniques and more recent thermodynamic and kinetic results have been used to refine this model, as illustrated in Figure 6.1.[19, 20]

Zerovalent metal atoms form during the embryonic stage of nucleation.[19] These zerovalent metal atoms collide with other metal atoms and metal ions to irreversibly form a stable metal nuclei that is known as a *seed*. Depending upon the strength of the metal-metal bond and redox potential of the metal salt and reducing agent, the diameter of the seed lies below 1 nm in size. Once the stable nuclei forms, stabilizer molecules are adsorbed on them and prevent further growth. Stabilization is governed by two main factors: electrostatic repulsion between particles and steric hindrance from stabilizer molecules (Figure 6.2).[21]

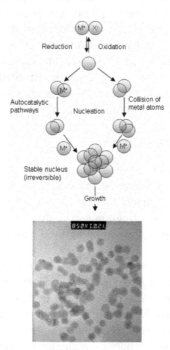

FIGURE 6.1 Formation of nanostructured gold colloids by the salt reduction method. TEM image is of gold nanocolloids of 20 ± 2 nm size prepared by citrate reduction method.

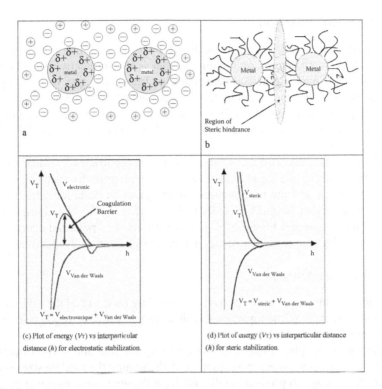

a

b

Region of
Steric hindrance

(c) Plot of energy (V_T) vs interparticular
distance (h) for electrostatic stabilization.

(d) Plot of energy (V_T) vs interparticular distance
(h) for steric stabilization.

FIGURE 6.2 (a) Electrostatic stabilization of nanostructured metal colloids, (b) steric stabilization of nanostructured metal colloids; (c) and (d) are plot of energy versus interparticular distances for electrostatic and steric stabilization ((c) and (d) adapted with permission from Reference 22).

6.3.2 Need for Stabilization

Nanoparticles are generally found in a metastable state. The force of attraction (van der Walls force) between any two particles is inversely proportional to the square of the distance between them. Therefore, in the absence of a barrier to overcome this attraction, particles attract each other and start to grow and eventually coagulate. Two primary types of stabilization exist with nanoclusters, namely electrostatic and steric stabilization.

6.3.3 Electrostatic Stabilization

Electrostatic stabilization (see Figure 6.2a) is based on the coulombic repulsion between the particles caused by the electrical double layer formed by ions adsorbed on the particle surface (e.g., halides) and the corresponding counter ions (tetraalkylammonium). Metal surfaces are generally electrophilic, so, in principle, any type of anion should be attracted towards it and able to stabilize it, for example, chloride, citrate, and hydroxyl ion. The classical example is gold sol prepared by the reduction of $[AuCl_4]^-$ with sodium citrate.[3,18]

6.3.4 STERIC STABILIZATION

Steric stabilization (Figure 6.2b) is achieved by the coordination of long chain organic molecules (polymers or oligomers) that act as protective shields on the metallic surface. Nanometallic cores are separated from each other with a long chain of organic moiety, and thus agglomeration is prevented. The main classes of protective groups predominant in the literature are macromolecules containing P, N, and S (e.g., phosphanes, amines, thioethers)[23–25] and solvents such as THF[23, 26] or THF/ MeOH.[27]

Molecules like ionic surfactants stabilize the nanoparticles by means of both electrostatic repulsion and steric hindrance. Their polar group forms an electrostatic layer around them, and the long organic moiety gives steric hindrance. Examples for this system include long-chain alcohols,[28–30] surfactants,[12, 31] and organometallics.[32, 33] Figure 6.2c and Figure 6.2d give the variation of potential energy versus interparticular distance for electrostatic and steric stabilization, respectively.

6.3.5 REDUCING AGENTS

Wet chemical reduction procedures have been widely used to stabilize the different types of transition metal nanoparticles with different kinds of organic as well as inorganic stabilizers.

Reducing agents play a major role in directing the size and shape of particles. In general, the stronger the reducing agent used, the smaller the particle size obtained. Stronger reducing agents produce smaller nuclei in the seed; these nuclei grow during the ripening process to yield colloidal metal particles.[19]

Schmid and coworkers synthesized $Au_{55}(PPh_3)_{12}Cl_6$ (1.4 nm), a full shell (magic number) nanocluster stabilized by phosphane ligands, by using diborane as a reducing agent.[24, 34] Homogeneously dispersed Au_{55} clusters formed during the reduction of Au(III) ions with B_2H_6. Aiken and Finke reviewed nanoclusters of $M_{55}L_{12}Cl_n$ synthesized by the diborane route.[35] Corbierre and Lennox recently synthesized gold nanoparticles by chemical reduction of gold thiolates using superhydride (lithium tri-Et borohydride) in suitable inert solvents and found that sizes and dispersity are comparable to gold nanoparticles synthesized by the conventional methods.[36]

Toshima and coworkers [37, 38] have prepared various nanoparticle systems stabilized by organic polymers such as poly(vinylpyrrolidone) (PVP), poly(vinyl alcohol) (PVA), and poly(methylvinyl ether) by alcohol reduction. During the reduction of metal ions, alcohols having α-hydrogen oxidize to the corresponding aldehyde. A similar approach has been applied for the synthesis of bimetallic nanoparticles, as well.[37]

Hydrogen is one of the best reducing agents and has been applied in the synthesis of various metal clusters.[39, 40] The advantage of using hydrogen lies in the avoidance of byproducts; the disadvantage is that it is a very weak reducing agent and unable to reduce many metal complexes. Polymer-stabilized hydrosols of Pd, Pt, Rh, and Ir[41, 42] have been successfully synthesized by using hydrogen. Moiseev prepared Pd nanoclusters (Figure 6.3a),[43–46]and Finke prepared polyoxoanion stabilized Ir, Rh

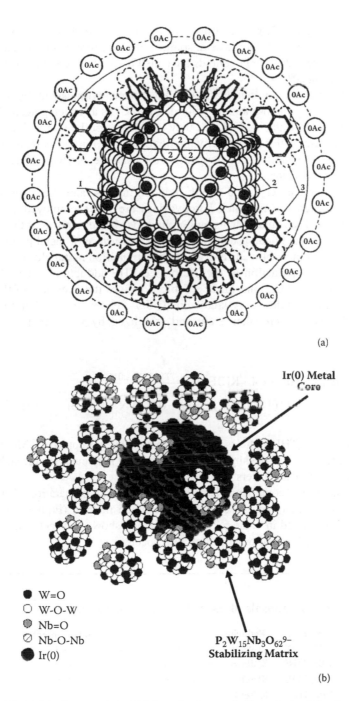

(a)

(b)

FIGURE 6.3 (a) Idealized model of Moiseev's giant palladium cluster $Pd_{\approx561}L_{\approx60}(OAc)_{\approx180}$ (phen = phenanthroline) (adapted from Reference 25); (b) idealized model of a Finke-type Ir^0 nanocluster $P_2W_{15}Nb_3O_{62}^{9-}$ and Bu_4N^+ stabilized $Ir^0_{\approx300}$ (adapted with permission from Reference 14).

nanoparticles (Figure 6.3b)[8, 47] by using hydrogen pathways. Pt hydrosols have been synthesized by using CO, formic acid, sodium formate, formaldehyde, and benzaldehyde as reductants.[4, 48, 49] Pt colloids have also been synthesized by using silanes as reducing agents.[50]

Ag, Cu, Pt, and Au nanoparticles have been synthesized by using tetrakis(hydroxymethyl) phosphonium chloride (THPC) as a reducing agent. Duff and coworkers introduced the method, and they could control the size and morphology of the nanoparticles in this particular synthetic route.[51, 52] Furthermore, hydrazine,[53] hydroxylamine,[54] and electrons trapped in, for example, $K^+[(crown)_2K]^-$ [55] have been applied as reductants.

Borohydrides are very good strong reducing agents, and there are numerous reports in the literature in which borohydrides are being employed as reducing agents.[56–58] The disadvantage in using borohydrides is the removal of their byproducts; the side product borate usually adsorbs on the surface of metal particles.[59, 60]

Tetrahydrofuran-stabilized zerovalent early transition metal clusters were prepared by $[BEt_3H]^-$ reduction of the preformed THF adducts of $TiBr_4$ (Equation 6.1), $ZrBr_4$, VBr_3, $NbCl_4$, and $MnBr_2$. Table 6.1 gives the overview of the results obtained by this synthesis. A study on Ti·0.5 THF[26] shows that the cluster contains Ti_{13} in the zerovalent state and stabilized by six THF molecules, as shown in Figure 6.4.

$$x \cdot TiBr_4 \; 2\; THF + x\; 4 \cdot K[BEt_3H] \xrightarrow{\text{THF, 2h, 20°C}}$$

$$[Ti \cdot 0.5\; THF]_x + x4 \cdot BEt_3 + x \cdot 4 \cdot KBr \downarrow + x4H_2 \uparrow \quad (6.1)$$

Similarly, $Mn_{0.3}THF$ particles of 1–2.5 nm have been prepared.[61] Moreover, the THF molecules were replaced by tetrahydrothiophene molecules for the stabilization of Mn, Pd, and Pt nanoclusters (Figure 6.5).[31]

Organoaluminum compounds have been used for the preparation of mono and bimetallic nanoparticles by a reductive stabilization method (Equation (6.2). The results are summarized in Table 6.2.[32, 62] The study shows that metal nanoparticles are stabilized by a layer of organoaluminum species.

TABLE 6.1
THF Stabilized Organosols of Early Transmission Metal

Product	Starting material	Reducing agent	T [°C]	t [h]	Metal content [%]	Size [nm]
[Ti·0.5THF]	TiBr$_4$·2THF	K[BEt$_3$H]	Room temp.	6	43.5	(<0.8)
[Zr·0.4THF]	ZrBr$_4$·2THF	K[BEt$_3$H]	Room temp.	6	42.0	—
[V·0.3THF]	VBr$_3$·3THF	K[BEt$_3$H]	Room temp.	2	51.0	—
[Nb·0.3THF]	NbCl$_4$·2THF	K[BEt$_3$H]	Room temp.	4	48.0	—
[Mn·0.3THF]	MnBr$_2$·2THF	K[BEt$_3$H]	50	3	70	1–2.5

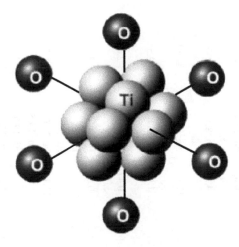

FIGURE 6.4 Ti_{13} cluster stabilized by six THF O atoms in an octahedral configuration.

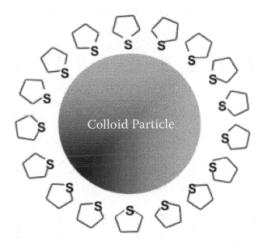

FIGURE 6.5 Organosols stabilized by tetrahydrothiophene; M = Ti, V (decomposition); M = Mn, Pd, Pt; stable colloids.

$$MX_n + AlR_3 \xrightarrow{\text{Toluene}} M[Alacac]_n + R_2Alacac \qquad (6.2),$$

Where
M = Metals of groups 6–11
X = Halogen, acetylacetonate, n = 2–4, R = C_1-C_8-alkyl, particle size = 1–12 nm

6.3.6 POLYOXOANION STABILIZED NANOCLUSTERS

A recent method demonstrated by Finke and coworkers involves the stabilization of nanosized metallic particles using polyoxoanions. They prepared Ir_{300} and Ir_{900}

TABLE 6.2
Mono- and Bimetallic Nanocolloids Prepared by the Organoaluminum Route (See Reference 62)

Metal salt	[g]	[mmol]	Reducing agent	[g]	[mmol]	Toluene [mL]	Conditions T [°C]	t [h]	Product m [g]	Metal content [wt-%]	Particle size F [nm]
Ni(acac)$_2$	0.275	1	Al(iBu)$_3$	0.594	3	100	20	10	0.85	Ni: 13.8	2–4
Fe(acac)$_2$	2.54	10	Al(Me)$_3$	2.1	30	100	20	3	2.4	n.d.	
RhCl$_3$	0.77	3.1	Al(oct)$_3$	4.1	11.1	150	40	18	4.5	Rh: 8.5	2–3
Al: 6.7											
Ag neodecanoate	9.3	21.5	Al(oct)$_3$	8.0	21.8	1000	20	36	17.1	Ag: 11.8	8–12
Al: 2.7											
Pt(acac)$_2$	1.15	3	Al(Me)$_3$	0.86	7.6	150	20	24	1.45	Pt: 35.8	2.5
Al: 15.4											
PtCl$_2$	0.27	1	Al(Me)$_3$	0.34	3	125	40	16	0.47	Pt: 41.1	2.0
Al: 15.2											
Pd(acac)$_2$	0.54	1.8	Al(Et)$_3$	0.46	4	500	20	2	0.85	P: 22	3.2
Pt(acac)$_2$	0.09	0.24								Pt: 5.5	
										Al: 12.7	
Pt(acac)$_2$	7.86	20	Al(Me)$_3$	8.64	120	400	60	21	17.1	Pt: 20.6	1.3
Ru(acac)$_3$	7.96	20								Ru: 10.5	
										Al: 19.6	
Pt (acac)$_2$	1.15	2.9	Al(Me)$_3$	0.86	12	100	60	2	1.1	Pt: 27.1	
SnCl$_2$	0.19	1								Al: 14.4	

polyoxoanion/Bu_4N^+-stabilized, organic solvent soluble nanoclusters [Ir_{300}-($P_4W_{30}Nb_6O_{123}^{-16}$)$_{33}$]$_{533}$ and [Ir_{900}-($P_4W_{30}Nb_6O_{123}^{-16}$)$_{60}$]$_{960}$ in mg scale.[14] In addition, they studied the mechanism of the formation of nanoclusters with a focus on LaMer's classic work and the mechanistic studies of the formation of novel [$P_2W_{15}Nb_3O_{62}$]$^{9-}$ stabilized $Ir_{190-450}$ nanoclusters.[14, 47] They also investigated nanocluster size control and magic number cluster synthesis and its reaction conditions using Ir-[$P_2W_{15}Nb_3O_{62}$]$^{9-}$.[63] They were able to synthesize Rh(0) stabilized by [$P_2W_{15}Nb_3O_{16}$]$^{9-}$, which was the second example for polyoxoanion stabilized nano-particles with a particle size of 4 ± 0.6 nm.[64,65] The same group was also able to scale up [$P_2W_{15}Nb_3O_{62}$]$^{-9}$ stabilized Ir(0) nanoparticles to 1 g and optimize the reaction conditions.[66] The Papaconstantinou group has made a significant contribution in this research area by synthesizing silver and gold metal nanoparticles by using polyoxo-metalates as both photocatalysts and stabilizers.[67]

Cabuil et al. synthesized gold nanoparticles using the prefunctionalized poly-oxometalate γ-[$SiW_{10}O_{36}(RSi)_2O$]$^{4-}$ (R = HSC_3H_6), in which the organic ligand forms a covalent link to gold particles through the thiol group and a covalent link with the polyanion through the bridging oxo groups.[68]

This field of polyanion stabilized transition metal nanoclusters has also facilitated intensive mechanistic investigations through which researchers have been able to establish repeatable synthesis of monodisperse systems.[63, 64, 69, 70] The synthesis involved hydrogen reduction under completely air- and water-free conditions using the polyoxoanion-supported Ir(I) organometallic complex [(1,5-COD)Ir $P_2W_{15}Nb_3O_{62}$]$^{8-}$ as the molecular precursor. A zero-valent transition metal core, e.g., Ir(0), is doubly stabilized by a polyoxoanion (e.g., [$P_2W_{15}Nb_3O_{62}$]$^{9-}$) and by a cationic surfactant (e.g., NR_4^+). When this precursor is treated with H_2 in acetone and excess cyclohexene, the reduction results in the formation of nearly monodisperse (i.e., ±15% size dispersion, 2 ± 0.3 nm Ir(0)$_{\approx300}$) nanoclusters.[14] In the absence of cyclohexene, the same procedure yields larger Ir(0)$_{\approx900}$ nanoclusters (3 ± 0.4 nm). The electron diffraction patterns of Ir(0) nanoparticles exactly match that of bulk Ir metal powder, which revealed that the Ir(0) core is uncharged and has cubic close packing (ccp) or hexagonal close packing (hcp) atomic arrangement and that the particles have an extremely clean, exposed, chemically reactive surface. The degree to which these clusters have been characterized is directly related to their monodispersity and has allowed the researchers to collect a wealth of information on the size-dependent properties exhibited by these nanoclusters. The catalytic properties demonstrated by these systems were unprecedented in terms of stability and catalytic turnovers. Further studies have suggested that these extraordinary properties are most likely a result of the type of anion used to support the colloid. This work has opened the door to a wide range of metal combinations of both core-shell and alloy particles.

6.3.7 Synthesis of Bimetallic Nanoparticles

Nanoscale bimetallic particles are comprised of two different metals but may be arranged in a variety of forms including alloys, core/shell, and decorated surfaces. Unique surfaces are obtained whenever a second metal is added to a conventional metal colloid. These combinations may create new properties that may not be achieved by the monometallic components. Toshima and coworkers[71] reported the preparation of polymer protected Pt-Pd alloy colloids by simultaneously reducing the metal ions in water-ethanol solutions and studied their structure by EXAFS. Pt-Pd alloy colloids in water-in-oil microemulsions were prepared by Boutonnet-Kizling and coworkers,[72] and their catalytic behavior for isomerization reactions was studied. Rousset and coworkers reported clusters of Pt-Pd on amorphous carbon.[73] Other bimetallic systems, Cu-Ru[74, 75], Pd-Ni[76], Cu-Ni[77], Pd-Ag,[78] Pt-Pd,[58] and Co-Pd[79] combinations in silicate matrices have been reported. The literature on Pt-Pd bimetallic systems[71, 80] reveals that the preparation conditions employed varies from polymer protection of the bimetallic colloid in aqueous media to laser vaporization of bulk alloys.[73] Rao and coworkers synthesized fcc structured Ag-Pd and Cu-Pd nanoscale alloys in bulk quantities and characterized them with various techniques such as TEM and XRD[112]. Doudna et al.[81] synthesized Ag-Pt bimetallic nanoparticles with high aspect ratios by radiolytic means. The particles were found to be filament-like nanostructures, which were several microns long and a few nanometers in diameter, with an Ag core and a Pt shell structure. Recently, Sastry and coworkers[82] reported the synthesis of Au core-Ag shell nanoparticles by a photochemical reduction process using polyoxometalates and stated that their approach provides a new strategy for realizing bimetallic core-shell structures using Keggin ions with potential applications in nanomaterial synthesis and catalysis.

6.3.8 MICROEMULSIONS

This is a relatively new technique for the preparation of nanoparticles and has received considerable interest in recent years.[83–86] A literature survey depicts that the ultrafine nanoparticles in the size range between 5 and 50 nm can be easily prepared by this method. This technique uses an inorganic phase in water-in-oil microemulsions, which are isotropic liquid media with nanosized water droplets that are dispersed in a continuous oil phase. In general, microemulsion consists of at least a ternary mixture of water, a surfactant, and a mixture of surface-active agents and oil. The classical examples for emulsifiers are sodium dodecyl sulfate (SDC) and aerosol bis(2-ethylhexyl)sulfosuccinate (AOT). The surfactant (emulsifier) molecule stabilizes the water droplets, which have polar heads and nonpolar organic tails. The organic (hydrophobic) portion faces towards the oil phase, and the polar (hydrophilic) group towards water. In diluted water (or oil) solutions, the emulsifier dissolves and exists as monomer, but when its concentration exceeds a certain limit called the critical micelle concentration (CMC), the molecules of emulsifier associate spontaneously to form aggregates called micelles. These water microdroplets then form nanoreactors for the formation of nanoparticles. The nanoparticles formed usually have monodisperse properties.

One method of formation consists of mixing two microemulsions or macroemulsions and aqueous solutions carrying the appropriate reactants in order to obtain the desired particles. The interchange of the reactants takes place during the collision of the water droplets in the microemulsions. The interchange of the reactant is very fast, so that for the most commonly used microemulsions, it occurs just during the mixing process. The reduction, nucleation, and growth occur inside the droplets, which controls the final particle size. The chemical reaction within the droplet is very fast, so the rate-determining step will be the initial communication step of the microdroplets with different droplets. The rate of communication has been defined by a second-order communication-controlled rate constant and represents the fastest possible rate constant for the system. The reactant concentration has a greater influence on the reduction rate. The rate of both nucleation and growth are determined by the probabilities of the collisions between several atoms, between one atom and a nucleus, and between two or more nuclei. Once a nucleus forms with the minimum number of atoms, the growth process starts. For the formation of monodisperse particles, all of the nuclei must form at the same time and grow simultaneously and with the same rate.

The method for the preparation of metal nanoparticles within micells consists of forming two microemulsions, one with the metal salt of interest and the other with the reducing agent, and mixing them together. A schematic diagram is shown in Figure 6.6.

When two different reactants mix, the interchange of the reactants takes place due to the collision of water microdroplets. The reaction (reduction, nucleation, and growth) takes place inside the droplet, which controls the final size of the particles. The interchange of nuclei between two microdroplets does not take place due to the special restrictions from the emulsifier. Once the particle inside the droplets attains its full size, the surfactant molecules attach to the metal surface, thus stabilizing and preventing further growth. A diverse range of metal nanoparticles have been prepared by this method including Fe,[88] Fe/Au,[89] Pt,[90] Ag,[91, 92] CdS,[93, 94] Pd,[95, 96] Cu,[97] Ni,[98] and Au.[99]

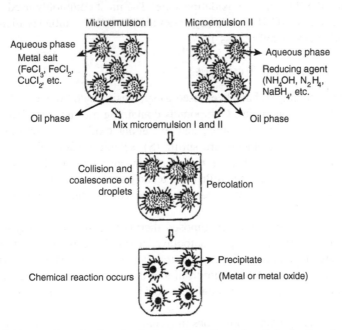

FIGURE 6.6 Proposed mechanism for the formation of metal particles by the microemulsion approach (adapted with permission from Reference 87).

6.3.9 DECOMPOSITION OF LOW-VALENT TRANSITION METAL COMPLEXES

Low-valent organometallic complexes and several organic-modified derivatives of the transition metals decompose to give short-lived nucleation particles of zerovalent metals in solution, which may be stabilized by colloidal protecting agents. This decomposition is typically initiated by an external force, such as sonication or thermolysis,[100, 101] or by introduction of a chemical agent such as H_2 or CO. PVP stabilized, 45 nm Co nanoparticles can be prepared by decomposition of $Co_2(CO)_8$ at 130–170°C in decaline or ethylene glycol solvents, as shown in Equation 6.3.[102]

$$Co_2(CO)_8 \rightarrow 2Co\ (s) + 8CO\uparrow \qquad (6.3)$$

Similarly, styrene-4-vinyl pyridine and styrene-N-vinyl pyrolidone stabilized Fe nanoparticles have been prepared starting from $Fe(CO)_5$.[103]

Chaudret and coworkers [104–109] have demonstrated the use of low-valent transition metal olefin complexes as a very clean source for the preparation of nanostructured mono- and bimetallic colloids (Co, Ni, Ru, Pd, Pt, CoPt, CoRh, and RuPt). Syntheses were carried out in the presence of suitable stabilizers using CO or H_2 as reducing agents at room or slightly elevated temperature. A number of nanoparticulate metal oxide systems have also been successfully developed by this method.[6, 110] Olefin complexes are similar to metal carbonyl complex, except the metal is in either low or zero oxidation state. The most commonly used ligands are 1,5-cyclooctadiene (COD), 1,3,5-cyclooctatriene (COT), dibenzylidene acetone (DBA), and cyclooctenyl ($C_8H_{13}^-$).

6.3.10 SOLVOTHERMAL TECHNIQUE

Solvothermal techniques have been used to synthesize metal oxide[111–114] and chalcogenide[115–118] nanoparticles. Compounds containing a chalcogen element, that is, in group 16 of the periodic table, and excluding oxides, are commonly termed *chalcogenides*. These elements are sulfur (S), selenium (Se), tellurium (Te), and polonium (Po). Common chalcogenides contain one or more of S, Se, and Te, in addition to other elements. Solvothermal techniques have been used to synthesize metal oxides and mixed metal oxides, which are usually spinels, and semiconductor chalcogenides.[119, 120]

The metal complexes are decomposed thermally either by boiling the contents in an inert atmosphere or using an autoclave. A suitable capping agent or stabilizer such as a long chain amine, thiol, or trioctylphospine oxide (TOPO) is added to the reaction contents at a suitable point to hinder the growth of the particles and hence provide their stabilization. The stabilizers also help in dissolution of the particles in different solvents.

Some of the important innovations in recent years are the decomposition of metal alkoxides for the synthesis of TiO_2[121] nanoparticles, TOPO capped autoclave synthesis of TiO_2 by metathetic reaction,[122] decomposition of metal N-nitroso N-phenyl hydroxylamine complex to get metal oxide[123, 124], synthesis of CdE (E = S, Se, Te) nanoparticles using organometallic precursor in TOPO[125] as solvent toluene[117] as solvent and in nujol.[126]

Solvothermal processes have been also used to synthesize metal nanoparticles other than metal oxide nanoparticles. Palladium nanoparticles[127, 128] of about 1 μm have been synthesized by a solvothermal route, as shown by following reaction:

$$2\ CH_2OH\text{-}CH_2OH \xrightarrow[160°C/18hrs]{-2\ H_2O} 2\ CH_3CHO$$

$$\xrightarrow{Pd\ (II)} CH_3\text{-}CO\text{-}CH_3 + Pd + H_2O \qquad (6.4)$$

First, acetyl acetone forms a ring-like complex with Pd(II), which may lead to Pd^{2+} being released slowly and reduced slowly.[128] A similar method was applied for synthesizing Ag nanoparticles. Rosemary and Pradeep[120] synthesized the Ag nanoparticles as follows: First, an Ag-thiolate complex was precipitated by adding a thiol-toluene solution to AgNO$_3$, then the precipitate was separated by centrifugation, washed, and vacuum-dried. The second step was dispersion of Ag-complex in toluene and heating overnight in an autoclave at 200°C to obtain Ag nanoparticles (Figure 6.7).

The solvothermal synthesis of Fe and Ni nanoparticles have also been reported.[119] Fe can be prepared by reacting Fe(acac)$_3$ or Fe(acac)$_2$ with 85% N$_2$H$_5$OH in the presence of poly-(N-vinyl-pyrrolidone). The reaction was carried out in a

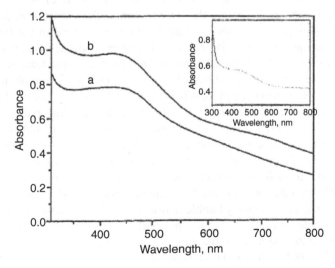

FIGURE 6.7 UV/Vis absorption spectrum of (a) Ag-octanethiolate and (b) Ag-octadecanethiolate in toluene at 200°C. Inset shows the absorption spectrum of Ag-octanethiolate at 190°C in toluene (adapted with permission from Reference 120).

basic pH medium and in an autoclave at a temperature of 120–140°C. The reaction can be written as

$$M(acac)_n + N_2H_5OH + OH^- \rightarrow M + N_2 + H_2O \qquad (6.5)$$

Thus, obtained Fe and Ni nanoparticles had an average particle size of 5 and 15 nm, respectively. The polymer on the surface of the nanoparticles hindered the oxidation of Fe or Ni surface atoms. The obtained Fe nanoparticles showed soft ferromagnetism, and Ni showed superparamagnetism.

6.3.11 SONOCHEMISTRY

Sonochemical methods for the preparation of nanoparticles were pioneered by Suslick et al. in 1991.[129] They prepared Fe nanoparticles by sonication of $Fe(CO)_5$ in a decaline solution, which gave them 10–20 nm amorphous iron nanoparticles. Sonochemical decomposition methods have been further developed by Suslick et al.[130] and Gedanken and coworkers[131, 132] and have produced Fe, Mo_2C, Ni, Pd, and Ag nanoparticles in various stabilizing environments.

The sonochemical method has been found useful in many areas of material science, starting from the preparation of amorphous products[133, 134] and insertion of nanomaterials into mesoporous materials,[135, 136] to deposition of nanoparticles on ceramic and polymeric surfaces.[137, 138]

The principle of sonochemistry is breaking the chemical bond with the application of high-power ultrasound waves, usually between 20 kHz and 10 MHz. The physical phenomenon responsible for the sonochemical process is acoustic cavitation. According to published theories for formation of nanoparticles by sonochemistry; the main events that occur during the preparation are creation, growth, and collapse of the solvent bubbles that are formed in the liquid. These bubbles are in the nanometer size range. The solute vapors diffuse into the solvent bubble, and when the bubble reaches a certain size, its collapse takes place. During the collapse, very high temperatures of 5,000—25,000 K [129] are obtained, which is enough to break the chemical bonds in solute. The collapse of the bubble takes place in less than a nanosecond,[139, 140] hence a high cooling rate (1011 K/s) is also obtained. This high cooling rate hinders the organization and crystallization of the products. Because the breaking of bonds in the precursor occurs in the gas phase, amorphous nanoparticles are obtained. Though the reason for formation of amorphous products is well understood, the formation of nanostructures is not. The possible explanations are the fast kinetics does not stop the growth of the nuclei, and in each collapsing bubble a few nucleation centers are formed whose growth is limited by the collapse; the other possibility is the precursor is a nonvolatile compound and the reaction occurs in a 200 nm ring surrounding the collapsing bubble.[141] In the latter case, the sonochemical reaction occurs in a liquid phase, and the products could be either amorphous or crystalline, depending on the temperature in the ring region of the bubble. Suslick has estimated the temperature of the ring region as 1900°C.[129]

Chalcogenides are commonly synthesized by this method. The metal sulphide syntheses have been carried out in ethanol,[142] water,[143] and ethylenediamine,[144] whereas the sources of metal ions have been acetates[142, 144] or the chlorides.[143] The precursor for sulfur is usually thioacetamide or thiourea. Nanoparticle synthesis of d-block elements have also been carried out, for example, platinum,[145] gold,[146] cobalt,[147] iron,[147] palladium,[148, 149] gold,[150] nickel,[151] and bimetallic alloys such as Co/Cu,[152] Pt/Ru,[153] Au/Pd,[154] Fe/Co.[155]

A typical synthesis of $Fe_{40}Co_{60}$ is as follows[156]: 3.0 ml of 0.15 mol.l-1 $Fe(CO)_5$ and 1.5 ml of 0.15 mol.l-1 $Co(NO)(CO)_3$ were dissolved in 100 ml of diphenyl methane. The sonolysis of the solution was performed for 3 h at 20–30°C. The Sonics and Material ultrasonic device is direct immersion titanium horn type (working frequency 20 kHz, electric power of generator 600 W, irradiation surface area of the horn 1 cm^2). After the reaction, a black solid product was separated by centrifugation, washed with pentane, and dried under vacuum at room temperature. The solid was finally annealed at 600°C for 5 h in an argon atmosphere to obtain $Fe_{40}Co_{60}$ nanoparticles. During the annealing process, protective amorphous carbon film forms from the decomposition of diphenyl methane and prevents particle from agglomeration, as demonstrated by XPS. The protective carbon film thickness varied in the range 5–10 nm. The TEM micrograph of the particle is shown in the Figure 6.8, and the particles are of 20–40 nm in size.

6.3.12 GAS PHASE SYNTHESIS

Metal vapor techniques have provided chemists with a very versatile route for the production of a wide range of nanostructure metal colloids on a preparative laboratory scale.[157–159] The use of metal vapor techniques is limited because the

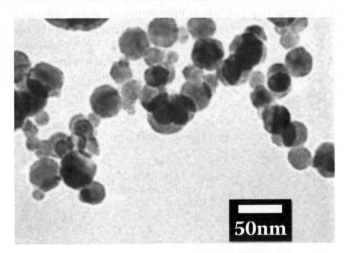

FIGURE 6.8 TEM image of $Fe_{40}Co_{60}$ annealed at 600°C for 5 h (adapted from Reference 155 with permission).

operation of the apparatus is demanding, and it is difficult to obtain narrow particle size distributions. However, this is the cleanest method of preparing gram quantities of compounds, because this does not involve the purification of counter ions like nitrate, chlorides, and sulfates, which are common in reduction synthesis methods.

The reactor consists of a four-necked thick-walled glass chamber, two water-cooled copper electrodes, a glass tube with small nozzle shower heads for spraying organic solvents, an alumina crucible with a resistive heating system, and a vacuum line, as shown in Figure 6.9. The reactor is immersed in liquid nitrogen (–196°C) to condense the metal vapors and substrate (liquid ligand) that enter through the shower head. The whole reactor is maintained under high vacuum (typically 5×10^{-3} torr.) in order to prevent undesired reaction of metal vapor with the atmospheric contents and facilitate the continuous evaporation of the substrate. The water-cooled electrode extends down into the reactor chamber, and a heating element is connected across the base of these electrodes. The metal sample is placed in an alumina crucible with an internal tungsten wire heater. The bottom of the crucible-heater assembly should be covered with refractory wool to increase the heating efficiency of the source and cooling of the reactor. A high electric power of up to 300 Amp is needed to reach the temperature of up-metal vaporization.

Vaporized metal atoms are highly reactive due to the input of the heat of vaporization and lack of steric hindrance. The metal of interest is heated in a crucible at elevated temperature and cocondensed with a specified amount of liquid ligand substrate on the cold walls of the reactor vessel. In a laboratory experiment, about 50 mmol of ligand to about 3 mmol of metal is introduced in 1 h. The reaction chamber should be filled with an inert gas before removing the electrode assembly. The glass reactor is removed from the liquid nitrogen, and the reactor is allowed to warm slowly. The resulting solution is then removed, along with excess substrate, by means of a syringe or other inert atmosphere technique. After the warming stage, particles can be stabilized either sterically (by solvation) or electrostatically (by incorporation of negative charge). Because metal atomic vapor is often pyrophoric, extra care should be taken while transferring the resulting solution and disassembly of the setup.

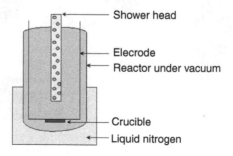

FIGURE 6.9 Schematic of a metal atom vapor reactor.

$$\text{Au}_{\text{atoms}} + \text{acetone vapor} \xrightarrow{\ 77\ \text{K}\ } \text{Au-acetone matrix}$$

$$\xrightarrow[\text{melt}]{\text{warm}} (\text{Au})_n\text{-acetone colloid}$$

$$\xrightarrow[\text{RSH}]{\text{toluene}} (\text{Au})_n\text{-acetone-toluene-RSH mixture}$$

$$\xrightarrow{\text{remove acetone under vacuum}} (\text{RSH})_x(\text{Au})_y \text{ toluene colloid} \qquad (6.6)$$

Monodisperse thiol stabilized gold colloids have been prepared by SMAD preparation followed by a digestive ripening.[160] Instead of depositing metal vapor and stabilizer simultaneously on the cold walls of the glass reactor, first dodecanethiol and toluene were placed at the bottom of the glass reactor and frozen to −196°C; gold and acetone vapors were then deposited on the sides of the glass reactor. Because acetone has a high degree of solvation with gold particles, it functions as a preliminary stabilizing agent in this system. After removal of the liquid nitrogen coolant from the glass reactor, the Au-acetone matrix melts into the toluene-thiol mixture in the bottom of the reactor. After complete melting of the matrix, the mixture was stirred for 45 min, resulting in Au-thiol colloid formation. Then, the acetone was removed and pure thiol stabilized gold colloids were obtained. The TEM image of the thus obtained colloid is shown in Figure 6.10, which shows that particle size ranges from 5–40 nm with no definite geometrical shapes and very similar to particles obtained in pure acetone solvent.

Similar methods have been applied for the synthesis of metal sulfides, metal oxides, and other types of dispersed compounds in different solvents.

6.3.13 HIGH-GRAVITY REACTIVE PRECIPITATION

High-gravity reactive precipitation (HGRP) is a new method for synthesizing nanoparticles introduced by Chen et al.[161] They originally applied this technique for the preparation of metal carbonates and hydroxides. Using this method, they could control and adjust the particle size of $CaCO_3$ particles in the region of 17–36 nm by adjusting the reaction parameters such as high-gravity levels, fluid flow rate, and reactant concentrations. The synthesis of nanofibrils of $Al(OH)_3$ of 1–10 nm in diameter and 50–300 nm in length, as well as $SrCO_3$ with a mean size of 40 nm in diameter, was also demonstrated by the same group.

HGRP is based on Higee technology;[162] it consists of rotating a packed bed under a high-gravity environment. This technology is a novel technique to intensify mass transfer and heat transfer in multiphase systems. The rate of mass transfer between a gas and liquid in a rotating packed bed is 1–3 orders of magnitude larger than that in a conventional packed bed. This is very helpful in the generation of higher supersaturated concentrations of the product in the gas-/liquid-phase reaction and precipitation process. Due to the high shear field experienced by the passing fluid, formation of fine droplets, threads, and thin films takes place in the rotating

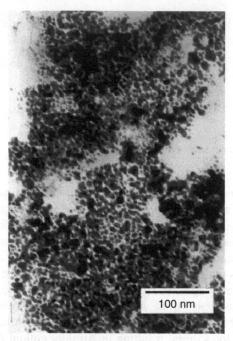

FIGURE 6.10 Synthetic steps for preparation of nanocrystal superlattice and TEM image of thus obtained gold nanoparticle (adapted from Reference 160 with permission).

packed bed. This effect facilitates the intense micromixing between the fluid components.[161]

The experimental apparatus for synthesis by high-gravity reactive precipitation is shown schematically in Figure 6.11. The key part of the RPB (Higee machine) is a packed rotator (6). The inner diameter of the rotator is d_{in} = 50 mm, and the outer diameter is d_{out} = 150 mm. The axial width of the rotator is 50 mm. The distributor (5) consists of two pipes (10 and 1.5 mm), each having a slot of 1 mm in width and 48 mm in length, which just covers the axial length of the packing section in the rotator. The rotator is installed inside the fixed casing and rotates at the speed of several hundred to thousands of rpm. Liquid (including slurry) is introduced into the eye space of the rotator from the liquid inlet pipe and then sprayed by the slotted pipe distributor (5) onto the inside edge of the rotator. The liquid on the bed flows in the radial direction under centrifugal force, passing through the packing, and emerges outside the space between the rotator and the shell, finally leaving the equipment through the liquid exit (7) and collects in (1). The gas is introduced from an outside source (gas cylinder) through the gas inlet (10), flows inward counter-currently to the liquid in the packing of the rotator, and finally goes out through the gas exit under the force of pressure gradient.

The HGRP approach has not evolved further because this was originally applied for metal carbonates and hydroxides. This method has to be modified for metal nanoparticle synthesis to gain more scope.

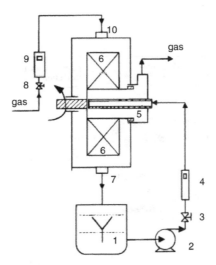

FIGURE 6.11 Schematic of experimental setup. 1. Stirred tank, 2. pump, 3. valve, 4. rotor flowmeter, 5. distributor, 6. packed rotator, 7. outlet, 8. valve, 9. rotor flow meter, 10. inlet.

6.3.14 ELECTROCHEMICAL SYNTHESIS

This method uses a sacrificial anode of the metal of interest and a cathode, which can be composed of any other metal. Under the suitable applied current density, the anode sacrificially dissolves in the electrolyte, the metal ions migrate towards the cathode, and reduction occurs. The nucleation and growth of reduced metal atoms occurs at the electrode surface. The stabilizing agent in the reaction vessel arrests the growth of the nanoparticles and facilitates the formation of stable nanostructures. The final step is diffusion of nanoparticles from the electrode surface into the bulk of the solution. Equation 6.7 gives the reactions taking place at the electrode surfaces. Figure 6.12 is a schematic diagram of reaction vessel and the main processes occurring during the electrochemical synthesis of nanoparticles.

$$\text{Anode} \qquad M_{bulk} \rightarrow Mn^+ + n\ e^-$$

$$\text{Cathode} \qquad M^{n+} + n\ e^- + \text{stabilizer} \rightarrow M_{coll/stabilizer}$$

$$\text{Sum:} \qquad M_{bulk} + \text{stabilizer} \rightarrow M_{coll/stabilizer} \qquad (6.7)$$

The advantages of the electrochemical pathway are that the contamination with byproducts resulting from chemical reducing agents is avoided and that the products are easily isolated from the precipitate. Further, the electrochemical preparation allows for size-selective particle formation. The electrochemical technique for preparing nanostructured mono- and bimetallic colloids has been further developed by Reetz and his group since 1994.[7]

The particle size obtained by the electrochemical route depends on many factors: The distance between the electrodes, reaction time, temperature, and polarity of the

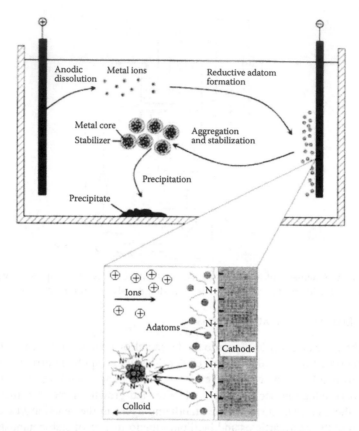

FIGURE 6.12 Electrochemical formation of NR_4^+ Cl^- stabilized nanometal (adapted with permission from Reference 7).

solvent contribute to the particle size. Experiments have also shown that the applied current density also has a major influence on the particle size. For example, experiments using Pd as the sacrificial anode in the electrochemical cell to give $(C_8H_{17})_4N^+Br^-$ stabilized Pd^0 particles have demonstrated that the particle size depends on the current density applied: High current densities lead to small Pd particles (1.4 nm), whereas low current densities give larger particles (4.8 nm).[163]

Pd,[164] Ni, Co, Fe, Ti, AgO,[165] Ag,[166, 167] Cu,[168] and Au[169], on a scale of several hundred milligrams, have been successfully prepared by electrochemical means.[163] In the case of Pt, Rh, Ru, and Mo, which are anodically less readily soluble, the corresponding metal salts were electrochemically reduced at the cathode (see the bottom part of Figure 6.12 and Table 6.3). Solvent (propylene carbonate) stabilized Pd nanoparticles have also been prepared.[164] It has further been demonstrated that bimetallic particles (Pd/Ni, Fe/Co, Fe/Ni) can be prepared by using two sacrificial anodes in a single electrolysis step.[170]

Tetraalkylammonium acetate has been used as both the supporting electrolyte and the stabilizer in a Kolbe electrolysis at an anode [see Equation (8)]. [171] Bimetallic nanoparticles can be prepared by combining the Equation 7 and 8 (see Table 4). [171]

TABLE 6.3
Electrochemical Colloid Synthesis[171]

Metal salt	d [nm]	EA[a]
PtCl$_2$	2.5[b]	51.21% Pt
PtCl$_2$	5.0[c]	59.71% Pt
RhCl$_3$H$_2$O	2.5	26.35% Rh
RuCl$_3$H$_2$O	3.5	38.55% Ru
OsCl$_3$	2.0	37.88% Os
Pd(OAc)$_2$	2.5	54.40% Pd
Mo$_2$(OAc)$_4$	5.0	36.97% Mo
PtCl$_2$ + RuCl$_3$H$_2$O	2.5	41.79% Pt + 23.63% Rh[d]

[a] Based on stabilizer-containing material. [b] Current density: 5.00 mA/cm^2. [c] Current density: 0.05 mA/cm^2. [d] Pt-Ru bimetallic cluster.

TABLE 6.4
Bimetallics Prepared Electrochemically[171]

Anode	Metal salt	d [nm]	Stoichiometry (EDX)
Sn	PtCl$_2$	3.0	Pt$_{50}$Sn$_{50}$
Cu	Pd(OAc)$_2$[a]	2.5	Cu$_{44}$Pd$_{56}$
Pd	PtCl$_2$	3.5	Pd$_{50}$Pt$_{50}$

[a] Electrolyte: 0.1 M [(n-octyl)4N]OAc/THF.

$$\text{Cathode:} \quad Pt^{2+} + 2\ e^- \rightarrow Pt^0$$

$$\text{Anode:} \quad 2\ CH_3CO_2^- \rightarrow 2\ CH_3CO_2 + 2\ e^- \tag{6.8}$$

Layered bimetallic nanocolloids have been prepared by modifying the electrochemical method (Pt/Pd).[12, 172] (Oct)$_4$NBr stabilized Pt/Pd bimetallic nanocolloids have been prepared by the electrolysis of preformed (Oct)$_4$NBr stabilized Pt colloid core (size 3.8 nm) in 0.1 M (Oct)$_4$NBr solution in THF with Pd as the sacrificial anode (Figure 6.13). Here, the preformed Pt core acts as a living-metalpolymer, on which the Pd atoms are deposited to give onion-type bimetallic nanoparticles (5 nm).

One problem with electrochemical synthesis is that the reactions are generally carried out in an aqueous media, in which noble metal ions can be coated on the cathode more easily, and when reduced they remain on just the electrode, giving an electroplated coat instead of nanoparticles.[173] Usually, ultrasonic agitation is used to transfer the nanoparticles formed to the bulk of the solution. Even in the presence of ultrasonic and mechanical agitation, the chances of electroplating cannot be overcome completely.

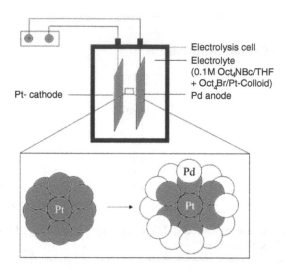

FIGURE 6.13 Modified electrolysis cell for the preparation of layered bimetallic Pt/Pd nano-colloids (adapted with permission from Reference 172).

This problem was overcome by using poly N-vinyl pyrrolidone (PVP) as a stabilizing agent.[174] The PVP was found to accelerate the formation of silver particles and lower the silver deposition on the cathode. Consequently, an external excitation system such as ultrasonication is no longer required.

The same group has recently synthesized gold nanoparticles using PVP as stabilizers.[169] The protective mechanism of PVP can be explained on the basis of its structural features. PVP has a structure of a polyvinyl skeleton with polar pyrolidone groups, as is shown in Figure 6.14. The donated lone pairs of both nitrogen and oxygen atoms in the polar groups of one PVP unit may occupy two sp orbitals of silver ions to form a complex compound.

The mechanism of silver nanoparticle formation in the electrochemical/PVP process can be explained in four steps. The first step is the formation of coordinative bonding between PVP and silver ions, producing the Ag_m^{m+}-PVP complex (see Figure 6.15), in which m is the number of silver ions anchored at a PVP molecule. This case also occurs in the chemical reduction process of silver ions in the presence of PVP. The second step is the electrochemical reduction of the Ag_m^{m+}-PVP complex at the cathode/electrolyte interface, producing silver adatoms (Ag_m^0-PVP) protected by PVP. In this step, the influence of the chemical bond must be considered. Because the ligand of C-N and C = O in PVP contributes more electronic density to the sp orbital of silver ions than H_2O does, the silver ions in the Ag_m^{m+}-PVP complex may obtain electrons more easily from the cathode than those in the Ag^+-H_2O complex. Thus, the presence of PVP ensures that the Ag_m^{m+}-PVP complex rather than single Ag^+ ion is reduced. In this way, the silver deposition trend at the cathode surface will be effectively lowered. The subsequent step is PVP-accelerated formation of large amounts of silver nuclei. In a fourth step, coalescence of the silver clusters during the growth progress is prevented by the PVP, which has the role to promote

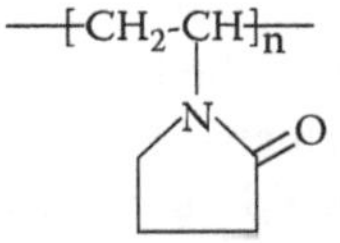

$$-\!-\!\{CH_2\text{-}CH\}_{\overline{n}}\!-$$

FIGURE 6.14 Structural scheme of the PVP molecule. Here, n represents the polymerization number. n = 360 for PVPK30 (adapted from Reference 174).

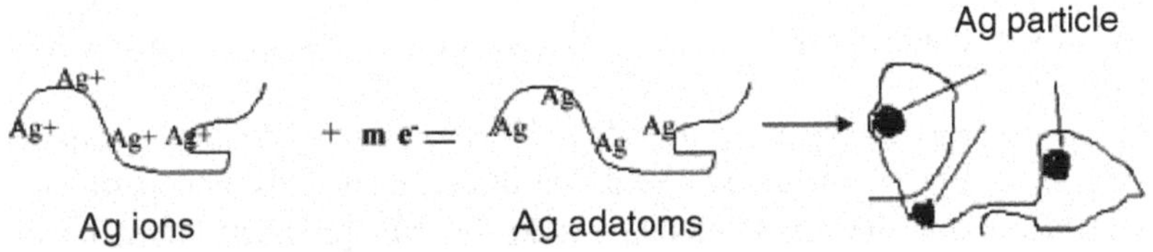

FIGURE 6.15 Schematic diagram showing formation of electrochemically produced PVP-stabilized silver clusters (adapted from Reference 174).

silver nucleation and to prohibit grain growth and particle aggregation. The steric effect arising from the long polyvinyl chains of PVP on the surface of silver particles can contribute to antiagglomeration, whereas the chemical bond between the PVP and silver powders may prohibit aggregation of silver particles.[174]

Nanocomposites of gold/polypyrrole have also been prepared by electrochemical methods.[175] The composite system can be prepared by simultaneous reduction of $AuCl_4^-$ and autopolymerization of pyrrole. The polymer coating can be removed by subjecting the composite to ultrasonic waves to obtain elemental gold nanoparticles of 2 nm in diameter.

6.4 PARTICLE SIZE CONTROL

The size distribution is an extremely important aspect in colloidal chemistry, and control over this parameter is necessary for the development of nanotechnology as a whole. It is only through studying the properties of monodisperse samples that we can begin to understand the size-related phenomena that makes nanotechnology so exciting. A standard deviation σ of 20% is considered narrow size distribution, and a monomodal size distribution is favored to bimodal or trimodal size distributions. The nucleation and growth of the particles (which determine the size and dispersity) depend on many factors: strength of the metal-metal bond,[176] the molar ratio of metal salt to stabilizer and reducing agent,[3] the extent of conversion or the reaction time,[48] the applied temperature,[3, 177] and the pressure.[177] Achieving reproducible narrow and monomodal size distribution by wet chemical preparation for nanoparticles is very difficult, though many groups have demonstrated significant progress in the last decade. The kinetics of the particle nucleation from atomic units and of the subsequent growth process can be directly observed by physical methods, e.g., light-scattering method[178] or *in situ* catalytic processes.[64]

The presently available options for obtaining narrow size distribution from broadly distributed samples are size-selective precipitation (SPP) method, [179] crystallization, [180]and size-selective ultra-centrifuge separation.[181] However, by far the most successful method for obtaining monodisperse samples is to make the initial preparation as monodisperse as possible (size-selective preparation). Turkevich could prepare colloidal Pd between 0.55 and 4.5 nm by size-selective synthetic methods. The crucial synthetic parameters are pH of the medium and reducing agent concentration.[3, 4] One promising technique is the electrochemical preparation, which was developed by Reetz and coworkers, who demonstrated that by controlling the current densities and other key parameters, they could synthesize monodisperse Pd and Ni nanoparticles in the size range between 1 and 6 nm.[12, 170, 182]

Size control has also been reported for the sonochemical decomposition method and γ-radiolysis.[183, 184] The domain of preparation methods using constrained environments offers control over the metal particle shape by predetermining the size and morphology of the products in nanoscale reaction chambers.

6.5 PREPARATION OF HETEROGENEOUS CATALYSTS BY THE PRECURSOR METHOD

Heterogeneous catalysts are traditionally obtained by impregnation methods consisting of immersing a solid support like Al_2O_3, SiO_2, or ZrO_2 in a solution of metal salts of interest and inducing deposition. After some specified time, the solid substance is filtered and dried. The dried product is calcined at high temperature in the presence of gaseous reducing agents such as H_2 or CO. The problem with impregnation methods is the difficulty in controlling the particle size; also, the metal particle formed after the reduction may be located where it is not accessible for further application.

An alternative method for the preparation of heterogeneous catalysis is the precursor concept, which was originally developed by Turkevich in 1970[4] and was further developed in the 1990s by many groups.[23] This method consists of depositing the preprepared nanocolloid on a solid support; by doing this, one can control the size and shape of the nanoparticles. Application of the precursor concept also allows full pre-characterization of the catalyst components (support and colloid) prior to deposition, thus facilitating the possibilities for comprehensive study of heterogeneous catalyst support/metal effects. Moreover, almost all of the particles on the inert support are accessible because the preformed colloids generally do not fill cracks and other deformities as molecular precursors often do. This has been demonstrated for supports such as charcoal, various oxidic support materials, and even low surface area materials such as sapphire, quartz, and highly oriented pyrolitic graphite (HOPG), for which no subsequent calcination is required (see Figure 6.16). Researchers at Degussa have demonstrated the applicability of this method for industrial-level catalytic applications.

Larpent et al. performed a combination of AFM, STM, and XPS[185] study to understand the interaction of platinum hydrosols with oxides (sapphire, quartz) and graphite single-crystal substrates (Figure 6.17). It has been observed that when a support was dipped into an aqueous Pt colloid at 20°C, the metal core was immediately

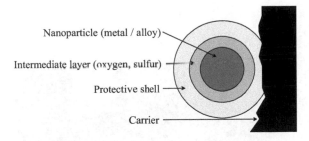

FIGURE 6.16 The precursor concept.

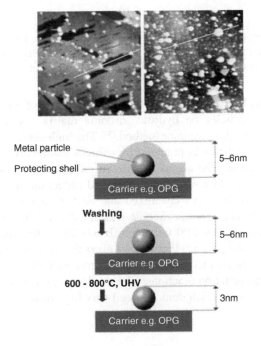

FIGURE 6.17 Top: AFM investigation of a sulfobetaine-12 stabilized Pt hydrosol (3 nm) absorbed on highly oriented pyrolytic graphite (HOPG) after dipping (left) and after additional washing (right); bottom: scheme of the Pt hydrosol adsorbtion on HOPG derived from a combined STM and XPS study (Reference 186).

adsorbed on the support surface. It was further observed that the organic stabilizer decomposed at 280°C during an annealing step. In addition, there was no change observed in the particle size up to 800°C, although above this temperature, sintering of the particles occurred.

Reetz et al. have used (octyl)$_4$NBr stabilized Pd colloids (of typical size 3 nm) as precursors of so-called *cortex catalysts*, where the active metal forms an extremely fine shell less than 10 nm thick on the support (e.g., Al$_2$O$_3$).[187] The impregnation of Al$_2$O$_3$ pellets by dispersed nanostructured metal colloids occurs with a time-dependent penetration of the support over the first 1–4 s and is complete after 10 s (see Figure 6.18a–c).

a)   b)   c)

FIGURE 6.18 Heterogeneous Pd colloid catalysts of the cortex-type (adapted with permission from Reference 187): (a) Al_2O_3 pellets (3.2 mm) after penetration with a 0.5 M solution of $(octyl)_4NBr$ stabilized colloidal Pd in THF for 1 s, (b) for 4 s, (c) for 10 s.

Recently, it has been demonstrated that the deposition of around 12 nm acetate stabilized Pd nanoparticles on hydrous zirconia matrix could be accomplished through application of the precursor method.[39] The high surface area zirconia with enhanced thermal stability was prepared by using a tetramethyl ammonium anion-directed synthetic route. The acetate stabilized palladium nanoparticles were prepared in acetone media in a single-step chemical reduction method. The colloidal solution and hydrous zirconia were mixed and heat-treated at 110°C by refluxing the mixture for 12 h. The hydrous zirconia was transformed into zirconia by annealing in an inert atmosphere at 500°C for 5 h. It was found that the Pd nanoparticles were not sintered in this process. Figure 6.19 gives the TEM micrograph of Pd/ZrO_2 obtained by the precursor method. Thus, prepared catalysts were found to be very active towards hydrogenation reactions. The catalyst was used in hydrogenation of 1-hexene hydrogenation, which demonstrated very high turnover frequency of up to 20,000 s^{-1}.

6.6 APPLICATIONS

There are numerous reports of nanoparticles applied in catalytic processes. Table 6.5 gives an overview of the important applications of d-block metal nanoparticles, especially Pd, Pt, and Au reported in recent decades. The applications vary from standard hydrogenations to selective hydrogenation, hydrosilylation, oxidation, reduction, and C-C coupling reactions. For details, please refer to literature cited. A very recent review on gold nanoparticles[188] depicts the important applications of this metal nanoparticle in different fields, such as biological sensing, supramolecular chemistry, molecular recognition, catalysis, and nonlinear optics.

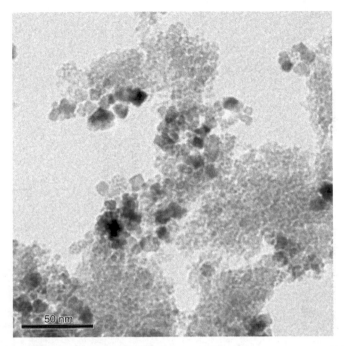

FIGURE 6.19 TEM micrograph of Pd/ZrO$_2$ prepared by a precursor concept, Pd nanoparticles of ≈12nm dispersed in ZrO$_2$ matrix.

6.7 CONCLUSIONS

This chapter covers the important methods for the preparation of nanoparticles in both gas and liquid phase, with particular attention given to wet-chemical preparations. Though electrochemical synthesis offers a narrow size distribution of nanoparticles and seems promising, the field is not growing as initially expected. Alternatively, chemical reductions of metal salts and organometallic compounds is growing rapidly, as evidenced from the number of publications in this field. There is a great deal of research still required to generalize the preparative methods to reach industrial-scale applications.

Furthermore, it should be noted that whereas nanomaterials produced by the methods discussed here are particularly useful in catalysis, significant advancements have taken place in biological and electronic applications, which are not covered in this chapter. Finally, nanotechnology is still in a developing stage, and research is particularly hindered by expensive instrumentation involved in this research. The future of nanotechnology is still somewhat unclear, but all indications are that developments in the field will be of particular relevance to catalytic applications.

TABLE 6.5
Catalytic Application Survey of Metal Nanoparticles[22, 171, 188]

Reaction	Nanoparticles/colloid	References
I. Hydrosilylation reactions		
Catalytic hydrosilylation reaction	Co, Ni, Pd, or Pt	[189, 190]
$R\!-\!\!=\ +\ R_3'SiH \xrightarrow{\ Catalyst\ } R\!\!\sim\!\!SiR_3'$		
Hydrosilylation of Oct-1-ene with colloid metal particles	Pt, Pt/Au, Pt/Pd/Al$_2$O$_3$	[191]
Oct-1-ene $+$ (Me)$_3$SiO$-$Si(H)(Me)$-$OSi(Me)$_3$ (HMTS) $\xrightarrow[60\,^{\circ}C]{\text{Colloids/Al}_2\text{O}_3}$ (Me)$_3$SiO$-$Si(CH$_2$$-$(CH$_2$)$_6$$-CH_3$)(Me)$-$OSi(Me)$_3$		
II. Oxidation catalyzed by metallic nanoparticles		
Catalytic Oxidation of Cyclohexane with Co and O$_2$	Co	Industrial catalytic reaction for the synthesis of adipic acid, which is an intermediate for the preparation of Nylon 6 and Nylon 6.6. [22]
Cyclohexane $\xrightarrow[160\,^{\circ}C,\ 15\ \text{bars}]{O_2,\ Co}$ cyclohexanone $+$ cyclohexanol $\xrightarrow{HNO_3}$ Adipic acid (CO$_2$H, CO$_2$H)		
Catalytic oxidation of cyclooctane with t-BHP and colloidal metals	Fe, Ru	[192]
Cyclooctane $\xrightarrow[20\,^{\circ}C-50\,^{\circ}C]{Fe_{coll.}\ Ru_{coll}\ \ tBHP}$ cyclooctanone $+$ cyclooctanol (OH) $+$ (OOtBu)		

Catalytic oxidation of ethene

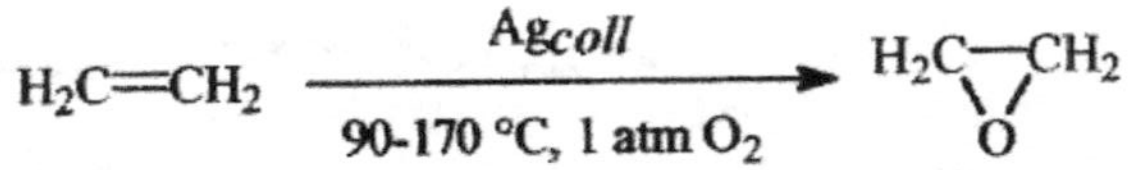

$$H_2C{=}CH_2 \xrightarrow[\text{90-170 °C, 1 atm } O_2]{Ag_{coll}} \begin{array}{c} H_2C{-}CH_2 \\ \diagdown O \diagup \end{array}$$

Ag   [193]

Oxidation by $Co(NH_3)_5Cl_2$+ of p-phenylene diamine derivatives such as N,N,N',N'-tetramethyl (TMPPD) or N,N-dimethyl (DMPPD)

Pd   [194, 195]

Oxidation of D-glucose

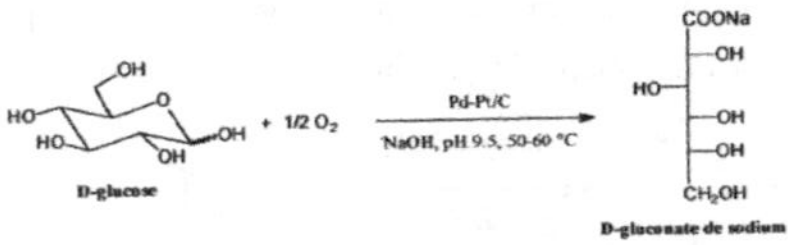

Degussa, [196]

Pt/Pd

III. C-C coupling reaction catalyzed by metallic nanoparticles
Carbonylation of methanol

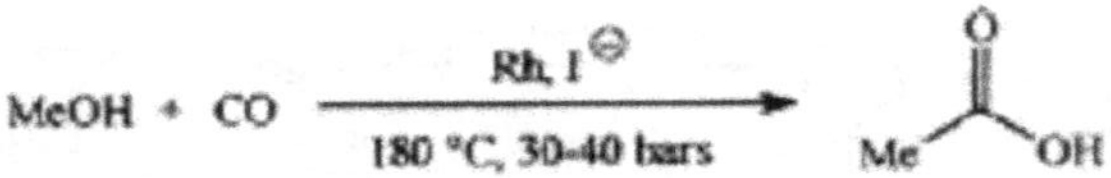

$$MeOH + CO \xrightarrow[\text{180 °C, 30-40 bars}]{Rh, I^{\ominus}} \begin{array}{c} O \\ \| \\ Me{-}\!\!-OH \end{array}$$

Rh   [80, 197]

Heck coupling reaction

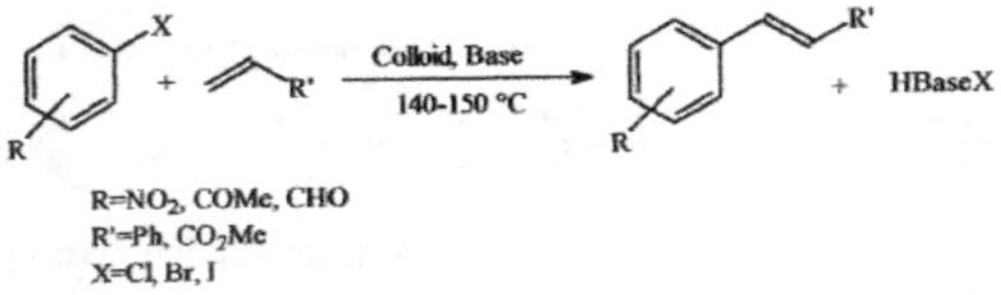

R=NO$_2$, COMe, CHO
R'=Ph, CO$_2$Me
X=Cl, Br, I

Pd   [198–200]

(Continued)

TABLE 6.5 (*Continued*)
Catalytic Application Survey of Metal Nanoparticles[22, 171, 188]

Reaction	Nanoparticles/colloid	References
Suzuki coupling reaction	Pd, Pd/Ni	[164, 201, 202]
R=NO₂, CN, COCH₃, CF₃, OCH₃ X=Cl, Br (Pd_coll or Pd/Ni_coll, Base, DMA)		
[3+2] Cycloaddition Reaction (Ni_coll, Toluene, reflux; + CO₂CH₃)	Ni	[203]
McMurry reaction catalyzed by Ti colloids (2 Ph₂C=O → Ti_coll, THF, reflux → Ph₂C=CPh₂ + Ph₂CHOH + Ph₂CHCHPh₂)	Ti	[204]
IV. Hydrogenation catalyzed by metal nanoparticles Selective hydrogenation of unsaturated aldehydes (Cinnamaldehyde, Cinnamic alcohol)	Pt, Pt/Co	[205, 206]

Hydrogenation of chloronitrobenzene

Pt, Ru, Ru/Pt, Ru/Pd, Pd/Pt [207–209]

Regioselective hydrogenations: Hydrogenation of 1,5-cyclooctadiene

Pd, Pd/Pt, Au/Pd,Cu/Pt [210–212]

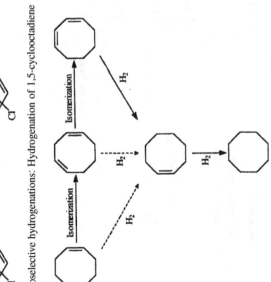

(Continued)

TABLE 6.5 (Continued)
Catalytic Application Survey of Metal Nanoparticles[22, 171, 188]

Reaction	Nanoparticles/colloid	References
Selective hydrogenation of dehydrolinalol	Pd	[213, 214]
Stereo- and enantioselective reactions		
Stereoselective hydrogenation of alkynes	Pd, Au/Pd, Pt	[215, 216]

Dehydrolinalol $\xrightarrow{H_2}$ Linalol

Structures referenced:

Dehydrolinalol, Linalol

hex-3-yn-1-ol $\xrightarrow[H_2]{Pd_{coll}}$ cis-hex-3-en-1-ol +

hex-2-yne $\xrightarrow[H_2]{Au/Pd_{coll}}$ cis-hex-2-ene +

Stereoselective hydrogenation of DB18C6

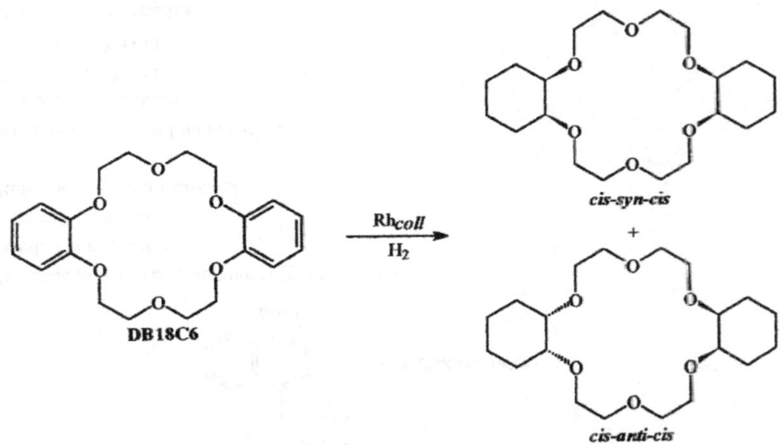

Rh

[217, 218]

Enantioselective hydrogenations: Enantioselective reduction of o-cresol

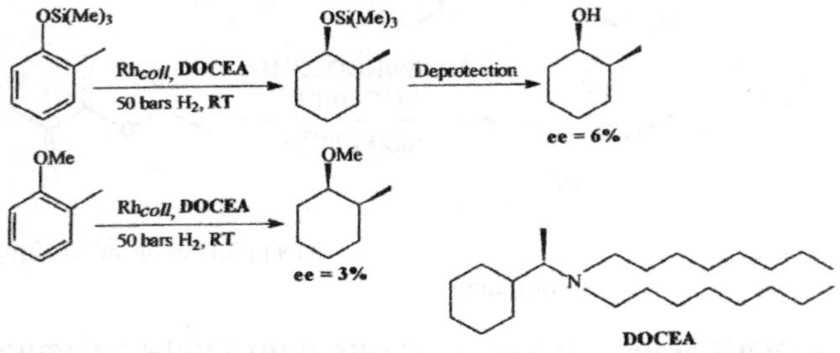

Rh

[219]

(Continued)

TABLE 6.5 (Continued)
Catalytic Application Survey of Metal Nanoparticles[22, 171, 188]

Reaction	Nanoparticles/colloid	References
Ethyl pyruvate to ethyl lactate	Pt, Pd	[6, 220, 221]
Benzene derivatives compounds hydrogenation		
Arene hydrogenation	Pd, Ni, Ir, Pt, and Ru	[219, 222, 223]
Fuel cell catalysts	Pt, Pt alloy, $Pt_{50}Co_{30}Cr_{20}$, Ru, Pt/Ru, Pt_{50}/Ru_{50}	[224–231]
Reformer gas into methanol		
Applications of gold nanoparticles		
CO and H_2 oxidation	Au	[232–236]
$CO + \frac{1}{2} O_2 \rightarrow CO_2$		
$H_2 + \frac{1}{2} O_2 \rightarrow H_2O$		
Water gas shift reaction	Au	[237]
$CO + H_2O \rightarrow CO_2 + H_2$		

CO$_2$ hydrogenation CO$_2$ + H$_2$ → CO + H$_2$O	Au	[238]
Catalytic combustion of methanol CH$_3$OH + O$_2$ → CO$_2$ + H$_2$O	Au	[239]
Electrochemical oxidation of CO and CH$_3$OH CO + 2OH$^-$ − 2e$^-$ → H$_2$O + CO$_2$ (CO$_3^{2-}$ in alkaline) CH$_3$OH + 6OH$^-$ − 6e$^-$ → 5H$_2$O + CO$_2$ (CO$_3^{2-}$ in alkaline)	Au	[240, 241]
Regioselective hydrogenation of acrolein to allylic alcohol Acrolein → allyl alcohol, propionaldehyde, n-propanol, and C2 and C3 hydrocarbons	Au	[242–245]
Hydrogenation of cyclohexene, cycloocta-1,3-diene	Au	[246]

Reduction of eosin by NaBH$_4$ C$_2$OH$_6$Br$_4$Na$_2$O$_5$ = eosin	Au	[247]
Epoxidation of propene using H$_2$ and O$_2$	Au	[248]

REFERENCES

1. Helcher, H. H. *J. Herbord Klossen* 1718, *31*, 309, 105.
2. Faraday, M. *Philos. Trans. R. Soc.* 1857, *147*, 145.
3. Turkevich, J.; Stevenson, P. C.; Hillier, J. *Diss. Fara. Soc.* 1951, *11*, 55–75.
4. Turkevich, J.; Kim, G. *Science* 1970, *169*, 873.
5. Schmid, G. *Aspects of Homogeneous Catalysis,* Kluwer: Dordrecht, 1990.
6. Bönnemann, H.; Braun, G.; Brijoux, W.; Brinkmann, R.; Schulze Tilling, K.; Seevogel, K.; Siepen, K. *J. Organomet. Chem.* 1996, *520*, 143–162.
7. Reetz, M. T.; Helbig, W.; Quaiser, S. A. *Active Metals,* VCH: Weinheim, 1996.
8. Aiken III, J. D.; Finke, R. G. *J. Mol. Catal. A* 1999, *145*, 1–44.
9. Schmid, G. *Applied Homogeneous Catalysis with Organometallic Compounds,* Wiley-VCH: Weinheim, 1996.
10. Herrmann, W. A.; Cornils, B. *Applied Homogeneous Catalysis with Organometallic Compounds,* Wiley-VCH: Weinheim, 1996.
11. Nieuwenhuys; B. E. In King, D. A.; Woodruff, D. P. (Eds.) *The Chemical Physics of Solid Surfaces and Heterogeneous Catalysis,* Elsevier: Amsterdam, 1993.
12. Kolb, U.; Quaiser, S. A.; Winter, M.; Reetz, M. T. *Chem. Mater.* 1996, *8*, 1889–1894.
13. Schmid, G. *Endeavour* 1990, *14*, 172–178.
14. Lin, Y.; Finke, R. G. *J. Am. Chem. Soc.* 1994, *116*, 8335–8353.
15. Teo, B. K.; Sloane, N. J. A. *Inorg. Chem.* 1985, *24*, 4545–4558.
16. Gaffet, E.; Tachikart, M.; El Kedim, O.; Rahouadj, R. *Mater. Charact.* 1996, *36*, 185–190.
17. Amulyavichus, A.; Daugvila, A.; Davidonis, R.; Sipavichus, C. *Fiz. Met. Metalloved.* 1998, *85*, 111–117.
18. Turkevich, J. *Gold Bull* 1985, *18*, 86–91.
19. Leisner, T.; Rosche, C.; Wolf, S.; Granzer, F.; Wöste, L. *Surf. Rev.Lett.* 1996, *3*, 1105–1108.
20. Michaelis, M.; Henglein, A. *J. Phys. Chem.* 1992, *96*, 4719–4724.
21. Bradley, J. S. *Clusters and Colloids,* VCH: Weinheim, 1994.
22. Roucoux, A.; Schulz, J.; Patin, H. *Chem. Rev.* 2002, *102*, 3757–3778.
23. Bönnemann, H.; Brijoux, W.; Brinkmann, R.; Fretzen, R.; Th. Joussen; Köppler, R.; Neiteler, P.; Richter, J. *J. Mol. Catal.* 1994, *86*, 129–177.
24. Schmid, G.; Pfeil, R.; Boese, R.; Bandermann, F.; Meyer, S.; Calis, G. H. M.; van der Velden, J. A. W. *Chem. Ber.* 1981, *114*, 3634–3642.
25. Vargaftik, M. N.; Zargorodnikov, V. P.; Stolarov, I. P.; Moiseev, I. I.; Kochubey, D. I.; Likholobov, V. A.; Chuvilin, A. L.; Zarnaraev, K. I. *J. Mol. Catal.* 1989, *53*, 315–349.
26. Franke, R.; Rothe, J.; Pollmann, J.; Hormes, J.; Bönnemann, H.; Brijoux, W.; Hindenburg, T. *J. Am. Chem. Soc.* 1996, *118*, 12090–12097.
27. Vidoni, O.; Philippot, K.; Amiens, C.; Chaudret, B.; Balmes, O.; Malm, J.-O.; Bovin, J.-O.; Senocq, F.; Casanove, M.-J. *Angew. Chem. Int. Ed.* 1999, *38*, 3736–3738.
28. Tanori, J.; Pileni, M. P. *Langmuir* 1997, *13*, 639–646.
29. Pileni, M. P. *Supramol. Sci.* 1998, *5*, 321–329.
30. Förster, S. *Ber. Bunsenges. Phys. Chem.* 1997, *101*, 1671–1678.
31. Bönnemann, H.; Brijoux, W. *Advanced Catalysts and Nanostructured Materials,* Academic Press: New York, 1996.
32. Sinzig, J.; de Jongh, L. J.; Bönnemann, H.; Brijoux, W.; Köppler, R. *Appl. Organomet. Chem.* 1998, *12*, 387–391.
33. Bradley, J. S.; Hill, E. W.; Leonowicz, M. E.; Witzke, H. *J. Mol. Catal.* 1987, *41*, 59–74.

34. Tominaga, T.; Tenma, S.; Watanabe, H.; Giebel, U.; Schmid, G. *Chem. Lett.* 1996, 1033.
35. Aiken III, J. D.; Finke, R. G. *J. Mol. Catal. A* 1999, *145*, 1244.
36. Corbierre, M. K.; Lennox, R. B. *Chem. Mater.* 2005, *17*, 5691–5696.
37. Toshima, N.; Yonezawa, T. *New J. Chem.* 1998, 1179–1201.
38. Lu, P.; Teranishi, T.; Asakura, K.; Miyake, M.; Toshima, N. *J. Phys. Chem. B* 1999, *103*, 9673–9682.
39. D'Souza, L.; Saleh-Subaie, J.; Richards, R. *J. Colloid Interface Sci.* 2005, *292*, 476–485.
40. Finke, R. G.; Ozkar, S. *Coord. Chem. Rev.* 2004, *248*, 135–146.
41. Rampino, L. D.; Nord, F. F. *J. Am. Chem. Soc.* 1941, *63*, 2745–2749.
42. Hernandez, L.; Nord, F. F. *J. Colloid. Sci* 1948, *3*, 363–375.
43. Vargaftik, M. N.; Zagorodnikov, V. P.; Stolarov, I. P.; Moiseev, I. I.; Kochubey, D. I.; Likholobov, V. A.; Chuvilin, A. L.; Zamaraev, K. I. *J. Mol. Catal.* 1989, *53*, 315–348.
44. Vargaftik, M. N.; Zagorodnikov, V. P.; Stolarov, I. P.; Moiseev, I. I.; Likholobov, V. A.; Kochubey, D. I.; Chuvilin, A. L.; Zaikovski, V.; Zamaraev, K. I.; Timofeeva, G. I. *J. Chem. Soc., Chem. Comm.* 1985, 937–939.
45. Volkov, V. V.; van Tendeloo, G.; Tsirkov, G. A.; Cherkashina, N. V.; Vargaftik, M. N.; Moiseev, I. I.; Novotortsev, V. M.; Kvit, A. V.; Chuvilin, A. L. *J. Cryst. Growth* 1996, *163*, 377.
46. Moiseev, I. I.; Vargaftik, M. N.; Chernysheva, T. V.; Stromnova, T. A.; Gekhman, A. E.; Tsirkov, G. A.; Makhlina, A. M. *J. Mol. Catal. A* 1996, *108*, 77.
47. Watzky, M.; Finke, R. G. *J. Am. Chem. Soc.* 1997, *119*, 10382–10400.
48. Meguro, K.; Nakamura, Y.; Hayashi, Y.; Torizuka, M.; Esumi, K. *Bull. Chem. Soc. Jpn.* 1988, *61*, 347.
49. Mucalo, M. R.; Cooney, R. P. *J. Chem. Soc., Chem. Commun.* 1989, 94–95.
50. Lewis, L. N.; Lewis, N. *Chem. Mater.* 1989, *1*, 106–114.
51. Curtis, A. C.; Duff, D. G.; Edwards, P. P.; Jefferson, D. A.; Johnson, B. F. G.; Kirkland, A. I.; Logan, D. E. *Angew. Chem. Int. Ed. Engl.* 1987, *26*, 676.
52. Duff, D. G.; Curtis, A. C.; Edwards, P. P.; Jefferson, D. A.; Johnson, B. F. G.; Logan, D. E. *J. Chem. Soc., Chem. Commun.* 1987, 1264.
53. Van Rheenen, P. R.; McKelvy, M. J.; Glaunsinger, W. S. *J. Solid State Chem.* 1987, *67*, 151–169.
54. Duff, D. G.; Baiker, A. *Preparation of Catalysts VI*, Elsevier Science: Amsterdam, 1995.
55. Tsai, K.-L.; Dye, J. L. *Chem. Mater.* 1993, *5*, 540–546.
56. Nayak, B. B.; Vitta, S.; Nigam, A. K.; Bahadur, D. *Thin Solid Films* 2006, *505*, 109–112.
57. D'Souza, L.; Bera, P.; Sampath, S. *J. Colloid Interface Sci.* 2002, *246*, 92–99.
58. D'Souza, L.; Sampath, S. *Langmuir* 2000, *16*, 8510–8517.
59. van Wonterghem, J.; Mørup, S.; Koch, C. J. W.; Charles, S. W.; Wells, S. *Nature* 1986, *322*, 622.
60. Glavee, G. N.; Klabunde, K. J.; Sorensen, C. M.; Hadjipanayis, G. C. *Inorg. Chem.* 1993, *32*, 474–477.
61. Franke, R.; Rothe, J.; Becker, R.; Pollmann, J.; Hormes, J.; Bönnemann, H.; Brijoux, W.; Köppler, R. *Adv. Mater.* 1998, *10*, 126–131.
62. Bönnemann, H.; Brijoux, W.; Brinkmann, R.; Hofstadt, W.; Angermund, K. *Rev. Roum. Chim.* 1999, *44*, 1003–1010.
63. Watzky, M. A.; Finke, R. G. *Chem. Mater.* 1997, *9*, 3083–3095.
64. Aiken III, J. D.; Finke, R. G. *J. Am. Chem. Soc.* 1998, *120*, 9545–9554.

65. Widegren, J. A.; Aiken, J. D.; Özkar, S.; Finke, R. G. *Chem. Mater.* 2001, *13*, 312–324.
66. Hornstein, B. J.; Finke, R. G. *Chem. Mater* 2003, *15*, 899–909.
67. Troupis, A.; Hiskia, A.; Papaconstantinou, E. *Angew. Chem. Int. Ed.* 2002, *41*, 1911–1912.
68. Mayer, C.; Neveu, S.; Cabuil, V. *Angew. Chem. Int. Ed.* 2002, *41*, 501–503.
69. Widegren, J. A.; Bennett, M. A.; Finke, R. G. *J. Am. Chem. Soc.* 2003, *125*, 10301–10310.
70. Özkar, S.; Finke, R. G. *Langmuir* 2002, *18*, 7653–7662.
71. Toshima, N.; Harada, M.; Yonezawa, T.; Kushihashi, K.; Asakuri, K. *J. Phys. Chem.* 1991, *95*, 7448.
72. Touroude, R.; Girard, P.; Maire, G.; Kizling, J.; Boutonnet-Kizling, M.; Stenius, P. *Colloids Surf.* 1992, *67*.
73. Rousset, J. L.; Renouprez, A. J.; Cadrot, A. M. *Phys. Rev. B.* 1998, *58*, 2150.
74. Schubert, U. *New J. Chem.* 1994, *18*, 1049.
75. Schubert, U.; Breitscheidel, B.; Buhler, B.; Egger, C.; Urbaniak, W., *Better Ceramics Through Chemistry V*, MRS Proc, San Francisco, 271, 1992.
76. Morke, W.; Lamber, R.; Schubert, U.; Breitscheidel, B. *Chem. Mater.* 1994, *6*, 1639.
77. Kaiser, A.; Gorsmann, C.; Schubert, U. *J. Sol-Gel Sci. Technol.* 1997, *2*, 795.
78. Heinrichs, B.; Delhez, P.; Schoebrechts, J.-P.; Pirard, J.-P. *J. Catal.* 1997, *172*, 322.
79. Guczi, L.; Schay, A.; Stefler, G.; Mizukami, F. *J. Mol. Catal.* 1999, *141*, 177.
80. Wang, Q.; Liu, H.; Han, M.; Li, X.; Jiang, D. *J. Mol. Catal. A: Chem.* 1997, *118*, 145.
81. Doudna, C. M.; Bertino, M. F.; Blum, F. D.; Tokuhiro, A. T.; Lahiri-Dey, D.; Chattopadhyay, S.; Terry, J. *J. Phys. Chem. B* 2003, *107*, 2966–2970.
82. Mandal, S.; Selvakannan, P.; Pasricha, R.; Sastry, M. *J. Am. Chem. Soc.* 2003, *125*, 8440–8441.
83. Fernandez, C. A.; Wai, C. W. *J. Nanosci. Nanotechnol.* 2006, *6*, 669–674.
84. Rivas, J.; Garcia-Bastida, A. J.; Lopez-Quintela, M. A.; Ramos, C. *J. Magn. Magn. Mater.* 2006, *300*, 185–191.
85. Wang, W.; Tian, X.; Chen, K.; Cao, G. *Colloids Surf., A* 2006, *273*, 35–42.
86. Feng, J.; Zhang, C.-P. *J. Colloid Interface Sci.* 2006, *293*, 414–420.
87. Capek, I. *Adv. Colloid Interface Sci.* 2004, *110*, 49–74.
88. Lopez-Quintela, M. A.; Rivas, J. *J. Colloid Interface Sci.* 1993, *158*, 446.
89. Seip, C. T.; O'Connor, R. J. *Nanostruct. Mater.* 1999, *12*, 183.
90. Ingelsten, H. H.; Bagwe, R.; Palmqvist, A.; Skoglundh, M.; Svanberg, C.; Holmberg, K. *J. Colloid Interface Sci.* 2001, *241*, 104.
91. Barnickel, P.; Wokaun, A.; Sager, W.; Eicke, H. F. *J. Colloid Interface Sci.* 1992, *148*, 90.
92. Manna, A.; Kulkarni, B. D. *Chem. Mater.* 1997, *9*, 3032.
93. Dutta, P.; Fendler, J. *J. Colloid Interface Sci.* 2002, *247*, 47.
94. Henglein, A. *Chem. Rev.* 1989, *89*, 1861–1873.
95. Iida, M.; Ohkawa, S.; Er, H.; Asaoka, N.; Yoshikawa, H. *Chem. Lett.* 2002, 1050.
96. Hamada, K.; Hatanaka, K.; Kawai, T.; Konno, K. *Shikizai Kyokaishi* 2000, *73*, 385.
97. Karpov, S. V.; Popov, A. K.; Slatko, V. V.; Shevnina, G. B. *Colloid J.* 1988, *57*, 199.
98. Lufimpadio, N.; Nagy, J. B.; Derouane, E. G. *Surfactant in Solution*, Plenum: New York, 1984.
99. Nagy, J. B. *Colloid Surf.* 1989, *35*, 201.
100. Hess, P. H.; Parker, P. H. *J. Appl. Polym. Sci.* 1966, *10*, 1915–1927.
101. Thomas, J. R. *J. Appl. Phys.* 1966, *37*, 2914–2915.
102. Kato, Y.; Sugimoto, S.; Shinohara, K.; Tezuka, N.; Kagotani, T.; Inomata, K. *Mater. Trans., JIM* 2002, *43*, 406.

103. Smith, T. W.; Wychick, D. *J. Phys. Chem.* 1980, *84*, 1621.
104. Amiens, C.; de Caro, D.; Chaudret, B.; Bradley, J. S. *J. Am. Chem. Soc.* 1993, *115*, 11638–11939.
105. de Caro, D.; Wally, H.; Amiens, C.; Chaudret, B. *J. Chem. Soc., Chem. Commun.* 1994, 1891–1892.
106. Bradley, J. S.; Hill, E. W.; Behal, S.; Klein, C.; Chaudret, B.; Duteil, A. *Chem. Mater.* 1992, *4*, 1234–1239.
107. Duteil, A.; Que´au, R.; Chaudret, B.; Mazel, R.; Roucau, C.; Bradley, J. S. *Chem. Mater.* 1993, *5*, 341–347.
108. de Caro, D.; Agelou, V.; Duteil, A.; Chaudret, B.; Mazel, R.; Roucau, C.; Bradley, J. S. *New J. Chem.* 1995, *19*, 1265–1274.
109. Dassenoy, F.; Philippot, K.; Ould Ely, T.; Amiens, C.; Lecante, P.; Snoeck, E.; Mosset, A.; Casanove, M.-J.; Chaudret, B. *New J. Chem.* 1998, *19*, 703–711.
110. Reetz, M. T.; Quaiser, S.; Winter, M.; Becker, J. A.; Schaefer, R.; Stimming, U.; Marmann, A.; Vogel, R.; Konno, T. *Angew. Chem. Ind. Ed. Engl.* 1996, *35*, 2092–2094.
111. Yu, D. B.; Sun, X. Q.; Bian, J. T.; Tong, Z. C.; Qian, Y. T. *Physica E-Low-Dimensional Systems & Nanostructures* 2004, *23*, 50–55.
112. Ghosh, M.; Seshadri, R.; Rao, C. N. R. *J. Nanosci. Nanotechnol.* 2004, *4*, 136–140.
113. Zhang, W. X.; Yang, Z. H.; Liu, Y.; Tang, S. P.; Han, X. Z.; Chen, M. *J. Cryst. Growth* 2004, *263*, 394–399.
114. Kim, C. S.; Moon, B. K.; Park, J. H.; Choi, B. C.; Seo, H. J. *J. Cryst. Growth* 2003, *257*, 309–315.
115. Kar, S.; Panda, S. K.; Satpati, B.; Satyam, P. V.; Chaudhuri, S. *J. Nanosci. Nanotechnol.* 2006, *6*, 771–776.
116. Li, B.; Xie, Y.; Huang, J. X.; Qian, Y. T. *Adv. Mater.* 1999, *11*, 1456–1459.
117. Gautam, U. K.; Rajamathi, M.; Meldrum, F.; Morgan, P.; Seshadri, R. *Chem. Commun.* 2001, 629–630.
118. Hai, B.; Tang, K. B.; Wang, C. R.; An, C. H.; Yang, Q.; Shen, G. Z.; Qian, Y. T. *J. Cryst. Growth* 2001, *225*, 92–95.
119. Hou, Y. L.; Gao, S. *J. Alloys Compd.* 2004, *365*, 112–116.
120. Rosemary, M. J.; Pradeep, T. *J. Colloid Interface Sci.* 2003, *268*, 81–84.
121. Chemseddine, A.; Moritz, T. *Eur. J. Inorg. Chem.* 1999, 235.
122. Trentler, T. J.; Denler, T. E.; Bertone, J. F.; Agrawal, A.; Colvin, V. L. *J. Am. Chem. Soc.* 1999, *121*, 1613.
123. Rockenberger, J.; Scher, E. C.; Alivisatos, P. A. *J. Am. Chem. Soc.* 1999, *121*, 11595.
124. Thimmaiah, S.; Rajamathi, M.; Singh, N.; Bera, P.; Meldrum, F.; Chandrasekhar, N.; Seshadri, R. *J. Mater. Chem.* 2001, *11*, 3215–3221.
125. Murray, C. B.; Norris, D. J.; Bawendi, M. G. *J. Am. Chem. Soc.* 1993, *115*, 8706.
126. Mitchell, P. W. D.; Morgan, P. E. D. *J. Am. Ceram. Soc.* 1974, *57*, 278.
127. Li, Q.; Shao, M. W.; Zhang, S. Y.; Liu, X. M.; Li, G. P.; Jiang, K.; Qian, Y. T. *J. Cryst. Growth* 2002, *243*, 327–330.
128. Cominos, V.; Gavriilidis, A. *Euro. Phy. J* 2001, *15*, 69.
129. Suslick, K. S.; Choe, S. B.; Cichowlas, A. A.; Grinstaff, M. W. *Nature* 1991, *353*, 414.
130. Suslick, K. S.; Hyeon, T.; Fang, M.; Cichowlas, A. *Advanced Catalysts and Nanostructured Materials,* Chapter 8, Academic Press: San Diego, 1996.
131. Dhas, A.; Gedanken, A. *J. Mater. Chem.* 1998, *8*, 445–450.
132. Salkar, R. A.; Jeevanandam, P.; Aruna, S. T.; Koltypin, Y.; Gedanken, A. *J. Mater. Chem.* 1999, *9*, 1333–1335.
133. Livage, J. *J. Phys.* 1981, *42*, 981.

134. Sugimoto, M. *J. Magn. Magn. Mater.* 1994, *133*, 460.
135. Landau, M. V.; Vradman, L.; Herskowitz, M.; Koltypin, Y.; Gedamken, A. *J. Catal.* 2001, *201*, 22.
136. Perkas, N.; Wang, Y.; Koltypin, Y.; Gedanken, A.; Chandrasekhar, N. *Chem. Commun.* 2001, 988.
137. Ramesh, S.; Koltypin, Y.; Prozorov, R.; Gedanken, A. *Chem. Mater.* 1997, *9*, 546.
138. Pol, V. G.; Reisfeld, R.; Gedanken, A. *Chem. Mater.* 2002, *14*, 3920.
139. Hiller, R.; Putterman, S. J.; Barber, B. P. *Phy. Rev. Lett.* 1992, *69*, 1182.
140. Barber, B. P.; Putterman, S. J. *Nature* 1986, *352*, 414.
141. Suslick, K. S.; Hammerton, D. A.; Cline, R. E. *J. Am. Chem. Soc.* 1986, *108*, 5641.
142. Wang, G. Z.; Geng, B. Y.; Huang, Y. W.; Wang, Y. W.; Li, G. H.; Zhang, L. D. *Appl. Phys. A-Mater. Sci. Process.* 2003, *77*, 933–936.
143. Mukaibo, H.; Yoshizawa, A.; Momma, T.; Osaka, T. *Power Sources* 2003, *119*, 60–63.
144. Li, Q.; Ding, Y.; Shao, M. W.; Wu, J.; Yu, G. H.; Qian, Y. T. *Mater. Res. Bull.* 2003, *38*, 539–543.
145. Mizukoshi, Y.; Oshima, R.; Maeda, Y.; Nagata, Y. *Langmuir* 1999, *15*, 2733–2737.
146. Qiu, X. F.; Zhu, J. J.; Chen, H. Y. *J. Cryst. Growth* 2003, *257*, 378–383.
147. Yu, Y.; Zhan, Q. Y.; Li, X. G. *Acta Phys. Chim. Sin.* 2003, *19*, 436–440.
148. Kijima, N.; Takahashi, Y.; Akimoto, J.; Tsunoda, T.; Uchida, K.; Yoshimura, Y. *Chem. Lett.* 2005, *34*, 1658–1659.
149. Qui, X. F.; Zhu, J. J.; Chin, J. *Inorg. Chem.* 2003, *19*, 766–770.
150. Xu, X.; Li, J.; Liu, X.; Hao, Z.; Zhao, W. *J. Nanosci. Nanotechnol.* 2006, *6*, 872–874.
151. Wu, S. H.; Chen, D. H. *J. Colloid Interface Sci.* 2003, *259*, 282–286.
152. Jia, Y.; Niu, H.; Wu, M.; Ning, M.; Zhu, H.; Chen, Q. *Mater. Res. Bull.* 2005, *40*, 1623–1629.
153. Vinodgopal, K.; He, Y.; Ashokkumar, M.; Grieser, F. *J. Phys. Chem. B* 2006, *110*, 3849–3852.
154. Kan, C. X.; Cai, W. P.; Li, C. C.; Zhang, L. D.; Hofmeister, H. *J. Phys. D-Appl. Phys.* 2003, *36*, 1609–1614.
155. Li, Q. L.; Li, H. L.; Pol, V. G.; Bruckental, I.; Koltypin, Y.; Calderon-Moreno, J.; Nowik, I.; Gedanken, A. *New J. Chem.* 2003, *27*, 1194–1199.
156. Li, Q.; Li, H.; Pol, V. G.; Bruckental, I.; Koltypin, Y.; Calderon-Moreno, J.; Nowik, I.; Gedanken, A. *New J. Chem.* 2003, *27*, 1194–1199.
157. Schalnikoff, A.; Roginsky, R. *Kolloid Z* 1927, *43*, 67–70.
158. Blackborrow, J. R.; Young, D. *Metal Vapor Synthesis,* Springer Verlag: New York, 1979.
159. Klabunde, J. K. *Free Atoms and Particles,* Academic Press: New York, 1980.
160. Stoeva, S.; Klabunde, K. J.; Sorensen, C. M.; Dragieva, I. *J. Am. Chem. Soc.* 2002, *124*, 2305–2311.
161. Chen, J.-F.; Wang, Y.-H.; Guo, F.; Wang, X.-M.; Zheng, C. *Ind. Eng. Chem. Res.* 2002, *39*, 948–954.
162. Ramshaw, C.; Mallinson, R. *European Patent 2568B,* 1979; *U.S. Patent 4263255,* 1981.
163. Reetz, M. T.; Helbig, W. *J. Am. Chem. Soc.* 1994, *116*, 7401–7402.
164. Reetz, M. T.; Breinbauer, R.; Wanninger, K. *Tetrahedron Lett.* 1996, *37*, 4499.
165. Murray, B. J.; Li, Q.; Newberg, J. T.; Menke, E. J.; Hemminger, J. C.; Penner, R. M. *Nano Lett.* 2005, *5*, 2319–2324.
166. Plieth, W.; Dietz, H.; Anders, A.; Sandmann, G.; Meixner, A.; Weber, M.; Kneppe, H. *Surf. Sci.* 2005, *597*, 119–126.
167. Starowicz, M.; Stypula, B.; Banas, J. *Electrochem. Commun.* 2006, *8*, 227–230.

168. Huang, L.; Jiang, H.; Zhang, J.; Zhang, Z.; Zhang, P. *Electrochem. Commun.* 2006, *8*, 262–266.
169. Ma, H.; Huang, S.; Feng, X.; Zhang, X.; Tian, F.; Yong, F.; Pan, W.; Wang, Y.; Chen, S. *ChemPhysChem* 2006, *7*, 333–335.
170. Reetz, M. T.; Helbig, W.; Quaiser, S. A. *Chem. Mater.* 1995, *7*, 2227–2228.
171. Bönnemann, H.; Richards, R. M. *Eur. J. Inorg. Chem.* 2001, *83*, 1–27.
172. Winter, M. A. Ph.D. dissertation, Verlag Mainz, 1998.
173. Rodriguez-Sanchez, L.; Blanco, M. C.; Lopez-Quintela, M. A. *J. Phys. Chem. B* 2000, *104*, 9683–9688.
174. Yin, B.; Ma, H.; Wang, S.; Chen, S. *J. Phys. Chem. B* 2003, *107*.
175. Liu, Y.-C.; Chuang, T. C. *J. Phys. Chem. B* 2003, *107*, 12383–12386.
176. Hirai, H.; Nakao, Y.; Toshima, N. *J. Macromol. Sci. Chem.* 1979, *A13*, 727.
177. Papirer, E.; Horny, P.; Balard, H.; Anthore, R.; Pepitas, C.; Martinet, A. *J. Colloid Interface Sci.* 1983, *94*, 220–228.
178. D'Souza, L.; Suchopar, A.; Richards, R. *J. Colloid Interface Sci.* 2004, *279*, 458–463.
179. Pileni, M. *Pure Appl. Chem.* 2000, *72*, 53–65.
180. Schmid, G.; Bäumle, M.; Beyer, N. *Angew. Chem. Int. Ed.* 2000, *39*, 182–184.
181. Cölfen, H.; Pauck, T. *Colloid Polym. Sci.* 1997, *275*, 175–180.
182. Reetz, M. T.; Helbig, W.; Quaiser, S. Studiengesellschaft Kohle: U.S., 1999.
183. Jeon, J.-S.; Yeh, C.-S. *J. Chin. Chem. Soc.* 1998, *45*, 721–726.
184. Stietz, F. Träger, F. *Physikalische Blätter,* 55(9) 5–57, 1999.
185. Larpent, C.; Brisse-le-Menn, F.; Patin, H. *J. Mol. Catal.* 1991, *65*, 35–40.
186. Shaikhutdinov, S. K.; Möller, F. A.; Mestl, G.; Behm, R. J. *J. Catal.* 1996, *163*, 492–495.
187. Reetz, M. T.; Quaiser, S. A.; Breinbauer, R.; Tesche, B. *Angew. Chem. Int. Ed.* 1995, *34*, 2728–2730.
188. Daniel, M.-C.; Astruc, D. *Chem. Rev.* 2004, *104*, 293–346.
189. Lewis, L. N. *J. Am. Chem. Soc.* 1990, *112*, 5998.
190. Onopchenko, A.; Sabourin, E. T. *J. Org. Chem.* 1987, *52*, 4118.
191. Schmid, G.; West, H.; Mehles, H.; Lehnert, A. *Inorg. Chem.* 1997, *36*, 891.
192. Launay, F.; Roucoux, A.; Patin, H. *Tetrahedron Lett.* 1998, *39*, 1353.
193. Shiraishi, Y.; Toshima, N. *Colloids Surf., A* 2000, *169*, 59.
194. Spiro, M.; De Jesus, D. M. *Langmuir* 2000, *16*, 2464.
195. De Jesus, D. M.; Spiro, M. *Langmuir* 2000, *16*, 4896.
196. Bönnemann, H.; Brijoux, W.; Schulze Tilling, A.; Siepen, K. *Top. Catal.* 1997, *4*, 217.
197. Parshall, G. W.; Ittel, D. D. *Homogeneous Catalysts,* Wiley and Sons: New York, 1992.
198. Beller, M.; Fischer, H.; Ku¨hlein, K.; Reisinger, C.-P.; Herrmann, W. A. *J. Organomet. Chem.* 1996, *520*, 257.
199. Klingelhöfer, S.; Heitz, W.; Greiner, A.; Oestreich, S.; Förster, S.; Antonietti, M. *J. Am. Chem. Soc.* 1997, *119*, 10116.
200. Le Bars, J.; Specht, U.; Bradley, J. S.; Blackmond, D. G. *Langmuir* 1999, *15*, 7621.
201. Li, Y.; Hong, X. M.; Collard, D. M.; El-Sayed, M. A. *Org. Lett.* 2000, *2*, 2385.
202. Li, Y.; El-Sayed, M. A. *J. Phys. Chem. B* 2001, *105*, 8938.
203. Reetz, M. T.; Breinbauer, R.; Wedemann, P.; Binger, P. *Tetrahedron Lett.* 1998, *1233*, 1233.
204. Reetz, M. T.; Quaiser, S. A.; Merk, C. *Chem. Ber.* 1996, 741.
205. Yu, W.; Liu, H.; Tao, Q. *Chem. Commun.* 1996, 1773.
206. Feng, H.; Liu, H. *J. Mol. Catal. A: Chem.* 1997, *126*, L5–L8.
207. Yang, X.; Liu, H. *Appl. Catal. A: General* 1997, *164*, 197.

208. Liu, M.; Yu, W.; Liu, H.; Zheng, J. *J. Colloid Interface Sci.* 1999, *214*, 213.
209. Yang, X.; Liu, H.; Zhong, H. *J. Mol. Catal. A: Chem.* 1999, *147*, 55.
210. Hirai, H.; Komatsuzaki, S.; Toshima, N. *Bull. Chem. Soc. Jpn.* 1984, *57*, 488.
211. Toshima, N.; Kushihashi, K.; Yonezawa, T.; Hirai, H. *Chem. Lett.* 1989, 1769.
212. Gittins, D. I.; Caruso, F. *Angew. Chem., Int. Ed. Engl* 2001, *40*, 3001.
213. Sulman, E.; Bodrova, Y.; Matveeva, V.; Semagina, N.; Cerveny, L.; Kurtc, V.; Bronstein, L.; Platonova, O.; Valetsky, P. *Appl. Catal. A: Gen.* 1999, *176*, 75.
214. Sulman, E.; Matveeva, V.; Usanov, A.; Kosivtov, Y.; Demidenko, G.; Bronstein, L.; Chernyshov, D.; Valetsky, P. *J. Mol. Catal. A: Chem.* 1999, *146*, 265.
215. Schmid, G.; Maihack, V.; Lantermann, F.; Peschel, S. *J. Chem. Soc., Dalton Trans.* 1996, 591–594.
216. Lange, C.; D. De Caro; A, G.; Storck, S.; Bradley, J. S.; Maier, W. F. *Langmuir* 1999, *15*, 5333–5338.
217. Drognat-Landre, P.; Lemaire, M.; Richard, D.; Gallezot, P. *J. Mol. Catal.* 1993, *78*, 257.
218. Drognat-Landre, P.; Richard, D.; Draye, M.; Gallezot, P.; Lemaire, M. *J. Catal.* 1994, *147*, 214.
219. Nasar, K.; Fache, F.; Lemaire, M.; Beziat, J. C.; Besson, M.; Gallezot, P. *J. Mol. Catal.* 1994, *87*, 107.
220. Köhler, J.; Bradley, J. S. *Langmuir* 1998, *14*, 2730.
221. Collier, P. J.; Iggo, J. A.; Whyman, R. *J. Mol. Catal. A: Chem.* 1999, *146*, 149.
222. Fache, F.; Lehuede, S.; Lemaire, M. *Tetrahedron Lett.* 1995, *36*, 369.
223. Widegren, J. A.; Finke, R. G. *Inorg. Chem.* 2002, *41*, 1558–1572.
224. Bönnemann, H.; Endruschat, U.; Tesche, B.; Rufinska, A.; Lehmann, C. W.; Wagner, F. E.; Filoti, G.; Pãˆrvulescu, V.; Pãˆrvulescu, V. I. *Eur. J. Inorg. Chem.* 2000, 819–822.
225. Kordesch, K.; Simader, G. *Fuel Cells and their Applications,* VCH: Weinheim, 1996.
226. Petrow, H. G.; Allen, R. J. Prototech Comp.: US, 1977, Vol. 4.
227. Cameron, D. S. *Platinum Met. Rev.* 1999, *43*, 149–154.
228. Luczak, F. J.; Landsman, D. A. United Technologies Corp., 1986.
229. Bönnemann, H.; Brinkmann, R.; Britz, P.; Endruschat, U.; Mörtel, R.; Paulus, U. A.; Feldmeyer, G. J.; Schmidt, T. J.; Gasteiger, H. A.; Behm, R. J. *J. New Mat. Electrochem. Syst.* 2000, *3*, 199–206.
230. Itoh, T.; Sato, J. N. E. Chemcat Corporation, 1999.
231. Narayanan, S.; Surampudi, S.; Halpert, G. California Institute of Technology: U.S., 1999.
232. Haruta, M.; Kobayashi, T.; Sano, H.; Yamada, N. *Chem. Lett.* 1987, 405–406.
233. Haruta, M.; Yamada, N.; Kobayashi, T.; Ijima, S. *J. Catal.* 1989, *115*, 301–309.
234. Haruta, M. *Catal. Today* 1997, *36*, 153–166.
235. Kozlova, P.; Kozlov, I.; Sugiyama, S.; Matsui, Y.; Asakura, K.; Iwasawa, Y. *J. Catal.* 1999, *181*, 37–48.
236. Wagner, F. E.; Galvano, S.; Milone, C.; Visco, A. M. *J. Chem. Soc., Faraday Trans.* 1997, *93*, 3403–3409.
237. Ueda, A.; Oshima, T.; Haruta, M. *Appl. Catal. B* 1997, *12*, 81–93.
238. Andreeva, D.; Tabakova, T.; Idakiev, V.; Chistov, P.; Giovanoli, R. *Appl. Catal. A* 1998, *169*, 9–14.
239. Sakurai, H.; Haruta, M. *Catal. Today* 1996, *29*, 361–365.
240. Maye, M. M.; Lou, Y.; Zhong, C.-J. *Langmuir* 2000, *16*, 7520–7523.
241. Nakaso, K.; Shimada, M.; Okuyama, K.; Deppert, K. *Aerosol Sci.* 2002, *33*, 1061–1074.
242. El-Deab, M. S.; Ohsaka, T. *Electrochem. Commun.* 2002, *4*, 288–292.

243. Claus, P.; Brückner, A.; Mohr, C.; Hofmeister, H. *J. Am. Chem. Soc.* 2000, *122*, 11430–11439.
244. Mohr, C.; Hofmeister, H.; Claus, P. *J. Catal.* 2002, *213*, 86–94.
245. Mohr, C.; Hofmeister, H.; Radnik, J.; Claus, P. *J. Am. Chem. Soc.* 2003, *125*, 1905–1911.
246. Harada, M.; Asakura, K.; Toshima, N. J. *J. Phys. Chem.* 1993, *97*, 5103.
247. Mukherjee, P.; Patra, C. R.; Ghosh, A.; Kumar, R.; Sastry, M. *Chem. Mater.* 2002, *14*, 1678–1684.
248. Porta, F.; Prati, L.; Rossi, M.; Scari, G. *Colloids Surf.* 2002, *211*, 43–48.

Part II

Synthesis of Heterogeneous Catalysts

7 Microwave-Assisted Synthesis of Nanolayer Carbides and Nitrides

Justin Bender, Jennifer Dunn, and Ken Brezinsky

CONTENTS

7.1 INTRODUCTION

We have combined the advantages of microwave processing of metal powders with the excellent gas-solid reactant contact of fluidized beds to prepare carbide and nitride nanolayers on early transition metals. Carbides and nitrides are promising catalysts because they are both technically and economically competitive with traditional noble metal catalysts for applications ranging from producing hydrogen for fuel cells to cleaning hydrocarbon fuels.[1, 2] In this work, nanolayers of these compounds were prepared on transition metal powders by fluidizing the powders either alone in a reactant gas (e.g., nitrogen or ethylene) or together with carbon black in argon. The resulting products were characterized with several microscopy techniques and assessed for catalytic activity. The compounds that were prepared include Cr_2N, MoN, Cr_2C_3, Mo_2C, and WC. Of these products, Mo_2C had the highest catalytic activity in the water-gas shift reaction.

Demonstrations of microwave-assisted metals processing in the literature[3, 4] primarily focus on sintering. In the limited reports of microwave-assisted reactions between metals and gases, the reactant gas typically must filter through a porous compact of metal powders.[5] Microwave-assisted fluidized bed synthesis (MAFBS) offers several advantages to this approach. First, gas-solid reactions in fluidized beds do not experience macroscopic transport limitations. Furthermore, a fluidized bed allows particles access to microwave irradiation in what would otherwise be an optically thick environment. Once heated, these particles will not lose heat to surrounding, cold particles via conduction.[6] Despite these advantages, only this work[7, 8] and that of Whittaker and Mingos[6] have harnessed MAFBS to process metal powders.

7.2 MICROWAVE PROCESSING OF METALS

This section briefly describes microwave-material interactions and provides a general understanding of the phenomena that occur during MAFBS. A number of references offer an in-depth explanation of microwave processing.[3, 9–11]

The response of a material to microwave processing depends on the abundance and surroundings of the charged particles it contains. These particles experience a force acting upon them in an electric field. If they move in response, a current develops and the material is a conductor. If the particles are fixed in place or can only move sparingly, they adjust their position until a counterforce balances the electric field's force. The net result of charged particles' movement in response to an electric field is termed *dielectric polarization*. When polarization cannot exactly follow the rapid reversals of the field, the material absorbs energy, and dielectric heating results.

Several quantities describe the extent of dielectric polarization that will occur in a material. The dielectric constant, ε' is the polarizability of a molecule. Polar molecules like water that have localized charges have greater ε' than nonpolar molecules like benzene. The second quantity, ε'', is the dielectric loss. It reflects how efficiently energy from an electromagnetic field is converted into heat within the material. As ε'' increases, the material will heat more effectively in a microwave.

The temperature rise in a solid or liquid exposed to microwave heating is dependent on the dielectric loss. Equation 7.1 relates ε'' to temperature rise (ΔT), processing time (Δt), and properties of the material and electric field[9]:

$$\frac{\Delta T}{\Delta t} \alpha \frac{\varepsilon'' f E_{r.m.s}^2}{\rho C_P} \qquad [7.1],$$

where $E_{r.m.s.}$ is the average electric field, f is the frequency of the field, ρ is the density of the material, and C_P is its heat capacity.

7.2.1 MICROWAVE-METAL INTERACTIONS

Materials fall into three classes of microwave-material interaction: reflectors (bulk metals), transmitters (quartz), and absorbers (water). The reflectivity of metals presents both drawbacks and benefits to processing them in a microwave. On the one hand, a bulk reaction is hard to achieve because microwave energy is generally absorbed only in a thin shell of a metal. This region is called the *skin depth*, which Equation 5.2 defines in terms of the physical properties of the material and electric field[3],

$$d = \frac{1}{\sqrt{\pi f \varepsilon_0 \sigma}} \qquad [7.2],$$

where σ is the total effective conductivity and ε_0 is the permittivity of free space. On the other hand, if it is desirable to limit a reaction to the metal's surface, microwave processing is perhaps a more appropriate technology than conventional heating, which is more difficult to control with an on/off switch.

7.2.2 BENEFITS OF MICROWAVE TECHNOLOGY

In addition to the unique ability to promote reactions at a metal's surface, microwave technology can offer reduced processing time, costs, and energy consumption[3] over conventional heating methods. For example, microwave sintering of alumina and boron carbide saved 90% of the energy and time, respectively, consumed in conventional sintering processes.[11] Depending on the competing technology, the working environment of a microwave-based process may be safer because exterior parts to the microwave oven do not become excessively hot. Energy and time savings, the increasing interest in microwave processing[12] for both organic synthesis[13] and material processing,[3, 10, 14] and the ongoing development of industrial-scale microwave technology[15, 16] make microwave research both exciting and practical.

7.3 TRANSITION METAL CARBIDES AND NITRIDES: INTRIGUING CATALYSTS

Early transition metal carbides and nitrides are the focus of this research for several reasons. First, they mimic the properties of Pt[1, 2] and have been touted as potential replacements for expensive and increasingly rare[1] noble metal catalysts (Pt, Pd, Ru, Rh) for several industrially significant reactions including hydrogenation,[2, 17] hydrogenolysis,[2] methane activation,[1, 18] and isomerization.[19] Moreover, Patt and coworkers[20] demonstrated that from 240 to 300°C, the CO consumption rate in the WGS reaction, which generates H_2 for PEM fuel cells, was significantly higher for Mo_2C than for commercial Cu-Zn-Al. Figure 7.1 illustrates these promising results. Carbides also exhibit thermal stability in the WGS reaction[21] and in other applications.[2] Another attractive property of carbides and nitrides, their tolerance of sulfur and nitrogen,[2, 22] makes them candidate catalysts for hydroprocessing,[23] which produces cleaner-burning hydrocarbon fuels. Additionally, Bej and Thompson[24] recently demonstrated the activity of carbides and nitrides for acetone condensation, a key characterization reaction for acid and base catalysts.

The utility of carbide and nitride catalysts has prompted numerous studies of their reactivity that use carbide and nitride overlayers as the catalyst rather than bulk carbides or nitrides. This approach permits careful manipulation of the surface metal/nonmetal stoichiometry,[2] which is crucial to probing reactivity. These studies consistently reveal the catalytic activity of carbide and nitride overlayers[25–28] and, in several cases, the similarities between their behavior and that of noble metal catalysts.[29, 30] For example, the same benzene yield and reaction pathway for the dehydrogenation of cyclohexane was observed for both p(4x4)-C/Mo(110) and Pt(111) surfaces.[31] Furthermore, carbon-modified tungsten may be a more desirable catalyst for direct methanol fuel cells than Pt or Ru surfaces because the transition metal carbide exhibits higher activity toward methanol and water dissociation and is more CO-tolerant.[32]

Although carbide and nitride overlayers on transition metals have been used to probe the reactivity of bulk carbides and nitrides, they have not received much attention as catalysts in their own right, despite repeated demonstration of their

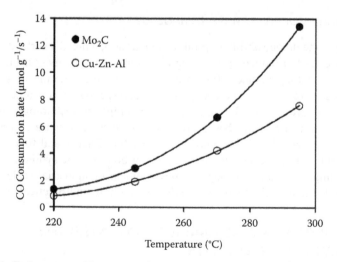

FIGURE 7.1 Carbon monoxide consumption rates for the Mo_2C (61 m^2 g^{-1}) and commercial Cu-Zn-Al (60 m^2 g^{-1}) catalysts.[20]

catalytic activity. Yet successful demonstrations of overlayers modifying the activity and selectivity of a parent transition metal exist. For example, titanium nanolayers on ruthenium augmented the turnover frequency for CO hydrogenation over that of the parent metal.[33] This example is indicative of the increased activity or selectivity that nanoscale catalysts, including Mo_2C,[34, 35] can exhibit when compared to conventional catalysts.[33] Nanoscale carbide and nitride overlayers on transition metals may display these advantages, as well.

Clearly, carbide and nitride overlayers on transition metals deserve attention as promising catalysts for important reactions that impact environmental quality. Current methods for preparing these compounds, however, have significant drawbacks. Laboratory techniques to prepare carbide or nitride overlayers on single transition metal crystals[2] would not be feasible on an industrial scale. On the other hand, controlling high-temperature industrial processes for carburization or nitridation to limit the reaction to the metal's surface is difficult. Microwave processing of transition metals to prepare carbide and nitride overlayers, however, may be a viable technology.

7.4 EXPERIMENTAL

In this section, we present a description of the MAFBS apparatus, an account of challenges encountered during its development, a brief description of the materials characterization techniques used in this research, and an explanation of the flow reactor used to test the catalytic activity of the prepared carbides and nitrides.

7.4.1 MATERIALS

Tungsten, chromium, molybdenum, nickel, and carbon black powders from Atlantic Equipment Engineers were between 1 and 5 μ in diameter. Bulk Mo_2C from Sigma-Aldrich had a 44 μ nominal diameter. Cu-Zn-Al was purchased from Sud-Chemie.

7.4.2 MAFBS APPARATUS

Figure 7.2 is a schematic of the MAFBS apparatus. The microwave, a 1.6 kW Panasonic NE-1757R, contains a quartz tube held in an aluminum base. A steel sieve in the base serves as a distributor for the fluidizing gas. From the base, a copper tube extends through a hole in the bottom of the microwave. It connects to a solenoid that receives a sine wave input signal and vibrates the base vertically to improve fluidization. A gas (e.g., ethylene, nitrogen) flows through the copper tube into the base and fluidizes the metal powders inside the quartz tube, which extends out the top of the microwave. A metal sieve caps the tube to prevent metal powders from escaping and to permit the fluidizing gas to vent into a fume hood.

The experimental procedure begins with loading a measured amount of metal powders into the quartz tube. The aluminum cap is then secured to the top of the tube. Next, the fluidizing gas is allowed into the tube at a flowrate that achieves satisfactory fluidization of the powders. Then, the microwave is operated either

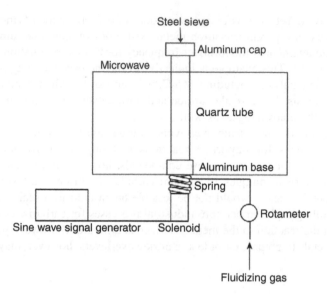

FIGURE 7.2 MAFBS apparatus.

continuously or in pulsed intervals. In some cases, pulsed operation replaces continuous operation when a reaction begins.

7.4.3 APPARATUS DEVELOPMENT

Designing experimental equipment to withstand chemical reactions in a microwave is challenging.[36] Three challenging aspects of developing the MAFBS apparatus were selecting a durable base material, effectively trapping the fluidized powders in the quartz tube, and preventing fires inside the microwave. To begin, three base materials were tested. A Teflon base degraded under reaction conditions, and a ceramic base cracked after several uses. An aluminum base, however, proved durable under reaction conditions. When using a microwave-reflecting material such as aluminum in a microwave, an adequate amount of microwave-absorbing material must also be present. Otherwise, the interior temperature of the oven can exceed the shut-off limit. In MAFBS experiments, if the height of the fluidized bed was sufficient, the metal powders absorbed enough microwave energy to avoid microwave overheating.

Retaining metal powders in the quartz tube presented another challenge. At first, a ceramic adhesive secured a metal sieve to the top of the tube. The adhesive cured in an oven for 2 h at 250°C to solidify. The curing process, however, often caused the quartz tube to crack. Additionally, the adhesive frequently burned during MAFBS experiments. These disadvantages prompted us to use an aluminum cap, rather than the ceramic adhesive, to secure the sieve. Using this cap eliminates tube cracks, the time-intensive curing step, and fires originating from the adhesive.

In addition to the adhesive, ethylene gas also burned in several experiments and formed a diffusion flame at the top of the quartz tube. In an earlier apparatus design, the tube terminated inside the microwave, and one diffusion flame caused irreparable damage to the microwave oven. Using a longer tube that extends outside the microwave minimizes damage in the event of an ethylene diffusion flame.

7.4.4 Materials Characterization and Catalytic Activity Testing

The microscopy techniques used to characterize the nanolayers that form on the metal powders include energy dispersive x-ray spectrometry (EDS), x-ray diffraction (XRD), and transmission electron microscopy (TEM). The EDS system was an Oxford Inca (Oxon, U.K.); the X-ray diffractometer was a Siemens D5000 (Karlsruhe, Germany). We analyzed XRD data with MDI Jade (version 5.1) software. The TEM is a JEOL 3010 (Tokyo). To prepare powder samples for TEM analysis, the samples were first crushed with a mortar and pestle, then mixed with ethanol. This slurry was applied to a holey-carbon grid with a micropipette and placed in the microscope's chamber. We examined TEM images for uniformity, composition, and depth of nanolayers. The interplanar or d-spacing of the nanolayer was used to determine its chemical identity. This parameter was calculated by counting the number of planes in a length of sample in a TEM micrograph and through XRD analysis. In some cases, the nanolayer was a small fraction of the total sample mass, and its d-spacing did not contribute to those detected in XRD analysis.

The WGS reaction (Equation 5.3) was used to evaluate the catalytic activity of the MAFBS-prepared carbides and nitrides.

$$CO + H_2O \leftrightarrow CO_2 + H_2 \qquad [7.3]$$

This reaction is a sensible test because CO is a common probe of the reactivity of carbides and nitrides.[37] In addition, the WGS reaction is of practical interest because it is a means to produce H_2 fuel for PEM fuel cells. As described earlier, Mo_2C is competitive with commercial WGS reaction catalysts.

Catalytic activity tests are conducted in a flow reactor at 275°C. In this apparatus, an electric heater contains a glass tube in which glass wool supports the powder catalyst. The feed to the reactor is 10% CO and 90% H_2. It passes through a bubbler to acquire water, the coreactant. The composition of the reactor effluent is determined with a gas chromatograph (GC). The CO conversion, an important measure of the catalyst's activity, is then calculated. The error associated with the CO conversion values reported herein, which was determined by calculating the standard deviation of the conversions from four replicate samples,[38] is approximately 0.1%.

7.5 RESULTS AND DISCUSSION

Both early transition metal nitrides and carbides were prepared with MAFBS. Nitrides were prepared exclusively from reactions between metal powders and fluidizing N_2 gas. Carbides were prepared either through fluidization of metal and carbon black powders with Ar gas or from reaction between metal powders and fluidizing ethylene gas. In this section, we present a description of the structural properties, catalytic activity, and visual reaction effects for the transition metal carbides and nitrides that we have prepared thus far. All standard d-spacing values that we compare to measured values are from the Powder Diffraction File of the Joint Committee of Powder Diffraction Standards (JCPDS). In catalytic activity tests, conditions were selected that would restrict CO conversions to below 7% to facilitate kinetic analysis.[38]

7.5.1 NITRIDES

7.5.1.1 Chromium Nitride

In these experiments, 5 g of between 1 and 5 μ chromium powder were fluidized in the quartz tube with N_2. During microwave irradiation of this mixture, the energy required for a reaction to begin was attained after roughly 10 s of irradiation. During the reaction, blue and white streaks were observed and arcs formed between the metal sieves below and atop the quartz tube. After a processing time of 1 min, the powder at the base of the tube glowed red.

Two samples, one prepared under continuous microwave irradiation for 5 min and one prepared during five 6-s pulses, were analyzed by XRD, EDS, and TEM. The precursor chromium powder was also analyzed by XRD (Figure 7.3) to provide a benchmark. The XRD pattern of the sample prepared under pulsed irradiation, presented in Figure 7.4, contained Cr_2N peaks in addition to the Cr peaks that are present in Figure 7.3. The processed material was also analyzed by EDS. Although the pattern from analysis at 30 kV in Figure 7.5 did not detect nitridation, the EDS pattern at 5 kV revealed that a nitride compound was present on the surface.

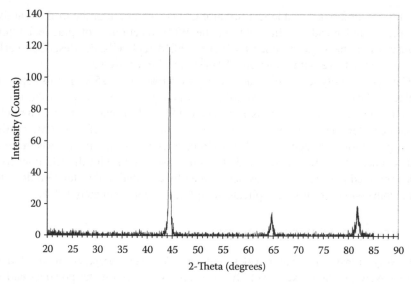

FIGURE 7.3 XRD pattern of precursor chromium powder.

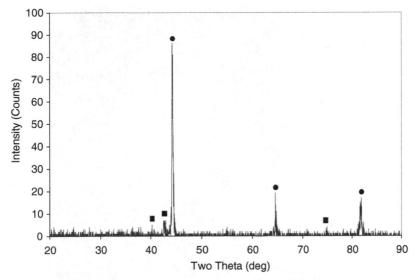

FIGURE 7.4 XRD pattern of chromium processed in nitrogen fluidizing gas for six 5 s pulses;
●: Cr, ■: Cr_2N.

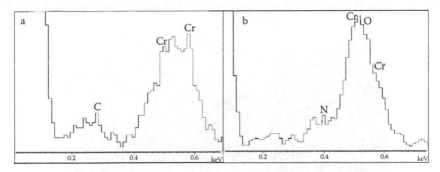

FIGURE 7.5 EDS patterns at (a) 5 kV or (b) 30 kV of chromium processed in nitrogen fluidizing gas for six 5 s pulses.

The TEM micrograph in Figure 7.6 reveals the presence of a nitride nanolayer. The d-spacing of this nanolayer, 1.64 Å, matches a JCPDS value for Cr_2N.

The XRD pattern of the sample prepared under continuous irradiation did not contain any nitride peaks. Two alternate explanations could account for this result: Either no nitridation occurred or the amount of chromium nitride that formed was not detectable with XRD. Further analysis by TEM was necessary to determine which explanation applied. The TEM micrograph of this sample (Figure 7.7) confirmed that a nanolayer had formed on the powder. The d-spacing of the material constituting the 3- to 5-nm-deep nanolayer was 2.24 Å, which corresponds with a standard value for Cr_2N. The d-spacing of the darker material just below the nanolayer was 2.04 Å, which agrees with the literature value for Cr. The depth and coverage of the nanolayer in Figure 7.6 appears more extensive than in Figure 7.7.

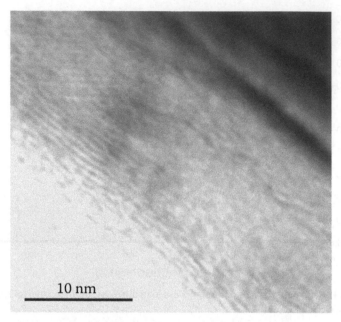

FIGURE 7.6 TEM micrograph of chromium processed in nitrogen fluidizing gas for six 5 s pulses.

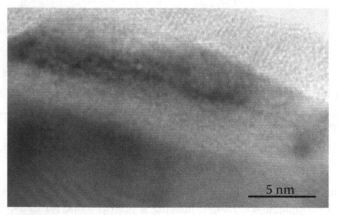

FIGURE 7.7 TEM micrograph of chromium processed in nitrogen fluidizing gas for 5 min.

This result may indicate that nitridation was more extensive during the pulsed irradiation experiments than under continuous irradiation, which is somewhat surprising.

Figure 7.8 and Figure 7.9 display the results of catalyst activity tests of both Cr_2N and MoN at 275°C for two ratios of catalyst weight (W) to initial CO flowrate (F_{CO_o}). In Figure 7.8, the CO conversion for Cr_2N and Cr were roughly equal, indicating that Cr_2N is not significantly more catalytically active than the parent metal.

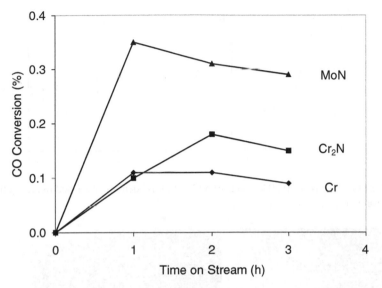

FIGURE 7.8 Variation in CO conversion with time on stream for MAFBS-prepared MoN and Cr_2N and unprocessed Cr at $\left(\frac{W}{F_{CO_o}}\right) = 13.5.$[38]

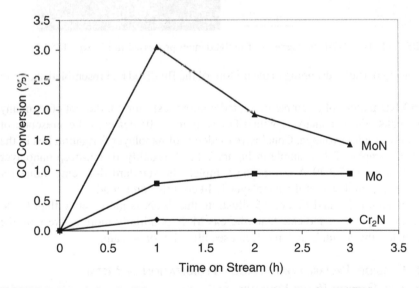

FIGURE 7.9 Variation in CO conversion with time on stream for MAFBS-prepared MoN and Cr_2N and unprocessed Mo at $\left(\frac{W}{F_{CO_o}}\right) = 4.2.$[38]

7.5.1.2 Molybdenum Nitride

The experimental procedure to prepare molybdenum nitride was similar to that for preparing chromium nitride. During microwave processing of the fluidized molybdenum powders, yellow and orange streaks appeared after an induction time of about

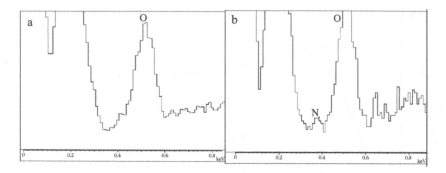

FIGURE 7.10 EDS patterns at (a) 30 kV and (b) 5 kV of molybdenum processed in nitrogen fluidizing gas for 1 h.

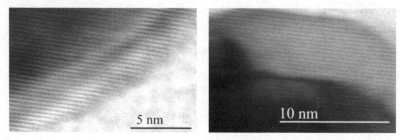

FIGURE 7.11 Two TEM micrographs of molybdenum processed in nitrogen for 1 h.

8 s. Glowing particles dropping to the bottom of the fluidized bed resembled shooting stars.

An XRD pattern of a sample irradiated continuously for 1 h did not contain any nitride peaks. Further analysis with EDS (Figure 7.10) detected the presence of nitride only at a low voltage. Conclusive evidence of nanolayer formation lies in the TEM micrographs of this sample in Figure 7.11. A roughly 10-nm-deep nanolayer with a d-spacing of 2.79 Å covers the sample. The standard d-spacing value for MoN, 2.80 Å, implies that the nanolayer is likely this compound.

Both Figure 7.8 and Figure 7.9 illustrate that MoN is more active than Cr_2N. The data in Figure 7.9 indicate that whereas MoN is initially more active than the parent metal, the nitride's activity decreases after 1 h on stream.

7.5.2 Carbides Prepared from Argon Fluidization of Metal and Carbon Black Powders

In initial experiments to prepare early transition metal carbides with MAFBS, argon gas fluidized mixtures of metal and carbon black powders. The extent of reaction in these experiments was limited by the number of high-temperature collisions between the two types of powders. Nonetheless, the results of these experiments are instructive in the development of this preparation technique. The catalytic activity of carbides prepared with this method was not assessed.

7.5.2.1 Chromium Carbide

Preliminary experiments with unfluidized mixed carbon and chromium powders irradiated for 10 min resulted in carburization of the chromium, as evidenced by the XRD pattern (Figure 7.12) of a sample from these tests. Unreacted chromium was also detected, which indicated that the powders were not entirely converted to chromium carbide.

In experiments to prepare chromium carbide via MAFBS, a total of 5 g of between 1 and 5 μ carbon black and chromium powders mixed in a chromium:carbide molar ratio of 3:2 were fluidized with argon. Within 15–40 seconds, dramatic sparks and yellow streaks formed within the tube. Blue arcs that likely resulted from argon degradation also formed. EDS analyses of the products from these experiments were not performed because the free, unreacted carbon could not be distinguished from carbon incorporated into the metal powder.

An XRD pattern for an MAFBS-prepared sample that was irradiated for fifteen 10 s pulses did not contain any carbide peaks. The TEM micrograph of this sample, in Figure 7.13, revealed that a nanolayer had formed. The d-spacing of this nanolayer was 2.76 Å, which matches a standard value for Cr_2C_3. The depth and coverage of this nanolayer were not uniform. In fact, TEM analysis of several samples from this experiment revealed that some samples did not experience carburization, whereas amorphous carbon was deposited on the surface of others.

To assess whether further heat treatment was a potential route to both sublime unreacted free carbon and to convert amorphous carbon on the surface of metal powders to chromium carbide, the MAFBS products were baked in an oven. The oven temperature increased to 800°C in 50°C intervals every 20 min. The final

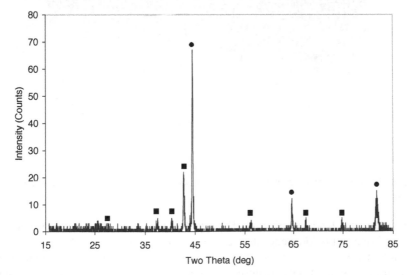

FIGURE 7.12 XRD pattern of contacting chromium and carbide powders processed for 10 min; ■: Cr_2C, ●: Cr.

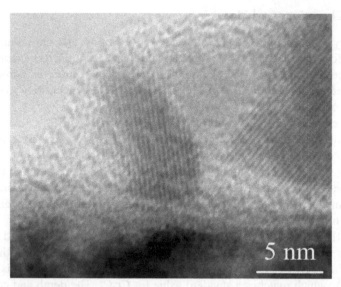

FIGURE 7.13 TEM micrograph of chromium processed with carbon black powder in argon for 15 pulses that were 10 s long.

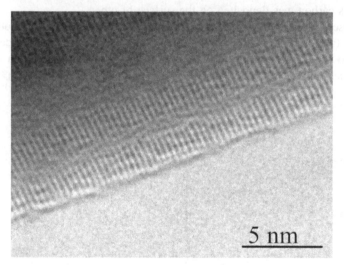

FIGURE 7.14 TEM micrograph of chromium processed in argon with carbon black powder for 1 h.

temperature was held for 30 min. During this treatment, the powders yellowed, indicating that oxidation rather than carburization occurred.

Another sample that was irradiated continuously for 1 h exhibited carbide and chromium peaks in its XRD pattern. In Figure 7.14, the d-spacing over the sample area depicted in the TEM micrograph of this sample is constant and consistent with chromium carbide formation. No unreacted chromium is observed. These results may indicate that bulk rather than surface reaction occurred.

7.5.2.2 Molybdenum Carbide

In MAFBS experiments to produce molybdenum carbide, molybdenum and carbon black powders were mixed in a 2:1 molar ratio. Five grams of this mixture were added to the quartz tube and fluidized with argon. Visible heating effects in these experiments were similar to those in experiments to prepare chromium carbide.

Again, samples were prepared under either pulsed or continuous radiation, and several samples were selected for characterization. The first was processed for 10 pulses that were 10 s long. Carbide peaks were absent from the XRD pattern of this sample. A TEM micrograph of the sample, shown in Figure 7.15, revealed a layer of amorphous carbon covering the metal powder. It is likely that the temperature reached by the metal powders was too low to prompt a carburization reaction between the carbon and the metal powder.

Another sample that was processed continuously for 1 h also did not exhibit a carbide peak in its XRD pattern. TEM analysis of a sample from this experiment (Figure 7.16), however, revealed a nanolayer roughly 2 nm deep with a d-spacing that matches a literature value for Mo_2C (3.05 Å).

The samples prepared by pulsed and continuous radiation were heat-treated with the method described in the previous section. Although oxidation of the powders occurred, carburization did, as well, because carbide peaks appeared in the XRD patterns for each type of powder after heat treatment. To maximize the competitiveness of MAFBS with traditional technologies, however, sufficient carburization should occur during MAFBS so that no secondary heating step is necessary.

FIGURE 7.15 TEM micrograph of molybdenum processed in argon with carbon black powder for ten pulses that were 10 s long.

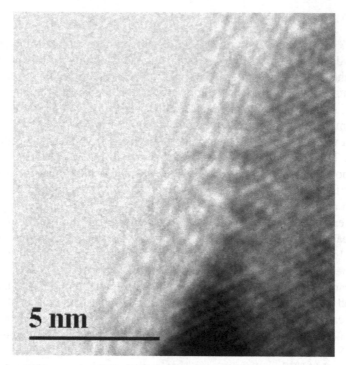

FIGURE 7.16 TEM micrograph of molybdenum processed in argon with carbon black powder for 1 h.

7.5.2.3 Nickel Molybdenum Carbide

In these experiments, nickel, molybdenum, and carbon black powders were mixed in a 1:1:1 molar ratio. Five gramsof this mixture were placed in the quartz tube and fluidized with argon. After 1 h of continuous processing, two distinct types of powders resided in the tube. The first type was black and settled at the base. The second type was light gray and adhered to the mesh at the top of the tube.

Figure 7.17 contains the XRD pattern for the precursor molybdenum powder. Although the XRD pattern for the black powder in Figure 7.18 contained peaks for unreacted molybdenum that correspond to those in Figure 7.17, no unreacted nickel was detected. Molybdenum and nickel carbide peaks were both present. The XRD pattern of the light gray powder is shown in Figure 7.19. Although nickel and molybdenum carbides are again present, less unreacted molybdenum was present than in the black powder. The cause of reaction of molybdenum with the carbide powder in the presence of nickel is unclear. Recall that in the absence of nickel, molybdenum processed for 1 h reacted only minimally at the surface.

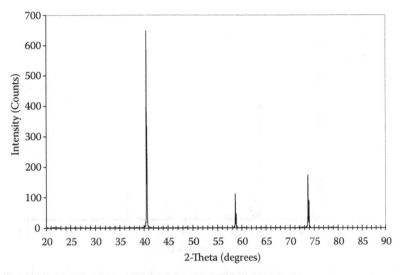

FIGURE 7.17 XRD pattern of precursor molybdenum powder.

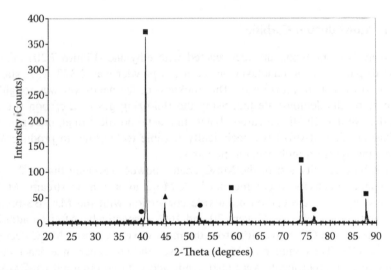

FIGURE 7.18 XRD pattern of black product from hour-long reaction among nickel, molybdenum, and carbon black powder in argon; ●: Mo_2C, ■: Mo, ▲: Ni_3C, ◆: Ni_3Mo.

7.5.3 CARBIDES PREPARED FROM ETHYLENE FLUIDIZATION OF METAL POWDERS

To overcome transport limitations associated with using a solid carbon source, ethylene was adopted as both the carbon source and fluidizing gas in several experiments. Metal powders in these experiments were either tungsten or molybdenum.

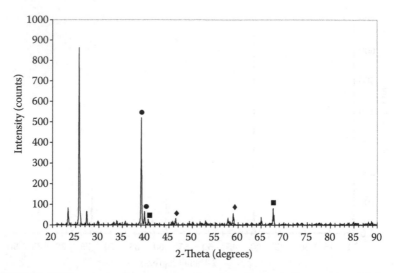

FIGURE 7.19 XRD pattern of gray product from hour-long reaction among nickel, molybdenum, and carbon black powder in argon; •: Mo_2C, : Mo, ♦: Ni_3Mo.

7.5.3.1 Molybdenum Carbide

TEM analysis of molybdenum that reacted with ethylene (Figure 7.20) indicated that the d-spacing of the nanolayer on the metal powder was 2.378 Å. A standard value for Mo_2C matches this result. The thickness of the nanolayer was roughly 20 nm. These results demonstrate that using the fluidizing gas as a carbon source is more effective than fluidizing carbon black and metal powders in an inert gas. They also illustrate that MAFBS is a technically feasible technology to produce Mo_2C carbide nanolayers on molybdenum powders.

A catalytic activity test of the Mo_2C-coated powder revealed that at 275°C, it had comparable activity to commercial bulk Mo_2C after 1 h on stream. At these conditions $\left(W/F_{CO_0} = 4.6 \right)$, the carbon monoxide conversion with the MAFBS-prepared product and commercial Mo_2C were 20 and 16%, respectively. After an additional hour on stream, the CO conversion from the microwave-processed Mo_2C decreased from 20 to 11%. The commercial Mo_2C, however, did not experience deactivation.

Flow reactor experiments were also conducted with commercial Cu-Zn-Al, an effective WGS catalyst,[20] and commercial bulk Mo_2C. The nominal reaction rates from the Mo_2C and Cu-Zn-Al catalysts were 4.2 μmol/[g min] and 240 μmol/[g min], respectively. The surface area of the Mo_2C (0.18 m²/g), however, was significantly less than that of the Cu-Zn-Al (11 m²/g). The rates on the basis of surface area for Mo_2C and Cu-Zn-Al were 2.3×10^{-3} mol^{-6}/[cm² min] and 2.2×10^{-3} mol^{-6}/[cm² min], respectively. This analysis indicates that if the surface area of the Mo_2C were increased, bulk Mo_2C may be competitive with Cu-Zn-Al. Although sufficient reaction rate data has not been gathered for MAFBS-prepared Mo_2C, this analysis and the report of Patt et al.[20] convey that it holds promise as a catalyst for the WGS reaction.

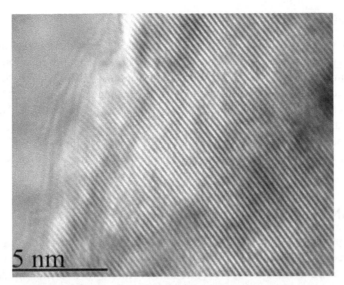

FIGURE 7.20 TEM micrograph of molybdenum processed in ethylene.

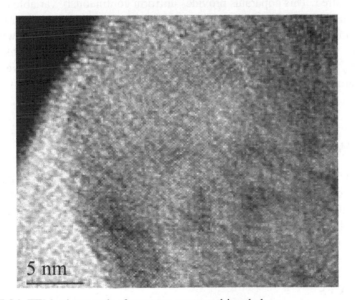

FIGURE 7.21 TEM micrograph of tungsten processed in ethylene.

7.5.3.2 Tungsten Carbide

Figure 7.21 is a sample TEM image from an experiment in which tungsten was fluidized in ethylene. The d-spacing of the nanolayer on W matched a standard value for WC, 2.52 Å. The depth of the nanolayer is approximately 15–20 nm. The sample showed very little catalytic activity. It is possible that a W_2C nanolayer may be more reactive or that a less thick WC layer would exhibit greater activity.

7.6 CONCLUSIONS

This work clearly demonstrates that it is possible to produce carbide and nitride nanolayers on early transition metals with MAFBS. The extent of reaction, however, varied from metal to metal. Cr_2N was prepared more readily than MoN, but Cr_2N was less catalytically active. Using ethylene rather than carbon black as the carbon source was a more effective technique to prepare carbides. Mo_2C and WC nanolayers were prepared with this method. Although none of the MAFBS-prepared catalysts were competitive with the commercial Cu-Zn-Al catalyst for the WGS reaction, Mo_2C had the most promising catalytic properties of the compounds that were prepared.

7.7 PLANS FOR FUTURE RESEARCH

Incorporating two new pieces of equipment will move this research program towards achieving controlled synthesis of catalytically active carbide and nitride nanolayers on early transition metals. Whereas the current microwave is multimode, like all domestic microwave ovens, a single axial-mode cylindrical cavity microwave oven will be acquired. This apparatus provides uniform continuously variable microwave heating that will eliminate the hot and cold spots that likely exist in the current oven.

Additionally, in future experiments, a noncontact optical pyrometer will be used to measure the temperature of the reacting particles. Knowledge of the temperature attained by the particles will promote an understanding of how different experimental parameters (e.g., fluidizing gas flowrate, processing time) affect the extent of reaction.

ACKNOWLEDGMENTS

The authors acknowledge the significant contributions to this research by participants in the National Science Foundation Research Experience for Undergraduates program at UIC from 2002–2004: Juan Gonzalez, Laura Sauber, Anthony New, Naja Joseph, and Jason Burbey. Mark Liska, a former graduate student of Professor John Regalbuto, conducted the catalytic activity tests.

REFERENCES

1. A. York. Magic Catalysts. *Chemistry in Britain. 55*, 25–27, 1999.
2. J. G. Chen. Carbide and Nitride Overlayers on Early Transition Metal Surfaces: Preparation, Characterization, and Reactivities. *Chemical Reviews. 96*, 1477–1498, 1996.
3. K. J. Rao; B. Vaidhyanathan; M. Ganguli; P. A. Ramakrishnan. Synthesis of Inorganic Solids Using Microwaves. *Chemistry of Materials. 11*, 882–895, 1999.
4. R. Roy; D. Agrawal; J. Cheng; S. Gedevanishvili. Full Sintering of Powdered-Metal Bodies in a Microwave Field. *Nature. 399*, 668–670, 1999.

5. H. W. Dandekar; J. A. Puszynski; V. Hlavacek. Two-Dimensional Numerical Study of Cross-Flow Filtration Combustion. *AIChE Journal. 36*, 1649–1660, 1990.

6. A. G. Whittaker; M. P. Mingos. Microwave-Assisted Solid-State Reactions Involving Metal Powders and Gases. *Journal of the Chemical Society Dalton Transactions. 16*, 2541–2543, 1993.

7. A. Jain; K. Brezinsky. Microwave-Assisted Combustion Synthesis of Chromium Nitride in a Fluidized Bed. *Proceedings of the Combustion Institute. 29*, 1109–1113, 2003.

8. A. Jain; K. Brezinsky. Microwave-Assisted Combustion Synthesis of Tantalum Nitride in a Fluidized Bed. *Journal of the American Ceramic Society. 86*, 222–226, 2003.

9. D. M. P. Mingos; D. R. Baghurst. Applications of Microwave Dielectric Heating Effects to Synthetic Problems in Chemistry. In *Microwave-Enhanced Chemistry*, Kingston, H. M., Haswell, S. J. (Eds.), American Chemical Society: Washington, DC, 1997.

10. D. E. Clark; D. C. Folz; J. K. West. Processing Materials with Microwave Energy. *Materials Science and Engineering. A287*, 153–158, 2000.

11. National Research Council. *Microwave Processing of Materials*, National Academy Press: Washington, DC, 1994.

12. D. Adam. Out of the Kitchen. *Nature. 421*, 571–572, 2003.

13. M. Nüchter; B. Ondruschka; W. Bonrath; A. Gum. Microwave-Assisted Synthesis— A Critical Technology Overview. *Green Chemistry. 6*, 128–141, 2004.

14. K. C. Patil; S. T. Aruna; T. Mimani. Combustion Synthesis: An Update. *Current Opinion in Solid State & Materials Science. 6*, 507–512, 2002.

15. R. Bierbaum; M. Nüchter; B. Ondruschka. Microwave-Assisted Reaction Engineering: Miniplant-Scale Microwave Equipment with On-Line Analysis. *Chemie Ingenieur Technik, 76*, 961–965, 2004.

16. N. S. Wilson; C. R. Sarko; G. P. Roth. Development and Applications of a Practical Continuous Flow Microwave Cell. *Organic Process Research & Development. 8*, 535–538, 2004.

17. Y. Z. Li; Y. N. Fan; J. He; B. L. Xu; H. P. Yang; J. W. Miao; Y. Chen. Selective Liquid Hydrogenation of Long Chain Linear Alkadienes on Molybdenum Nitride and Carbide Modified by Oxygen. *Chemical Engineering Journal. 99*, 213–218, 2004.

18. M. V. Iyer; L. P. Norcio; A. Punnoose; E. L. Kugler; M. S. Seehra; D. B. Dadyburjor. Catalysis for Synthesis Gas Formation from Reforming of Methane. *Topics in Catalysis. 29*, 197–200, 2004.

19. M. J. Ledoux; C. Pham-Huu; R. Chianelli. Catalysis with Carbides. *Current Opinion in Solid State & Materials Science. 1*, 96–100, 1996.

20. J. Patt; D. J. Moon; C. Phillips; L. Thompson. Molybdenum Carbide Catalysts for Water-Gas Shift. *Catalysis Letters. 65*, 193–195, 2000.

21. D. J. Moon; J. W. Ryu. Molybdenum Carbide Water-Gas Shift Catalyst for Fuel Cell-Powered Vehicles Applications. *Catalysis Letters. 92*, 17–24, 2004.

22. P. Da Costa; C. Potvin; J. Manoli; B. Genin; G. Djega-Mariadassou. Deep Hydrodesulphurization and Hydrogenation of Diesel Fuels on Alumina-Supported and Bulk Molybdenum Carbide Catalysts. *Fuel. 83*, 1717–1726, 2004.

23. E. Furimsky. Metal Carbides and Nitrides as Potential Catalysts for Hydroprocessing. *Applied Catalysis A: General. 240*, 1–28, 2003.

24. S. K. Bej; L. T. Thompson. Acetone Condensation Over Molybdenum Nitride and Carbide Catalysts. *Applied Catalysis A: General. 264*, 141–150, 2004.

25. L. Bugyi; A. Oszko; F. Solymosi. The Interaction of 1-Butyl Iodide with the Mo₂C/Mo(1 0 0) Surface. *Surface Science. 561*, 57–68, 2004.

26. J. Eng Jr.; J. G. Chen. Reaction Pathways of cis- and trans-2-butene on Mo(110) and C/Mo(110): Selective Activation of Alpha and Beta C-H Bonds. *Surface Science. 414*, 374–388, 1998.

27. H. H. Hwu; B. Frühberger; J. G. G. Chen. Different Modification Effects of Carbidic and Graphitic Carbon on Ni Surfaces. *Journal of Catalysis. 221*, 170–177, 2004.

28. M. H. Zhang; H. H. Hwu; M. T. Buelow; J. G. Chen; T. H. Ballinger; P. J. Anderson; D. R. Mullins. Decomposition Pathways of NO on Carbide and Oxycarbide-Modified W(111) Surfaces. *Surface Science. 522*, 112–124, 2003.

29. H. H. Hwu; J. G. Chen. Chemical Properties of Carbon-Modified Titanium: Reaction Pathways of Cyclohexene and Ethylene over Ti(0 0 0 1) and C/Ti (0 0 0 1). *Surface Science. 557*, 144–158, 2004.

30. B. Frühberger; J. G. Chen. Reaction of Ethylene with Clean and Carbide-Modified Mo(110): Converting Surface Reactivities of Molybdenum to Pt-Group Metals. *Journal of the American Ceramic Society. 118*, 11599–11609, 1996.

31. J. G. Chen; B. Frühberger; J. Eng Jr.; B. E. Bent. Controlling Surface Reactivities of Transition Metals by Carbide Formation. *Journal of Molecular Catalysis A: Chemical. 131*, 285–299, 1998.

32. H. H. Hwu; J. G. Chen. Potential Application of Tungsten Carbides as Electrocatalysts. *Journal of Vacuum Science and Technology A. 21*, 1488–1493, 2003.

33. A. T. Bell. The Impact of Nanoscience on Heterogeneous Catalysis. *Science. 299*, 1688–1691, 2003.

34. T. Hyeon; M. Fang; K. S. Suslick. Nanostructured Molybdenum Carbide: Sonochemical Synthesis and Catalytic Properties. *Journal of the American Ceramic Society. 118*, 5492–5493, 1996.

35. G. Dantsin; K. S. Suslick. Sonochemical preparation of a nanostructured bifunctional catalyst. *Journal of the American Ceramic Society. 122*, 5214–5215, 2000.

36. H. M. Kingston; P. J. Walter; W. G. Engelhart; P. J. Parsons. Laboratory Microwave Safety. In *Microwave-Enhanced Chemistry,* Kingston, H. M., Haswell, S. J. (Eds.), American Chemical Society: Washington, DC, 1997.

37. T. P. St. Clair; S. T. Oyama; D. F. Cox. CO and O₂ Adsorption on alpha-Mo₂C (0001). *Surface Science. 468*, 2000.

38. M. Liska. Alternative Catalysts and Synthesis Methods for Fuel Cell Reformers. M.S. thesis, University of Illinois at Chicago, 2003.

8 Sol-Gel Synthesis of Supported Metals

Benoît Heinrichs, Stéphanie Lambert, Nathalie Job, and Jean-Paul Pirard

CONTENTS

8.1 INTRODUCTION

The preparation of a supported metal catalyst with adequate physico-chemical characteristics leading to the properties required by a practical application is always a real challenge. Indeed, many characteristics relative to active metal particles, including high dispersion, adequate composition when formed of an alloy, low mobility, ... and to the support, including high and well distributed porosity, adequate acidity or basicity, thermal stability, leading to desired properties such as high activity, high selectivity, chemical, thermal and mechanical stability, rapid mass and heat transfers, ... have to be met in the same material. Moreover, as mentioned by LePage [1], one must avoid creating a structure that is only a laboratory curiosity which for technical or economic reasons cannot be manufactured on industrial scale.

In the seventies, a one-step sol-gel method to synthesize metal containing inorganic gels has been developed that consists in dissolving an adequate precursor of the metal in an alcoholic solution containing precursors of the support [2–4]. In order to maintain the high porosity of the wet gel in the final dry material, those authors used the supercritical drying imagined by Kistler as a way to remove the solvent from a gel without causing the collapse of the latter by avoiding the destructive effects of capillary pressure [5–7]. However, thanks to the use of an alcohol as solvent, the pressure and temperature needed to reach supercritical conditions were much lower than in the work of Kistler who performed such a drying with aqueous gels (critical constants: water: $T_c = 647$ K, $P_c = 22.1$ MPa; methanol: $T_c = 513$ K, $P_c = 8.1$ MPa; ethanol: $T_c = 514$ K, $P_c = 6.1$ MPa; 1-propanol: $T_c = 537$ K, $P_c = 5.2$ MPa; n-butanol: $T_c = 563$ K, $P_c = 4.4$ MPa [8]). By that method, aerogel catalysts like Ni/Al_2O_3, $Ni/Al_2O_3-SiO_2$, Cu/Al_2O_3 were prepared. Since then, the one-step sol-gel method in alcoholic medium has been used to prepare numerous supported metal catalysts including Pd/Al_2O_3 [9], Ru/SiO_2 [10–13], Pt/SiO_2 [14], Pd/SiO_2 [15, 16], Pt/Al_2O_3 [17], etc. Often, gels prepared in those studies are xerogels dried by a simple evaporative drying. Aerogel and xerogel catalysts prepared in one single step in alcohol are reviewed in several papers [18–21].

In the eighties and nineties, the use of functionalized ligands allowing to anchor metal complexes to a silica gel in the making has been proposed to prepare metal containing silica xerogels by a one-step processing [22–31]. In addition to obtaining uniform nanometer-sized metal particles homogeneously distributed in a silica matrix, several studies showed that the use of such functionalized ligands brings both significant improvements in the morphology of the final material and a considerable simplification of their manufacture since the advantageous aerogel-like morphology can be obtained even after an easy evaporative drying thus allowing to avoid technical difficulties inherent in supercritical drying [32–35]. The use of mono- or bimetallic catalysts prepared in such a way and named cogelled catalysts, in some first reactions of industrial interest, has demonstrated high activities and selectivities, excellent mass transfer, reasonable stability and easy regenerability in comparison with more conventional catalysts [33, 34, 36–41].

Beside metal catalysts supported on inorganic oxides, a new class of sol-gel materials consisting of one-step-made metal containing carbon aerogels or xerogels have been developed for about five years [42–50]. That development has been based on the knowledge acquired since the early nineties on the preparation of carbon aerogels and xerogels from the drying and pyrolysis of organic gels obtained from the polycondensation of resorcinol and formaldehyde in aqueous medium [51–66]. Although they are only in the early stages of their development and although characteristics as fundamental as metal dispersion and accessibility must be improved, metal containing carbon xerogels prepared in one single step and dried by a simple evaporative drying are promising materials as catalysts because, in comparison with metal catalysts supported on active carbons produced from natural sources and that are mainly microporous, their texture and morphology can be widely tailored according to the specific needs of the application considered.

Beyond presenting a review on the sol-gel preparation, mainly by one-step methods, and on the characteristics of metallic catalysts supported on inorganic oxide or carbon gels, the aim of the present chapter is to show that, since the initial works more than thirty years ago, significant discoveries and progresses have been made that allow to consider those materials no more as laboratory curiosities but as promising catalysts for which it is worth to try to find solutions to overcome the remaining technical obstacles in order to envisage their use in large scale processes.

8.2 PREPARATION

8.2.1 FUNDAMENTALS OF SOL-GEL PROCESS

A *sol* is a colloidal suspension of solid particles in a liquid. A *colloid* is a suspension in which the dispersed phase is so small (~1 nm-1 μm) that gravitational forces are negligible and interactions are dominated by short-range effects, such as van der Waals attraction and electrostatic forces resulting from surface charges. A *gel* can be interpreted to consist of continuous solid and liquid phases of colloidal dimensions. Continuity means that one could travel through the solid phase from one side of the sample to the other without having to enter the liquid; conversely, one could make the same trip entirely within the liquid phase. Since both phases are of colloidal dimension, a line segment originating in a pore and running perpendicularly into the nearest solid surface must re-emerge in another pore less than 1 μm away. Similarly, a segment originating within the solid phase and passing perpendicularly through the pore wall must re-enter the solid phase within a distance of 1 μm [67].

Hence, in principle, *sol-gel* designates a process in which a gel is formed from the particles of a sol when attractive forces cause them to stick together in such a way as to form a network. In other words, sol-gel process should imply the formation of a gel by aggregation of particles in a sol. However, the term *sol-gel* is often used in literature to designate a process that leads to a gel (or sometimes merely to a slurry) from a homogeneous solution of soluble monomer precursors whatever the underlying physico-chemical mechanisms. Indeed, mechanisms other than aggregation have been suggested in various cases. For example [67], if a monomer can make more than two bonds, then there is no limit on the size of the molecule that can fill the space. If one molecule reaches macroscopic dimensions so that it extends throughout the solution, the substance is said to be a gel. Another example concerns the mechanism based on the separation of the initial homogeneous solution into two immiscible continuous and interconnected liquid phases, reminiscent of the structure of a gel, and called *spinodal phase separation* [68–70].

The present review focuses on gels, as supported metallic catalyst precursors or catalyst support precursors, obtained from a homogeneous solution of monomer precursors. Gels obtained from the destabilization of a stable colloidal suspension, while representing an important class of sol-gel materials, are not presented here.

Depending on whether the liquid in the wet gel is removed by evaporative drying or by supercritical drying, that is in pressure and temperature conditions beyond the critical point of the liquid, the resulting dry material is named *xerogel* or *aerogel* respectively. A third class of materials are *cryogels* dried by freeze-drying or lyophilization [20].

8.2.1.1 Inorganic Gels

In theory, almost all metal oxides can be synthesized by sol-gel process. Among others, an abundant literature can be found on SiO_2, Al_2O_3, TiO_2, ZrO_2 as well as on the corresponding mixed oxides [19, 67, 71–73]. Thus all porous oxide materials used in heterogeneous catalysis as catalyst support or precursors of catalyst supports can be prepared by sol-gel process. In the field of inorganic gels, the present review will focus mainly on silica gels.

Probably the best starting materials for sol-gel preparations of metal oxides are the class of metalorganic compounds known as *metal alkoxides* [71, 74], which are focused on in the present paper. All metals form alkoxides which have the following general formula:

$$M(OR)_x$$

where M is a metal (Si, Zr, Al, Ti, …), R is an alkyl group (most often $-CH_3$, $-C_2H_5$, $-C_3H_7$ or $-C_4H_9$), and x is the valence state of the metal (for example, x = 4 with Si, Zr and Ti, and x = 3 with Al). Metal alkoxides, which are soluble in alcohol, are popular precursors because they are rapidly hydrolyzed to the corresponding hydroxide, except the most thoroughly studied among them, silicon tetraethoxide (or tetraethoxysilane, or tetraethyl orthosilicate, TEOS), $Si(OC_2H_5)_4$, which requires an acid or basic catalyst for hydrolysis. The reaction is called *hydrolysis*, because a hydroxyl ion becomes attached to the metal atom, as in the following reaction:

$$M(OR)_x + H_2O \leftrightarrow HO\text{-}M(OR)_{x-1} + ROH$$

where, again, R is an alkyl group and therefore ROH is an alcohol. Depending on the amount of water and catalyst present, hydrolysis may go to completion, so that all of the OR groups are replaced by OH,

$$M(OR)_x + x \ H_2O \leftrightarrow M(OH)_x + x \ ROH$$

or stop while the metal is only partially hydrolyzed, $M(OR)_{x-n}(OH)_n$. Let us mention that inorganic precursors as chlorides or nitrates can also be hydrolyzed [67, 75].

Two partially or completely hydrolyzed molecules can link together in a *condensation* reaction, such as

$$(OR)_{x-1}M\text{-}OH + HO\text{-}M(OR)_{x-1} \leftrightarrow (OR)_{x-1}M\text{-}O\text{-}M(OR)_{x-1} + H_2O$$

or

$$(OR)_{x-1}M\text{-}OR + HO\text{-}M(OR)_{x-1} \leftrightarrow (OR)_{x-1}M\text{-}O\text{-}M(OR)_{x-1} + ROH$$

By definition, condensation liberates a small molecule, such as water or alcohol. This type of reaction can continue to build larger and larger silicon or aluminum or titanium or zirconium or ... containing molecules by a polymerization process in alcoholic medium.

To summarize, the basic chemistry associated with formation of a typical sol-gel material such as silica gel is given in very simplified form in Figure 8.1 [76].

In the case of silicon alkoxides, hydrolysis occurs by the nucleophilic attack of the oxygen contained in water on the silicon atom and is most rapid and complete when acid (e.g. HCl, CH_3COOH, HF, ...) or basic (e.g. NH_3, KOH, amines, ...) catalysts are employed. Under acidic conditions, it is likely that an alkoxide group is protonated in a rapid first step. Electron density is withdrawn from silicon, making it more electrophilic and thus more susceptible to be attacked by water. Under basic conditions it is likely that water dissociates to produce nucleophilic hydroxyl anions in a rapid first step. The hydroxyl anion then attacks the silicon atom. Polymerization

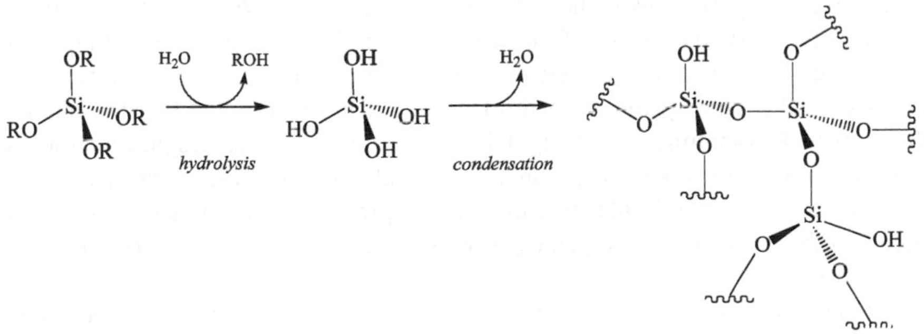

FIGURE 8.1 Basic steps of a typical sol-gel process from a metal alkoxide. Hydrolysis of alkylorthosilicates affords silicon hydroxides that undergo condensation reactions to form a silicate network (reproduced from [76] with permission from Wiley).

to form siloxane bonds occurs by either a water-producing condensation reaction or an alcohol-producing condensation reaction (see both condensation equations above). Depending on conditions, a complete spectrum of structures ranging from molecular networks to colloidal particles may result. Although the condensation of silanols can proceed thermally without involving catalysts, the use of acid or basic catalysts, similar to those used for hydrolysis, is often helpful and various mechanisms have been suggested to explain the role of the catalyst [67, 77].

An important advantage of sol-gel chemistry over conventional oxide preparation techniques is the possibility it offers to prepare mixed oxides, such as SiO_2-ZrO_2, SiO_2-Al_2O_3, SiO_2-TiO_2, ..., with an excellent control of mixing because of its ability to adjust the relative precursor reactivity [19, 35, 73, 78–80]. Mixed oxides are interesting catalysts or catalyst supports because they often display acid strengths that are significantly higher than either of the component oxides [19, 81, 82]. In a mixed oxide MO_x-$M'O_y$, the acidic properties are related to the homogeneous mixing of the two component oxides in terms of the number of available M-O-M' linkages. The key condition to obtain molecularly homogeneous mixed oxides is to adjust the relative reactivity of the precursors in such a way that they have similar rates of hydrolysis and/or condensation. The matching of precursor reactivity can be accomplished using various strategies including the use of a precursor containing a different alkoxy group [19], the prehydrolysis of a less reactive precursor [83–85], the slowing down of a more reactive precursor by replacing some of its alkoxy groups with different ligands or by stabilizing it by using an appropriate solvent [86–91], and the modification of synthesis temperature [19]. In a recent paper, Rupp *et al.* [92] prepared SiO_2-TiO_2 mixed oxides with low titania crystallization tendency by using single-source precursors containing both -$Si(OR)_3$ and -$Ti(OR)_x$ (x = 2 or 3) moieties linked by an organic group.

From the hydrolysis and condensation reactions presented above as starting point, various theories corresponding to various physico-chemical mechanisms have been suggested to explain the formation of the wet gel. According to several authors who investigated how inorganic polymer systems, such as SiO_2 [93–95], TiO_2 [96], and ZrO_2 [97, 98], are formed, the materials are obtained *via* the formation of elementary building blocks that, in a secondary stage, aggregate until the resulting clusters fill the space. This is the *aggregation* theory. According to that theory, the large variety in structures results from differences in the details of the building blocks and in their aggregation mechanism, which are governed by the nature of the precursor and by the synthesis conditions [67]. It must be noted that the building block notion is sometimes confusing in literature since it designates now a monomer molecule, now a silica, alumina, titania, ... particle of about several tens nanometer, that is a large macromolecule. However, the aggregation mechanism implies most often first the formation of inorganic particles by polycondensation followed by their aggregation.

On the other hand, in some cases, a *phase separation* mechanism, which is often proposed to explain the formation of organic gels (see Section 8.2.1.2. below), agrees better with experimental results. If the initial homogeneous solution is separated into two immiscible continuous and interconnected liquid phases, the phase separation is named *spinodal phase separation* or *spinodal decomposition*. The two phases are an oligomer-rich or polymer-rich phase which, after polycondensation, will constitute the porous inorganic skeleton of the gel, and a solvent-rich phase [70, 99, 100].

Let us mention that both clusters formation followed by aggregation and phase separation mechanisms can take place simultaneously and competitively [100].

8.2.1.2 Organic Gels

Beside gels composed of a metal oxide skeleton, sol-gel process is also a powerful tool for synthesizing porous organic three-dimensional networks in a solvent. The increasing popularity of these organic materials is largely due to their unique and controllable properties [65]. They open up new horizons for obtaining highly porous tailor-made carbon catalyst supports since, after drying, organic gels can be converted into carbon gels by pyrolysis (see 8.2.4.2.).

Porous carbon gels prepared by polycondensation of hydroxylated benzene (phenol, catechol, resorcinol, hydroquinone or phloroglucinol, ...) and aldehyde (formaldehyde, furfural, ...) in a solvent followed by drying and pyrolysis have been extensively studied for the past fifteen years [51–66, 101]. Various carbon materials whose texture depends on the nature of the precursors, the gelation conditions and the drying method can be obtained. The most common precursors are resorcinol (1,3-dihydroxybenzene) and formaldehyde (CH_2O) and the polymer is usually synthesized using water as solvent and sodium carbonate (Na_2CO_3) as catalyst.

The organic sol-gel polymerization chemistry involving a reaction between phenol and formaldehyde has been widely studied for a long time [102]. The reaction of resorcinol with formaldehyde is similar to that of phenol, except that resorcinol is more reactive because of the electron-donating and *ortho*- and *para*- directing effects of the attached hydroxyl groups [59]. As illustrated in Figure 8.2, the resorcinol-formaldehyde polymerization mechanism includes: the formation of resorcinol anions by hydrogen abstraction (enhanced by OH^-) (step 1a), the formaldehyde addition to obtain hydroxymethyl derivatives ($-CH_2OH$) of resorcinol (mono-, di- and tri-substituted methylolphenols) (step 2), and the condensation of the hydroxymethyl derivatives (catalyzed by H^+) to form methylene ($-CH_2-$) and methylene ether ($-CH_2OCH_2-$) bridged compounds which leads to cluster growth (step 3) [54, 55, 66, 103, 104].

The mechanism of step 1a is questionable in acidic solutions, because of the very low amount of charged species (resorcinol anions) present at the beginning of the polymerization reaction. In this case, the charged species that activates the addition reaction is probably the protonated form of formaldehyde, $(CH_2\text{-}OH)^+$ (step 1b), which is more sensitive to resorcinol addition than molecular formaldehyde.

According to Yu *et al.* [103] who examined the polymerization of 1,4-benzene-dimethanol ($HOCH_2\text{-}ph\text{-}CH_2OH$) in a concentrated H_2SO_4 medium, the condensation of hydroxymethyl derivatives is catalyzed by H^+ according to the presumed following mechanism: in the presence of a proton, such molecules lose their -OH groups to form a benzyl-type cation $-ph\text{-}CH_2^+$. The cation then undergoes an electrophilic reaction with the benzene ring of another molecule to connect the two benzene rings with a methylene bridge ($-CH_2-$) [102, 103]. Pekala *et al.* [54] also showed that the

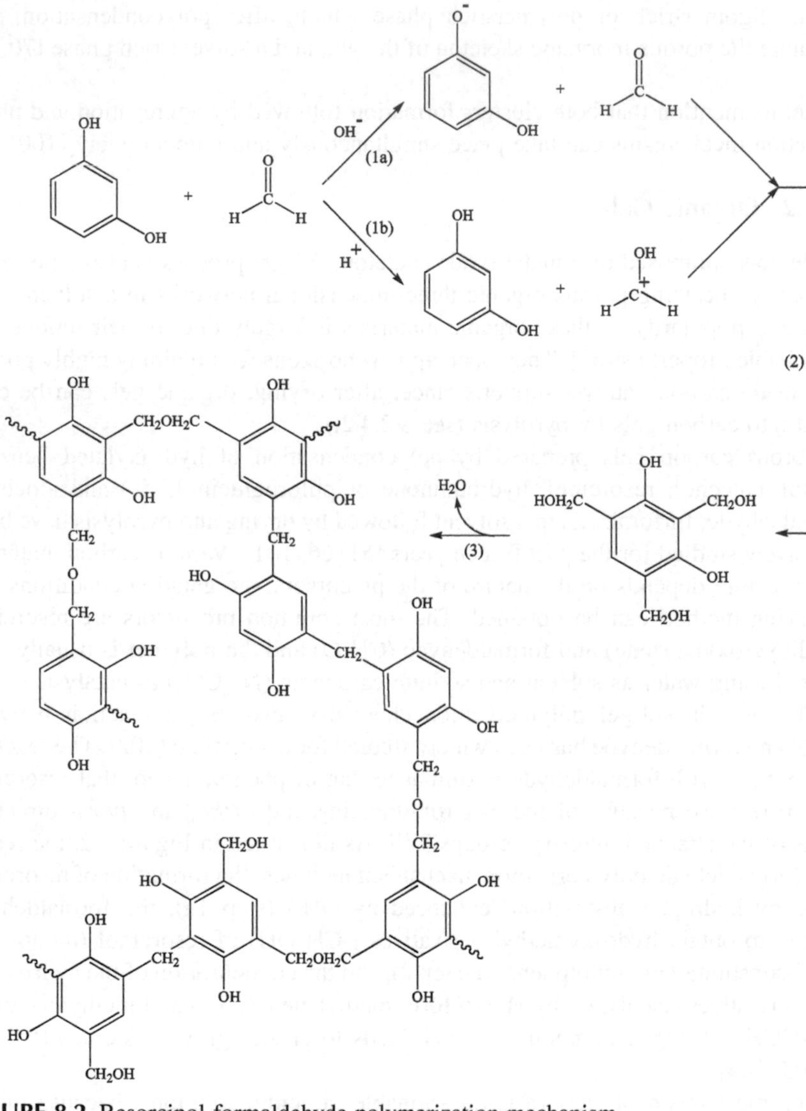

FIGURE 8.2 Resorcinol-formaldehyde polymerization mechanism.

cation can react with a hydroxymethyl group of another molecule to form a methylene ether bridge. In the case of the hydroxymethyl derivatives of resorcinol, successive repetitions of such a condensation reaction lead to the three-dimensional cross-linked polymer.

Note that the often used sodium carbonate is not a catalyst strictly speaking. Indeed, its primary role is to increase the acidic pH of the resorcinol-formaldehyde aqueous solution by increasing the OH^-/H^+ ratio. Sodium cations Na^+ have no direct role in the polymerization reaction whereas carbonate anions CO_3^{2-} basify the solution. In fact, the pH change can be achieved by addition of any base that does not react with resorcinol or formaldehyde such as sodium hydroxide, NaOH [49, 66].

In all cases, the concentration of activated species and the condensation rate are controlled by the pH of the solution.

Since the texture of the final material obtained after drying and pyrolysis is strongly influenced by the reaction mechanism which is itself influenced by pII, the latter is a key variable for controlling texture. Indeed, depending on the pH interval used for synthesis, micro-macroporous, micro-mesoporous, only microporous or totally non-porous carbon materials can be obtained [54, 59, 66].

The various mechanisms proposed in literature to explain the formation of inorganic gels are also encountered in the case of organic gels. However, in the latter case, the mechanism based on phase separation and particularly spinodal phase separation is most often suggested: as the polymerization proceeds, the molecular weight of clusters consisting of branched polymeric species increases which makes them immiscible and induces a phase separation, the polymer-rich phase becoming the precursor of the final organic gel skeleton [69, 104]. Beside the phase separation mechanism, the clusters formation followed by aggregation mechanism is suggested by some authors [105].

8.2.2 Sol-Gel Methods for Preparing Supported Metals

The common industrially important preparation methods of supported metal catalysts are multisteps processes consisting of [106, 107]:

(a) preparation of the support;
(b) distribution of the active component precursor over the support surface (by impregnation, ion exchange, anchoring, ...);
(c) drying, calcination or pyrolysis and possibly reduction of the catalyst to obtain the active species (often metal or metal oxide nanoparticles) dispersed in the porosity of the support.

An important advantage of sol-gel process over traditional methods for the preparation of catalysts is the possibility its offers to prepare a solid from a homogeneous solution which includes not only the support precursor(s), but also the metal(s) precursor(s). Steps (a) and (b) corresponding to classical methods are then gathered in one single step [19, 21, 72, 108, 109].

Although the present review is mainly focused on one-step synthesis methods, the two-steps method, which consists in preparing a xerogel or aerogel support which will be next loaded with one or several metal(s) processor(s), is briefly presented below.

8.2.2.1 Two-Steps Method

Although sol-gel process is particularly attractive in catalyst preparation because it allows the preparation of both support and active sites in one single-step, the two-steps method consisting in dispersing the active metal(s) precursor(s) in the porosity of a xerogel or aerogel support prepared in advance leads in some particular cases to catalysts with higher performances than one-step made catalysts (examples in

[110–112]). Indeed, while it often results in highly dispersed supported metal catalysts, the one-step method leads also sometimes to poor dispersion of metallic species [50, 113]. Furthermore, depending on the conditions of catalyst preparation and/or operation such as temperature, pressure or composition of gaseous atmosphere, there is a risk of active metal occlusion in the bulk of the support of one-step-made catalysts which would make it partly or completely inaccessible for the reactants in a catalytic process. Such an occlusion has been observed in a Pd-Ag/SiO_2 aerogel [33], in Pt/SiO_2 xerogels [114], in Ru/SiO_2 xerogels [12, 115], in Pd/SiO_2 xerogels [116], as well as in Ni/C aerogels [47]. It must be however mentioned that, while leading to a loss of active metallic surface, partial occlusion can be beneficial when it corresponds to a partial burying of metal particles in the support. Indeed, such more strongly anchored metal particles are more resistant to sintering, which makes the catalyst more stable at high temperature [19, 114, 115].

In consequence, it is sometimes preferable to use the sol-gel method in a two-steps process rather than in a one-step one in order to obtain an efficient catalyst.

Due to their particularly high specific surface area and porous volume, sol-gel materials are attractive candidates as catalysts supports. Conventional methods such as impregnation and ion exchange [107, 117] have been used to introduce one or several metal(s) precursor(s) into a xerogel or aerogel support. For example, such methods have been applied to Pt/SiO_2 aerogels [118, 119], Pd/SiO_2 xerogels and aerogels [16, 32], Ru/SiO_2 xerogel [12], Co/Al_2O_3 xerogels [112], Pt/ZrO_2 aerogel [120], Fe/TiO_2 and Fe/Al_2O_3 xerogels [121], Pt/TiO_2 xerogels [122], Rh/SiO_2-TiO_2 aerogels [123, 124], Ni/SiO_2-TiO_2 aerogels [124], Ni/Al_2O_3-TiO_2 xerogels [125], Co-Mo/SiO_2 xerogels [111], Ag-Co/Al_2O_3 xerogels [110], Ru/C aerogels [126, 127], Pt/C aerogels [101], and Pd-Ag/C xerogels [128].

In the case of inorganic supports, it is important to calcine xerogel and aerogel supports in an oxidizing atmosphere (most often in air) at high temperature (e.g. 400°C) before impregnation. Indeed, such a calcination increases drastically the microporosity, and thus the specific surface area, of the support because its allows to burn the many organic residues which block the pores, and particularly the micropores, in xerogels and aerogels just dried [32]. In the case of carbon supported catalysts, the organic gel, most often obtained from aqueous polycondensation of resorcinol and formaldehyde, is, after drying, pyrolyzed at high temperature (e.g. 800–1000°C) in an inert atmosphere to be converted into a high surface area carbon support [65].

When a monomodal metal particle size distribution as narrow as possible is required, an impregnation of xerogel or aerogel supports with chemical anchoring is useful [32, 76]. The anchoring method for obtaining homogeneous dispersion, i. e., narrow particle size distribution with small diameters, has been known for a long time and has been used successfully to prepare Ni, Pd, and Pt catalysts supported on inorganic supports [129–131]. The metal precursor used in these studies is an organometallic compound and the reactant are the -OH surface groups [15]. Such a method has been proposed by Heinrichs *et al.* [32] to prepare impregnated Pd/SiO_2 aerogel and xerogel catalysts. A Pd^{2+} amino complex containing methoxy groups (-OCH_3) which can condense with silica surface hydroxyl groups ($\equiv$Si-OH) has been used. Anchoring of palladium on the silica surface occurs as shown in Figure 8.3.

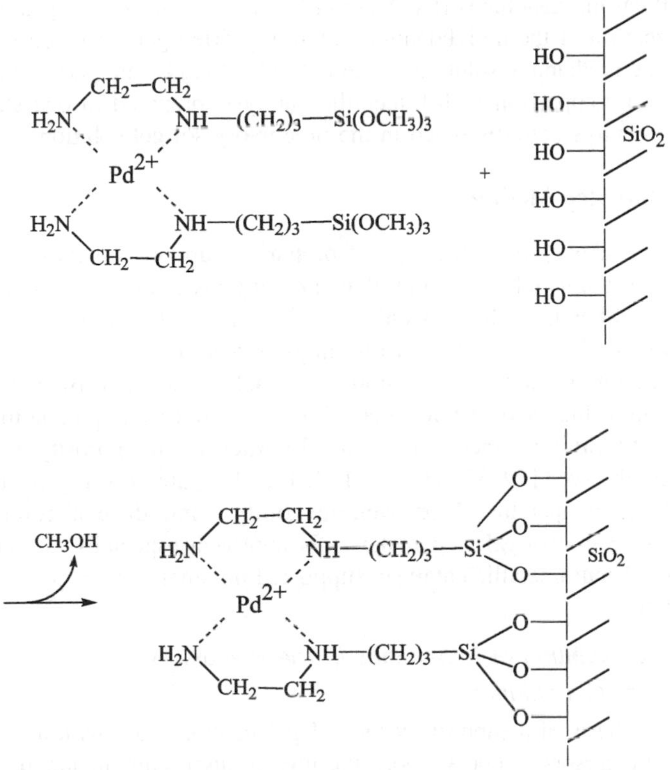

FIGURE 8.3 Impregnation, with anchoring, of silica (or any hydroxylated support) with a Pd^{2+} amino complex containing methoxy groups able to condense with surface hydroxyls (adapted from [32] with permission from Elsevier).

Note that a SiO_2 support prepared by the sol-gel process is highly hydroxylated even after thermal treatment [14].

A comparison of Pd/SiO_2 xerogel and aerogel catalysts prepared by impregnation with and without palladium anchoring ($Pd^{2+}(NH_3)_4$ complex used in the latter case) showed that, while not strongly influencing the Pd particle mean size in the final catalyst, the impregnation with anchoring method leads to a narrower metal particle size distribution.

The two-steps method has also been combined with a one-step method (described below) to synthesize bimetallic catalysts. In those cases, monometallic samples prepared by a one-step method have been impregnated with a solution of a second metal. With the aims to avoid the formation of Pt-Sn alloys and to promote the formation of 'tin-aluminate' compounds in order to obtain a $Pt-Sn/Al_2O_3$ bimetallic catalyst with high performance in the reforming of naphtas, Gómez *et al.* [132] added tetrabutyltin to a homogeneous solution containing aluminum tri-sec-butoxide to form a Sn/Al_2O_3 gel, which, after drying and calcination, was impregnated with an hexachloroplatinic acid solution. A second example concerns a comparison of different preparation methods of $Fe-Mo/Al_2O_3$ sol-gel catalysts for the synthesis of single wall carbon nanotubes (SWNTs) [133]. In that study, catalysts prepared by the impregnation of the molybdenum containing dried gel, prepared in one single step, with the methanolic solution of the $Fe(NO_3)_3.9H_2O$ were particularly selective in SWNTs in comparison with bimetallic catalysts prepared in one step only (Fe and Mo precursors both dissolved in the precursory sol-gel solution).

8.2.2.2 One-Step Methods

As mentioned above, one advantage of preparing supported metals by the sol-gel method is that the metal can be introduced *during* instead of *after* the formation of the support. Other than the convenience of saving a step, this feature allows to prepare supported metal catalysts with unique structures that are inaccessible with other preparation methods. Often, supported catalysts prepared by one-step sol-gel methods exhibit higher dispersion than those prepared by impregnation [32, 111, 121, 134, 135]. Stronger anchoring of metal particles in the porosity of the support is also often observed [19, 32, 33, 112, 114, 115, 135] and this may benefit a higher stability at high temperature. Moreover, the ability to introduce all components into solution during the sol-gel step makes this approach especially attractive for the preparation of multimetallic catalysts supported on single or mixed oxides as well as on carbon.

8.2.2.2.1 *Dissolution of Metal Salts in the Precursory*
Sol-Gel Solution

Since the synthesis of a support by the sol-gel method starts with a homogeneous solution of precursors of that support, the idea of dissolving an adequate precursor of the active species in that solution comes naturally to the mind. Many catalysts have been synthesized in that way. Let us cite Pd/SiO_2 xerogels [15, 16], Pt/SiO_2 xerogels [14], Ru/SiO_2 xerogels [10–13], Fe/SiO_2 xerogels [136–138], Ni/SiO_2 xerogels [136, 139], Co/SiO_2 xerogels [136, 140], Pd/Al_2O_3 aerogels [9], Pt/Al_2O_3 aerogels and xerogels [17], Ni/Al_2O_3 aerogels [9, 134, 141], Co/Al_2O_3 xerogel [112], Pd/AlF_3 and Ru/AlF_3 xerogels [142], Ni/SiO_2-Al_2O_3 aerogels and xerogels [9, 139], Pd/V-Zr-Al oxide xerogels [143], Pd-Sn/SiO_2 xerogels [144], Co-Mo/SiO_2 xerogels [111], Co-Fe/SiO_2, Co-Ni/SiO_2, and Fe-Ni/SiO_2 xerogels [136], Fe-Mo/Al_2O_3 xerogels and aerogels [133, 145], Co-In/Al_2O_3 xerogels [146], Pt-$Sn/MgAl_2O_4$ xerogels [147], Ni/C xerogels and aerogels [46–50], Fe/C xerogels and aerogels [46, 47, 50], Pd/C xerogels [44, 48–50], Pt/C aerogels [44], Ru/C xerogels [43], Ag/C aerogels [44], Ti/C aerogels [46], Cr/C, Mo/C and W/C aerogels [42, 46], Co/C aerogels [46], Ce/C and Zr/C aerogels [45].

The underlying purpose is to try to include the active metal precursor in the porous growing gel network without making it inaccessible in the final catalyst. An essential condition to obtain a final catalyst in which the metal is highly dispersed is to maintain the metal precursor in solution throughout the sol-gel step, which is sometimes difficult with noble metals which are easily reduced and then precipitate.

In a study of the synthesis of transition metal doped carbon xerogels by solubilization of metal salts in resorcinol-formaldehyde aqueous solution, Job *et al.* [50] used complexing agents when necessary to render the metal ions soluble. Indeed, some salts can be dissolved easily and kept dissolved during the gel synthesis (nickel acetate for example). But in many cases, the addition of a complexing agent proved to be essential for two reasons: i) some salts are not very soluble in the resorcinol-formaldehyde aqueous solution: the addition of a complexing agent improves their solubility. In particular, many metal ions (Fe^{3+}, Cu^{2+}, for instance) precipitate as hydroxides under the pH conditions chosen; ii) formaldehyde is a powerful reductant and often causes the metal to precipitate under metallic state before gelation, leading to non homogeneous gels and metal agglomerates. In that study, Ni/C, Fe/C and Pd/C xerogel samples were prepared. As significant amounts of nickel acetate tetrahydrate can be dissolved in resorcinol-formaldehyde aqueous solutions without precipitation, Ni/C samples were prepared by simple addition of nickel acetate tetrahydrate. The Fe/C samples were synthesized using iron acetate and hydroxy-ethyl-ethylenediaminetriacetic acid (HEDTA) as complexing agent to increase the solubility of the metal salt. The Pd/C xerogels were prepared with palladium acetate. Again, significant amounts of palladium acetate cannot be dissolved without additional complexing agent. Preliminary results showed that though HEDTA improves palladium acetate dissolution, this complexing agent cannot prevent the reduction of palladium cations during gelation. HEDTA was then replaced by diethylenetri-aminepentaacetic acid (DTPA). Unfortunately, despite those precautions, the study showed that nickel, iron and palladium are under metallic state after drying without needing further reduction treatment, which indicates that the metal complexes are not stable throughout the polymerization reaction. Though the xerogels seem homogeneous, the transition metals are reduced by formaldehyde and precipitate locally. As a result, the metal particles obtained are rather big (15–20 nm and more) in the final catalysts. Investigations have therefore to be conducted to improve the preparation method.

In another study, Cr, Mo and W oxides loaded carbon aerogel catalysts have been prepared by Moreno-Castilla et al. [42] by dissolving metal salts in aqueous solutions of resorcinol and formaldehyde. According to X-ray diffraction results, the metal oxide or carbide was well dispersed inside the organic matrix.

In the case of metals supported on inorganic gels, it is sometimes possible to find a salt of the desired metal which is soluble in the initial sol-gel solution whose solvent most often is an alcohol [136, 138, 145]. It is also possible to realize a dissolution of the metal salt in an adequate solvent and to add that metal precursor solution to the sol-gel solution [9, 16, 17, 143]. The preliminary dissolution of the metal salt in the water which will be used to hydrolyze the support precursor alkoxyde is often encountered [14, 15, 111, 112, 137, 139, 144].

Unlike the cogelation method which is presented below, metal precursors in the dissolution method do not participate directly in the sol-gel chemistry involving the oxide precursors which are usually alkoxides. In the dissolution method, the metal precursor is often simply encapsulated in a growing gel network, but its presence can still indirectly influence the sol-gel process [19] and leads to modifications in the structure of the final catalyst. For example, in a study of spectroscopic characterization of Pd/SiO_2 xerogel catalysts prepared by hydrolysis and polycondensation of tetraethoxysilane ($Si(OC_2H_5)_4$) in a solution containing $PdCl_2$, López et al. [15] showed that the preparation method leads to the formation of $[SiO_2]_{-[PdCl_xOH_y]}^{-OH}$ surface species with an unusual interaction detected by UV-VIS and FT-IR spectroscopies. The interaction of palladium with the silica gel has an important effect on the specific surface of the final catalyst: whereas the SiO_2 support prepared without palladium shows an area of 110 m2/g only, the Pd/SiO_2 catalysts exhibit an much higher area which decreases from 889 m2/g down to 630 m2/g when the metal loading increases from 0.1 to 3 wt.%. In another study on the synthesis of Ni/C, Fe/C and Pd/C xerogels, Job et al. [50] observed that the solubilization of transition metal salts in the resorcinol-formaldehyde initial aqueous solution does not prevent the texture regulation, even though this texture control is influenced: the limits of the pH interval leading to micro-mesoporous carbon materials can slightly differ when a metal salt and/or a complexing agent are added. The pH range shift depends mainly on the amount and nature of the complexing agent, but also slightly on the nature of the metal ion.

A serious drawback that can arise with the one-step dissolution method is the possible partial occlusion of the active metal in the bulk of the support leading to a loss of metallic surface available for the catalytic reaction. Cases in which that problem occurred have been mentioned in Section 8.2.2.1. devoted to the two-steps method which represents a possible alternative to avoid occlusion. However, because of the above mentioned advantages of the one-step methods, it is really worth to try to solve the problem of metal occlusion by improving the one-step methods. In the following Section, a second one-step method that allows to avoid occlusion and leads to unique structural characteristics is presented.

8.2.2.2.2 Cogelation of Metal Chelates with Support Precursors

Since the eighties, a particularly attractive method to prepare nanometer-sized uniform metal particles in a silica matrix by sol-gel processing of metal complexes has been developed. Numerous metals and alloys supported on SiO_2 including Ag, Co, Cu, Ni, Pd, Pt, Cu-Ru, Pd-Ni, Cu-Ni, and Pt-Cu were prepared [23–31, 148]. That method has been applied to prepare Pd/SiO_2, Ag/SiO_2, Cu/SiO_2, $Pd-Ag/SiO_2$ and $Pd-Cu/SiO_2$ xerogel and aerogel catalysts for various applications [32, 33, 36, 37, 39–41, 109, 149–152]. The method consists in using an alkoxysilane-functionalized ligand of the type $(RO)_3Si-X-L$ in which the ligand L, able of forming a complex - LNM with a metal M (M = Pd, Ni, Ag, Cu, etc.), is connected to the alkoxide moiety $(RO)_3Si$- via an inert and hydrolytically stable tethering organic group X. The concomitant hydrolysis and condensation of such molecules with a network-forming reagent such as $Si(OC_2H_5)_4$ (TEOS), i.e. their cogelation, result in materials in which the catalytic metal is anchored to the SiO_2 matrix [22, 24, 32, 76]. The process is illustrated in Figure 8.4 with [3-(2-aminoethyl)aminopropyl]trimethoxysilane as alkoxysilane-functionalized ligand and palladium as metal.

In general, the alkoxysilane-functionalized metal complex is formed from a metal precursor and an alkoxysilane-functionalized ligand in a solvent (most often in ethanol) before adding the silica precursor and water for gel formation. Examples of metal precursors are acetates $(Pd(OAc)_2$, $Co(OAc)_2$, $Cu(OAc)_2$, $Ni(OAc)_2$, Ag(OAc), ...), acetylacetonates $(Pd(acac)_2$, $Pt(acac)_2$, ...), and nitrates $(AgNO_3$, ...). Examples of alkoxysilane-functionalized ligands are given in Figure 8.5. It must be noted that only a few among these ones are commercially available (e.g. 3-(aminopropyl)triethoxysilane,

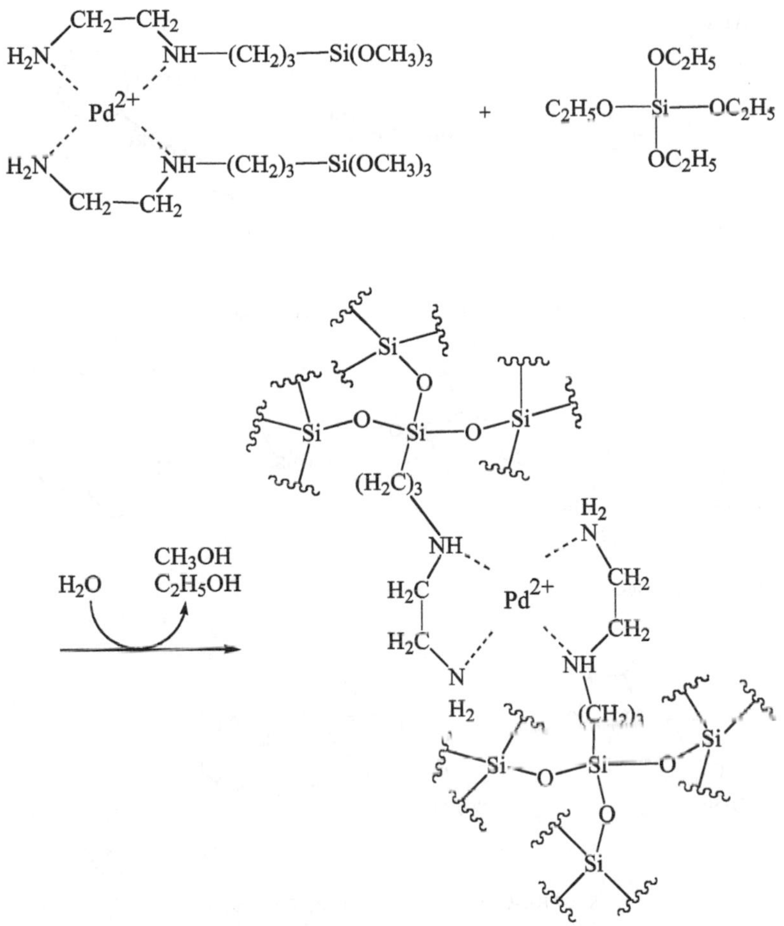

FIGURE 8.4 Anchoring of alkoxysilane-functionalized complexes of catalytically active metals to a silica matrix by cogelation (adapted from [32] with permission from Elsevier).

$H_2N(CH_2)_3Si(OC_2H_5)_3$ (Figure 5.1), [3-(2-aminoethyl)aminopropyl]trimethoxysilane, $H_2NCH_2CH_2NH(CH_2)_3Si(OCH_3)_3$ (Figure 5.1), and 4-(triethoxysilyl)butyronitrile, $NC(CH_2)_3Si(OC_2H_5)_3$ (Figure 5.2)). An excellent review on coordination complexes in sol-gel silica materials, including silylated ligands and silylated metal complexes, has been recently published by Watton *et al.* [76]. Note that alkoxysilane-functionalized organic molecules have been also synthesized to prepare new organo-ceramic materials [153]. A variety of routes have been developed for preparing metal-binding ligands with silylated side chains [22, 76, 150, 154].

Thus, contrary to the metal precursor in the dissolution method presented in the previous Section, once formed, the alkoxysilane-functionalized metal complex is directly involved in the sol-gel chemistry. Indeed, as the main SiO_2-network-forming reagent ($Si(OR)_4$), this complex undergoes hydrolysis and condensation. Besides its anchoring function, the alkoxysilane-functionalized ligand presents two particularly interesting advantages for preparing metal containing as well as pure

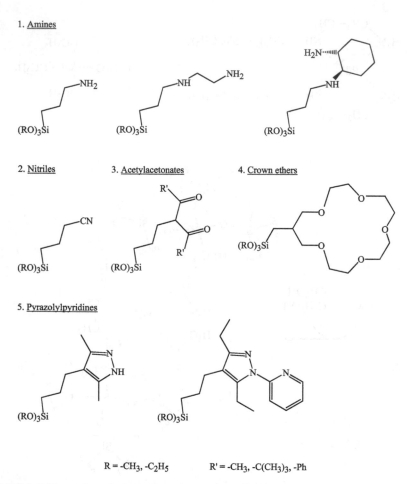

1. Amines

2. Nitriles　　　　3. Acetylacetonates　　　　4. Crown ethers

5. Pyrazolylpyridines

R = -CH₃, -C₂H₅　　　　R' = -CH₃, -C(CH₃)₃, -Ph

FIGURE 8.5 Examples of alkoxysilane-functionalized ligands ([22, 76, 150, 154]).

silica gels: i) it allows the solubilization of metal salts that are insoluble in alcohol (e.g. [3-(2-aminoethyl)aminopropyl]trimethoxysilane $H_2NCH_2CH_2NH(CH_2)_3Si(OCH_3)_3$ allows to solubilize $Pd(acac)_2$ in ethanol by forming the corresponding alkoxysilane-functionalized complex [32]), and ii) whereas attempts to synthesize a pure silica gel (without ligand and metal precursor) by the same method as cogelled Pd/SiO_2 samples failed [32], the introduction of ligands such as [3-(2-aminoethyl)aminopropyl]trimethoxysilane makes such a synthesis possible [151, 155, 156]. Therefore, the use of such ligands allows to prepare pure silica gels by another way than the usual acid-base method [157].

Physico-chemical characterization of cogelled catalysts indicates that the introduction of alkoxysilane-functionalized ligand in the sol-gel solution influences strongly the characteristics of the final material. These catalysts exhibit a very particular structure leading to remarkable properties which will be detailed throughout this review.

Before ending this Section on one-step methods, it is important to point out that in the case of samples prepared by one-step sol-gel methods, the actual metal loading can be sometimes significantly higher than the designed metal loading. When the TEOS is not completely converted into silica gel, the mass of silica of the resulting sample is lower than expected and actual metal loadings may exceed designed metal loadings. For this reason, calculation of metal loadings in final samples from elemental analysis (inductively coupled plasma atomic emission spectroscopy, ICP-AES for example) [119, 158] or from accurate measurements of weight losses [39, 40, 44, 150, 151] are essential.

8.2.3 DRYING OF WET GELS

Organic and inorganic wet gels are extremely porous materials whose porosity is completely filled with an interstitial liquid mainly constituted of the initial solvent. Since a high porosity is a key property for an efficient heterogeneous catalyst, it is crucial to maintain such an exceptional porosity as high as possible in the dried gel.

As the pore liquid is evaporated from a gel network, the capillary pressure associated with the liquid-vapor interface within a pore can become very large for small pores thus leading to a collapse in the pore structure and the corresponding shrinkage of the gel [72, 155]. The extent of shrinkage is controlled by a balance between the capillary pressure and the modulus of the solid network [159]. Various approaches can be envisaged to maintain the integrity of a gel network that are based on: i) minimizing the capillary pressure, ii) eliminating it altogether, and iii) increasing the stiffness of the wet gel.

The capillary pressure (P) associated with the liquid–vapor interface within a pore is roughly given by

$$P = -\frac{2\gamma \cos\theta}{r} \text{ ,}$$

where γ is the surface tension, θ is the contact angle between liquid and solid, and r is the pore radius [19, 159]. For a wetting fluid ($\theta < 90°$), P is negative indicating that the fluid is in tension. Obviously, using a solvent with a lower surface tension will reduce the capillary pressure and thus the extent of the pore structure collapse during evaporative drying. This approach has been demonstrated with silica gels by using various solvents with a range of surface tensions [160, 161]. Those studies established the feasibility of preparing high-surface area, low-density materials at ambient pressure, in contrast with the preparation of aerogels with supercritical drying. The method has also been applied to resorcinol-formaldehyde gels prepared in aqueous medium by exchanging water with acetone before evaporative drying [56, 59, 65]. A second way to reduce P consists in increasing the contact angle θ which can be done by modifying the surface of the solid skeleton, by methylation for example in the case of inorganic gels [162]. Besides increasing the contact angle, such a methylation causes a second effect which reduces extensively the shrinkage of inorganic gels during drying. Indeed, during drying, the shrinkage promotes condensation between hydroxyl (-OH) or between hydroxyl and alkoxy (-OR) groups present on the surface of the wet gel. The M-O-M (M = Si, Al, Ti, ...) bonds thus formed make the shrinkage irreversible. To prevent additional condensation during drying and so to let the 'springback' phenomenon take place, it is necessary to cover the surface with non-reactive species, like -CH$_3$, by means of surface modification with compounds such as trimethylchlorosilane or trimethylethoxysilane [162, 163].

In order to eliminate the liquid-vapor interface and the accompanying capillary pressure, two methods are available: supercritical drying and freeze drying. Supercritical drying has been extensively used to remove the usually used alcoholic solvent (containing water and other residues) in inorganic wet gels. To exceed the critical conditions (for ethanol, T_c = 514 K and P_c = 6.1 MPa), the wet gel is transferred into an autoclave. The high-pressure system is flushed with nitrogen, hermetically closed, and heated slightly above the critical point of the solvent used. After a period of thermal equilibration, the pressure is released and the autoclave is flushed with nitrogen to remove residual alcohol and subsequently cooled to ambient temperature [164]. The extremely porous materials obtained in such a way are aerogels [18, 20, 32, 36, 88, 164–166] and are reviewed by Pajonk in the present book. Note that, while maintaining very large pores, supercritical drying in alcohol can cause the closing of micropores and thus make an aerogel catalyst completely inactive because of active metal particle occlusion [33].

Because of the unlikely scaling-up at an industrial level of such a high-temperature process due to inherent risks, linked to hot alcohol vapors, and costs, a low-temperature supercritical drying in carbon dioxide has been developed. In this procedure, advantage is taken from the low critical temperature of CO_2 (T_c = 304 K, P_c = 7.4 MPa). Using this drying process involves a solvent exchange by liquid CO_2 before achieving the drying step [20, 42, 44, 47, 69, 104, 134, 141, 145, 164, 167]. Note that, in the case of resorcinol-formaldehyde gels prepared in aqueous medium, due to the very low solubility of water in liquid CO_2, a first solvent exchange from water to an organic such as acetone is required before the exchange by liquid CO_2 [20, 44, 47, 69, 104, 167].

One other approach to bypass the liquid-vapor interface is freeze drying, in which the pore liquid is frozen into a solid that subsequently sublimes to give an aerogel-like material called a cryogel [20, 72]. Freeze drying has been applied to organic gels synthesized in aqueous medium such as resorcinol-formaldehyde gels [60, 61, 63], as well as to inorganic gels prepared in alcoholic medium [133, 168].

As mentioned in point iii) above, a third method to maintain the integrity of a gel network during drying consists in increasing the stiffness of the wet gel. The strengthening of the gel structure can be achieved by additional precipitation of silica after the initial gelation [169–171]. In the case of cogelled catalysts presented in Section 8.2.2.2.2., it has been observed that even when they are submitted to an evaporative drying, the resulting xerogels exhibit a texture similar to that of aerogels dried under supercritical conditions with comparable pore volumes (up to 7 cm^3/g) and pore size distributions (from around 1 nm up to 1 μm) [33, 34, 39, 150–152, 172]. The reason of that aerogel-like structure has been explained by the role of nucleation agent played by the alkoxysilane-functionalized ligand or complex in the formation of the gel which allows to adjust the size of silica particles constituting the inorganic skeleton, the size of pores and thus the extent of shrinkage during evaporative drying [32, 155]. The role of the alkoxysilane-functionalized ligand or complex on gel structure is explained in more details in Section 8.3.1.1.

In consequence, the resistance of cogelled catalysts towards shrinkage induced by the capillary pressure makes them particularly attractive candidates for large-scale applications. Indeed, it allows the use of a standard evaporative drying which is much more easy and cheap to implement than the supercritical drying. That shrinkage resistance is a first remarkable property of cogelled catalysts.

In the field of carbon xerogels, it is also important to remark that, despite commonly accepted ideas, evaporative drying does not always completely destroy the pore texture of phenolic gels. Indeed, by choosing appropriate values of synthesis variables, in particular pH values, it is possible to maintain a large porosity in the dried xerogels. Pore volumes up to 2 cm³/g and pore size distributions from around 1 nm up to 200 nm have been observed [50, 66].

8.2.4 SUBSEQUENT THERMAL TREATMENTS

Before obtaining active supported metal catalysts, dried gels must still undergo one or several thermal treatment(s) in various gaseous atmospheres.

8.2.4.1 Calcination of Inorganic Gels in an Oxidative Atmosphere

Even when thoroughly dried, gels still contain a significant amount of residues that block up part of the porosity of the catalyst support. In order to burn off any residual organics (alcohol, ligands, …) present in dried inorganic gels and to convert organic precursors of the active metal into metal oxide or to burn off organic parts of alkoxysilane-functionalized complexes and produce metal oxide in cogelled catalysts, samples are calcined at high temperature (e. g. 400–500°C) in flowing air or oxygen. The calcination temperature should be high enough to ensure the complete removal of all organic parts but not higher than necessary to avoid sintering of the support (micropore closing) and/or of the metal compound. The metals act as oxidation catalysts because oxidation occurs at lower temperature when these ones are present in the gel [24, 32]. In cogelled materials, calcination results in a considerable increase of the support surface area corresponding to the unblocking of micropores that were completely filled with residues which makes active metal precursors (in the form of oxides) accessible [32, 150].

8.2.4.2 Pyrolysis of Organic Gels in an Inert Atmosphere

In order to convert resorcinol-formaldehyde xerogels or aerogels into the corresponding carbon xerogels or aerogels, the former have to be pyrolyzed at high temperature (e. g. 800–0–1000°C) in an inert atmosphere (e.g. N_2). Phenomena occurring during pyrolysis can be separated into the release of water and organic groups and the simultaneous formation of carbon entities followed by rearrangement of these carbonaceous structures to graphitic or semi-graphitic structures [173]. Note that transition metals present in the gel can induce graphitization of the material, since they act as catalysts in this process [47, 174–176]. According to Lin and Ritter [59], pyrolysis of low pH resorcinol-formaldehyde xerogels leads to nanocrystalline materials, which give rise to a highly porous structure compared to graphite, but not nearly as microporous as a reference activated carbon without further activation. Concerning texture evolution, Job *et al.* [66] observed that pyrolysis induces an increase of the specific surface area along with a decrease of the mesopore and macropore sizes.

For metal containing gels, it seems that it is sometimes possible to obtain highly dispersed metals or metal oxides in the final carbon gel provided that the pyrolysis temperature is not too high. Above a given temperature, which depends on the metal considered, pyrolysis leads to metal sintering. When metal oxides are present, a reduction by the carbon matrix can be observed at sufficiently high temperature [42, 47].

8.2.4.3 Reduction

When the metallic state is required for catalytic species to be active, the metal containing xerogels or aerogels are submitted to reduction in hydrogen at high temperature (e.g. 350–400°C) [39]. In the case of bimetallic catalysts, hydrogen treatment allows moreover the formation of alloy particles [33, 128, 152, 177].

Note that hydrogen treatment is not always necessary to obtain metals in their reduced state. Indeed, in the case of catalysts supported on inorganic aerogels, when alcohols are used as solvents, their chemical reducing properties combined with the supercritical conditions can be sufficient to convert oxidized species into metallic species [18]. As mentioned above, pyrolysis of metal containing organic xerogels or aerogels can also lead to reduced species on the carbon support [42, 47].

8.3 PHYSICO-CHEMICAL CHARACTERISTICS

As already mentioned, an important advantage of sol-gel process over traditional methods of catalyst preparation is the possibility it offers to prepare active metal particles and their support in one single step. Moreover, one-step made metal catalysts on inorganic xerogels prepared by the cogelation method presented in Section 2.2.2.2. are particularly attractive candidates for large-scale applications since their large porosity is maintained even after an easy and cheap evaporative drying.

For those reasons, physico-chemical characteristics of cogelled metal catalysts supported on inorganic xerogels and of one-step made metal catalysts supported on carbon xerogels or aerogels will be emphasized in the present Section.

8.3.1 Morphology and Texture of Catalysts

8.3.1.1 Metal Catalysts on Inorganic Xerogels

Cogelled xerogel catalysts exhibit very particular morphology and texture organized into a hierarchy and that are illustrated in Figs. 6–8 corresponding to various samples.

Figure 8.6 is a SEM (scanning electron microscopy) micrograph which gives us an idea of the 3-D morphology of the cogelled xerogels at the scale of the micrometer whereas Figure 8.7 is a TEM (transmission electron microscopy) micrograph corresponding to a much higher magnification and obtained from a slice of the material. At the scale of Figure 8.6, it appears that cogelled xerogel pellets contain silica aggregates separated by large pores. An approximate quantification indicates that the size of aggregates, as well as the width of pores between those aggregates, range from several tens to several hundreds of nanometers. The examination of such materials at higher magnification (Figure 8.7) shows that the silica aggregates look like an assembling of highly interpenetrated SiO_2 particles. While being difficult to evaluate accurately, the mean size of the latter ones in a given sample can strongly differ from one sample to another with values from 5 to 50 nm [32–34, 151, 152].

Figure 8.8 illustrates the texture of cogelled xerogel catalysts. In the micropore domain (width < 2 nm), the catalysts exhibit a very narrow pore size distribution centered around a mean value of about 0.8 nm that corresponds to the steep volume increase followed by a plateau. In the range of meso- (2 nm < width < 50 nm) and macropores (width > 50 nm), all samples exhibit a broad continuous distribution starting at about 2 nm and extending up to several hundred nanometers.

Let us remark that, apart from scarce cases [178], when given, pore size distributions presented in literature most often cover a very narrow range of sizes only, corresponding for example to the mesopore range (2–50 nm) calculated from nitrogen adsorption-desorption isotherms measured at 77 K [15, 17, 19, 21, 119, 125, 136, 179]. Such distributions give a truncated view of the catalyst pore size distribution which is probably often much broader. The knowledge of the pore size distribution over a range which is as broad as possible is however essential to understand the behavior of catalysts with regard to mass transfer, thermal stability, resistance towards deactivation, … (see Sections 8.3.3. and 8.4.1.). By using both

10^{-6} m

FIGURE 8.6 SEM micrograph of a 1.9wt.%Pd-3.7wt.%Ag/SiO$_2$ cogelled xerogel catalyst (reprinted from [34] with permission from AIChE-Wiley).

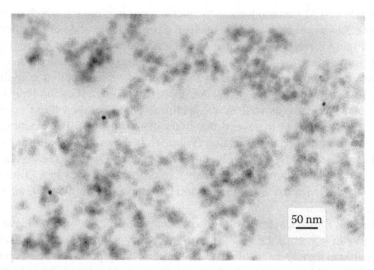

FIGURE 8.7 TEM micrograph of a 2.2 wt.%Pd-1.1 wt.%Ag/SiO$_2$ cogelled xerogel catalyst.

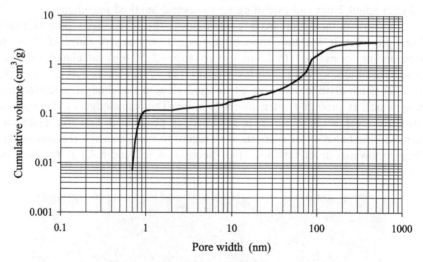

FIGURE 8.8 Pore size distribution of a 1.5wt.%Cu/SiO$_2$ cogelled xerogel catalyst (adapted from [151] with permission from Elsevier).

nitrogen adsorption-desorption at 77 K and mercury porosimetry and by applying appropriate models, it is possible to obtain a more complete view of pores present in a heterogeneous catalyst as shown in Fig. 8 [33, 34, 151, 152, 155, 180, 181]. Indeed, nitrogen adsorption-desorption allows to calculate the micropore size distribution by Brunauer's method [182–184] and the mesopore size distribution by the BJH method [185] or by the more elaborate Broekhoff and de Boer's method [184, 186–188]. When operating up to 200 MPa, mercury porosimetry allows to calculate size distribution of meso- and macropores larger than 7.5 nm. In order to derive correct pore size distributions from the latter technique, it is however essential to remark that highly porous xerogel catalysts prepared by the one-step cogelation method presented in Section 8.2.2.2.2. behave in a particular way during mercury compression that cannot be analyzed by the classical intrusion model of Washburn [184, 189] only. Indeed, submitted to an increasing mercury pressure, those materials exhibit two successive behaviors: at low pressure, they collapse under the isostatic pressure, and above a transition pressure, P_t, which is characteristic of the material composition and microstructure, mercury can enter into the network of small pores not destroyed during the compression at low pressure [190]. Two models must then be used in order to calculate the pore size distribution from mercury porosimetry: Washburn's model describing the mercury intrusion in small pores for pressures $P > P_t$ and Pirard's model describing the collapse of larger pores for pressures $P < P_t$ [190–197]. Note that such a collapse-intrusion mechanism as well as a collapse only mechanism be has been observed for various materials other than inorganic cogelled xerogels, such as polyurethane aero- and xerogels [198], carbon blacks [199, 200], precipitated silicas [194, 201], fumed silicas [202], silica gels [203] as well as SiO_2 ethylene polymerization catalysts [204].

In Figure 8.8, the distribution section below 2 nm has been calculated from N_2 adsorption and Brunauer's model, the section between 2 nm and 7.5 nm has been calculated from N_2 adsorption and Broekhoff and de Boer's model, the section between 7.5 nm and 94 nm has been calculated from Hg porosimetry and Washburn's model and the section beyond 94 nm has been calculated from Hg porosimetry and Pirard's model. The size of 94 nm corresponds to the transition from the collapse mechanism in pores larger than 94 nm to the intrusion mechanism in pores smaller than 94 nm ($P_t = 16$ MPa). The total cumulative porous volume is obtained by summing up the cumulative volumes corresponding those successive ranges.

According to Heinrichs et al. [32, 33], the 0.8 nm micropores are located inside the silica particles and the broad continuous meso- and macropore distribution is located in voids between those SiO2 particles and between aggregates constituted of those particles.

The size of metal particles in cogelled catalysts has been examined extensively [24, 28, 30, 32, 33, 39, 40, 109, 152]. Often, metal particles are distributed into two families of different size: a majority of small nanoparticles in the range 2–5 nm (small black points in Fig. 7) and a few large particles in the range 10–100 nm (four large black points in Fig. 7). Metal particle size depends on metal nature and loading as well as on many synthesis variables. By optimizing the latter ones, it is however possible to obtain small nanoparticles only with a narrow size distribution (2–3 nm) [40].

According to studies on the localization of metal particles in cogelled catalysts [32, 33, 39, 40, 109, 149–152], the small metal particles are located inside silica particles or aggregates and new results from 3D-TEM studies conducted for the moment confirm such a localization. Small metal particles would be located inside silica particles because of a role of nucleation agent played by the alkoxysilane-functionalized complex in the sol-gel process [32, 35, 151, 152, 156, 181]. Briefly, if the alkoxysilane-functionalized ligand used for metal complexation is more reactive towards hydrolysis and condensation than the main silica precursor, which has been demonstrated with the NH2(CH2)2NH(CH2)3Si(OCH3)3 / Si(OC2H5)4 couple [205], therefore, rapid hydrolysis and polycondensation of the former produce a metal complex containing core on which the main silica precursor condenses in a later stage to form a core-shell structure. The observation, in various cogelled catalysts, of a decreasing size of silica particles or aggregates when the ratio between the total amount of silica precursors and the amount of alkoxysilane-functionalized ligand decreases, is in agreement with a nucleation function of the latter.

Contrary to one-step made catalysts prepared by the dissolution method (Section 8.2.2.2.1). that leads sometimes to partial occlusion of metal particles [19, 72, 116], it has been demonstrated that metal particles in one-step made catalysts prepared by cogelation (Section 2.2.2.2). are completely accessible (provided that molecules are not too large) via the 0.8 nm micropores in silica particles since their sizes calculated from TEM and from CO chemisorption agree [32, 40]. Besides being completely accessible, a mass transfer study showed moreover that metal particles are easily accessible [34] (see Section 8.3.3.1). When present, the large metal particles are located outside silica particles.

Such a localization of metal nanocrystallites in very small pores of the support structure has been also observed by Pecchi et al. [16] in the case of catalysts prepared by the one-step dissolution method (Section 8.2.2.2.1). Likewise, those authors obtained quite similar metal particle sizes from TEM and chemisorption. In that case however, a lower activity per site for methane combustion in the catalysts prepared by one-step dissolution than in impregnated catalyst supported on xerogels has been attributed to that localization in the microporous structure.

In the case of impregnated catalysts supported on aerogels or xerogels, due to the molecular size of the metal precursor, metal particles are located outside SiO_2 particles [16, 32]. Heinrichs et al. [32] found a correlation between the size of those metal particles and the pore size distribution of the aerogel or xerogel support: metal particles are larger on supports characterized by larger pores.

8.3.1.2 Metal Catalysts on Carbon Xerogels and Aerogels

As mentioned in the introductory Section, carbon gels are interesting catalyst supports in relation to active carbons produced from natural sources because, unlike the latter ones that are mainly microporous, they can contain a broad distribution of pores spreading over micro-, meso- or even macroporous domains thus enhancing mass transfer [42, 47, 50]. However, in order to obtain such a highly porous texture, synthesis variables such as pH as well as subsequent treatments must be chosen carefully for fear of obtaining a completely non-porous material [44, 50]. Furthermore, Job et al. [50] showed that it is possible to obtain an aerogel-like texture even when wet gel are submitted to a simple evaporative drying.

As in the case of cogelled metal catalysts supported on inorganic xerogels, the morphology and texture of highly porous metal containing carbon xerogels and aerogels synthesized by a one-step method are illustrated by SEM and TEM micrographs in Figures 8.9 and 8.10 respectively as well as by a pore size distribution in Figure 8.11.

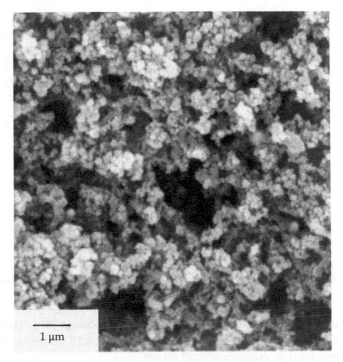

FIGURE 8.9 SEM micrograph of a one-step made 0.5wt.%Pt/C aerogel (reprinted from [44] with permission from Elsevier).

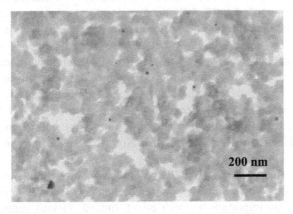

FIGURE 8.10 TEM micrograph of a one-step made 1.1wt.%Pd/C xerogel (reprinted from [50] with permission from Elsevier).

At the scale of Figure 8.9 which is very similar to Figure 8.6 in the case of catalysts supported on inorganic xerogels, it appears that the materials are composed of interconnected carbon aggregates separated by large pores. At higher magnification, Figure 8.10 shows that aggregates are made of interconnected particles. The word 'particles' is maybe inappropriate. Pictures taken with a higher magnification

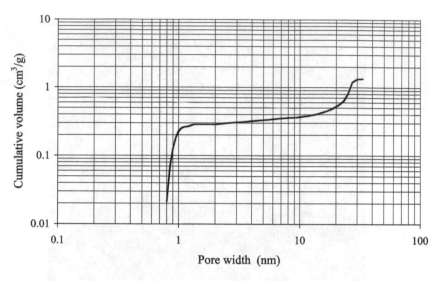

FIGURE 8.11 Pore size distribution of a 1.1wt.%Ni/C one-step made xerogel [50].

show that the carbon materials are in fact very continuous [50]. Though it is not easy to determine the limits of a 'particle', Figure 8.10 allows to determine an approximate particle diameter around 40 nm.

Again, as in the case of inorganic xerogels, Figure 8.11 shows that metal containing carbon gels exhibit a broad pore size distribution from micropores (width < 2 nm) to large mesopores (2 nm < width < 50 nm) or even macropores (width > 50 nm) in some cases [42, 44, 47, 50]. Here again, due to the presence of pores that are much smaller than the size of carbon particles (~40 nm), it can be reasonably assumed that those small pores are located inside the latter ones which are then (micro)porous and that larger pores are located outside carbon particles. Interestingly, Maldonado-Hódar *et al.* [44] found that activation of samples in steam results in a large increase of porosity and surface area, most probably due to a gasification mechanism including the heterogeneous water-gas reaction: $C + H_2O \leftrightarrow H_2 + CO$ [206]. Such a gasification in steam is at the root of the activation process in the manufacture of active carbon [207, 208]. Note that steam can be replaced with carbondioxide.

Examination of Figure 8.10 indicates that obtaining highly dispersed metal particles in one-step made metal containing carbon gels is a challenging task which will still require an important research effort. Indeed, various studies showed that after or even before pyrolysis at high temperature (e.g. 750–1000°C), when reduced in the metallic state (which often occurs during pyrolysis since carbon is a good reductant at high temperature), metal particles are usually rather large (>10 nm) [43, 47, 50]. In some cases, these particles are obviously located outside the carbon network and this suggests that the metal ions or metal complexes are rejected outside the organic phase during the polymerization. In such a case, in order to produce highly dispersed metal catalysts, a ligand able to form a stable complex with the metal to be incorporated in the gel should probably be chosen that react with resorcinol and formaldehyde precursors so as to anchor the metal to the polymeric network as carried out in cogelled metal catalysts supported on inorganic xerogels [50]. However, incorporating small metal particles in the polymeric network could lead, after pyrolysis to metal particles covered with a graphite layer [47] which would make them inaccessible. In such a case, post-treatments in mild oxidative conditions should be used to make metal particles free.

8.3.2 BIMETALLIC CATALYSTS: COMPOSITION OF METALLIC ALLOY PARTICLES

Thanks to the possibility its offers to synthesize supported metal catalysts in one single step, sol-gel process is particularly attractive for preparing bimetallic catalysts [72]. Since the obtaining of an alloy is often a key factor for a bimetallic catalyst to be selective [209–213], the aim of this Section is to show that cogelation (see Section 8.2.2.2.2) is an efficient tool for producing finely dispersed supported metallic alloy nanoparticles.

As listed in Section 8.2.2.2.2, cogelation has been used to prepare Cu-Ru, Pd-Ni, Cu-Ni, Pd-Ag, Pd-Cu, and Pt-Cu catalysts supported on silica. Heinrichs *et al.* [33, 158] and Lambert *et al.* [152, 177] examined the size and composition of metal particles in Pd-Ag/SiO$_2$ and Pd-Cu/SiO$_2$ cogelled xerogel catalysts by X-ray diffraction (XRD) and TEM. XRD results are illustrated in Figure 8.12 in the case of three Pd-Cu/SiO$_2$ catalysts.

Between the (111) Bragg lines of Pd and Cu, the three samples, which contain the same amount of Pd and a growing amount of Cu, exhibit a broad peak whose

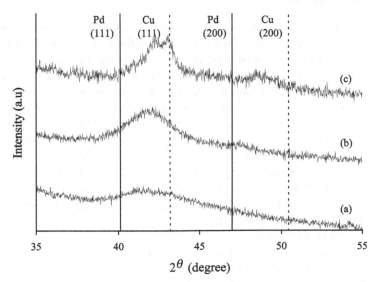

FIGURE 8.12 XRD patterns of Pd-Cu/SiO$_2$ cogelled xerogel catalysts: (a) 1.5 wt.%Pd-0.8 wt.%Cu/SiO$_2$, (b) 1.5wt.%Pd-1.5 wt.%Cu/SiO$_2$, (c) 1.4wt.%Pd-3.0wt.%Cu/SiO$_2$. (reprinted from [177] with permission from Elsevier].

maximum position moves away from the Pd line and comes closer to the Cu line when the Cu loading increases, that is from sample (a) to sample (c). This corresponds to the presence of small alloy particles with a size varying between 2 and 4 nm (confirmed by TEM) that contain an increasing atomic percentage of Cu that has been estimated to 49, 55 and 68 at.% from sample (a) to sample (c). Note that pattern (c) shows also the presence of pure Cu particles as indicated by the second peak on the (111) Cu line [152, 177]. Similar results have been obtained for Pd-Ag/SiO$_2$ samples [33, 152, 158]. Serykh *et al.* [148] who prepared Pt-Cu/SiO$_2$ samples by the cogelation technique confirm also the presence of small alloy nanocrystallites (4–8 nm) from TEM, XPS (X-ray photoelectron spectroscopy) and DRIFT (diffuse reflectance infrared spectroscopy of chemisorbed CO) characterization results. This confirms the ability of cogelation to produce metallic alloy particles highly dispersed on silica. Note that Hammoudeh and Mahmoud [144] report the preparation of Pd-Sn alloys on SiO$_2$ by the dissolution technique (Section 8.2.2.2.1.).

8.3.3 Remarkable Properties of Cogelled Catalysts

Physico-chemical characteristics of cogelled metal catalysts supported on silica presented in the previous Sections lead to two properties of those materials that are particularly useful for catalysts in operation: an easy diffusion of reactants and products inside the catalyst pellets, that is an easy mass transfer, and a high resistance of finely dispersed metal particles towards sintering. Those two properties are detailed below.

8.3.3.1 Mass Transfer

In a heterogeneous catalytic process, mass transfer limitations, that occur when the consumption of reactants on active sites is faster than the diffusive supply of reactants towards those active sites, should be as low as possible for fear of extensively decreasing the overall rate of the process and thus leading to a poor activity of the catalyst. Mass transfer of reactants and products to and from active sites in a porous catalyst support is intimately linked to the internal texture and morphology of the catalyst pellets constituting the catalytic bed in a reactor [214].

Results from the analysis of cogelled catalyst texture and morphology presented in Section 3.1.1. indicate that such catalysts are composed of overlapping entities exhibiting different textures: silica particles are basic blocks which constitute aggregates which themselves constitute the catalyst pellet. In order to reach the active metal particles, reactants have to diffuse through a continuous distribution of macro- and mesopores located in voids between aggregates of elementary SiO$_2$ particles and between those elementary particles and then through the 0.8 nm micropores located inside the elementary SiO$_2$ particles. This situation is schematically illustrated in Figure 8.13 which gives also morphological and textural characteristics of the three levels present in a cogelled catalyst pellet.

As indicated in Figure 8.13, in order to reach the active sites, reactants must diffuse in pores of decreasing width, w_0, (which makes mass transfer more difficult)

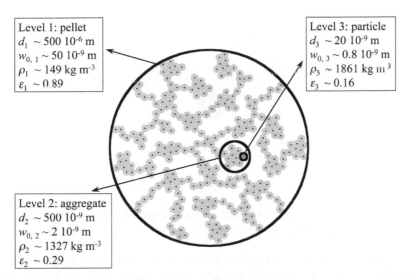

Level 1: pellet
$d_1 \sim 500\ 10^{-6}$ m
$w_{0,1} \sim 50\ 10^{-9}$ m
$\rho_1 \sim 149$ kg m^{-3}
$\varepsilon_1 \sim 0.89$

Level 3: particle
$d_3 \sim 20\ 10^{-9}$ m
$w_{0,3} \sim 0.8\ 10^{-9}$ m
$\rho_3 \sim 1861$ kg m^3
$\varepsilon_3 \sim 0.16$

Level 2: aggregate
$d_2 \sim 500\ 10^{-9}$ m
$w_{0,2} \sim 2\ 10^{-9}$ m
$\rho_2 \sim 1327$ kg m^{-3}
$\varepsilon_2 \sim 0.29$

FIGURE 8.13 Schematic representation of the texture and morphology of a cogelled xerogel catalyst divided in three levels (i = 1, 2, 3): pellet (i = 1), aggregate of elementary SiO$_2$ particles (i = 2), and elementary silica particle (i = 3) (reprinted from [34] with permission from AIChE-Wiley). d_i = characteristic diameter; $w_{0,i}$ = characteristic pore width; ρ_i = bulk density; ε_i = void fraction.

located in entities of decreasing size, d, (which makes mass transfer easier), increasing bulk density, ρ, and decreasing void fraction, ε (which makes mass transfer more difficult). Taken as a whole, those antagonistic effects lead to negligible diffusional limitations which indicates that a catalyst with such a 'funnel' structure exhibits very good mass-transfer properties [34].

8.3.3.2 Thermal Stability of Metal Particles

Resistance of supported metal particles towards sintering is essential for maintaining an acceptable activity in high temperature catalytic processes. Indeed, metal sintering leads to a considerable loss of exposed metal surface and thus to a drastic drop of activity. Sintering of supported metal catalysts has been reviewed in detail by Ruckenstein [215]. Ruckenstein and Pulvermacher [216] have proposed the generation of a chemical trap on the substrate to inhibit sintering. Irregular substrates have valleys and the atoms of the substrate may generate energy wells. In a theoretical study, they suggested that a narrow distribution of metal particle sizes could be attained through the use of these chemical traps [21]. Examples of such traps can be found in one-step made sol-gel catalysts prepared by the dissolution method (Section 8.2.2.2.1) where metal particles are sometimes partially buried in the support leading to an enhanced resistance towards sintering but also to a partial occlusion of the metal corresponding to a loss of accessible metal surface (see Section 8.2.2.1).

Without losing any surface developed by metal nanoparticles, the one-step cogelation method (Section 8.2.2.2.2) allows to build more than a chemical trap for active metal particles: its allows to build a real cage around the latter ones that

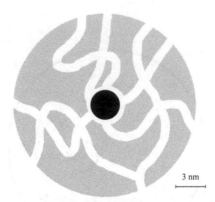

3 nm

FIGURE 8.14 Schematic representation of a metal nanoparticle caged inside a microporous silica particle in sol-gel catalysts prepared by cogelation.

prevent them to sinter by a migration and coalescence mechanism [215]. Indeed, cogelled catalysts contain metal particles in the range 2–5 nm located inside microporous silica particles containing micropores around 0.8 nm (see Section 8.3.1.1). Thus, because they are larger than the micropores of the silica particles in which they are located, the highly dispersed metal crystallites in cogelled catalysts are caged while being completely accessible. Therefore, these crystallites are sinter-proof during high temperature activation and reaction. This situation is illustrated in Figure 8.14.

8.4 USES

8.4.1 CATALYSIS

Because of the easiness of their preparation (one-step method) and their remarkable properties such as high dispersion of metal or alloy, high porosity, easy mass transfer and thermal stability, one-step made sol-gel supported metals are attractive candidates as catalysts for various process of practical interest. Some applications are briefly reported below.

8.4.1.1 Monometallic Catalysts

The ability of one-step sol-gel methods (dissolution and cogelation) to achieve high metal dispersion has been exploited by several authors to prepare efficient monometallic catalyst. For example, Kim *et al.* [134] synthesized Ni/Al$_2$O$_3$ aerogel catalysts in which very small metal particles (~ 3 nm) are evenly distributed over the alumina support. Such a high dispersion, which results from adequate texture, and excellent thermal stability during calcination at 500°C, allowed a remarkable low coking rate of the aerogel catalysts during CO$_2$ reforming of CH$_4$ in comparison with a Ni/Al$_2$O$_3$ catalyst prepared by conventional impregnation. Again for the CO$_2$ reforming of CH$_4$ to synthesis gas (CO/H$_2$), Ji *et al.* [112] showed that a one-step made sol-gel Co/Al$_2$O$_3$ xerogel catalyst possesses smaller metallic Co particles, richer surface OH species and stronger metal-support interaction which leads to a better coking resistivity in comparison with a Co/Al_2O_3 catalyst prepared by conventional impregnation. Lambert *et al.* [40] optimized synthesis variables to produce 2.5 nm Pd particles supported on SiO_2 xerogels by the cogelation method allowing to obtain high activities in 1,2-dichloroethane hydrodechlorination into ethane. Zou and Gonzalez [179] demonstrated the thermal stability in O_2 at 650°C of small Pd particles (2.4 nm) supported on a Pd/SiO_2 xerogel prepared by the dissolution method when the average particle size coincides with the average pore diameter.

In the case of catalysts supported on carbon, Moreno-Castilla *et al.* [42] prepared a WO_3/C aerogel by the one-step method. That catalyst was used in the skeletal isomerization of 1-butene for which it exhibited a good selectivity in isobutene, an easy diffusion of reactants and products, as well as a low deactivation with reaction time thanks to an adequate pore structure.

8.4.1.2 Bimetallic Catalysts

The cogelation method has been applied to the production of Pd-Ag and Pd-Cu alloy particles (~ 2–4 nm) finely dispersed in the porosity of a SiO_2 xerogel [33, 37, 38, 41, 177, 212]. In comparison with catalysts prepared by impregnation, those cogelled xerogels are more selective in olefins, the desired product, during hydrodechlorination of chlorinated alkanes [37]. Moreover, their unique texture and morphology allow an easy diffusion of reactants and products [34] and make them resistant to sintering and coking [38].

In the field of carbon nanotubes synthesis, widely studied for the moment, the one-step sol-gel method is examined as a way to prepare bimetallic catalysts leading to a high productivity in single-walled carbon nanotubes (SWNTs). For example, Su *et al.* [145] prepared a $Fe-Mo/Al_2O_3$ aerogel (dried in supercritical CO_2) and Méhn *et al.* [133] prepared a $Fe-Mo/Al_2O_3$ cryogel (freeze-dried) both with a good activity in SWNTs synthesis.

8.4.1.3 Heterogenization of Homogeneous Catalysts

Besides the preparation of supported metals, the cogelation method described in Section 8.2.2.2.2 is a promising way for the preparation of supported homogeneous catalysts [22, 150, 217]. Concomitant hydrolysis and condensation of alkoxysilane-functionalized metal complexes of the type $(RO)_3Si-X-L_nM$, where $-L_nM$ is a catalytically active metal complex, with a network-forming reagent such as $Si(OC_2H_5)_4$ (TEOS) leads to a heterogenized catalyst in which the homogeneous catalytic complex $-L_nM$ is covalently attached to the silica support. Covalent attachment *versus* entrapment of catalytic metal complex improves the stability of the complex towards leaching and thus enhances the recyclability of catalytic materials considerably [76]. Using that method, Panster *et al.* [22] prepared various immobilized complexes such as $RhCl_3\{S[(CH_2)_3-SiO_{3/2}]_2\}_3.xSiO_2$ hydroformylation catalyst for example. Schubert *et al.* [217] prepared a $Rh(CO)Cl(PPh_2CH_2CH_2SiO_{3/2})_2.xSiO_2$ catalyst by the polycondensation of $Rh(CO)Cl[PPh_2CH_2CH_2Si(OC_2H_5)_3]_2$ with TEOS. The catalyst was used in the hydrosilylation of 1-hexene with triphenylsilane, in CO oxidation as well as in the water-gas shift reaction. Lambert *et al.* [150] prepared acetylacetonate Pd complexes anchored to silica xerogels by hydrolysis and condensation of silylated acetylacetonate Pd complexes with TEOS. Such xerogels could be used in several organic reactions such as cyclopropanation and Kharasch reactions. With the aim of preparing heterogenized pyrazolylpyridine bidentate ligands coordinating a transition metal, Sacco *et al.* [154] developed the synthesis of a new series of trimethoxysilyl-tethered *N*-substituted 3,5-dialkylpyrazolylpyridines.

Let us remark that the sol-gel process can also be used for the preparation of covalent-attached [22] or entrapped [72, 165] enzymes and proteins.

Beside catalysis, other research fields could take benefit from sol-gel prepared supported metals.

8.4.2 ELECTROCHEMISTRY

There has been increasing interest in electrochemical capacitors as energy storage systems because of their high power density and long cycle life, compared to battery devices [43]. Miller *et al.* [126, 127] prepared electrode materials for supercapacitors, with Ru nanoparticles deposited within the pore of a carbon aerogel. With such materials, they obtained specific capacitances as high as 250 F/g. Following that idea, Lin *et al.* [43] developed a one-step sol-gel method to prepare carbon-ruthenium xerogels with high capacitances.

8.4.3 OPTICS

Non-linear optical materials are expected to play an essential role in the emerging technology of photonics. Metal quantum dots, including small metal particles embedded in glasses and gels, have been shown to exhibit strong non-linear optical responses [138, 218]. Following that idea, Selvan *et al.* [218] and Rebbouh *et al.* [138] prepared respectively Au/SiO$_2$ and Fe/SiO$_2$ xerogels by a one-step sol-gel method and showed that these materials exhibit non-linear optical properties.

Let us end this Section on applications of sol-gel prepared supported metal by emphasizing the potential that those materials offer in fields like magnetic resonance medical diagnosis or magnetic recording media. Indeed those fields are looking for encapsulated metal nanoparticles [219, 220] with specific magnetic properties for which the sol-gel methods described in the present paper could be an attractive way of synthesis.

8.5 CONCLUSIONS

One aim of the present review is to show that sol-gel process allows preparing in a very simple way (one step) supported metal catalysts exhibiting some remarkable characteristics in comparison with more conventional catalysts.

In the case of metals on inorganic supports, the cogelation method produces small metal or alloy nanoparticles caged, while being completely accessible, inside microporous silica particles. Such a localization makes those metallic particles sinter-proof during high temperature pretreatments or operation thus leading to thermally stable catalysts. The use of alkoxysilane-functionalized ligand allows obtaining a highly porous aerogel-like structure even after a simple evaporative drying. Moreover, cogelled catalysts are characterized by an easy diffusion of reactants and prod ucts in the porosity of the silica support due to their hierarchical morphology and texture in which elementary microporous silica particles trapping the active sites are the basic blocks that constitute larger mesoporous aggregates which are separated by large macropores and constitute the macroscopic pellets in a catalytic reactor.

Concerning metals on carbon supports, again, a one-step method followed by simple evaporative drying, or possibly a drying in supercritical CO_2, and a high temperature pyrolysis leads to an open and highly porous morphology that can be easily tailored by modifying synthesis variables.

However, in order to make one-step made sol-gel catalysts usable on a large scale in industrial processes, several fundamental and practical issues must still be addressed.

Among the practical issues, the large scale production of xerogel or aerogel catalysts in an adequate shape with, among others, sufficient mechanical strength and resistance towards abrasion, is a challenging task. Some attempts have been carried out with inorganic aerogels and xerogels such as encasement of the gel in a supporting structure (honeycombs, Rashig rings, mineral foams, ...) by addition of those structures in the precursor sol-gel solution [221–223], or the preparation of xerogel catalyst microspheres from a suspension of crushed catalyst containing a binder [224], but an important research effort has still to be performed. Let us remark that, in the case of carbon xerogels, a satisfactory mechanical strength can be obtained provided that macropores are avoided in the final material which can be obtained by adjusting pH at an adequate value during gel formation.

Among the fundamental issues, since supports other than silica are often needed for a catalyst to be efficient (needs for thermal stability at temperature as high as 1000°C, needs for acid or basic sites on the support, ...), studies should be performed to transpose the cogelation method to other supports such as Al_2O_3, TiO_2, mixed oxides, ... while keeping the advantageous properties obtained with SiO_2. Moreover, investigations should be carried out on a way to improve the metal dispersion in metal containing carbon xerogels. Transposing the cogelation mechanism from inorganic to organic gels by using adequate functionalized ligands able, on the one hand to form complexes with the metal under consideration and, on the other hand to link to the growing resorcinol-formaldehyde polymer, could help to obtain a high dispersion. In such a case however, as mentioned in Section 3.1.2., incorporating small metal particles in the polymeric network could lead, after pyrolysis to metal particles covered with a graphite layer which would make them inaccessible. If such an occlusion occurs, post-treatments in mild oxidative conditions will have to be examined with the aim of making metal particles free.

In their review in 1995 on the preparation of catalytic materials by the sol-gel method, Ward and Ko [19] mentioned four issues to be addressed in anticipation of the eventual industrial applications of those materials, one of them being the necessity to prepare high-surface-area, high-pore-volume materials without supercritical drying. It is encouraging for the future of sol-gel catalysts to note that this issue is now solved as demonstrated by the aerogel-like morphology of xerogel catalysts prepared by cogelation. Similarly, the morphology of carbon xerogel catalysts is close to the one of carbon aerogels.

ACKNOWLEDGMENTS

Many coworkers in the Laboratoire de Génie Chimique at the Université de Liège contributed to the progressive building of the knowledge presented in this paper. Many thanks to all of them, in particular to Dr. René Pirard, Dr. Christelle Alié, Dr. Cédric Gommes and Dr. Silvia Blacher. SL is 'collaborateur scientifique' of the Belgian Fonds National de la Recherche Scientifique (FNRS). The FNRS, the Communauté Française de Belgique, la Région Wallonne and the Fonds de Bay are gratefully acknowledged for financial support.

REFERENCES

1. J. F. LePage, "Developing Industrial Catalysts," Handbook of Heterogeneous Catalysis, **1** (G. Ertl, H. Knözinger, J. Weitkamp, Eds.), Wiley-VCH, Weinheim, pp. 49–53 1997.
2. G. Gardes and S. Teichner, "Methods for the Manufacture of Composite Catalysts Containing a Composition of a Transition Metal on a Support," U.S. Patent No. 3,963,646, 1976.
3. G. E. E. Gardes, G. Pajonk, and S. J. Teichner, "Preparation and Properties of Aerogels of Simple or Mixed Inorganic Oxides Containing Metallic Nickel," Bull. Soc. Chim. Fr. 9–10, pp. 1327–32, 1976.
4. M. Astier, A. Bertrand, D. Bianchi, A. Chenard, G. E. E. Gardes, G. Pajonk, M. B. Taghavi, S. J. Teichner, and B. L. Villemin, "Preparation and Catalytic Properties of Supported Metal or Metal-Oxide on Inorganic Oxide Aerogels," Stud. Surf. Sci. Catal. **1**, pp. 315–30, 1976.
5. S. S. Kistler, "Coherent Expanded Aerogels," J. Phys. Chem. 36, pp. 52–64, 1932.
6. S. S. Kistler, "Inorganic aerogel compositions," U.S. Patent No. 2,188,007, 1940.
7. S. S. Kistler, "Method of making aerogels," U.S. Patent No. 2,249,767, 1941.
8. R. C. Reid, J. M. Prausnitz, and B. E. Poling, "The Properties of Gases and Liquids," McGraw-Hill, New York, 1987.
9. J. N. Armor, E. J. Carlson, and P. M. Zambri, "Aerogels as Hydrogenation Catalysts," Appl. Catal., 19, pp. 339–48, 1985.
10. T. López, A. López-Gaona, and R. Gómez, "Synthesis, Characterization and Activity of Ru/SiO_2 Catalysts Prepared by the Sol-Gel Method," J. Non-Cryst. Solids, 110, pp. 170–74, 1989.
11. T. López, A. López-Gaona, and R. Gómez, "Deactivation of Ruthenium Catalysts Prepared by the Sol-Gel Method in Reactions of Benzene Hydrogenation and *n*-Pentane Hydrogenolysis," Langmuir, 6, pp. 1343–46, 1990.
12. T. López, P. Bosch, M. Asomoza, and R. Gómez, "Ru/SiO_2-Impregnated and Sol-Gel-Prepared Catalysts: Synthesis, Characterization, and Catalytic Properties," J. Catal., 133, pp. 247–59, 1992.

13. T. López, R. Gómez, O. Novaro, A. Ramírez-Solís, E. Sánchez-Mora, S. Castillo, E. Poulain, and J. M. Martínez-Magadán, "Sol-Gel Ru/SiO$_2$-Catalysts: Theoretical and Experimental Determination of the Ru-in-Silica Structures," J. Catal., 141, pp. 114–23, 1993.

14. T. López, A. Romero, and R. Gómez, "Metal-Support Interaction in Pt/SiO$_2$ Catalysts Prepared by the Sol-Gel Method," J. Non-Cryst. Solids, 127, pp. 105–13, 1991.

15. T. López, M. Asomoza, P. Bosch, E. Garcia-Figueroa, and R. Gómez, "Spectroscopic Characterization and Catalytic Properties of Sol-Gel Pd/SiO$_2$ Catalysts," J. Catal., 138, pp. 463–73, 1992.

16. G. Pecchi, P. Reyes, I. Concha, and J. L. G. Fierro, "Methane Combustion on Pd/SiO$_2$ Sol Gel Catalysts," J. Catal., 179, pp. 309–14, 1998.

17. K. Balakrishnan and R. D. Gonzalez, "Preparation of Pt/Alumina Catalysts by the Sol-Gel Method," J. Catal., 144, pp. 395–413, 1993.

18. G. M. Pajonk, "Aerogel Catalysts," Appl. Catal., 72, pp. 217–66, 1991.

19. D. A. Ward and E. I. Ko, "Preparing Catalytic Materials by the Sol-Gel Method," Ind. Eng. Chem. Res., 34, pp. 421–33, 1995.

20. G. M. Pajonk, "Catalytic Aerogels," Catal. Today, 35, pp. 319–37, 1997.

21. R. D. Gonzalez, T. López, and R. Gómez, "Sol-Gel Preparation of Supported Metal Catalysts," Catal. Today, 35, pp. 293–317, 1997.

22. U. Deschler, P. Kleinschmit, and P. Panster, "3-Chloropropyltrialkoxysilanes-Key Intermediates for the Commercial Production of Organofunctionalized Silanes and Polysiloxanes," Angew. Chem. Int. Ed. Engl., 25, pp. 236–52, 1986.

23. U. Schubert, S. Amberg-Schwab, and B. Breitscheidel, "Metal Complexes in Inorganic Matrixes. 4. Small Metal Particles in Palladium-Silica Composites by Sol-Gel Processing of Metal Complexes," Chem. Mater., 1, pp. 576–78, 1989.

24. B. Breitscheidel, J. Zieder, and U. Schubert, "Metal Complexes in Inorganic Matrixes. 7. Nanometer-Sized, Uniform Metal Particles in a SiO$_2$ Matrix by Sol-Gel Processing of Metal Complexes," Chem. Mater., 3, pp. 559–66, 1991.

25. U. Schubert, "Catalysts Made of Organic-Inorganic Hybrid Materials," New J. Chem., 18, pp. 1049–58, 1994.

26. W. Mörke, R. Lamber, U. Schubert, and B. Breitscheidel, "Metal Complexes in Inorganic Matrixes. 11. Composition of Highly Dispersed Bimetallic Ni, Pd Alloy Particles Prepared by Sol-Gel Processing: Electron Microscopy and FMR Study," Chem. Mater., 6, pp. 1659–66, 1994.

27. A. Kaiser, A. Görsmann, and U. Schubert, "Influence of the Metal Complexation on Size and Composition of Cu/Ni Nano-Particles Prepared by Sol-Gel Processing," J. Sol-Gel Sci. Technol., 8, pp. 795–99, 1997.

28. C. Lembacher and U. Schubert, "Nanosized Platinum Particles by Sol-Gel Processing of Tethered Metal Complexes: Influence of the Precursors and the Organic Group Removal Method on the Particle Size," New J. Chem., 22, pp. 721–24, 1998.

29. G. Trimmel and U. Schubert, "Sol-Gel Processing of Tethered Metal Complexes: Influence of the Metal and the Complexing Alkoxysilane on the Texture of the Obtained Silica Gels," J. Non-Cryst. Solids, 296, pp. 188–200, 2001.

30. G. Trimmel, C. Lembacher, G. Kickelbick, and U. Schubert, "Sol-Gel Processing of Alkoxysilyl-Substituted Nickel Complexes for the Preparation of Highly Dispersed Nickel in Silica," New J. Chem., 26, pp. 759–65, 2002.

31. U. Schubert, "Metal Oxide/Silica and Metal/Silica Nanocomposites from Organo-functional Single-Source Sol-Gel Precursors," Adv. Eng. Mater., 6, pp. 173–76, 2004.

32. B. Heinrichs, F. Noville, and J.-P. Pirard, "Pd/SiO$_2$-Cogelled Aerogel Catalysts and Impregnated Aerogel and Xerogel Catalysts: Synthesis and Characterization," J. Catal., 170, pp. 366–76, 1997.

33. B. Heinrichs, P. Delhez, J.-P. Schoebrechts, and J.-P. Pirard, "Palladium-Silver Sol-Gel Catalysts for Selective Hydrodechlorination of 1,2-Dichloroethane into Ethylene I. Synthesis and Characterization," J. Catal., 172, pp. 322–35, 1997.

34. B. Heinrichs, J.-P. Pirard, and J.-P. Schoebrechts, "Mass Transfer in Low-Density Xerogel Catalysts," AIChE J., 47, pp. 1866–73, 2001.

35. C. Alié, A. J. Lecloux, and J.-P. Pirard, "Control and Characterisation of the Morphology of Solids Prepared via Sol-Gel Chemistry," Recent Research Developments in Non-Crystalline Solids, 2 (S. G. Pardalai, Ed.), Transworld Research Network, Kerala, pp. 335–62, 2002.

36. B. Heinrichs, J.-P. Pirard, and R. Pirard, "Transition Metal Aerogel-Supported Catalyst," U.S. Patent No. 5,538,931, 1996.

37. P. Delhez, B. Heinrichs, J.-P. Pirard, and J.-P. Schoebrechts, "Process for the Preparation of a Catalyst and its Use for the Conversion of Chloroalkanes into Alkenes Containing less Chlorine," U.S. Patent No. 6,072,096, 2000.

38. B. Heinrichs, F. Noville, J.-P. Schoebrechts, and J.-P. Pirard, "Palladium-Silver Sol-Gel Catalysts for Selective Hydrodechlorination of 1,2-Dichloroethane into Ethylene IV. Deactivation Mechanism and Regeneration," J. Catal., 220, pp. 215–25, 2003.

39. S. Lambert, C. Cellier, P. Grange, J.-P. Pirard, and B. Heinrichs, "Synthesis of Pd/SiO$_2$, Ag/SiO$_2$, and Cu/SiO$_2$ Cogelled Xerogel Catalysts: Study of Metal Dispersion and Catalytic Activity," J. Catal., 221, pp. 335–46, 2004.

40. S. Lambert, J.-F. Polard, J.-P. Pirard, and B. Heinrichs, "Improvement of Metal Dispersion in Pd/SiO$_2$ Cogelled Xerogel Catalysts for 1,2-Dichloroethane Hydrodechlorination," Appl. Catal. B, 50, pp. 127–40, 2004.

41. S. Lambert, F. Ferauche, A. Brasseur, J.-P. Pirard, and B. Heinrichs, "Pd-Ag/SiO$_2$ and Pd-Cu/SiO$_2$ Cogelled Xerogel Catalysts for Selective Hydrodechlorination of 1,2-Dichloroethane into Ethylene," Catal. Today, 100, pp. 283–89, 2005.

42. C. Moreno-Castilla, F. J. Maldonado-Hódar, J. Rivera-Utrilla, and E. Rodríguez-Castellón, "Group 6 Metal Oxide-Carbon Aerogels. Their Synthesis, Characterization and Catalytic Activity in the Skeletal Isomerization of 1-Butene," Appl. Catal. A, 183, pp. 345–56, 1999.

43. C. Lin, J. A. Ritter, and B. N. Popov, "Development of Carbon-Metal Oxide Supercapacitors from Sol-Gel Derived Carbon-Ruthenium Xerogels," J. Electrochem. Soc., 146, pp. 3155–60, 1999.

44. F. J. Maldonado-Hódar, M. A. Ferro-García, J. Rivera-Utrilla, and C. Moreno-Castilla, "Synthesis and Textural Characteristics of Organic Aerogels, Transition-Metal-Containing Organic Aerogels and their Carbonized Derivatives," Carbon, **37**, pp. 1199–1205, 1999.

45. E. Bekyarova and K. Kaneko, "Structure and Physical Properties of Tailor-Made Ce, Zr-Doped Carbon Aerogels," Adv. Mater., 12, pp. 1625–28, 2000.

46. F. J. Maldonado-Hódar, C. Moreno-Castilla, J. Rivera-Utrilla, and M. A. Ferro-García, "Metal-Carbon Aerogels as Catalysts and Catalyst Supports," Stud. Surf. Sci. Catal., **130**, pp. 1007–12, 2000.

47. F. J. Maldonado-Hódar, C. Moreno-Castilla, J. Rivera-Utrilla, Y. Hanzawa, and Y. Yamada, "Catalytic Graphitization of Carbon Aerogels by Transition Metals," Langmuir, 16, pp. 4367–73, 2000.

48. N. Job, F. Ferauche, R. Pirard, and J.-P. Pirard, "Single Step Synthesis of Metal Catalysts Supported on Porous Carbon with Controlled Texture," Stud. Surf. Sci. Catal., 143, pp. 619–26, 2002.

49. J.-P. Pirard, R. Pirard, and N. Job, "Matériau Carboné Poreux," European Patent EP 1280215, 2003.

50. N. Job, R. Pirard, J. Marien, and J.-P. Pirard, "Synthesis of Transition Metal-Doped Carbon Xerogels by Solubilization of Metal Salts in Resorcinol-Formaldehyde Aqueous Solution," Carbon, 42, pp. 3217–27, 2004.

51. R. W. Pekala and F. M. Kong, "A Synthetic Route to Organic Aerogels: Mechanism, Structures and Properties," Rev. Phys. Appl., 24, pp. 33–40, 1989.

52. R. W. Pekala, C. T. Alviso, and J. D. LeMay, "Organic Aerogels: Microstructural Dependence of Mechanical Properties in Compression," J. Non-Cryst. Solids, 125, pp. 67–75, 1990.

53. R. W. Pekala, "Low Density Resorcinol-Formaldehyde Aerogels," U.S. Patent No. 4,997,804, 1991.

54. R. W. Pekala, C. T. Alviso, and J. D. LeMay, "Organic Aerogels: a New Type of Ultrastructured Polymer," Chemical Processing of Advanced Materials (L. L. Hench and J. K. West, Eds.), Wiley, New York, pp. 671–83, 1992.

55. R. W. Pekala, and C. T. Alviso, "Carbon Aerogels and Xerogels," Mater. Res. Soc. Symp. Proc., 270, pp. 3–14, 1992.

56. S. T. Mayer, J. L. Kaschmitter, and R. W. Pekala, "Method of Low Pressure and/or Evaporative Drying of Aerogel," U.S. Patent No. 5,420,168, WO 9422943, 1993.

57. R. W. Pekala, C. T. Alviso, X. Lu, J. Gross, and J. Fricke, "New Organic Aerogels Based upon a Phenolic-Furfural Reaction," J. Non-Cryst. Solids, 188, pp. 34–40, 1995.

58. J. L. Kaschmitter, S. T. Mayer, and R. W. Pekala, "Carbon Aerogel Electrodes for Direct Energy Conversion," U.S. Patent No. 5,601,938, WO 9520246, 1995.

59. C. Lin, and J. A. Ritter, "Effect of Synthesis pH on the Structure of Carbon Xerogels," Carbon, 35, pp. 1271–78, 1997.

60. B. Mathieu, S. Blacher, R. Pirard, J.-P. Pirard, B. Sahouli, and F. Brouers, "Freeze-Dried Resorcinol-Formaldehyde Gels," J. Non-Cryst. Solids, 212, pp. 250–61, 1997.

61. B. Mathieu, B. Michaux, O. Van Cantfort, F. Noville, R. Pirard, and J.-P. Pirard, "Synthesis of Resorcinol-Formaldehyde Aerogels by the Freeze-Drying Method," Ann. Chim. Fr., 22, pp. 19–29, 1997.

62. R. W. Pekala, J. C. Farmer, C. T. Alviso, T. D. Tran, S. T. Mayer, J. M. Miller, and B. Dunn, "Carbon Aerogels for Electrochemical Applications," J. Non-Cryst. Solids, 225, pp. 74–80, 1998.

63. R. Kocklenberg, B. Mathieu, S. Blacher, R. Pirard, J.-P. Pirard, R. Sobry, and G. Van den Bossche, "Texture Control of Freeze-Dried Resorcinol-Formaldehyde Gels," J. Non-Cryst. Solids, 225, pp. 8–13, 1998.

64. J. L. Kaschmitter, R. L. Morrison, S. T. Mayer, and R. W. Pekala, "Method for Making Thin Carbon Foam Electrodes," U.S. Patent No. 5,932,185, WO 9506002, 1999.

65. C. Lin, and J. A. Ritter, "Carbonization and Activation of Sol-Gel Derived Carbon Xerogels," Carbon, 38, pp. 849–61, 2000.

66. N. Job, R. Pirard, J. Marien, and J.-P. Pirard, "Porous Carbon Xerogels with Texture Tailored by pH control during Sol-Gel Process," Carbon, 42, pp. 619–28, 2004.

67. C. J. Brinker and G. W. Scherer, "Sol-Gel Science: The Physics and Chemistry of Sol-Gel Processing," Academic Press, San Diego, 1990.

68. P. Flory, "Principles of Polymer Chemistry," Cornell University Press, Ithaca, 1971.

69. D. W. Schaefer, R. Pekala, and G. Beaucage, "Origin of Porosity in Resorcinol-Formaldehyde Aerogels," J. Non-Cryst. Solids, 186, pp. 159–67, 1995.

70. C. Gommes, S. Blacher, B. Goderis, R. Pirard, B. Heinrichs, C. Alié, and J.-P. Pirard, "*In Situ* SAXS Analysis of Silica Gel Formation with an Additive", J. Phys. Chem. B, 108, pp. 8983–91, 2004.

71. I. M. Thomas, "Multicomponent Glasses from the Sol-Gel Process," Sol-Gel Technology for Thin Films, Fibers, Preforms, Electronics and Specialty Shapes (L. C. Klein, Ed.), Noyes Publications, Park Ridge, NJ, pp. 2–15, 1988.

72. E. I. Ko, "Sol-Gel Process," Handbook of Heterogeneous Catalysis (G. Ertl, H. Knözinger, and J. Weitkamp, Eds.), Wiley-VCH, Weinheim, pp. 86–94, 1997.

73. M. Schneider and A. Baiker, "Titania-Based Aerogels," Catal. Today, 35, pp. 339–65, 1997.

74. D. C. Bradley, R. C. Mehrotra, and D. P. Gaur, "Metal Alkoxides," Academic Press, London, 1978.

75. R. K. Iler, "The Chemistry of Silica," Wiley, New York, 1979.

76. S. P. Watton, C. M. Taylor, G. M. Kloster, and S. C. Bowman, "Coordination Complexes in Sol-Gel Silica Materials," Progress in Inorganic Chemistry, 51 (K. D. Karlin, Ed.), Wiley, New York, pp. 333–420, 2003.

77. C. J. Brinker, "Hydrolysis and Condensation of Silicates: Effects on Structure," J. Non-Cryst. Solids, 100, pp. 31–50, 1988.

78. A. J. Lecloux and J.-P. Pirard, "High-Temperature Catalysts Through Sol-Gel Synthesis," J. Non-Cryst. Solids, 225, pp. 146–52, 1998.

79. D. C. M. Dutoit, M. Schneider, and A. Baiker, "Titania-Silica Mixed Oxides I. Influence of Sol-Gel and Drying Conditions on Structural Properties," J. Catal., 153, pp. 165–76, 1995.

80. F. Legrand, L. Lerot, and P. De Bruycker, "Process for the Manufacture of a Powder of Mixed Metal Oxides, and Mixed Metal Oxide Powders," U.S. Patent No. 4,929,436, 1990.

81. K. Tanabe, "Solid Acid and Base Catalysts," Catalysis: Science and Technology, 2 (J. R. Anderson and M. Boudart, Eds.), Springer, Berlin, pp. 231–273, 1981.

82. W. K. Hall, "Acidity and Basicity - Introduction, Chemical Characterization," Handbook of Heterogeneous Catalysis (G. Ertl, H. Knözinger, and J. Weitkamp, Eds.), Wiley-VCH, Weinheim, pp. 689–98, 1997.

83. B. E. Yoldas, "Monolithic Glass Formation by Chemical Polymerization," J. Mater. Sci., 14, pp. 1843–49, 1979.

84. B. E. Yoldas, "Deposition and Properties of Optical Oxide Coatings from Polymerized Solutions," Appl. Opt., 21, pp. 2960–64, 1982.

85. M. Schraml-Marth, K. L. Walther, A. Wokaun, B. E. Handy, and A. Baiker, "Porous Silica Gels and Titania/Silica Mixed Oxides Prepared via the Sol-Gel Process: Characterization by Spectroscopic Techniques," J. Non-Cryst. Solids, 143, pp. 93–111, 1992.

86. C. Sanchez, J. Livage, M. Henry, and F. Babonneau, "Chemical Modification of Alkoxide Precursors," J. Non-Cryst. Solids, 100, pp. 65–76, 1988.

87. J. Livage and C. Sanchez, "Sol-Gel Chemistry," J. Non-Cryst. Solids, 145, pp. 11–19, 1992.

88. R. Pirard, D. Bonhomme, S. Kolibos, J.-P. Pirard, and A. J. Lecloux, "Textural Properties and Thermal Stability of Silica-Zirconia Aerogels," J. Sol-Gel Sci. Technol., 8, pp. 831–36, 1997.

89. O. Van Cantfort, A. Abid, B. Michaux, B. Heinrichs, R. Pirard, J.-P. Pirard, and A. J. Lecloux, "Synthesis and Characterization of Porous Silica-Alumina Xerogels," J. Sol-Gel Sci. Technol., 8, pp. 125–30, 1997.

90. A. Pirson, A. Moshine, P. Marchot, B. Michaux, O. Van Cantfort, J.-P. Pirard, and A. J. Lecloux, "Synthesis of SiO$_2$-TiO$_2$ Xerogels by Sol-Gel Process," J. Sol-Gel Sci. Technol., 4, pp. 179–85, 1995.

91. J.-P. Pirard, P. Petit, A. Moshine, B. Michaux, F. Noville, and A. J. Lecloux, "Silica-Zirconia Monoliths from Gels," J. Sol-Gel Sci. Technol., 2, pp. 875–80, 1994.

92. W. Rupp, N. Huesing, and U. Schubert, "Preparation of Silica-Titania Xerogels and Aerogels by Sol-Gel Processing of New Single-Source Precursors," J. Mater. Chem., 12, pp. 2594–96, 2002.

93. D. W. Schaefer and K. D. Keefer, "Fractal Geometry of Silica Condensation Polymers," Phys. Rev. Lett., 53, pp. 1383–86, 1984.

94. E. Blanco, M. Ramirez-del-Solar, N. De la Rosa-Fox, and A. F. Craievich, "SAXS Study of Growth Kinetics of Fractal Aggregates in TEOS-Water-Alcohol Solutions with Formamide," J. Non-Cryst. Solids., 147&148, pp. 238–44, 1992.

95. D. R. Vollet, D. A. Donatti, and A. Ibañez Ruiz, "A SAXS Study of Kinetics of Aggregation of TEOS-Derived Sonogels at Different Temperatures," J. Non-Cryst. Solids, 288, pp. 81–87, 2001.

96. S. Lebon, J. Marignan, and J. Appell, "Titania Gels: Aggregation and Gelation Kinetics," J. Non-Cryst. Solids, 147&148, pp. 92–96, 1992.

97. D. Chaumont, A. Craievich, and J. Zarzycki, "A SAXS Study of the Formation of Zirconia Sols and Gels," J. Non-Cryst. Solids, 147&148, pp. 127–34, 1992.

98. A. Lecomte, A. Dauger, and P. Lenormand, "Dynamical Scaling Property of Colloidal Aggregation in a Zirconia-Based Precursor Sol During Gelation," J. Appl. Crystallogr., 33, pp. 496–99, 2000.

99. H. Kaji, K. Nakanishi, N. Soga, T. Inoue, and N. Nemoto, "*In Situ* Observation of Phase Separation Processes in Gelling Alkoxy-Derived Silica System by Light Scattering Method," J. Sol-Gel Sci. Technol., 3, pp. 169–88, 1994.

100. K. Nakanishi, "Pore Structure Control of Silica Gels Based on Phase Separation," J. Porous Mater., 4, pp. 67–112, 1997.

101. G. M. Pajonk, A. V. Rao, N. Pinto, F. Ehrburger-Dolle, and M. B. Gil, "Monolithic Carbon Aerogels for Fuel Cell Electrodes," Stud. Surf. Sci. Catal., 118, pp. 167–74, 1998.

102. A. Streitwieser Jr. and C. H. Heathcock, "Introduction to Organic Chemistry," Macmillan Publishing Co., New York, 1976.

103. H. A. Yu, T. Kaneko, S. Yoshimura, and S. Otani, "A Turbostratic Carbon with High Specific Surface Area from 1,4-Benzenedimethanol," Carbon, 34, pp. 676–78, 1996.

104. R. W. Pekala and D. W. Schaefer, "Structure of Organic Aerogels. 1. Morphology and Scaling," Macromolecules, 26, pp. 5487–93, 1993.

105. H. Tamon and H. Ishizaka, "SAXS Study on Gelation Process in Preparation of Resorcinol-Formadehyde Aerogel," J. Colloid Interf. Sci., 206, pp. 577–82, 1998.

106. K. Foger, "Dispersed Metal Catalysts," Catalysis: Science and Technology, 6 (J. R. Anderson and M. Boudart, Eds.), Springer-Verlag, Berlin, pp. 227–305, 1984.

107. M. Che, O. Clause, and C. Marcilly, "Impregnation and Ion Exchange," Handbook of Heterogeneous Catalysis, 1 (G. Ertl, H. Knözinger, and J. Weitkamp, Eds.), Wiley-VCH, Weinheim, pp. 191–207, 1997.

108. B. Heinrichs, P. Delhez, J.-P. Schoebrechts, and J.-P. Pirard, "Pd-Ag/SiO$_2$ Sol-Gel Catalysts Designed for Selective Conversion of Chlorinated Alkanes into Alkenes," Stud. Surf. Sci. Catal., 118, pp. 707–16, 1998.

109. B. Heinrichs, S. Lambert, C. Alié, J.-P. Pirard, G. Beketov, V. Nehasil, and N. Kruse, "Cogelation: an Effective Sol-Gel Method to Produce Sinter-Proof Finely Dispersed Metal Catalysts Supported on Highly Porous Oxides," Stud. Surf. Sci. Catal., 143, pp. 25–33, 2002.

110. C. Guelduer and F. Balikci, "Catalytic Oxidation of CO over Ag-Co/Alumina Catalysts," Chem. Eng. Commun., 190, pp. 986–98, 2003.

111. A. M. Venezia, V. La Parola, G. Deganello, D. Cauzzi, G. Leonardi, and G. Predieri, "Influence of the Preparation Method on the Thiophene HDS Activity of Silica Supported CoMo Catalysts," Appl. Catal. A, 229, pp. 261–71, 2002.

112. L. Ji, S. Tang, H. C. Zeng, J. Lin, and K. L. Tan, "CO_2 Reforming of Methane to Synthesis Gas over Sol-Gel-Made Co/γ-Al_2O_3 Catalysts from Organometallic Precursors," Appl. Catal. A., 207, pp. 247–55, 2001.

113. P. Gronchi, A. Kaddouri, P. Centola, and R. Del Rosso, "Synthesis of Nickel Supported Catalysts for Hydrogen Production by Sol-Gel Method," J. Sol-Gel Sci. Technol., 26, pp. 843–46, 2003.

114. M. Asomoza, T. López, R. Gómez, and R. D. Gonzalez, "Synthesis of High Surface Area Supported Pt/SiO_2 Catalysts from $H_2PtCl_6·6H_2O$ by the Sol-Gel Method," Catal. Today, 15, pp. 547–54, 1992.

115. T. López, L. Herrera, R. Gómez, W. Zou, K. Robinson, and R. D. Gonzalez, "Improved Mechanical Stability of Supported Ru Catalysts: Preparation by the Sol-Gel Method," J. Catal., 136, pp. 621–25, 1992.

116. T. López, P. Bosch, J. Navarrete, M. Asomoza, and R. Gómez, "Structure of Pd/SiO_2 Sol-Gel and Impregnated Catalysts," J. Sol-Gel Sci. Technol., 1, pp. 193–203, 1994.

117. J. W. Geus and J. A. R. van Veen, "Preparation of Supported Catalysts," Stud. Surf. Sci. Catal., 123, pp. 459–85, 1999.

118. W. Zou and R. D. Gonzalez, "Pretreatment Chemistry in the Preparation of Silica-Supported Pt, Ru, and Pt-Ru Catalysts: An *in situ* UV Diffuse Reflectance Study," J. Catal., 133, pp. 202–19, 1992.

119. W. Zou and R. D. Gonzalez, "The Preparation of High-Surface-Area Pt/SiO_2 Catalysts with Well-Defined Pore Size Distributions," J. Catal., 152, pp. 291–305, 1995.

120. H. Kalies, D. Bianchi, and G. M. Pajonk, "Hydrogenation of Carbonaceous Adsorbed Species on a Zirconia Aero-Gel Catalyst in Presence of Platinum," Spillover and Migration of Surface Species on Catalysts (C. Li and Q. Xin, Eds.), Elsevier, Amsterdam, pp. 81–91, 1997.

121. G. Pecchi, P. Reyes, and J. Villasenor, "Fe Supported Catalysts Prepared by the Sol-Gel Method. Caracterization and Evaluation in Phenol Abatement," J. Sol-Gel Sci. Technol., 26, pp. 865–67, 2003.

122. J. A. Wang, A. Cuan, J. Salmones, N. Nava, S. Castillo, M. Moran-Pineda, F. Rojas, "Studies of Sol-Gel TiO_2 and Pt/TiO_2 Catalysts for NO Reduction by CO in an Oxygen-Rich Condition," Appl. Surf. Sci., 230, pp. 94–105, 2004.

123. M. A. Cauqui, J. J. Calvino, G. Cifredo, L. Esquivias, and J. M. Rodríguez-Izquierdo, "Preparation of Rhodium Catalysts Dispersed on TiO_2-SiO_2 Aerogels," J. Non-Cryst. Solids., 147&148, pp. 758–63, 1992.

124. J. J. Calvino, M. A. Cauqui, G. Cifredo, J. M. Rodríguez-Izquierdo, and H. Vidal, "Microstructure and Catalytic Properties of Rh and Ni Dispersed on TiO_2-SiO_2 Aerogels," J. Sol-Gel Sci. Technol., 2, pp. 831–36, 1994.

125. J. Escobar, J. Antonio De Los Reyes, and T. Viveros, "Nickel on TiO_2-Modified Al_2O_3 Sol-Gel Oxides. Effect of Synthesis Parameters on the Supported Phase Properties," Appl. Catal. A, 253, pp. 151–63, 2003.

126. J. M. Miller, B. Dunn, T. D. Tran, and R. W. Pekala, "Deposition of Ruthenium Nanoparticles on Carbon Aerogels for High Energy Density Supercapacitor Electrodes," J. Electrochem. Soc., 144, pp. L309–11, 1997.

127. J. M. Miller and B. Dunn, "Morphology and Electrochemistry of Ruthenium/Carbon Aerogel Nanostructures," Langmuir, 15, pp. 799–806, 1999.

128. N. Job, B. Heinrichs, F. Ferauche, F. Noville, J. Marien, and J.-P. Pirard, "Hydrodechlorination of 1,2-Dichloroethane on Pd-Ag Catalysts Supported on Tailored Texture Carbon Xerogels," Catal. Today, 102–103, pp. 234–41, 2005.

129. Y. I. Yermakov, "Supported Catalysts Obtained by Interaction of Organometallic Compounds of Transition Elements with Oxide Supports," Catal. Rev., 13, pp. 77–120, 1976.

130. Y. I. Yermakov and B. N. Kuznetzov, "Supported Metallic Catalysts Obtained via Organometallic Compounds of Transition Elements," Kinet. Catal., 18, pp. 955–65, 1977.

131. Y. I. Yermakov and B. N. Kuznetzov, "Supported Metallic Catalysts Prepared by Decomposition of Surface Organometallic Complexes," J. Mol. Catal., 9, pp. 13–40, 1980.

132. R. Gómez, V. Bertin, M. A. Ramirez, T. Zamudio, P. Bosch, I. Schifter, and T. López, "Synthesis, Characterization and Activity of Pt-Sn/Al$_2$O$_3$ Sol-Gel Catalysts," J. Non-Cryst. Solids, 147&148, pp. 748–52, 1992.

133. D. Méhn, A. Fonseca, G. Bister, and J. B. Nagy, "A Comparison of Different Preparation Methods of Fe/Mo/Al$_2$O$_3$ Sol-Gel Catalyst for Synthesis of Single Wall Carbon Nanotubes," Chem. Phys. Lett., 393, pp. 378–84, 2004.

134. J.-H. Kim, D. J. Suh, T.-J. Park, and K.-L. Kim, "Effect of Metal Particle Size on Coking During CO$_2$ Reforming of CH$_4$ over Ni-Alumina Aerogel Catalysts," Appl. Catal. A, 197, pp. 191–200, 2000.

135. P. Kim, Y. Kim, H. Kim, I. K. Song, and J. Yi, "Synthesis and Characterization of Mesoporous Alumina with Nickel Incorporated for Use in the Partial Oxidation of Methane into Synthesis Gas," Appl. Catal. A, 272, pp. 157–66, 2004.

136. Á. Kukovecz, Z. Kónya, N. Nagaraju, I. Willems, A. Tamási, A. Fonseca, J. B. Nagy, and I. Kiricsi, "Catalytic Synthesis of Carbon Nanotubes over Co, Fe and Ni Containing Conventional and Sol-Gel Silica-Aluminas," Phys. Chem. Chem. Phys., 2, pp. 3071–76, 2000.

137. M. Pérez-Cabero, I. Rodríguez-Ramos, and A. Guerrero-Ruíz, "Characterization of Carbon Nanotubes and Carbon Nanofibers Prepared by Catalytic Decomposition of Acetylene in a Fluidized Bed Reactor," J. Catal., 215, pp. 305–16, 2003.

138. L. Rebbouh, V. Rosso, Y. Renotte, Y. Lion, F. Grandjean, B. Heinrichs, J.-P. Pirard, J. Delwiche, M.-J. Hubin-Franskin, and G. J. Long, "The Nonlinear Optical, Magnetic, and Mössbauer Spectral Properties of Some Iron(III) Doped Silica Xerogels," J. Mater. Sci., 41, pp. 2839–2849, 2006.

139. C. Guimon, A. Auroux, E. Romero, and A. Monzon, "Acetylene Hydrogenation over Ni-Si-Al Mixed Oxides Prepared by Sol-Gel Technique," Appl. Catal. A, **251**, pp. 199–214, 2003.

140. B. C. Dunn, D. J. Covington, P. Cole, R. J. Pugmire, H. L. C. Meuzelaar, R. D. Ernst, E. C. Heider, E. M. Eyring, N. Shah, G. P. Huffman, M. S. Seehra, A. Manivannan, and P. Dutta, "Silica Xerogel Supported Cobalt Metal Fisher-Tropsch Catalysts for Syngas to Diesel Range Fuel," Energy & Fuels, 18, pp. 1519–21, 2004.

141. D. J. Suh, T.-J. Park, J.-H. Kim, and K.-L. Kim, "Nickel-Alumina Aerogel Catalysts Prepared by Fast Sol-Gel Synthesis," J. Non-Cryst. Solids, 225, pp. 168–72, 1998.

142. R. H. Hina and R. Kh. Al-Fayyoumi, "Conversion of Dichlorodifluoromethane with Hydrogen over Pd/AlF$_3$ and Ru/AlF$_3$ Prepared by Sol-Gel Method," J. Mol. Catal. A, 207, pp. 27–33, 2004.

143. D. H. Kim, S. I. Woo, J. Noh, and O-B. Yang, "Synergistic Effect of Vanadium and Zirconium Oxides in the Pd-Only Three-Way Catalysts Synthesized by Sol-Gel Method," Appl. Catal. A, 207, pp. 69–77, 2001.

144. A. Hammoudeh and S. Mahmoud, "Selective Hydrogenation of Cinnamaldehyde over Pd/SiO$_2$ Catalysts: Selectivity Promotion by Alloyed Sn," J. Mol. Catal. A, 203, pp. 231–39, 2003.

145. M. Su, B. Zheng, and J. Liu, "A Scalable CVD Method for the Synthesis of Single-Walled Carbon Nanotubes with High Catalyst Productivity," Chem. Phys. Lett., 322, pp. 321–26, 2000.

146. Z. Liu, J. Hao, L. Fu, T. Zhu, J. Li, and X. Cui, "Activity Enhancement of Bimetallic Co-In/Al$_2$O$_3$ Catalyst for the Selective Reduction of NO by Propene," Appl. Catal. B, 48, pp. 37–48, 2004.

147. J. Salmones, J.-A. Wang, J. A. Galicia, and G. Aguilar-Rios, "H$_2$ Reduction Behaviors and Catalytic Performance of Bimetallic Tin-Modified Platinum Catalysts for Propane Dehydrogenation," J. Mol. Catal. A, 184, pp. 203–13, 2002.

148. A. I. Serykh, O. P. Tkachenko, V. Y. Borovkov, V. B. Kazansky, K. M. Minachev, C. Hippe, N. I. Jaeger, and G. Schulz-Ekloff, "Characterization of Silica-Gel Supported Pt-Cu Alloy Particles Prepared *via* the Sol-Gel Technique," Phys. Chem. Chem. Phys., 2, pp. 2667–72, 2000.

149. C. Alié, S. Lambert, B. Heinrichs, and J.-P. Pirard, "Nucleation Phenomenon in Silica Xerogels and Pd/SiO$_2$, Ag/SiO$_2$, Cu/SiO$_2$ Cogelled Catalysts," J. Sol-Gel Sci. Technol., 26, pp. 827–30, 2003.

150. S. Lambert, L. Sacco, F. Ferauche, B. Heinrichs, A. Noels, and J.-P. Pirard, "Synthesis of SiO$_2$ Xerogels and Pd/SiO$_2$ Cogelled Xerogel Catalysts from Silylated Acetylacetonate Ligand," J. Non-Cryst. Solids, 343, pp. 109–20, 2004.

151. S. Lambert, C. Alié, J.-P. Pirard, and B. Heinrichs, "Study of Textural Properties and Nucleation Phenomenon in Pd/SiO$_2$, Ag/SiO$_2$ and Cu/SiO$_2$ Cogelled Xerogel Catalysts," J. Non-Cryst. Solids, 342, pp. 70–81, 2004.

152. S. Lambert, C. Gommes, C. Alié, N. Tcherkassova, J.-P. Pirard, and B. Heinrichs, "Formation and Structural Characteristics of Pd-Ag/SiO$_2$ and Pd-Cu/SiO$_2$ Catalysts Synthesized by Cogelation," J. Non-Cryst. Solids, 351, pp. 3839–53, 2005.

153. D. Avnir, L. C. Klein, D. Levy, U. Schubert, and A. B. Wojcik, "Organo-Silica Sol-Gel Materials," The Chemistry of Organic Silicon Compounds, 2 (Z. Rappoport and Y. Apeloig, Eds.), Wiley, Chichester, pp. 2317–62, 1998.

154. L. Sacco, S. Lambert, J.-P. Pirard, and A. F. Noels, "Synthesis of Pyrazolylpyridine Derivatives Bearing a Tethered Alkoxysilyl Group," Synthesis, 5, pp. 663–64, 2004.

155. C. Alié, R. Pirard, A. J. Lecloux, and J.-P. Pirard, "Preparation of Low-Density Xerogels Through Additives to TEOS-Based Alcogels," J. Non-Cryst. Solids, 246, pp. 216–28, 1999.

156. C. Alié, R. Pirard, A. J. Lecloux, and J.-P. Pirard, "The Use of Additives to Prepare Low-Density Xerogels," J. Non-Cryst. Solids, 285, pp. 135–41, 2001.

157. C. J. Brinker, K. D. Keefer, D. W. Schaefer, and C. S. Ashley, "Sol-Gel Transition in Simple Silicates," J. Non-Cryst. Solids, 48, pp. 47–64, 1982.

158. B. Heinrichs, F. Noville, J.-P. Schoebrechts, and J.-P. Pirard, "Palladium-Silver Sol-Gel Catalysts for Selective Hydrodechlorination of 1,2-Dichloroethane into Ethylene II. Surface Composition of Alloy Particles," J. Catal., 192, pp. 108–18, 2000.

159. D. M. Smith, G. W. Scherer, and J. M. Anderson, "Shrinkage During Drying of Silica Gel," J. Non-Cryst. Solids, 188, pp. 191–206, 1995.

160. D. M. Smith, R. Desphande, and C. J. Brinker, "Preparation of Low-Density Aerogels at Ambient Pressure," Better Ceramics through Chemistry V, 271 (M. J. Hampden-Smith, W. G. Klemperer, and C. J. Brinker, Eds.), Materials Research Society, Pittsburgh, pp. 567–72, 1992.

161. R. Desphande, D. M. Smith, and C. J. Brinker, "Pore Structure Evolution of Silica Gel during Aging/Drying: Effect of Surface Tension," Better Ceramics through Chemistry V, 271 (M. J. Hampden-Smith, W. G. Klemperer, and C. J. Brinker, Eds.), Materials Research Society, Pittsburgh, pp. 553–58, 1992.

162. R. Desphande, D. M. Smith, and C. J. Brinker, "Preparation of High Porosity Xerogels by Chemical Surface Modification," U.S. Patent No. 5,565,142, 1996.

163. D. M. Smith, D. Stein, J. M. Anderson, and W. Ackerman, "Preparation of Low-Density Xerogels at Ambient Pressure," J. Non-Cryst. Solids, 186, pp. 104–12, 1995.

164. M. Schneider and A. Baiker, "Aerogels in Catalysis," Catal. Rev.-Sci. Eng., 37, pp. 515–56, 1995.

165. A. C. Pierre and G. M. Pajonk, "Chemistry of Aerogels and Their Applications," Chem. Rev., 102, pp. 4243–65, 2002.

166. G. M. Pajonk, "Some Applications of Silica Aerogels," Colloid Polym. Sci., 281, pp. 637–51, 2003.

167. U. Klett, T. Heinrich, A. Emmerling, and J. Fricke, "Structural Changes upon Super-critical CO_2-Drying of Gels," Sol-Gel Processing and Applications (Y. A. Attia, Ed.), Plenum Press, New York, pp. 295–302, 1994.

168. E. Ponthieu, E. Payen, G. M. Pajonk, and J. Grimblot, "Comparison of Drying Procedures for the Preparation of Alumina Powders with the System Al-Alkoxide/Tertiary Butanol/Water," J. Sol-Gel Sci. Technol., 8, pp. 201–06, 1997.

169. S. Haereid, M. Dahle, S. Lima, and M.-A. Einarsrud, "Preparation and Properties of Monolithic Silica Xerogels from TEOS-Based Alcogels Aged in Silane Solutions," J. Non-Cryst. Solids, 186, pp. 96–103, 1995.

170. S. Haereid, E. Nilsen, and M.-A. Einarsrud, "Properties of Silica Gels Aged in TEOS," J. Non-Cryst. Solids, 204, pp. 228–34, 1996.

171. M.-A. Einarsrud, M. B. Kirkedelen, E. Nilsen, K. Mortensen, and J. Samseth, "Structural Development of Silica Gels Aged in TEOS," J. Non-Cryst. Solids, 231, pp. 10–16, 1998.

172. S. Blacher, B. Heinrichs, B. Sahouli, R. Pirard, and J.-P. Pirard, "Fractal Characterization of Wide Pore Range Catalysts: Application to Pd-Ag/SiO_2 Xerogels," J. Colloid Interf. Sci., 226, pp. 123–30, 2000.

173. J. Kuhn, R. Brandt, H. Mehling, R. Petricevic, and J. Fricke, "*In Situ* Infrared Observation of the Pyrolysis Process of Carbon Aerogels," J. Non-Cryst. Solids, 225, pp. 58–63, 1998.

174. A. Oberlin and J. P. Rouchy, "Transformation of Nongraphitable Carbon by Thermal Treatment in the Presence of Iron," Carbon, 9, pp. 39–46, 1971.

175. M. Audier, A. Oberlin, M. Oberlin, M. Coulon, and L. Bonnetain, "Morphology and Crystalline Order in Catalytic Carbons," Carbon, 19, pp. 217–24, 1981.

176. M. Inagaki, Y. Okada, V. Vignal, H. Konno, and K. Oshida, "Graphite Formation from a Mixture of Fe_3O_4 and Polyvinylchloride at 1000°C," Carbon, 36, pp. 1706–08, 1998.

177. S. Lambert, B. Heinrichs, A. Brasseur, A. Rulmont, and J.-P. Pirard, "Determination of Surface Composition of Alloy Nanoparticles and Relationships with Catalytic Activity in Pd-Cu/SiO$_2$ Cogelled Xerogel Catalysts," Appl. Catal. A, 270, pp. 201–08, 2004.
178. R. S. Cunningham and C. J. Geankoplis, "Effects of Different Structures of Porous Solids on Diffusion of Gases in the Transition Region," Ind. Eng. Chem. Fundam., 7, pp. 535–42, 1968.
179. W. Zou and R. D. Gonzalez, "Thermal Stability of Silica Supported Palladium Catalysts Prepared by the Sol-Gel Method," Appl. Catal. A, 126, pp. 351–64, 1995.
180. C. Alié, A. Benhaddou, R. Pirard, A. J. Lecloux, and J.-P. Pirard, "Textural Properties of Low-Density Xerogels," J. Non-Cryst. Solids, 270, pp. 77–90, 2000.
181. C. Alié, R. Pirard, and J.-P. Pirard, "The Role of the Main Silica Precursor and the Additive in the Preparation of Low-Density Xerogels," J. Non-Cryst. Solids, 311, pp. 304–13, 2002.
182. R. S. Mikhail, S. Brunauer, and E. E. Bodor, "Investigations of a complete pore structure analysis: I. Analysis of micropores," J. Colloid Interface Sci., 26, pp. 45–53, 1968.
183. R. S. Mikhail, S. Brunauer, and E. E. Bodor, "Investigations of a complete pore structure analysis: II. Analysis of four silica gels," J. Colloid Interface Sci., 26, pp. 54–61, 1968.
184. A. J. Lecloux, "Texture of Catalysts," Catalysis: Science and Technology, 2 (J. R. Anderson and M. Boudart, Eds.), Springer, Berlin, pp. 171–230, 1981.
185. E. P. Barrett, L. G. Joyner, and P. P. Halenda, "The Determination of Pore Volume and Area Distributions in Porous Substances. I. Computations from Nitrogen Isotherms," J. Amer. Chem. Soc., 73, pp. 373–80, 1951.
186. J. C. P. Broekhoff and J. H. de Boer, "Pore Systems in Catalysts. IX. Calculation of Pore Distributions from the Adsorption Branch of Nitrogen Sorption Isotherms in the Case of Open Cylindrical Pores. 1. Fundamental Equations," J. Catal., 9, pp. 8–14, 1967.
187. J. C. P. Broekhoff and J. H. de Boer, "Pore Systems in Catalysts. XI. Pore Distribution Calculations from the Adsorption Branch of a Nitrogen Adsorption Isotherm in the Case of 'Ink-Bottle' Type Pores," J. Catal., 10, pp. 153–65, 1968.
188. J. C. P. Broekhoff and J. H. de Boer, "Pore Systems in Catalysts. XII. Pore Distributions from the Desorption Branch of a Nitrogen Sorption Isotherm in the Case of Cylindrical Pores. 1. An Analysis of the Capillary Evaporation Process," J. Catal., 10, pp. 368–76, 1968.
189. E. W. Washburn, "Note on a Method of Determining Distribution of Pore Sizes in a Porous Material," Proc. Nat. Acad. Sci., 7, pp. 115–16, 1921.
190. R. Pirard, C. Alié, and J.-P. Pirard, "Specific Behavior of Sol-Gel Materials in Mercury Porosimetry: Collapse and Intrusion," Handbook of Sol-Gel Technology, 2, Characterization of Sol-Gel Materials and Products (S. Sakka, Ed.), Kluwer, Dordrecht, pp. 211–33, 2004.
191. R. Pirard, S. Blacher, F. Brouers, and J.-P. Pirard, "Interpretation of Mercury Porosimetry Applied to Aerogels," J. Mater. Res., 10, pp. 2114–19, 1995.
192. R. Pirard and J.-P. Pirard, "Aerogel Compression Theoretical Analysis," J. Non-Cryst. Solids, 212, pp. 262–67, 1997.
193. R. Pirard, B. Heinrichs, O. Van Cantfort, and J.-P. Pirard, "Mercury Porosimetry Applied to Low Density Xerogels; Relation Between Structure and Mechanical Properties," J. Sol-Gel Sci. Technol., 13, pp. 335–39, 1998.

194. R. Pirard and J.-P. Pirard, "Mercury Porosimetry Applied to Precipitated Silica," Stud. Surf. Sci. Catal., 128, pp. 603–11, 2000.
195. C. Alié, R. Pirard, and J.-P. Pirard, "Mercury Porosimetry Applied to Porous Silica Materials: Successive Buckling and Intrusion Mechanisms," Colloids Surf. A, 187–188, pp. 367–74, 2001.
196. C. Alié, R. Pirard, and J.-P. Pirard, "Mercury Porosimetry: Applicability of the Buckling-Intrusion Mechanism to Low-Density Xerogels," J. Non-Cryst. Solids, 292, pp. 138–49, 2001.
197. R. Pirard, C. Alié, and J.-P. Pirard, "Characterization of Porous Texture of Hyperporous Materials by Mercury Porosimetry Using Densification Equation," Powder Technol., 128, pp. 242–47, 2002.
198. R. Pirard, A. Rigacci, J. C. Maréchal, D. Quenard, B. Chevalier, P. Achard, and J.-P. Pirard, "Characterization of Hyperporous Polyurethane-Based Gels by Non-Intrusive Mercury Porosimetry," Polymer, 44, pp. 4881–87, 2003.
199. D. R. Milburn, B. D. Adkins, and B. H. Davis, "Comparison of Results from Nitrogen Adsorption and Mercury Penetration for Spherical-Particle Carbon Blacks," Stud. Surf. Sci. Catal., 39, pp. 501–08, 1988.
200. R. Pirard, B. Sahouli, S. Blacher, and J.-P. Pirard, "Sequentially Compressive and Intrusive Mechanisms in Mercury Porosimetry of Carbon Blacks," J. Colloid Interface Sci., 217, pp. 216–17, 1999.
201. S. M. Brown and E. W. Lard, "A Comparison of Nitrogen and Mercury Pore Size Distributions of Silicas of Varying Pore Volume," Powder Technol., 9, pp. 187–90, 1974.
202. D. M. Smith, G. P. Johnston, and A. J. Hurd, "Structural Studies of Vapor-Phase Aggregates via Mercury Porosimetry," J. Colloid Interface Sci., 135, pp. 227–37, 1990.
203. A. Y. Fadeev, O. R. Borisova, and G. V. Lisichkin, "Fractality of Porous Silicas: a Comparison of Adsorption and Porosimetry Data," J. Colloid Interface Sci., 183, pp. 1–5, 1996.
204. E. S. Vittoratos and P. R. Auburn, "Mercury Porosimetry Compacts SiO₂ polymerization Catalysts," J. Catal., 152, pp. 415–18, 1995.
205. C. Alié and J.-P. Pirard, "Preparation of Low-Density Xerogels from Mixtures of TEOS with Substituted Alkoxysilanes. I. ¹⁷O NMR Study of the Hydrolysis-Condensation Process," J. Non-Cryst. Solids, 320, pp. 21–30, 2003.
206. H.-J. Mühlen and K. H. van Heek, "Porosity and Thermal Reactivity," Porosity in Carbons (J. W. Patrick, Ed.), Edward Arnold, London, pp. 131–49, 1995.
207. F. Derbyshire, M. Jagtoyen, and M. Thwaites, "Activated Carbons - Production and Applications," Porosity in Carbons (J. W. Patrick, Ed.), Edward Arnold, London, pp. 227–52, 1995.
208. R. Schlögl, "Carbons," Handbook of Heterogeneous Catalysis (G. Ertl, H. Knözinger, and J. Weitkamp, Eds.), Wiley-VCH, Weinheim, pp. 138–91, 1997.
209. J. H. Sinfelt, "Bimetallic Catalysts – Discoveries, Concepts, and Applications," Wiley, New York, 1983.
210. V. Ponec and G. C. Bond, "Catalysis by Metals and Alloys," Elsevier, Amsterdam, 1995.
211. J. M. Thomas and W. J. Thomas, "Principles and Practice of Heterogeneous Catalysis," VCH, Weinheim, 1997.
212. B. Heinrichs, J.-P. Schoebrechts, and J.-P. Pirard, "Palladium-Silver Sol-Gel Catalysts for Selective Hydrodechlorination of 1,2-Dichloroethane into Ethylene III. Kinetics and Reaction Mechanism," J. Catal., 200, pp. 309–20, 2001.

213. V. Y. Borovkov, D. R. Luebke, V. I. Kovalchuk, and J. L. d'Itri, "Hydrogen-Assisted 1,2-Dichloroethane Dechlorination Catalyzed by Pt-Cu/SiO$_2$: Evidence for Different Functions of Pt and Cu Sites," J. Phys. Chem. B, 107, pp. 5568–74, 2003.
214. G. F. Froment and K. B. Bischoff, "Chemical Reactor Analysis and Design," Wiley, New York, 1990.
215. S. A. Stevenson, J. A. Dumesic, R. T. K. Baker, and E. Ruckenstein, "Metal-Support Interactions in Catalysis, Sintering, and Redispersion," Van Nostrand-Reinhold, New York, 1987.
216. E. Ruckenstein and B. Pulvermacher, "Growth Kinetics and the Size Distributions of Supported Metal Crystallites," J. Catal., 29, pp. 224–45, 1973.
217. U. Schubert, C. Egger, K. Rose , and C. Alt, "Metal Complexes in Inorganic Matrices. Part III. Catalytic Activity of Rh(CO)Cl(PR$_3$)$_2$ Heterogenised by the Sol-Gel Method," J. Mol. Catal., 55, pp. 330–39, 1989.
218. S. T. Selvan, M. Nogami, A. Nakamura, and Y. Hamanaka, "A Facile Sol-Gel Method for the Encapsulation of Gold Nanoclusters in Silica Gels and Their Optical Properties," J. Non-Cryst. Solids, 255, pp. 254–58, 1999.
219. T. Balzer, A. Mühler, and P. Reimer, "Use of Ferrites for Determining the Perfusion of Human Tissue by Magnetic Resonance Diagnosis," World Patent No. WO 9627394, 1996.
220. K. J. Klabunde, D. Zhang, and C. Sorensen, "Encapsulated Nanometer Magnetic Particles," World Patent No. WO 9907502, 1999.
221. J. N. Armor and E. J. Carlson, "A Novel Procedure for "Pelletizing" Aerogels," Appl. Catal., 19, pp. 327–37, 1985.
222. S. Blacher, A. Léonard, B. Heinrichs, N. Tcherkassova, F. Ferauche, M. Crine, P. Marchot, E. Loukine, and J.-P. Pirard, "Image Analysis of X-Ray Microtomograms of Pd-Ag/SiO$_2$ Xerogel Catalysts Supported on Al$_2$O$_3$ Foams," Colloids Surf. A, 241, pp. 201–06, 2004.
223. C. Alié, F. Ferauche, A. Léonard, S. Lambert, N. Tcherkassova, B. Heinrichs, M. Crine, P. Marchot, E. Loukine, J.-P. Pirard, "Pd-Ag/SiO$_2$ Xerogel Catalyst Forming by Impregnation on Alumina Foams," Chem. Eng. J., 117, pp. 13–22, 2006.
224. C. Alié, F. Ferauche, B. Heinrichs, R. Pirard, N. Winterton, and J.-P. Pirard, "Preparation and Characterization of Xerogel Catalyst Microspheres," J. Non-Cryst. Solids, 350, pp. 290–98, 2004.

9 Synthesis of Supported Metal Catalysts by Dendrimer-Metal Precursors

D. Samuel Deutsch, Christopher T. Williams, and Michael D. Amiridis

CONTENTS

9.1 INTRODUCTION

Starburst dendrimers have received considerable attention in the area of heterogeneous catalyst synthesis in the last decade. Although Tomalia et al. and Bosman et al.[1, 2] originally discovered these hyperbranched macromolecules and their host-guest properties in the mid-1980s, Crooks and coworkers were the first to demonstrate the ability of poly(amidoamine) (PAMAM) starburst dendrimers to act as metal nanoparticle stabilizers that could potentially aid in the synthesis of supported metal catalysts.[3] The benchmark work of Crooks et al. and subsequent literature reports have underscored the advantages of successfully utilizing PAMAM dendrimer-nanocomposite precursors over conventional catalyst preparation methods.

Conventional techniques for the synthesis of heterogeneous catalysts often involve the deposition of an inorganic salt precursor onto a porous support via

209

incipient wetness impregnation or coprecipitation techniques, followed by rigorous drying, calcination, and reduction steps. Control of the size of the resulting metal particles is often attempted through the optimization of precursor solution concentrations and subsequent time/temperature thermal treatments. Still, these arrangements result in poorly controlled particle morphologies and wide particle size distributions.[4] On the other hand, the potential benefits of using dendrimer-metal nanocomposites as precursors in the synthesis of supported metal catalysts include greater control of metal nanoparticle size, better distribution of the active phase on the support, controlled particle composition in the case of bimetallic catalysts, and the ability to deposit pre-formed metal nanoparticles with identical size and compositional character on different supports.

The principles behind obtaining such desired characteristics are directly related to the exploitation of the unique structure of dendrimers, and in particular the PAMAM dendrimers utilized in our work. These macromolecules consist of alternating tertiary qamines and secondary amides that are separated by ethylenic moieties, composing the branches of the starburst polymer. The amine nitrogen atoms can act as binding sites for metal precursors, whereas the ethylenic groups put distance between metal-dendrimer complexation sites. Furthermore, the terminal functional groups of the PAMAM dendrimer may be chosen to suit the needs of the user (i.e., to prevent undesired reactions at the dendrimer's periphery, increase solubility in organic or aqueous solutions, or facilitate dendrimer adhesion on a support). The structure of a second generation amine-terminated (G2NH$_2$) PAMAM dendrimer is shown in Figure 9.1, while some key parameters of PAMAM dendrimers are summarized in Tables 9.1 and 9.2.

The synthesis of PAMAM dendrimers is relatively straightforward. Michael addition of methylacrylate to an ethylenediame molecule is followed by amination by four additional ethylenediamine species. This process is repeated and yields the regular, hyperbranched structure of PAMAM dendrimers. However, control of the degree to which the dendrimer grows (i.e., the dendrimer generation) is a challenging process. Thus, commercially available dendrimers have been frequently used for the synthesis of dendrimer-metal nanocomposites (DMNs) in most catalyst synthesis studies.

The most common method of synthesizing supported metal catalysts from dendrimer-metal nanocomposites consists of four stages: complexation of metal precursors with PAMAM dendrimers, reduction of the metal-dendrimer complexes to form DMNs in solution, deposition of the DMNs onto a support, and the exposure of the supported active metal phase via removal of the dendrimer. A schematic of the general synthesis procedure is shown in Figure 9.2. It must be emphasized that this depiction of the synthesis of supported catalysts from dendrimer precursors represents an ideal case. In fact, there is growing evidence that contests the formation of zerovalent metal nanoparticles housed within PAMAM dendrimers.[5,6] Instead, the addition of a reducing agent to metal-dendrimer complexes in solution can result in the formation of structures that retain certain valence characteristics of the metal precursor, such as the presence of chloride ligands, solvation shell coordinations, and a net ionic charge. Generally speaking, the degree to which a particular metal will form a zerovalent dendrimer-metal nanocomposite will be strongly dependent upon the nature of the metal precursor and the strength of its interaction with the dendrimer (with the latter factor relying strongly upon solution conditions).[5]

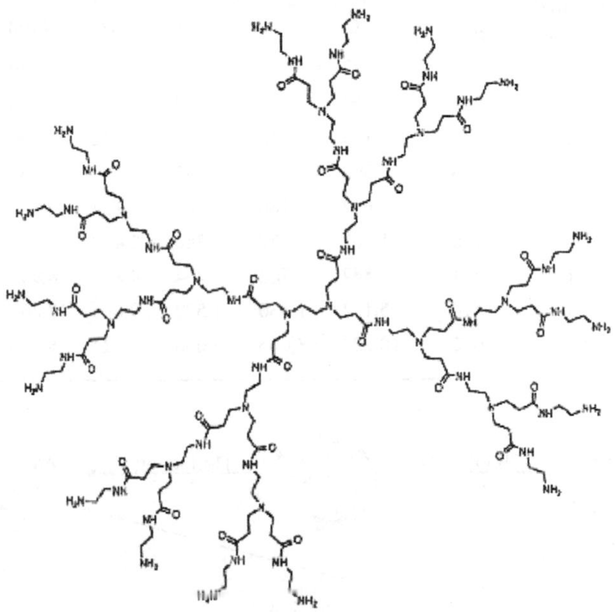

FIGURE 9.1 Structure of a 2nd generation amine-terminated (G2NH$_2$) poly(amidoamine) (PAMAM) starburst dendrimer. (Courtesy of Dendritech, Inc., http://www.dendritech.com/pamam.html.)

TABLE 9.1
Properties of Amine-Terminated PAMAM Dendrimers by Generation

Generation	NH$_2$ Terminal Groups	Tertiary Mmines	Number of Atoms					Molecular Weight
			C	N	O	H	Total	
0	4	2	22	10	4	48	84	517
1	8	6	62	26	12	128	228	1,430
2	16	14	142	58	28	288	516	3,256
3	32	30	302	122	60	608	1,092	6,909
4	64	62	622	250	124	1,248	2,244	14,214
5	128	126	1,262	506	252	2,528	4,548	28,825
6	256	254	2,542	1,018	508	5,088	9,156	58,046
7	512	510	5,102	2,042	1,020	10,208	18,372	116,489
8	1,024	1,022	10,222	4,090	2,044	20,448	36,804	233,374

TABLE 9.2
Properties of Hydroxyl-Terminated PAMAM Dendrimers by Generation

Generation	OH Terminal Groups	Tertiary Amines	Number of atoms					Molecular Weight
			C	N	O	H	Total	
0	4	2	22	6	8	44	80	521
1	8	6	62	18	20	120	220	1,438
2	16	14	142	42	44	272	500	3,272
3	32	30	302	90	92	576	1,060	6,940
4	64	62	622	186	188	1,184	2,180	14,277
5	128	126	1,262	378	380	2,400	4,420	28,951
6	256	254	2,542	762	764	4,832	8,900	58,298
7	512	510	5,102	1,530	1,532	9,696	17,860	116,993
8	1,024	1,022	10,222	3,066	3,068	19,424	35,780	234,382

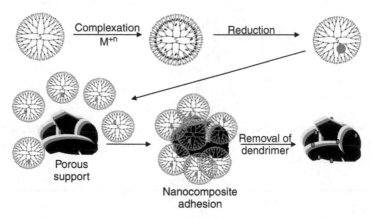

FIGURE 9.2 General procedure for the synthesis of supported metal catalysts from dendrimer-metal nanocomposites. This figure is meant to depict a series of synthetic steps and does not necessarily represent the actual configuration of the chemical species involved.

The numerous possible combinations of metal weight loadings, dendrimer generations, dendrimer terminal groups, metal-to-dendrimer ratios, and supports require a nomenclature system that is clear and distinct. Most dendrimer-prepared catalysts reported in the literature are listed in the form, $Wt\% M_nGXT/S$, with Wt, M, n, X, T, and S representing the percentage weight metal loading, the type of metal, the theoretical number of metal atoms per dendrimer during complexation, the generation of the dendrimer, the terminal group of the dendrimer, and the support, respectively. For example, a 2% Pt/SiO_2 catalyst prepared by fourth-generation amine-terminated dendrimers with a metal-to-dendrimer ratio of 40:1 is represented as 2% $Pt_{40}G4NH_2/SiO_2$.

9.2 METAL-DENDRIMER COMPLEXATION AND FORMATION OF DENDRIMER-METAL NANOCOMPOSITES

The first two stages of the synthesis of catalysts prepared by dendrimers are inextricably linked. Proper incorporation of the metal precursor within the PAMAM dendrimer is essential for the formation of dendrimer-metal nanocomposites and, eventually, nanoparticles with controlled particle sizes. Complications in the complexation stage, such as incomplete or inadequate incorporation of the metal precursor, will leave free metal cations or colloidal particles in the impregnating solution, resulting in the formation of supported catalysts that exhibit wide particle size distributions. In the case of bimetallic catalysts, the loss is twofold: In addition to an array of metal particle sizes, there will also be a significant loss of compositional control in the active phase. In short, if the complexation step is not tightly controlled, the dendrimer-prepared catalyst will not differ substantially from a catalyst prepared by wet impregnation.

The complexation stage relies solely on the ability of metal precursors to adhere to the interior of the dendrimer. In principle, the nitrogen-containing functional groups of the dendrimer (i.e., primary amines, tertiary amines, and secondary amides) can complex with metal precursors, thus providing excellent control of the number of metal atoms in the DMN and the eventual size of the metal particle formed at the end of catalyst synthesis. As one increases or decreases the dendrimer generation and, hence, the number of sites available for metal complexation, a greater or lesser number of metal cations can be incorporated into the dendrimer, ultimately yielding larger or smaller reduced metal nanoparticles, respectively. Factors such as the nature of the guest species, the chemical compositions of the dendrimer interior and periphery, and the cavity size of the dendrimer play a role in metal complexation.[3] Combinations of chemical interactions, such as covalent bonding, dipole interactions, steric effects, van der Waals forces, and hydrogen bonding, provide the driving force for metal incorporation within the dendrimer.[7]

Successful incorporation of the metal requires the selective complexation of the metal precursor with only the interior amine groups of the PAMAM dendrimer, and not with the primary amine surface groups. Two approaches are used to prevent such unselective complexation. The first involves the selective protonation of the primary amine surface groups, rendering them inactive for complexation. These primary amines are more alkaline than the internal tertiary amines (pK$_a$ of 9.23 and 6.30, respectively) and, thus, can be more readily protonated.[8] Consequently, it is possible to selectively protonate the primary amines while leaving the interior tertiary amines available for complexation simply by reducing the solution pH. This approach will be discussed further in Section 9.2.1. The second approach involves the substitution of the surface primary amines by noncomplexing functional groups, such as surface hydroxyl groups. This approach eliminates both the need for selective protonation and operation within a narrow pH range.

Finally, dialysis has proven to be a useful tool for removing metal precursors that do not complex with the dendrimer and remain in the solution. This is important because, should the problem of incomplete dendrimer-metal complexation arise, deposition of the uncomplexed metal precursor species will yield catalysts akin to those prepared by conventional wet impregnation techniques. However, dialysis results in the separation and removal of uncomplexed species and alleviates this problem. For example, Gu and coworkers[6] have observed incomplete complexation of platinum precursors with PAMAM dendrimers; they found that a substantial amount of platinum can remain in solution even after a period of 10 days. Dialysis of the metal-dendrimer complex solution with a cellulose dialysis membrane resulted ultimately in the formation of Pt nanoparticles with a narrower size distribution than a sample prepared without dialysis.[9] Others have also employed this technique, as dialysis has become a necessary step in the synthesis of many catalysts prepared by dendrimers.[10–14]

9.2.1 Synthesis of Monometallic Dendrimer-Metal Nanocomposites

Since their host-guest properties were discovered, PAMAM dendrimers have been used in the synthesis of dendrimer-metal nanocomposites containing various metals, including Cu, Pt, Pd, Au, Ag, Rh, Ru, Ni, Fe, Mn, Co, and Sn.[6, 8, 10, 15–34] The first five of these metals have received the most attention in the literature with regard to catalyst synthesis and will be the focus of this section.

Copper is perhaps the most studied metal for incorporation within PAMAM dendrimers. This is due to the relative ease of its complexation with the polymer, as well as to the readily interpretable UV-vis spectra obtained from aqueous Cu^{2+}/PAMAM solutions.[3] It has been shown by Zhao and coworkers[29] that Cu^{2+} cations bind strongly to both G4OH and G4NH$_2$ PAMAM dendrimers. A readily observable UV-vis band located at 810 nm corresponding to fully solvated $CuSO_4$ undergoes a blue shift (to 605 nm) and increases in intensity when the $CuSO_4$ solution is mixed with G4OH PAMAM dendrimers. Furthermore, a ligand-to-metal-charge-transfer (LMCT) band appears simultaneously at 300 nm, emphasizing the fact that the Cu^{2+} ions are interacting with the dendrimer. At a neutral pH, the number of copper cations that strongly bind to the PAMAM dendrimer is proportional to the number of available tertiary amine coordination sites.

However, at a pH of 1.3, hydronium cations displace the coordinated copper cations from the tertiary amine sites, yielding once again solvated copper. Upon chemical reduction of Cu-complexed dendrimers with excess $NaBH_4$, there is an immediate color change of the solution from blue to brown, as well as a disappearance of the UV-vis bands at 605 and 300 nm. A new band at 590 nm becomes visible and has been assigned to the formation of copper clusters in solution.[35] Transmission electron micrographs of these samples show particles with diameters of less than 1.8 nm. Although it has been shown that these nanoparticles remain stable for many days in an oxygen-free environment, exposure of the dendrimer-copper nanocomposite solution to air was found to result in the formation of intradendrimer Cu cations, as observed by UV-vis spectroscopy.[3]

Platinum is another metal that has attracted considerable attention with regard to catalyst synthesis via dendrimers. Pt^{2+} and Pt^{4+} from different precursors can be used to synthesize dendrimer-metal nanocomposites in solution, two of the most frequently used being $K_2Pt(II)Cl_4$ and $H_2Pt(IV)Cl_6$. Complexation of platinum precursors is different from that of copper precursors in that the initial hydrolysis product of the former still contains four chlorine atoms. These chlorinated Pt precursors undergo a relatively slow ligand exchange reaction with the dendrimer amine groups 25 monitored by UV-vis spectroscopy, which takes place over the course of 72 h.[3] More specifically, upon addition of K_2PtCl_4 to an aqueous solution of G4NH$_2$ PAMAM dendrimer, Ye et al.[23] were able to observe the disappearance of LMCT bands for Pt at 216 nm and the appearance of a band at 260 nm assigned to a Pt complex with interior amines. From these data, however, it is difficult to ascertain the exact number of Pt-N bonds present in each complex.

The ligand exchange process has also been monitored in our own group by the use of extended x-ray absorption fine structure (EXAFS) spectroscopy.[5] Our results indicate that the hydrolysis of H_2PtCl_6 and K_2PtCl_4 leads to the formation of $[PtCl_3(H_2O)_3]^+$ and $[PtCl_2(H_2O)_2]$ complexes, respectively. These species strongly interact with amine or amide groups within the dendrimer. The EXAFS results further suggest that the interaction of these two Pt precursors with the dendrimers leads on average to the replacement of approximately one Cl$^-$ anion and one aquo ligand in the coordination sphere of Pt^{n+} with two internal dendrimer nitrogen atoms. The platinum complexation process has also been investigated via the use of ^{195}Pt NMR.[36] The data in this case suggest that upon addition of aqueous solutions of K_2PtCl_4 to a solution of G2OH PAMAM dendrimer, various modes of Pt incorporation take place. Even after complexation for 10 days, 41% of the platinum remains in solution, while 15% is bound to one tertiary nitrogen, 5% is bound to one amide nitrogen, 10% is bound to two tertiary nitrogens, 3% is bound to one tertiary and one amide nitrogen, 17% is bound to three tertiary nitrogens, 7% is bound to two tertiary nitrogens and one amide nitrogen, and 2% is bound to three amide nitrogens. These ^{195}Pt NMR data are not necessarily in disagreement with the EXAFS data, as the latter technique measures only an average Pt coordination and does not provide information regarding the possible distribution of coordination states around this average.

As was stated earlier, complications that arise during dendrimer-metal complexation can be manifested later in the overall synthesis as a loss of control over metal particle size. After reduction with NaBH$_4$, Crooks and coworkers reported metal particle diameters for Pt$_{40}$G4OH and Pt$_{60}$G4OH DMNs measured by TEM of 1.4 ± 2 and 1.6 ± 2 nm, respectively.[3] In contrast, Pellechia and others reported a much wider distribution of Pt particle sizes (from 0.4 to 3.3 nm) for Pt$_{40}$G4OH DMNs prepared from the same precursors.[36, 6] These results suggest some uncertainties regarding the formation of Pt-DMNs. The EXAFS data clarify this imprecision and suggest that addition of NaBH$_4$ to Pt-containing DMNs results in incomplete reduction of the metal and the formation of only Pt dimers. Moreover, no changes in the Pt^{4+} or Pt^{2+} shell geometries (octahedral and square planar, respectively) were observed during this step, reinforcing the notion that Pt is not fully reduced by this treatment.[5] Incomplete reduction is most likely the result of strong Pt-N interactions; thus, hydronium ions introduced by the reducing agent cannot displace a Pt atom from a Pt-N bond as easily as it can remove a copper atom from a Cu-N complex. The EXAFS data further suggest that Pt nanoparticle formation takes place only upon deposition of the DMNs onto a solid support and the subsequent removal of solvent. Clearly, more work is needed in this area to understand how and under what conditions zerovalent platinum nanoparticles are formed.

More extensive studies have been conducted on the synthesis of palladium nanoparticles through dendrimer stabilization. The most widely used precursor in this case is K_2PdCl_4, which, upon hydrolysis, yields $PdCl_3(H_2O)^-$ and $PdCl_4^{2-}$. These species interact strongly with G4-PAMAM dendrimers, and precursor incorporation in this case takes only minutes rather than days. Palladium complexation with the dendrimer has been followed with UV-vis spectroscopy.[23] The results suggest that the water ligand in $PdCl_3(H_2O)^-$ is displaced during complexation, and a covalent bond with an amine nitrogen within the dendrimer is formed.[10] A corresponding absorbance band at 221 nm, assigned to an LMCT interaction, appears within seconds of mixing and can be readily observed at ambient conditions. Upon subsequent reduction with $NaBH_4$, a broad band develops at 230 nm that exhibits substantial tailing up to wavelengths of 500 nm. This band has been assigned to interband transitions of zerovalent Pd nanoparticles.[37]

Although the band at 230 nm initially overlaps with the LMCT band at 221 nm, it eventually replaces the latter and is of much lower intensity. An alternative method has also been developed for monitoring the complexation of Pd^{2+} within PAMAM dendrimers, if and when this interference becomes problematic. Prior to the addition of $PdCl_3(H_2O)^-$ to the dendrimer, addition of KCl to the precursor solution ensures that the dominant species in solution is $PdCl_4^{2-}$. UV-vis spectra of the resulting precursor-PAMAM mixture exhibit the original LMCT band at 221 nm, plus an additional LMCT band at 279 nm. This latter band decreases in intensity upon complexation with the dendrimer and is located at a wavelength that does not interfere with that of the reduced species. Furthermore, the $PdCl_4^{2-}$ precursor complexes with the dendrimer more slowly than the $PdCl_3(H_2O)^-$ precursor, allowing for more facile observation of this process with visible spectroscopy.[23] UV-vis spectra of solutions containing G4NH$_2$ PAMAM dendrimers and $PdCl_3(H_2O)^-$ and $PdCl_4^{2-}$ precursors are shown in Figure 9.3a and Figure 9.3b, respectively.

Although both of these palladium precursors can form complexes with both G4OH and G4NH$_2$ PAMAM dendrimers, different synthetic steps are required to complete this process, depending on the peripheral functional groups of the dendrimer. It was mentioned earlier that amine-terminated PAMAM dendrimers must undergo selective protonation in order to prevent metal complexation with the peripheral amine groups. For example, complexation of palladium precursors with amine-terminated dendrimers at a pH greater than 5 results in the formation of a white precipitate in solution. This is caused by the cross-linking of the dendrimer by palladium ions complexed at peripheral primary amine groups.[38, 39] This problem can be overcome by operating in a pH range between 2 and 5, where the palladium precursors form complexes with only the interior amines of the dendrimer, because the peripheral amines are all protonated under these conditions.

Further reduction in pH results in the additional protonation of internal dendrimer amine groups and eliminates any complexation of the palladium precursor with the dendrimer.[40, 41] Hydroxyl groups do not interact with most metal precursor species,

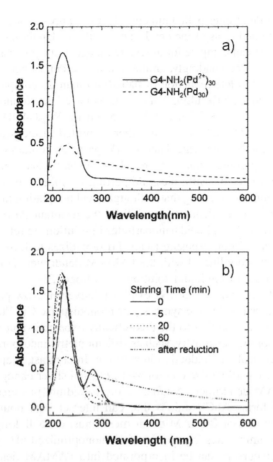

FIGURE 9.3 a) UV-vis spectra of 2.5 μM $Pd_{30}G4NH_2$ solutions before (pH 3) and after (pH 8) reduction. **b)** Time-resolved UV-vis spectra of 2.5 μM $Pd_{30}G4NH_2$ solutions in the presence of 0.1 M KCl before (pH 3) and after (pH 8) reduction. Reprinted with permission from H. Ye, R.W.J. Scott, R.M. Crooks, "Synthesis, Characterization, and Surface Immobilization of Platinum and Palladium Nanoparticles Encapsulated within Amine-Terminated Poly(amidomine) Dendrimers" Langmuir 20 (2004) 2915. Copyright 2004 American Chemical Society.

thus eliminating the need for controlling the pH during metal complexation in this case. In addition, G4OH dendrimers require less than 1 min for complete incorporation of $PdCl_3(H_2O)^-$ into their branches, whereas G4NH$_2$ dendrimers require at least 10 min for the same process to take place. This is to be expected, because G4NH$_2$ dendrimers require a relatively low pH for complexation, and the resulting hydronium ions in solution compete with the palladium precursors for the tertiary amine sites.[40] However, it has been shown that the maximum number of palladium atoms strongly incorporated within the dendrimer is the same regardless of the nature of the dendrimer periphery. In fact, G4 through G8 PAMAM dendrimers exhibit complexation with palladium precursors in a 1:1 tertiary amine-to-Pd ratio at equilibrium.[10]

Reduction of dendrimer-palladium complexes requires a large excess of $NaBH_4$ due to the acidic conditions of the dendrimer solutions. Under these conditions, the hydronium ions present compete for the reducing agent in a process termed *parasitic proton reduction*. Correspondingly, amine-terminated dendrimer complexes require more $NaBH_4$ for complete reduction of palladium than do complexes of hydroxyl-terminated dendrimers.[10] Resulting nanoparticles have been found to exhibit diameters on the order of 1.4 ± 0.4 and 1.7 ± 0.5 nm for $Pd_{40}G4OH$ and $Pd_{40}G4NH_2$, respectively. These values,[10] as well as others reported in the literature,[23, 42, 43] are larger than theoretical values calculated for Pd particles containing 40 atoms, indicating that some agglomeration is taking place beyond what is predicted within a single dendrimer molecule. Still, these Pd particles exhibit a relatively narrow particle size distribution, suggesting that, although ideal templating may have not taken place, the presence of the dendrimer stabilizes the resultant Pd particles and minimizes agglomeration. Metallic Pd nanoparticles in solution are relatively stable when kept under nitrogen. Upon exposure to air, Pd reoxidizes to form Pd^{2+} cations that once again become coordinated with the PAMAM dendrimer. This reaction can be reversed by exposure of the liquid solution to hydrogen.[10]

All of the aforementioned DMNs exhibit relatively narrow particle size distributions, thus suggesting that the synthesis of monodisperse Cu, Pt, and Pd particles with average diameters similar to the theoretically expected values can be achieved through precise solution chemistry, such as stoichiometric incorporation of the metal precursors, pH control, and reduction in solution. In contrast, it appears to be more difficult to exercise particle size control over gold and silver nanoparticles in DMNs involving PAMAM dendrimers. Similar to the palladium precursors, $HAuCl_4$ dissolved in water forms $AuCl_4^-$ anions that can interact with both the tertiary and primary amine groups of PAMAM dendrimers; thus, cross-linking of the polymer at the peripheral amine sites is possible under nonoptimized pH conditions.[27, 44] In addition, gold precursors can be incorporated into PAMAM dendrimers in a 1:1 tertiary amine-to Au ratio, just as in the case of palladium. The difference is in the mechanism of metal incorporation; rather than forming covalent bonds to tertiary amines, $AuCl_4^-$ is thought to migrate into the dendrimer's interior through electrostatic interactions.[26] Rapid reduction of Au-dendrimer complexes with concentrated $NaBH_4$ leads to the formation of large polydisperse particles, whereas insufficient reduction leads to the formation of a red-violet precipitate.

Grohn and coworkers[26] tried to optimize the conditions utilizing $NaBH_4$ in 0.3 M NaOH solution and obtained monodisperse Au nanoparticles 2–5 nm in diameter. Both the location and the mechanism of formation of the resultant Au nanoparticles are also dependent on the dendrimer generation. Dendrimers of lower generations tend to limit the growth of gold nanoparticles through a process called *arrested precipitation*. Instead of gold agglomerates forming inside the dendrimers, peripheral groups of many dendrimers appear to surround gold clusters formed outside the dendrimers. The use of G2 through G4 amine-terminated dendrimers results in the formation of such interdendrimer Au particles, whereas G6 through G10 dendrimers stabilize Au within the polymeric structure. Samples prepared with G6 through G8 dendrimers give rise to a single gold particle per dendrimer, whereas two or three Au nanoparticles are sometimes observed in G10 dendrimers following reduction.[26]

Because all of the samples mentioned above pertain to the 1:1 tertiary amine- to-Au complexation stoichiometry, it is possible that there is insufficient flexibility in the large G10NH$_2$ PAMAM structure to allow for the formation of single gold particles.

Another method of reducing Au-containing DMNs is via bombardment by UV radiation. Light of the proper wavelength (i.e., 532 nm) produced by an Nd/YAG laser has been shown to reduce these gold-dendrimer complexes.[45] This technique has enjoyed only limited success, however, because the fully reduced gold particles formed a range in diameter between 5 and 30 nm, clearly not adhering to the templating model. However, refinement of this technique may prove useful, because it does not involve the use of a strong reducing agent.

More recently, Oh and coworkers[46] developed PAMAM dendrimers with both primary amines and quaternary ammonium groups at the dendrimer periphery, in an attempt to facilitate templating of Au precursors within the dendrimer. Upon reduction by NaBH$_4$, gold nanoparticles on the order of 1 to 2 nm were formed for DMNs containing between 55 and 300 Au atoms. Utilizing this polycationic dendrimer precursor results in the formation of gold particles with smaller and more controlled particle diameters than those previously reported.[8, 46–48] Just as in the case of Pt, however, more work needs to be performed to elucidate the mechanism of Au nanoparticle formation.

All of the metal precursors that have been discussed up to this point readily form complexes with the interior amine groups of PAMAM dendrimers that can be subsequently reduced. However, this is not the case for silver, because the interaction between Ag$^+$ cations and PAMAM dendrimers is weak and does not yield a stable complex. Still, it is possible to form zerovalent silver-dendrimer nanocomposites through an intradendrimer redox displacement route.[49] This technique consists of synthesizing dendrimer-Cu nanocomposites, as outlined earlier in this section, followed by exposure of the Cu-DMNs to a solution of silver cations. The ensuing redox reaction is fast, leaving silver in a metallic state within the PAMAM dendrimer while the resulting Cu^{2+} cations may either remain trapped within the dendrimer or extracted back into the solution, depending on the solution pH.[3] This displacement process can be followed via UV-vis spectroscopy.[35]

Upon addition of Ag$^+$ cations to a solution of Cu$_{55}$G6OH dendrimers, a band appears at 400 nm that is assigned to reduced silver. Upon neutralization of the solution, the band representative of the LMCT of Cu^{2+} is also observed (300 nm), indicating that metallic silver and cationic copper coexist in the solution under these conditions.[3] Although a UV-vis band at 400 nm is visible after reduction of an Ag$^+$-PAMAM dendrimer solution (i.e., an attempt to directly complex Ag$^+$ cations within the dendrimer), its presence can be attributed to the formation of large silver colloids in this solution. Many of these particles exhibit diameters greater than 60 nm, clearly indicating that the presence of the dendrimer in this case had no effect on particle formation.[19] Silver nanoparticles synthesized through redox displacement of Cu are stable in air for months. This approach can be extended to other noble metals with more positive standard potentials of half-reaction as well, such as Au, Pt, or Pd. These metals can displace Ag within the dendrimer in a second redox reaction, just as copper was displaced by silver. A schematic representation of this process is shown in Figure 9.4.

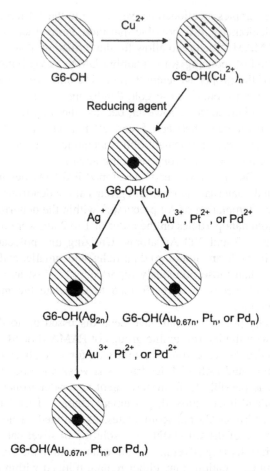

FIGURE 9.4 A schematic representation of intradendrimer redox displacement of Cu to form zerovalent dendrimer–metal nanocomposites. Further displacement of Ag can also occur by Au, Pt, or Pd. Reprinted with permission from R.M. Crooks, M. Zhao, L. Sun, V. Chechik, L.K. Yeung, "Dendrimer-Encapsulated Metal Nanoparticles: Synthesis, Characterization, and Applications to Catalysis" Accounts of Chemical Rsearch 34 (2001) 181. Copyright 2001 American Chemical Society.

9.2.2 Synthesis of Bimetallic Dendrimer-Metal Nanocomposites

PAMAM dendrimers can also be used as templating agents and nanoparticle stabilizers for the synthesis of bimetallic particles. The unique ability of dendrimers to host various metal precursors enables the simultaneous complexation of multiple metallic species at its various internal functional groups. The three primary methods of bimetallic nanoparticle synthesis through dendrimer stabilization are partial displacement, co-complexation, and sequential complexation.

Bimetallic nanoparticle synthesis via partial displacement is a natural extension of the intradendrimer redox displacement described earlier for the case of silver nanoparticle synthesis. If, instead of adding silver cations to fully reduced copper-dendrimer nanocomposites in a 2:1 Ag^+-to-Cu ratio, substoichiometric amounts of Ag^+ are used, they only partially displace the metallic Cu, resulting in the formation of a bimetallic Ag-Cu nanoparticle.[3] Just as is the case with the complete intradendrimer redox displacement reaction, this approach can be used with any pair of metals in which the standard potential of half-reaction of the latter is stronger than that of the former.

Co-complexation is the most straightforward technique for synthesizing bimetallic particles via the dendrimer route. The simultaneous addition of two metal precursors to a solution of PAMAM dendrimer results in a competition between the two metal precursors for complexation sites within the dendrimer. Subsequent reduction of the complexed species could, in principle, lead to the formation of bimetallic particles. Still, the exact mechanism of nanoparticle formation, from dendrimer-metal complex to zerovalent nanoparticle, is not well understood.

Finally, the sequential loading technique takes advantage of the fact that PAMAM dendrimers retain their polymeric structure following mild reduction treatment in solution. In other words, the formation of a dendrimer-metal nanocomposite with the first precursor does not affect the templating ability of the dendrimer. Therefore, the same host polymer may form a second complex in a subsequent step with the second metal precursor, which can be further reduced to form a second metallic phase in close proximity to the first one. Ideally, this technique could also enable the user to create *core-shell* type bimetallic particles. Although the partial displacement method should theoretically yield nanoparticles with similar core-shell type structures, sequential loading is different in that it allows the user to choose the core and shell metals simply by selecting which species will be first in the overall sequence. A schematic representation of these three techniques is shown in Figure 9.5.

All of the above processes, in principle, afford the benefit of controlling the particle sizes of bimetallic systems that are eventually formed, whereas it may be more difficult to synthesize bimetallic colloids via traditional methods. Depending on the thermodynamic properties of the two metals, conventional catalyst preparation techniques (i.e., co-impregnation, co-precipitation) may result in the formation of a homogeneous alloy, segregation into pure monometallic phases, or a combination thereof. On the other hand, templating by dendrimers at least ensures that both metals are in close proximity with each other during complexation and after reduction.

Only limited information has been published with regard to synthesis of bimetallic particles through dendrimer stabilization. Among the existing literature reports in this area, bimetallic particles containing palladium have received the most attention. Because Pd is a very active catalyst for hydrogenation reactions,[50] liquid-phase Pd-X bimetallic dendrimer nanocomposites were used for proof-of-concept experiments for the synergy of two metals in the dendrimer environment. Chung and Rhee, for example, have synthesized bimetallic Pd-Rh nanocomposites with G4OH dendrimers through co-complexation/co-reduction.[51] The metal particle diameters observed in this case were in the range of 1.5 to 2.8 nm, and these materials showed substantial activity for the hydrogenation of 1, 3-cyclooctadiene in the liquid phase.

A later report from the same group also documents the synthesis of Ag-Pd bimetallic particles prepared via a slightly different co-complexation/co-reduction

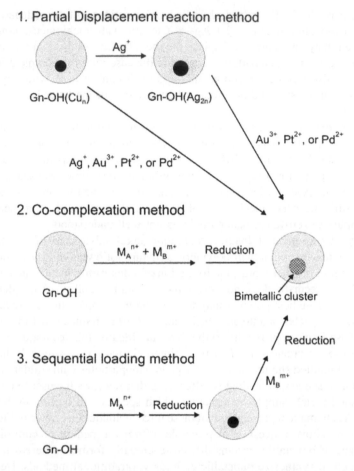

FIGURE 9.5 A schematic representation of three different approaches for the synthesis of bimetallic dendrimer-metal nanocomposites. Reprinted with permission from R.M. Crooks, M. Zhao, L. Sun, V. Chechik, L.K. Yeung, "Dendrimer-Encapsulated Metal Nanoparticles: Synthesis, Characterization, and Applications to Catalysis" Accounts of Chemical Research 34 (2001) 181. Copyright 2001 American Chemical Society.

method.[52] In this case, K_2PdCl_4 was dissolved with $K_2C_2O_4$ at 60°C to form a $K_2[Pd(C_2O_4)_2] \cdot 4H_2O$ precipitate that was easily recovered from a solution containing the remaining K^+ and Cl^- ions. Mixing of a new solution containing the redissolved precipitate and $AgNO_3$ led to the formation of $Ag_2[Pd(C_2O_4)_2] \cdot 3H_2O$. This bimetallic Ag-Pd precursor was complexed within PAMAM dendrimers, an approach that eliminated the possibility of silver loss due to the formation of AgCl. Finally, Crooks and coworkers have synthesized a bimetallic Pd-Pt hydrogenation catalyst via cocomplexation with G4OH dendrimers and observed metal particle diameters ranging from 2 to 5 Nm.[53]

Other bimetallics synthesized via dendrimers include Ag-Au and Pd-Au systems.[54, 55] In the latter system, it was shown that the Pd-Au catalysts prepared via cocomplexation and sequential complexation routes exhibited different catalytic performance for the liquid-phase hydrogenation of allyl alcohols, although they exhibited similar particle diameters in the 1–3 nm range, suggesting that fundamental differences in nanoparticle structure can be achieved through these different synthetic routes.

Although the main focus of this chapter is the synthesis of heterogeneous catalysts prepared via dendrimers, it should be noted that both mono- and bimetallic dendrimer nanocomposites have exhibited substantial activity in homogeneous catalytic systems.[3, 8, 47, 53, 55, 56] In some cases, the dendrimer can act as a selective barrier, permitting only reactants that are sterically favorable to reach the reactive nanoparticle within, whereas other applications include using nonreduced dendrimer-metal complexes for catalysis at single metal cation sites.[57–60]

9.3 DEPOSITION OF DENDRIMER-METAL NANOCOMPOSITES ONTO HIGH SURFACE AREA SUPPORTS

The next step in the synthesis of supported metal catalysts via dendrimers is the immobilization of dendrimer-metal nanocomposites onto a solid support. An array of techniques exists for achieving this task. Wet impregnation and sol-gel incorporation of dendrimer-metal nanocomposites may lead to strongly adhered metal particles. Other techniques, such as functionalization of the support to facilitate dendrimer growth or adhesion, provide a route for deposition of empty dendrimers that can subsequently undergo complexation with metal precursors to form dendrimer-metal complexes and eventually zerovalent nanoparticles. Whereas the complexation and reduction phases of catalyst synthesis via dendrimers can be fairly complicated, most methods of dendrimer deposition are rather straightforward.

Extensive hydrogen bonding and electrostatic interactions are present following deposition by wet impregnation of empty dendrimers and DMNs on model surfaces such as mica.[13] These interactions are sufficiently strong to force both OH and NH_2-terminated dendrimers to adopt a flattened shape on the surface and inhibit surface mobility.[61, 62] Following deposition on a mica surface, the Pt particles formed from Pt-G4OH nanocomposites appear to exhibit a quasihemispherical structure. Furthermore, HR-TEM micrographs do not show any lattice fringes and exhibit a speckled appearance, suggesting the formation of a disordered array of Pt atoms and not a crystalline Pt particle.[6] Results from EXAFS spectroscopy appear to agree with this model.[5]

More specifically, when Pt_{40}G4OH was deposited on γ-Al_2O_3, Pt-N and Pt-O contributions were still observed in the EXAFS spectra, suggesting that the Pt atoms remain, to a large extent, coordinated to the dendrimer even after reduction in solution. Pt particles are formed only upon room-temperature drying of the deposited Pt_{40}G4OH/Al_2O_3 material, as dendrimers lose a considerable percentage of their volume upon dehydration,[63] resulting in both the collapse of the dendrimer and the aggregation of nearby platinum atoms, along with their adhering dendrimer functional groups.[21, 64] As of yet, there is not enough information to ascertain whether this phenomenon is unique to supported Pt-DMNs. One would assume that using any metal precursor that forms a complex with the dendrimer interior that is strong enough to resist reduction in solution would result in behavior similar to that of the Pt-DMNs.

Dendrimer-metal nanocomposites can also be incorporated into sol gels. While wet impregnation requires evaporation of the solvent, thus increasing the concentration of DMNs in solution, low dendrimer concentrations can be maintained during sol-gel synthesis, minimizing the possibility of particle agglomeration in solution. Additionally, evenly distributed DMNs within the hardened gel should reduce the probability of metal sintering upon dendrimer removal. Platinum DMNs can be incorporated into sol-gel silica at various stages of overall catalyst development. For example, in a sol composed of acetic acid and tetramethoxysilane, $Pt_{50}G5OH$ can be added either at, or 24 h after, the beginning of gel formation (the gel is completely formed after 48 h), resulting in the formation of platinum nanoparticles housed within porous silica.[65] Similarly, nanocomposites of gold and palladium with amine-terminated PAMAM dendrimers can be incorporated within titania sols to form supported catalysts.[66] During this process, Scott and coworkers observed a negligible change in the average diameter of the gold nanoparticles throughout the gel formation, from 1.9 ± 0.5 to 2.0 ± 0.6 nm.[47, 54]

When particularly concerned with securely anchoring DMNs onto supports, it is also possible to immobilize dendrimers or their precursor reagents on surfaces that have been chemically pretreated to facilitate this process. Such functionalized supports should contain surface moieties that can interact strongly with the peripheral groups of the dendrimer or its precursor species. The techniques that make use of such deposition methods can be divided into two groups: direct immobilization of intact dendrimers and stepwise propagation of dendrimers within the pores of a support material. The former can be performed by grafting an epoxide linker onto the support, followed by addition of $GXNH_2$. Chung and Rhee,[67, 68] for example, have used 3-glycidoxypropyltrimethoxysilane (GPTMS) to treat a silica support under reflux toluene to yield surface epoxides, which in turn can easily bond with the primary amines present at the periphery of the dendrimer.

On the other hand, the stepwise propagation of dendrimers inside silica pores involves a repeating sequence of polymerization reactions to form the supported PAMAM dendrimer. Various research groups have synthesized these materials by adding 3-aminopropyltriethoxysilane to porous silica in boiling toluene to form surface-bound propylamine moieties. Subsequent Michael addition of methylacrylate, followed by amidation with ethylenediamine, yields a first-generation surface-bound PAMAM dendrimer. Repetition of the Michael addition and amination steps, leads to the growth of higher-generation dendrimers on the silica surface.[58–60, 68] A schematic of these direct immobilization and stepwise propagation methods is shown in Figure 9.6. It should be pointed out that these routes are often used for synthesizing heterogenized homogeneous catalysts (i.e., supported metal complexes or materials with a polymeric active phase).

9.4 REMOVAL OF PAMAM DENDRIMERS FROM SUPPORTED DENDRIMER-METAL NANOCOMPOSITES

Although DMNs deposited on a solid support may allow liquid reagents at ambient temperatures to access metal active sites within their structures, following a drying

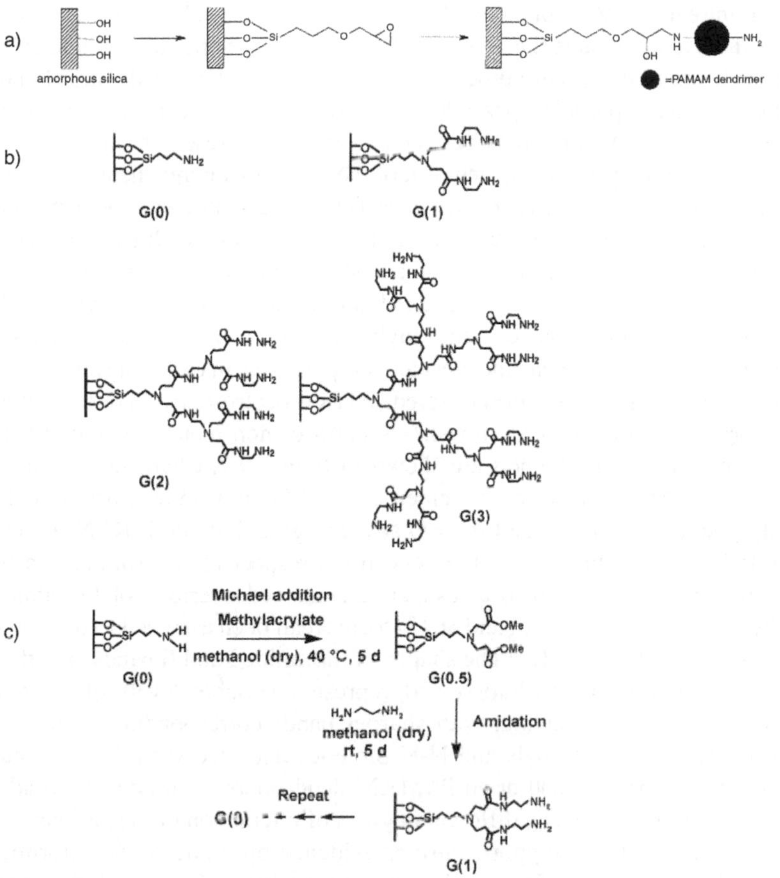

FIGURE 9.6 Schematic representations of different methods used for immobilizing PAMAM dendrimers on functionalized supports. (a) Direct immobilization of an amine-terminated PAMAM dendrimer on silica. Reprinted from Figure 1 of Y.M. Chung, H.K. Rhee, "Design of Silica-Supported Dendritic Chiral Catalysts for the Improvement of Enantioselective Addition of Diethylzinc to Benzaldehyde" Catalysis Letters 82 (2002) 249, with kind permission from Springer Science and Business Media. (b) Structures of amine-terminated PAMAM dendrimers of different generations. (c) Stepwise propagation of PAMAM dendrimers on a silica support. The dendrimer generation in parentheses indicates polymer growth within the pores of the support. Reprinted with permission from J.P.K. Reynhardt, Y. Yang, A. Sayari, H. Apler, "Periodic Mesoporous Silica-Supported Recyclable Rhodium-Complexed Dendrimer Catalysts" Chemistry of Materials 16 (2004) 4095. Copyright 2004 American Chemical Society.

step, the catalytically active metal sites are no longer accessible by even very small gas-phase molecules, such as CO.[20, 64] It is suspected that this is due to covering of these metal sites by an abundance of collapsed polymer ligands.

Thus, removal of the dendrimer component needs to be performed on supported DMNs to render active metal sites accessible for catalyzing gas-phase reactions.[20]

Supported untreated DMNs often exhibit small average particle sizes. While exposure to elevated temperatures (i.e., 400–500°C) for prolonged periods of time ensures removal of the dendrimer, these processes can lead to sintering of the metal and, in the case of bimetallic particles, possible segregation, as well. Thus, optimization of the necessary activation conditions has been an important research question.

Dendrimer thermolysis, i.e., the removal of the dendrimer component via exposure to high temperatures in oxidative, reducing, or neutral gas-phase environments, is the technique that has been so far exclusively employed for this task. Other potential low-temperature removal methods, such as chemical leaching of the dendrimer or treatment under plasma conditions, have been suggested but not demonstrated experimentally.

We have previously conducted a thorough study of the thermolysis of supported PAMAM dendrimers at ambient and elevated temperatures in oxidizing and reducing environments via Fourier-transform infrared (FTIR) spectroscopy, in order to understand the process of dendrimer removal.[69] Spectra of non-metal-containing G4OH dendrimers on an alumina support are shown in Figure 9.7. Characteristic infrared bands of the PAMAM dendrimers centered at 1647 and 1548 cm^{-1} have been previously assigned to the amide I (C = O stretching) and amide II (C-N stretching and C-N-H bending) vibrations of the dendrimer, respectively.[20] Similarly, a band at 3080 cm^{-1} can be assigned to a resonance-enhanced overtone of the amide II band, whereas a relatively weak band at 1245 cm^{-1} can been assigned to the opening of the C-N-H group (amide III).[70] The shapes of the amide I and II bands are affected by the degree of polymer hydration and represent variable levels of inter- and intramolecular hydrogen bonding, with sharper bands corresponding to more uniformly oriented amide carbonyls and N-H moieties and vice versa.[71] These bands in the spectra of supported and dried PAMAM dendrimers are relatively broad and exhibit some tailing, suggesting different degrees of interaction of the various amide functional groups with the support. Further evidence of amide Non-Uniformity is also provided by the breadth of the band assigned to hydrogen-bonded N-H stretching of the monosubstituted amide groups, centered at approximately 3300 cm^{-1}.[70]

Additional bands observed in the room-temperature spectra can be assigned to the C-H deformations and C-C backbone skeletal vibrations of the dendrimer. More specifically, asymmetric and symmetric stretching vibrations of methylene C-H bonds are observed at 2936 and 2825 cm^{-1}, respectively.[70] Dendrimer reconfiguration or bond scission on the support can take place upon deposition, as indicated by a shoulder at 2972cm^{-1} assigned to asymmetric methyl C-H stretching. Additional low wavenumber bands at 1464 and 1435 cm^{-1} are observed in the spectra in accordance with the presence of methylene groups.[72] Bands observed in the 1400–1300 cm^{-1} region are more difficult to assign. Although not always readily discernible with infrared spectroscopy, they can generally be assigned to methylene rocking, wagging, and twisting deformations.[70] Finally, C-C skeletal stretching can sometimes be observed at ~1160–1130 cm^{-1}. Worthy of mention is also the absence of any tertiary amine vibrations in the room-temperature spectra of supported poly(amido)amine dendrimers. Such groups are indeed present; however, their characteristic vibrations often overlap with the stronger features induced by other functionalities. For example, the N-C$_3$ stretching vibration is weak and coincides with the dendrimer's C-C deformation modes.[73] In addition, the C-H stretches of the amine N-CH$_2$ groups are

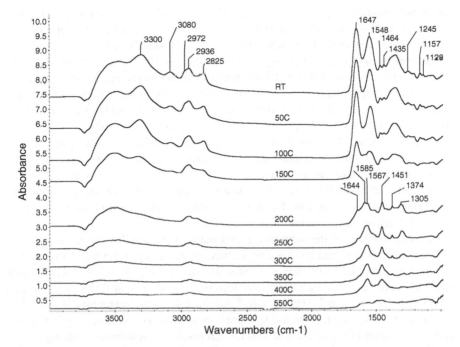

FIGURE 9.7 FTIR spectra of a G4OH PAMAM dendrimer supported on γ-alumina obtained at different temperatures in flowing H₂. From D.S. Deutsch, A. Siani, C.T. Williams, and M.D. Amiridis, "FT-IR Investigation of the Thermal Decomposition fo Poly(amiduamide) (PAMAM) Dendrimers and Dendrimer-Metal Nanocomposites Supported on Al₂O₃ and ZrO₂," Journal of Physical Chemistry B, Submitted 2006.

almost identical to those of the dendrimer's methylene groups (C-CH₂) observed at 2825 cm⁻¹.[70]

FTIR spectra collected at elevated temperatures have been used to explore the conditions necessary to remove the dendrimer and its fragments from the support following decomposition.[69] The characteristic vibrational modes of the supported dendrimer remain unchanged from room temperature up to approximately 100°C. At 150°C, both the amide I and II bands decrease in intensity, suggesting that dendrimer decomposition is initiated in this temperature range. The FTIR results further suggest that, at 200°C, the PAMAM dendrimer has decomposed to form an assemblage of smaller adsorbed fragments. The most identifiable products of this decomposition process are surface-bound carboxylates. The observed FTIR bands suggest that the majority of these species are of the formate and acetate types.[74-76] Upon further temperature increase above 200°C, bands corresponding to formate vibrations decrease in intensity until they finally disappear at approximately 350°C. Surface acetates appear to be more stable and remain bound to alumina up to 550°C finally, when the removal of these species from the support takes place.

Thermogravimetric analysis measurements have also provided information regarding dendrimer decomposition.[9] Thermolysis of G4OH dendrimers on a gold surface in both air and argon environments has shown that the greatest weight loss occurs between room temperature and 250°C. The sample weight remained relatively constant over the next 100°C, until a sharp decrease in mass loss was observed at 400°C. Metal-containing $Pt_{40}G4OH$ DMNs exhibit a steadier mass loss. Although weight loss is first observed at higher temperatures in this case, thermolysis of Pt-DMNs results in complete dendrimer removal at temperatures lower than those required for empty dendrimers, indicating that the presence of Pt catalyzes the removal of the resulting carboxylate species.[9, 20] Similar results have also been observed for many other supported DMNs.[77] Thus, the overall process of removal of supported dendrimers via thermolysis may be considered to take place in two stages. The first stage is dendrimer decomposition, followed by a second stage involving further reaction and desorption of the decomposition products. The first phase begins at approximately 150°C, whereas further reaction and desorption of the surface species formed in the first stage may occur over a broad temperature range, depending on the nature of these species.[69]

Characterization of the supported metal nanoparticles following the removal of the dendrimer component has been performed in our group using carbon monoxide as a probe molecule. FTIR measurements conducted with $Pt_{40}G4OH/SiO_2$ indicate that CO adsorption can be maximized following oxidation at 425°C for 1 h and subsequent reduction at 200°C for 1 h.[20] Additionally, this treatment did not affect the mean Pt particle size, whereas exposure to nonoptimal conditions (e.g., oxidation at 425°C followed by reduction at 400°C) can result in substantial sintering.

Dendrimer thermolysis experiments conducted on Pt/Al_2O_3 catalysts show that 400°C is the lowest temperature that will result in the removal of adsorbed carboxylates. It is not until these carboxylates are removed from the surface that the platinum nanoparticles exhibit significant CO uptake, indicating that the dendrimer decomposition products, possibly including undetectable nitrogen moieties, interfere with the metal active sites (Figure 9.8).[77]

Conventional supported ruthenium catalysts are known to undergo sintering if an oxidative pretreatment is used before a reduction step.[79–81] Thus, supported Ru- DMNs were treated directly in flowing hydrogen for the removal of the dendrimer component. Indeed, reduction for 2 h at 300°C is sufficient for exposure of the metalsites in a 1% $Ru_{40}G4OH/\gamma$-Al_2O_3 catalyst. When compared to a catalyst of the same metal weight loading prepared by incipient wetness impregnation, the dendrimer-prepared catalyst shows a much more narrow particle size distribution as well as a smaller average particle diameter (2.0 versus 3.4 nm), as shown in Figure 9.9.[24, 25]

A variety of temperature-time combinations has been used by other groups to remove the dendrimer component of supported DMNs. Supported FeG6OH DMNs calcined at 800°C for 5 min, for example, yield iron oxide particles with an average feature height of 1.3 nm, as measured by atomic force density (AFM).[16] In contrast, Chandler and coworkers have developed a low-temperature activation protocol for the activation of supported monometallic Pt and bimetallic Pt-Au catalysts.[78, 82] Pretreatment in an oxidizing atmosphere at 300°C for 2 h effectively decomposes the PAMAM dendrimer; however, an additional 10 to 14 h of treatment is required for obtaining a catalyst free of adsorbed carboxylates that exhibits high activity for CO oxidation. A potential advantage of such lower temperature activation protocols is that sintering of the metal particles is kept to a minimum.[21, 78]

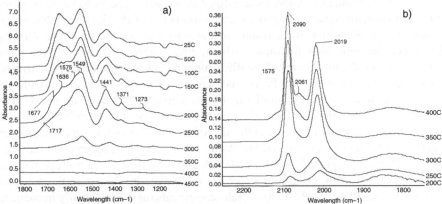

FIGURE 9.8 (a) Thermolysis of PtG4OH/Al$_2$O$_3$ in a flowing oxidizing environment. From D.S. Deutsch, A. Siani, C.T. Williams, and M.D. Amiridis, "FT-IR Investigation of the Thermal Decomposition of Poly(amidoamide) (PAMAM) Dendrimers and Dendrimer-Metal Nanocomposites Supported on Al$_2$O$_3$ and ZrO$_2$," Journal of Physical Chemistry B, Submitted 2006. (b) Room-temperature FT-IR spectra of CO adsorbed on PtG4OH/SiO$_2$ after (a) helium flow at 100°C for 1 hour, and oxidation at (b) 275, (c) 350, and (d) 425°C for 1 hour and subsequent reduction at 200°C for 1 hour. The band in the 2080 cm^{-1} region is assigned to CO linearly boned to zerovalent Pt.

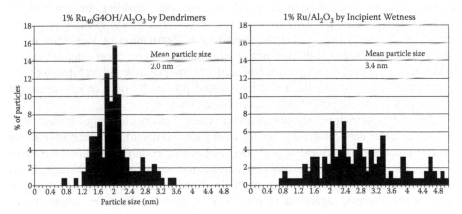

FIGURE 9.9 Metal particle size distributions of a 1%wt Ru/Al$_2$O$_3$ catalyst prepared from Ru$_{40}$G4OH DMNs and a 1%wt Ru/Al$_2$O$_3$ catalyst prepared via wet impregnation with RuCl$_3$. Both samples were reduced for 2 h at 300°C before HR-TEM imaging. Reprinted from Figure 3 of G. Lafaye, C.T. Williams, M.D. Amiridis, "Synthesis and Microsocopic Characterization of Dendrimer-Derived Ru/Al$_2$O$_3$ Catalysts" Catalysis Letters 96 (2004) 43, with kind permission of Springer Science and Business Media.

Supported bimetallic Pt-Au catalysts with 1:1 stoichiometry formed via displacement of Cu^0 within G5OH DMNs can also be activated under similar conditions.[82] The majority of these metal nanoparticles are bimetallic in nature and smaller than 3 nm after activation. Although there is some evidence of CO-induced restructuring of the active phase, this material has been shown to exhibit high catalytic activity for CO oxidation near room temperature.

Finally, activation of DMNs immobilized in sol gels appears to be more problematic. Low-temperature activation, such as oxidation of $Pt_{50}G5OH$ DMNs within a silica gel at 300°C for 16 h, results in the poisoning of the Pt active sites by dendrimer decomposition fragments, which, presumably, is due to slower mass transport within the small pores of the silica support.[65] On the other hand, dendrimer-derived Pd, Au, and bimetallic Pd-Au nanoparticles incorporated within titania sol gels resist sintering more than similar materials prepared from deposition of DMNs onto titania via wet impregnation. Calcination of these materials at 500°C for 3 h in air results in a fourfold increase in the size of the particles of the wet impregnation catalysts, whereas the Au, Pd, and bimetallic Pd-Au samples prepared via incorporation within the titania gels exhibit a 40, 80, and 80% increase in average particle size, respectively. Still, even after sintering, the bimetallic Pd-Au particles within the sol gel retain their 1:1 Pd-to-Au stoichiometric ratio as was originally introduced during co-complexation with the dendrimer.

9.5 CONCLUSION

The development of supported metal catalysts from dendrimer-metal nanocomposite precursors is a promising new approach that, in principle, allows for much better control of the active metal phase during catalyst synthesis. This technique can be applied to many different types of metals and supports, enabling the synthesis of custom-tailored catalysts for specific reactions. Among the most exciting prospects are the ability to tightly control the particle size distribution of supported metal catalysts, with averages in the 1–3 nm or even subnanometer range, and the ability to synthesize bimetallic nanoparticles of uniform composition and architecture (i.e., well-mixed or core-shell). Such materials could potentially have unique chemical properties that can be exploited for catalytic applications.

Still, this approach is very new and there are still many challenges remaining if dendrimer-metal nanocomposites are to reach their full potential as supported metal catalyst precursors. A critical step in the synthesis procedure that must be investigated is the formation of the dendrimer-metal nanocomposites in solution. Until recently, a generally accepted (and relatively simple) mechanism has been used to describe this process, regardless of the metal in question. However, recent mechanistic studies using *in situ* techniques such as EXAFS suggest that this step is more complicated and is likely to be dependent on the metal in question. Success in synthesizing bimetallic nanocomposites of controlled composition will depend greatly on developing a more complete understanding of the mechanism and kinetics of dendrimer-metal complexation.

The second major concern is removal of the dendrimer and its decomposition products from the support while limiting the sintering of the resulting supported nanoparticles. Although recent studies have shown clearly that the dendrimer itself can serve as a sintering control agent on the support surface, the exact mechanism for this control is not yet elucidated. Furthermore, this control may vary depending on the type of support material that is used. Finally, there is a great need to perform comparative catalytic studies with traditionally prepared supported catalysts, especially for bimetallic systems. Of particular importance are kinetic studies of structure-sensitive reactions that may reveal unique properties that are only accessible via the dendrimer approach.

REFERENCES

1. D.A. Tomalia, H. Baker, J. Dewald, M. Hall, G. Kallos, S. Martin, J. Roeck, J. Ryder, P. Smith, "Dendritic Macromolecules: Synthesis of Starburst Dendrimers," Macromolecules 19, 2466, 1986.
2. A.W. Bosman, H.M. Janssen, E.W. Meiger, "About Dendrimers: Structure, Physical Properties, and Applications," Chemistry Reviews 99, 1665, 1999.
3. R.M. Crooks, M. Zhao, L. Sun, V. Chechik, L.K. Yeung, "Dendrimer-Encapsulated Metal Nanoparticles: Synthesis, Characterization, and Applications to Catalysis" Accounts of Chemical Research 34, 181, 2001.
4. G. Ertl, H. Knozinger, J. Weitkamp, *Handbook of Heterogeneous Catalysis*, VCH: Weinheim, 1997.
5. O.S. Alexeev, A. Siani, G. Lafaye, C.T. Williams, H.J. Ploehn, M.D. Amiridis, "EXAFS Characterization of Dendrimer-Pt Nanocomposites Used for the Preparation of Pt/γ-Al$_2$O$_3$ Catalysts," Journal of Physical Chemistry B, Submitted 2006.
6. Y. Gu, H. Xie, J. Gao, D. Liu, C.T. Williams, C.J. Murphy, H.J. Ploehn, "AFM Characterization of Dendrimer-Stabilized Platinum Nanoparticles," Langmuir 21, 3122, 2005.
7. J.F.G.A. Jansen, D.M.M. de Brabander-van den Berg, E.W. Meijer, "Encapsulation of Guest Molecules into a Dendritic Box," Science 266, 1226, 1994.
8. Y. Niu, R.M. Crooks, "Dendrimer-Encapsulated Metal Nanoparticles and their Applications to Catalysis," Comptes Rendus Chimie 6, 1049, 2003.
9. O. Ozturk, T.J. Black, K. Perrine, K. Pizzolato, C.T. Williams, F.W. Parsons, J.S. Ratliff, J. Gao, C.J. Murphy, H. Xie, H.J. Ploehn, D.A. Chen, "Thermal Decomposition of Generation-4 Polyamidoamine Dendrimer Films: Decomposition Catalyzed by Dendrimer-Encapsulated Pt Particles," Langmuir 21, 3998, 2005.
10. R.W.J. Scott, H. Ye, R.R. Henriquez, R.M. Crooks, "Synthesis, Characterization, and Stability of Dendrimer-Encapsulated Palladium Nanoparticles," Chemistry of Materials 15, 3873, 2003.
11. M.R. Knecht, J.C. Garcia-Martinez, R.M. Crooks, "Hydrophobic Dendrimers as Templates for Au Nanoparticles," Langmuir 21, 11981, 2005.
12. M. Zhao, R.M. Crooks, "Intradendrimer Exchange of Metal Nanoparticles," Chemistry of Materials 11, 3379, 1999.
13. L. Sun, R.M. Crooks, "Dendrimer-Mediated Immobilization of Catalytic Nanoparticles on Flat, Solid Supports" Langmuir 18, 8231, 2002.
14. S.K. Oh, Y. Niu, R.M. Crooks, "Size-Selective Catalytic Activity of Pd Nanoparticles Encapsulated within End-Group Functionalized Dendrimers," Langmuir 21, 10209, 2005.
15. M. Zhao, R.M. Crooks, "Homogeneous Hydrogenation Catalysis using Monodisperse, Dendrimer-Encapsulated Pd and Pt Nanoparticles," Angewandte Chemie International Edition 38, 364, 1999.
16. H.C. Choi, W. Kim, D. Wang, H. Dai, "Delivery of Catalytic Metal Species onto Surfaces with Dendrimer Carriers for the Synthesis of Carbon Nanotubes with Narrow Diameter Distribution," Journal of Physical Chemistry B 106, 12361, 2002.

17. M.F. Ottaviani, S. Bossmann, N.J. Turro, D.A. Tomalia, "Characterization of Starburst Dendrimers by the EPR Technique. 1. Copper Complexes in Water Solution," Journal of the American Chemical Society 116, 661, 1994.
18. M.F. Ottaviani, F. Montalti, N.J. Turro, D.A. Tomalia, "Characterization of Starburst Dendrimers by the EPR Technique. Copper(II) Ions Binding Full-Generation Dendrimers," Journal of Physical Chemistry B 101, 158, 1997.
19. M.F. Ottaviani, R. Valluzzi, L. Balogh, "Internal Structure of Silver-Poly(amidoamine) Dendrimer Complexes and Nanocomposites," Macromolecules 35, 5105, 2002.
20. D.S. Deutsch, G. Lafaye, D. Liu, B.D. Chandler, C.T. Williams, M.D. Amiridis, "Decomposition and Activation of Pt-Dendrimer Nanocomposites on a Silica Support," Catalysis Letters 97, 139, 2004.
21. H. Lang, R.A. May, B.L. Iverson, B.D. Chandler, "Dendrimer-Encapsulated Nanoparticle Precursors to Supported Platinum Catalysts," Journal of the American Chemical Society 125, 14832, 2003.
22. M. Zhao, R.M. Crooks, "Intradendrimer Exchange of Metal Nanoparticles," Chemistry of Materials 11, 3379, 1999.
23. H. Ye, R.W.J. Scott, R.M. Crooks, "Synthesis, Characterization, and Surface Immobilization of Platinum and Palladium Nanoparticles Encapsulated within Amine-Terminated Poly(amidoamine) Dendrimers," Langmuir 20, 2915, 2004.
24. G. Lafaye, C.T. Williams, M.D. Amiridis, "Synthesis and Microscopic Characterization of Dendrimer-Derived Ru/Al$_2$O$_3$ Catalysts," Catalysis Letters 96, 43, 2004.
25. G. Lafaye, A. Siani, P. Marecot, M.D. Amiridis, C.T. Williams, "Particle Size Control in Dendrimer-Derived Supported Ruthenium Catalysts," Journal of Physical Chemistry B, 110, 7725, 2006.
26. F. Grohn, B.J. Bauer, Y.A. Akpalu, C.L. Jackson, E.J. Amis, "Dendrimer Templates for the Formation of Gold Nanoclusters," Macromolecules 33, 6042, 2000.
27. J. Zheng, M.S. Stevenson, R.S. Hikida, P.F. Van Patten, "Influence of pH on Dendrimer-Protected Nanoparticles," Journal of Physical Chemistry B 106, 1252, 2002.
28. M.F. Ottavaiani, F. Montalti, M. Romanelli, N.J. Turro, D.A. Tomalia, "Characterization of Starburst Dendrimers by EPR. 4. Mn(II) as a Probe of Interphase Properties," Journal of Physical Chemistry 100, 11033, 1996.
29. M. Zhao, L. Sun, R.M. Crooks, "Preparation of Cu Nanoclusters within Dendrimer Templates," Journal of the American Chemical Society 120, 4877, 1998.
30. R.M. Crooks, B.I. Lemon, L.K. Yeung, M. Zhao, "Dendrimer-Encapsulated Metals and Semiconductors: Synthesis, Characterization, and Applications," Topics in Current Chemistry 212, 81, 2000.
31. M.R. Knecht, J.C. Garcia-Martinez, R.M. Crooks, "Hydrophobic Dendrimers as Templates for Au Nanoparticles," Langmuir 21, 11981, 2005.
32. A. Manna, T. Imae, K. Aoi, M. Okada, T. Yogo, "Synthesis of Dendrimer-Passivated Noble Metal Nanoparticles in a Polar Medium: Comparison of Size between Silver and Gold Particles," Chemistry of Materials 13, 1674, 2001.
33. M. Zhao, R.M. Crooks, "Dendrimer-Encapsulated Pt Nanoparticles: Synthesis, Characterization, and Applications to Catalysis," Advanced Materials 11, 217, 1999.
34. G. Larsen, S. Noriega, "Dendrimer-Mediated Formation of Cu-CuO$_x$ Nanoparticles on Silica and their Physical and Catalytic Characterization," Applied Catalysis A: General 278, 73, 2004.
35. U. Kreibig, M. Vollmer, *Optical Properties of Metal Clusters,* Springer: Berlin, 1995.
36. P.J. Pellechia, J. Gao, Y. Gu, H.J. Ploehn, C.J. Murphy, "Platinum Ion Uptake by Dendrimers: An NMR and AFM Study," Inorganic Chemistry 43, 1421, 2004.

37. A. Henglein, "Colloidal Palladium Nanoparticles: Reduction of Pd(II) by H_2; $Pd_{Core}Au_{Shell}$ Particles," Journal of Physical Chemistry B 104, 6683, 2000.

38. S. Watanabe, S.L. Regen, "Dendrimers as Building Blocks for Multilayer Construction," Journal of the American Chemical Society 116, 8855.

39. V. Chechik, M. Zhao, R.M. Crooks, "Self-Assembled Inverted Micelles Prepared from a Dendrimer Template: Phase Transfer of Encapsulated Guests," Journal of the American Chemical Society 121, 4910, 1999.

40. Y. Niu, L. Sun, R.M. Crooks, "Determination of the Intrinsic Proton Binding Constants for Poly(amidoamine) Dendrimers via Potentiometric pH Titration," Macromolecules 36, 5725.

41. L. Sun, R.M. Crooks, "Interactions between Dendrimers and Charged Probe Molecules. A. Theoretical Methods for Simulating Proton and Metal Ion Binding to Symmetric Polydentate Ligands," Journal of Physical Chemistry B 106, 5864, 2002.

42. M. Ooe, M. Murata, T. Mizuzaki, K. Bitani, K. Kaneda, "Dendritic Nanoreactors Encapsulating Pd Particles for Substrate-Specific Hydrogenation of Olefins," Nano Letters 2, 999, 2000.

43. Y. Li, M.A. El-Sayed, "The Effect of Stabilizers on the Catalytic Activity and Stability of Pd Colloidal Nanoparticles in the Suzuki Reactions in Aqueous Solution," Journal of Physical Chemistry B 105, 8938, 2001.

44. J. Fink, C.J. Kiely, D. Bethell, D.J. Schiffrin, "Self-Organization of Nanosized Gold Particles," Chemistry of Materials 10(3), 922, 1998.

45. K. Hayakawa, T. Yoshimura, K. Esumi, "Preparation of Gold-Dendrimer Nanocomposites by Laser Irradiation and Their Catalytic Reduction of 4-Nitrophenol," Langmuir 19, 5517, 2003.

46. S.K. Oh, Y.G. Kim, H. Ye, R.M. Crooks, "Synthesis, Characterization and Surface Immobilization of Metal Nanoparticles Encapsulated within Bifunctionalized Dendrimers," Langmuir 19, 10420, 2003.

47. R.W.J. Scott, O.M. Wilson, R.M. Crooks, "Synthesis, Characterization, and Applications of Dendrimer-Encapsulated Nanoparticles," Journal of Physical Chemistry B 109, 692, 2005.

48. Y.G. Kim, S.K. Oh, R.M. Crooks, "Preparation and Characterization of 1–2 nm Dendrimer-Encapsulated Gold Nanoparticles Having Very Narrow Size Distributions," Chemistry of Materials 16, 167, 2004.

49. M. Zhao, R.M. Crooks, "Intradendrimer Exchange of Metal Nanoparticles," Chemistry of Materials 11, 3379, 1999.

50. J. Silvestre-Albero, G. Rupprechter, H.J. Freund, "Atmospheric Pressure Studies of Selective 1,3-Butadiene Hydrogenation on Pd Single Crystals: Effect of CO Addition," Journal of Catalysis 235, 52, 2005.

51. Y.M. Chung, H.K. Rhee, "Partial Hydrogenation of 1,3-Cyclooctadiene using Dendrimer-Encapsulated Pd-Rh Bimetallic Catalysts," Journal of Molecular Catalysis A: Chemical 206, 291, 2003.

52. Y.M. Chung, H.K. Rhee, "Dendrimer-Templated Ag-Pd Bimetallic Nanoparticles," Journal of Colloid and Interface Science 271, 131, 2004.

53. R.W.J. Scott, A.K. Datye, R.M. Crooks, "Bimetallic Palladium-Platinum Dendrimer-Encapsulated Catalysts," Journal of the American Chemical Society 125, 3708, 2003.

54. O.M. Wilson, R.W.J. Scott, J.C. Garcia-Martinez, R.M. Crooks, "Synthesis, Characterization, and Structure-Selective Extraction of 1–3 nm Diameter AuAg Dendrimer-Encapsulated Bimetallic Nanoparticles," Journal of the American Chemical Society 127, 1015, 2005.

55. R.W.J. Scott, O.M. Wilson, S.K. Oh, E.A. Kenik, R.M. Crooks, "Bimetallic Palladium-Gold Dendrimer-Encapsulated Catalysts," Journal of the American Chemical Society 126, 15583, 2004.
56. R. Narayanan, M.A. El-Sayed, "Effect of Colloidal Catalysis on the Nanoparticle Size Distribution: Dendrimer-Pd vs. PVP-Pd Nanoparticles Catalyzing the Suzuki Coupling Reaction," Journal of Physical Chemistry B 108, 8572, 2004.
57. S.M. Lu, H. Apler, "Hydroformylation Reactions with Recyclable Rhodium-Complexed Dendrimers on a Resin," Journal of the American Chemical Society 125, 13126, 2003.
58. J.P.K. Reynhardt, Y. Yang, A. Sayari, H. Apler, "Periodic Mesoporous Silica-Supported Recyclable Rhodium-Complexed Dendrimer Catalysts," Chemistry of Materials 16, 4095, 2004.
59. S.M. Lu, H. Apler, "Intramolecular Carbonylation Reactions with Recyclable Palladium-Complexed Dendrimers on Silica: Synthesis of Oxygen, Nitrogen, or Sulfur-Containing Medium Ring Fused Heterocycles," Journal of the American Chemical Society 127, 14776, 2005.
60. S. Antebi, P. Arya, L.E. Manzer, H. Apler, "Carbonylation Reactions of Iodoarenes with PAMAM Dendrimer-Palladium Catalysts Immobilized on SiO_2," Journal of Organic Chemistry 67, 6623, 2002.
61. H. Tokuhisa, M. Zhao, L.A. Baker, V.T. Phan, D.L. Dermody, M.E. Garcia, R.R. Peez, R.M. Crooks, T.M. Mayer, "Preparation and Characterization of Dendrimer Monolayers and Dendrimer-Alkanethiol Mixed Monolayers Adsorbed to Gold," Journal of the American Catalysis Society 120, 4492, 1998.
62. A. Hierlemann, J.K. Campbell, L.A. Baker, R.M. Crooks, A.J. Ricco, "Structural Distortion of Dendrimers on Gold Surfaces: A Tapping-Mode AFM Investigation," Journal of the American Chemical Society 120, 5323, 1998.
63. P.K. Maiti, T. Cagin, S.T. Lin, W.A. Goddard, "Effect of Solvent and pH on the Structure of PAMAM Dendrimers," Macromolecules 38, 979, 2005.
64. D. Liu, J. Gao, C.J. Murphy, C.T. Williams, "*In Situ* Attenuated Total Reflection Infrared Spectroscopy of Dendrimer-Stabilized Platinum Nanoparticles Adsorbed on Alumina," Journal of Physical Chemistry B 108, 12911, 2004.
65. L.W. Beakley, S.E. Yost, R. Cheng, B.D. Chandler, "Nanocomposite Catalysts: Dendrimer Encapsulated Nanoparticles Immobilized in Sol-Gel Silica," Applied Catalysis A: General 292, 124, 2005.
66. R.W.J. Scott, O.M. Wilson, R.M. Crooks, "Titania-Supported Au and Pd Composites Synthesized from Dendrimer-Encapsulated Metal Nanoparticle Precursors," Chemistry of Materials 16, 5682, 2004.
67. Y.M. Chung, H.K. Rhee, "Design of Silica-Supported Dendritic Chiral Catalysts for the Improvement of Enantioselective Addition of Diethylzinc to Benzaldehyde," Catalysis Letters 82, 249, 2002.
68. Y.M. Chung, H.K. Rhee, "Silica-Supported Dendritic Chiral Auxiliaries for Enantioselective Addition of Diethylzinc to Benzaldehyde," Comptes Rendus Chimie 6, 695, 2003.
69. D.S. Deutsch, A. Siani, C.T. Williams, and M.D. Amiridis, "An FTIR Investigation of the Thermal Decomposition of Poly(amidoamine) (PAMAM) Dendrimers and Dendrimer-Metal Nanocomposites supported on Al_2O_3 and ZiO_2," Journal of Physical Chemistry B, submitted 2006.
70. D. Lin-Vien, N.B. Colthup, W.G. Fateley, J.G. Grasselli, *The Handbook of Infrared and Raman Characteristic Frequencies of Organic Molecules*, San Diego: Academic Press Inc., 1991.

71. A. Pevsner, M. Diem, "Infrared Spectroscopic Studies of Major Cellular Components. Part I: The Effect of Hydration on the Spectra of Proteins," Applied Spectroscopy 55, 788, 2001.
72. G. Socrates, *Infrared and Raman Characteristic Group Frequencies: Tables and Charts, 3rd Ed.*, West Sussex, U.K.: John Wiley & Sons, 2001.
73. C.J. Pouchert, *The Aldrich Library of FT-IR Spectra*, Aldrich Chemical Company, Inc., Vol. I, p. 279, 1986.
74. V.S. Escribano, G. Busca, V. Lorenzelli, "Fourier Transform Infrared Spectroscopic Studies of the Reactivity of Vanadia-Titania Catalysts Towards Olefins. 1. Propylene," Journal of Physical Chemistry 94, 8939, 1990.
75. G.R. Bamwenda, A. Ogata, A. Obuchi, J. Oi, A. Mizuno, J. Skrzypek, "Selective Reduction of Nitric Oxide with Propene over Platinum-Group-Based Catalysts: Studies of Surface Species and Catalytic Activity," Applied Catalysis B 6, 311, 1995.
76. M. Xin, I.C. Hwang, S.I. Woo, "FTIR Studies of the Reduction of Nitric Oxide by Propene on Pt/ZSM-5 in the Presence of Oxygen," Journal of Physical Chemistry B 101, 9005, 1997.
77. D.S. Deutsch, A. Siani, C.T. Williams, and M.D. Amiridis, "Synthesis and Characterization of Dendrimer-Stabilized Rhodium Nanoparticles on a Zirconia Support," *manuscript in preparation.*
78. A. Singh, B.D. Chandler, "Low-Temperature Activation Conditions for PAMAM Dendrimer Templated Pt Nanoparticles," Langmuir 21, 10776, 2005.
79. C. Elmasides, D.I. Kondarides, W. Grunert, X.E. Verykios, "XPS and FTIR Study of Ru/Al2O3 and Ru/TiO2 Catalysts: Reduction Characteristics and Interaction with a Methane-Oxygen Mixture," Journal of Physical Chemistry B 103, 5227, 1999.
80. V. Mazzieri, F. Coloma-Pascual, A. Arcoya, P.C. L'Argentiere, N.S. Figoli, "XPS, FTIR and TPR Characterization of Ru/Al2O3 Catalysts," Applied Surface Science 210, 222, 2003.
81. G.C. Bond, J.C. Slaa, "Catalytic and Structural Properties of Ruthenium Bimetallic Catalysts: Effects of Pretreatment on the Behaviour or Various Ru/Al2O3 Catalysts in Alkane Hydrogenolysis," Journal of Molecular Catalysis A 96, 163, 1995.
82. H. Lang, S. Maldonado, K.J. Stevenson, B.D. Chandler, "Synthesis and Characterization of Dendrimer-Templated Supported Bimetallic Pt-Au Nanoparticles," Journal of the American Chemical Society 126, 12949, 2004.

10 Synthesis of Oxide- and Zeolite-Supported Catalysts from Organometallic Precursors

Bruce C. Gates

CONTENTS

ABSTRACT

Supported metal complexes and clusters with well-defined structures offer the advantages of catalysts that are selective and structures that can be understood in depth. Such catalysts can be synthesized precisely with organometallic precursors, as illustrated in this review. Synthetic methods are illustrated with examples, including silica-supported chromium and titanium complexes for alkene polymerization; rhodium carbonyls bonded predominantly at crystallographically specific sites in a zeolite; and metal clusters, including Ir_4, Rh_6, Os_5C, and bimetallics.

10.1 INTRODUCTION

Supported metals are among the most important industrial catalysts. They are usually made from metal salts and porous supports such as metal oxides. Typical preparation routes include impregnation of the support or ion exchange with an aqueous solution of a metal salt, followed by calcination and reduction. The well-known preparation routes are often efficient and economical, being widely used in technology, but they typically produce highly nonuniform materials. Because of this nonuniformity, it is difficult to determine incisive relationships between catalyst structure and performance; it is also difficult to prepare highly selective catalysts because the variety of surface structures corresponds to a range of active sites.

The motivation to prepare structurally well-defined supported catalysts leads to the recognition of the value of precursors that react precisely with the functional groups of supports. Thus, metal compounds with reactive ligands (typically, organometallics) have been used frequently to make supported catalysts, both mononuclear (single-metal-atom) metal complexes and metal clusters. Some of the catalysts made in this way are structurally rather simple and well defined, and they are providing routes to fundamental understanding of supported metal catalysts that extends to the more complex materials made conventionally. This understanding has approached a level that justifies use of the term *synthesis* rather than just preparation of supported catalysts. However, it is emphasized that some degree of non-uniformity is unavoidable in all catalyst preparations when the support surface is nonuniform—as almost all of them are.

The goal of this chapter is to summarize methods for synthesis of supported catalysts from organometallics and to illustrate them with examples. Recent work is emphasized, and reviews[1, 2, 3, 4, 5] provide connections to the older literature. For brevity, details are largely omitted here, and readers are directed to the cited literature to find them. Many of the inferences about the structures and therefore the synthesis chemistry of these catalysts are based on physical characterization methods such as spectroscopy and microscopy; the details of these methods are also glossed over here and available in the cited references.

10.2 ORGANOMETALLIC PRECURSORS OF SUPPORTED CATALYSTS

When the goals of catalyst preparation are to make nearly uniform supported metal-containing species that are so small as to be nearly molecular in character, a good strategy is to choose precursors with reactive groups bonded to the metals (as ligands) and to use known molecular chemistry to guide the synthesis. If a precursor reacts with the functional groups on a support (e.g., OH groups and oxygen atoms on oxides and zeolites), then the resultant surface species may be simply bonded to the support by metal-oxygen bonds, which can be characterized experimentally and theoretically.

These surface species may be catalyst precursors that are converted in subsequent treatments into active catalysts. One of the synthesis goals is often to avoid remnants on the catalyst that remain from the precursor, such as chloride, nitrate, or phosphines, because these often affect catalytic activity (e.g., as poisons) and complicate characterization. To meet this goal, it is advantageous to use precursors with ligands that readily react away in the preparation or subsequent treatment steps. Thus, organometallic compounds with small, highly reactive ligands such as alkyl, allyl, and cyclopentadienyl are good choices, because they often give volatile products, such as small alkanes or alkenes. Other attractive precursor ligands include hydride and small oxygen-containing organics such as carbonyl and acetylacetonate.

Carbonyls may be problematic in the sense that CO poisons metal catalysts and is not so easily removed as, say, alkyls, but metal carbonyls are nonetheless frequently chosen because there are so many of them.

Because they are almost all air- or moisture-sensitive compounds, organometallics require handling in gloveboxes, vacuum lines, and the like. Thus, the resultant catalysts themselves are regarded as air- and moisture-sensitive materials that also require such handling. The relative complexity of the methods and the lack of the available equipment in many catalyst preparation laboratories seem to have contributed to a relative lack of application of organometallics in catalyst preparation. Another detriment of these precursors is cost; most organometallics are expensive in comparison with salts of the same metal. Nonetheless, organometallic catalyst precursors are being used increasingly, apparently even for industrial catalyst preparation (supported metallocenes—single-site catalysts for alkene polymerization[6]) when specific molecular surface sites are needed for selective catalysis.

Examples given below illustrate synthesis of alkene polymerization catalysts, but these catalysts are simpler than the supported metallocenes used in industry, because they lack the promoter methylaluminoxane (MAO), an ill-defined material that greatly complicates characterization. Other examples given below illustrate (a) details of the surface chemistry of conversion of an organometallic precursor into a supported catalyst; (b) synthesis of metal clusters of various sizes and compositions on a family of supports from metal carbonyl precursors; and (c) synthesis of supported bimetallic clusters with combinations of noble (e.g., Pt) and oxophilic (e.g., W) metals that give quite stable catalysts with extremely high metal dispersions.

10.3 OXIDE AND ZEOLITE SUPPORTS

Oxides and zeolites are the dominant catalyst supports, because many of them are inexpensive and can be prepared with wide ranges of pore structures that, to some degree, can be tailored. The supports considered here are, therefore, largely restricted to oxides and zeolites. The surface chemistry of the supports is important in the catalyst synthesis, but the understanding of this chemistry is usually highly simplified and barely begins to take account of the minority surface sites, although they may dominate the chemistry. Often, the only guidelines are vaguely defined acid/base properties of the support. For example, MgO is basic, γ-Al_2O_3 is more or less neutral, and SiO_2-Al_2O_3 is acidic. Simplified as they are, these representations of the properties go a long way in affording predictions of surface reactivity for catalyst synthesis.

Zeolites and other molecular sieves offer additional advantages: They are crystalline and available with a wide range of almost uniform pores of various sizes, and their chemical properties (such as acidity and basicity) can be varied by changing the composition—for example, of the exchange cations and the Si:Al ratio or the incorporation of atoms other than Si or Al in the framework. The pore structures of molecular sieves offer another advantage: The cages within the pore structure offer opportunities for "ship-in-a-bottle" synthesis for conversion of small metal complexes into metal clusters that are too large to leave through the cage windows.

The resulting entrapment, at least in prospect, may stabilize the supported metals by hindering their migration and sintering.

An illustration of a ship-in-a-bottle synthesis is given below. Other examples that follow illustrate (a) transition metal complexes bonded in nearly uniform sites in a zeolite, constituting one of the best-defined supported catalysts, and (b) transition metal clusters in a zeolite that are simple and uniform enough to allow precise characterization by x-ray absorption spectroscopy and density functional theory, with results that provide fundamental new, and apparently general, understanding of the metal-support interface in supported metals and the chemistry of hydrogen spillover.

10.4 EXAMPLES

Supported metal complex catalysts for alkene polymerization. Supported chromium complexes on silica have been used for many years in the Phillips process for ethylene polymerization, and promoters are not required. Like these supported complexes, the classical $TiCl_3$ Ziegler polymerization catalysts have also long been viewed as presenting surface catalytic sites that are well described as molecular analogues.

The Phillips catalyst is prepared from relatively inexpensive chromium salts; it is robust, but structurally complex, and the catalytic sites are not identified. To make a structurally simpler silica-supported alkene polymerization catalyst, Ajjou et al.[7, 8, 9] used the precursor bis(neopentyl)chromium(IV). The synthesis chemistry was represented as follows:

$$\{SiO\}_2Cr(CH_2CMe_3)_2 \rightarrow \{SiO\}_2Cr=CHCMe_3 + CMe_4 \qquad (1),$$

where Me is methyl and the braces denote groups terminating the silica (the chromium is bonded to oxygen). This catalyst was found to be active for ethylene polymerization without a promoter, and its relative simplicity allowed the authors to follow the chemistry in enough depth to infer details of the polymerization mechanism.

The description of the catalyst given above does not distinguish the various sites on the silica surface where the chromium complex could be bonded. In an attempt to better define sites for bonding of a supported metal complex for alkene polymerization and to ensure that they would be isolated from each other (i.e., site isolated, not present on neighboring surface sites), McKittrick and Jones[10] used a patterning molecule to be anchored to the silica that incorporated large trityl groups (Structure 1, Figure 10.1) that allowed for a spacing of aminosilane groups in the anchored species (Structure 2), preventing them from being immediately adjacent to each other. The support was then treated with a capping agent (hexamethyldisilazane) to remove remaining silanol groups (Structure 3), and then the trityl groups were removed (Figure 10.1). The resulting tethered amine ligands then reacted with an organotitanium precursor to form the catalyst (Figure 10.2), which was found to be highly active for ethylene polymerization. This strategy of multistep synthesis using protective groups indicates how methods of organic chemistry are being implemented in synthesis of supported catalysts.

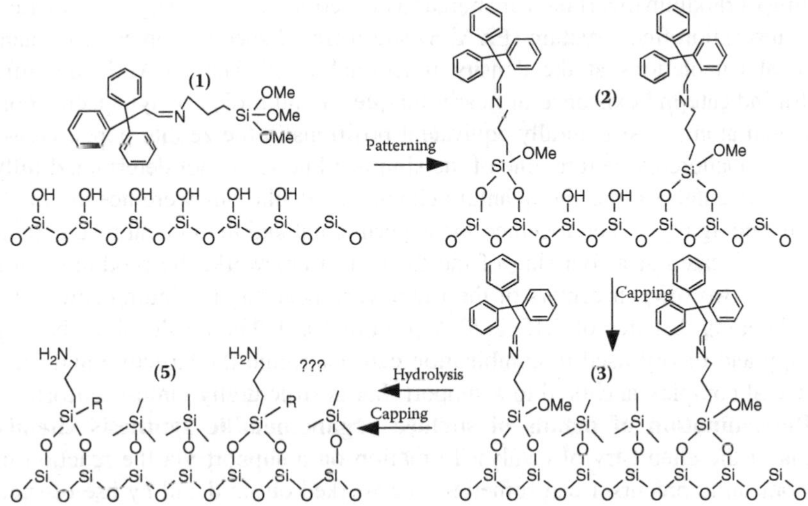

FIGURE 10.1 Schematic representation of synthesis of spaced amine groups on the surface of silica.[10]

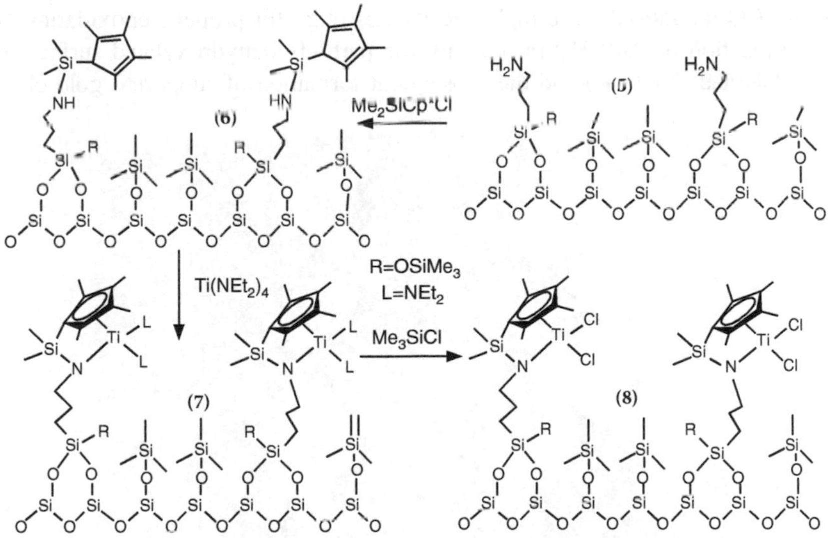

FIGURE 10.2 Schematic representation of anchoring of organotitanium complexes to silica support functionalized with spaced amine groups.[10]

A different approach to the synthesis of structurally uniform metal complexes on a surface was reported by Goellner et al.,[11] who used a crystalline support, attempting to prepare a complex bonded predominantly at a particular crystallographic site. The support was dealuminated Y zeolite, and the catalyst was prepared by its reaction with Rh(CO)₂(acac) (acac is acetylacetonate). The structure and bonding of rhodium dicarbonyl anchored to the zeolite were investigated by extended x-ray absorption fine structure (EXAFS) and infrared spectroscopies, and quantum chemical calculations at the density functional level. The EXAFS and infrared spectra indicate the existence of nearly unique rhodium dicarbonyl species bonded to oxygen atoms at structurally equivalent positions in the zeolite pores. However, even this anchored structure, one of the simplest known, is not determined fully by the experimental results, and quantum chemical calculations were needed to eliminate the ambiguity. Taken together, the experimental and theoretical results indicate $Rh^+(CO)_2$ located at a four-ring of the faujasite framework; the rhodium center is bonded to two oxygen centers of the framework near an aluminum center with an Rh-O bonding distance of 2.15–2.20 Å (Figure 10.3). The results show how spectroscopy and theory used in combination can determine the structure and location of a metal complex anchored to a support that is structurally almost uniform.

Determination of details of surface organometallic synthesis chemistry. Details of the chemistry of catalyst formation on a support via the reaction of an organometallic precursor can sometimes be worked out in detail by use of spectroscopic methods to characterize the surface species combined with analysis of gas-phase products. A number of examples are given by the Copéret group in a recent review.[2] The example given here involves supported gold, as there is intense interest now in such catalysts, because some of them have been found to have high activities (e.g., for CO oxidation[12]) and high selectivities (e.g., for propene epoxidation[13]).

The reaction of $Au(CH_3)_2(acac)$ with the partially dehydroxylated surface of γ-Al_2O_3 (calcined at 673 K) and the subsequent formation of supported gold clusters

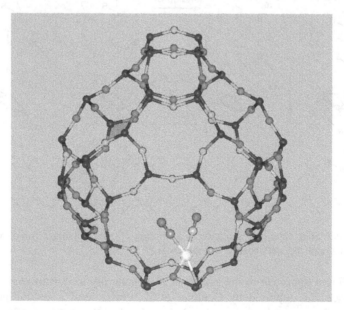

FIGURE 10.3 Structural model representing the location of a rhodium dicarbonyl complex at a faujasite four-ring; the structure was determined by EXAFS and infrared spectroscopies and calculations with density functional theory.[11]

upon treatment of the supported species in He at various temperatures and atmospheric pressure were investigated with infrared, EXAFS, and x-ray absorption near edge structure (XANES) spectroscopies.[14] The infrared spectra show that $Au(CH_3)_2(acac)$ reacts readily at room temperature with both OH groups and coordinatively unsaturated aluminum sites of the γ-Al_2O_3 surface, forming supported mononuclear gold complexes and {Al}-acac species. EXAFS data demonstrate the formation of mononuclear gold complexes as the predominant surface gold species, confirmed by the lack of Au-Au contributions in the EXAFS spectra. The mononuclear complex is represented as ${AlO}_2$-$Au^{III}(Me)_2$, with the Au-O bonding distance being 2.15 Å. Treatment of ${AlO}_2$-$Au^{III}(Me)_2$ in He at 323 K and 1 atm led to formation of supported Au_2O_3 in the form of highly dispersed clusters. Further treatment led to the formation of metallic gold clusters of increasing size, ultimately those with an average diameter of about 30 Å. The {Al}-acac groups were converted to {Al}-acetate groups upon treatment in He at 423 K; further treatment at increasing temperatures removed all the organic ligands from the surface. The results provide the first evidence of the reaction of a metal acetylacetonate complex with an oxide surface that is characterized by essentially all the resultant surface species.

Metal clusters on supports. Methods for the synthesis of structurally well-defined supported metal clusters include bonding of precursor metal carbonyl clusters to the support or reaction of precursors to form metal carbonyl clusters on the support. In subsequent steps, the carbonyl ligands are removed. The methods are reviewed elsewhere,[3, 5] and only a few details are presented here.

Metal carbonyl clusters, including $Ir_4(CO)_{12}$, $Ir_6(CO)_{16}$, and $Rh_6(CO)_{16}$, are adsorbed intact from alkane solution onto neutral supports such as γ-Al_2O_3 or TiO_2. Adsorption on basic supports such as MgO or La_2O_3 causes formation of surface carbonylate ions such as $[HIr_4(CO)_{11}]^-$ and $[Ir_6(CO)_{15}]^{2-}$ from $Ir_4(CO)_{12}$ and $Ir_6(CO)_{16}$, respectively.

Alternatively, supported metal carbonyl clusters are formed by surface-mediated synthesis[4, 15] from mononuclear precursors. For example, $[HIr_4(CO)_{11}]^-$ is formed from $Ir(CO)_2(acac)$ on MgO, and $Rh_6(CO)_{16}$ is formed from $Rh(CO)_2(acac)$ on MgO or γ-Al_2O_3 in the presence of CO. Synthesis of metal carbonyl clusters on oxide supports apparently often involves OH groups or water on the support surface; analogous chemistry occurs in solution.[5] The synthesis from a mononuclear metal complex is likely to occur with a yield less than that associated with simple adsorption of a preformed metal cluster, and so performed cluster precursors are preferred, except when they do not fit into the pores of the support (e.g., a zeolite).

Quantitative characterization of the formation of $[Os_5C(CO)_{14}]^{2-}$ from a smaller precursor on MgO was characterized by ^{13}C NMR spectroscopy.[16] The multistep synthesis from $Os_3(CO)_{12}$ gave $[Os_5C(CO)_{14}]^{2-}$ in a yield of about 65%; other products included tri- and tetra-osmium carbonyl clusters (Figure 10.4). The clusters on MgO have been observed by high-resolution transmission electron microscopy (Figure 10.5).[17]

Because CO is a poison of many catalytic reactions, supported metal carbonyl clusters are usually treated to remove these ligands. The decarbonylation of supported metal carbonyl clusters sometimes occurs almost without changes in the metal

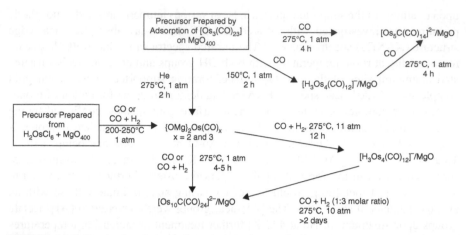

FIGURE 10.4 Synthesis of osmium carbonyl clusters on the surface of MgO; the basic properties of the support are essential for the synthesis.[16a] A yield of 65% of $[Os_5C(CO)_{14}]^{2-}$ was obtained, as indicated by ^{13}C NMR spectroscopy.[16b]

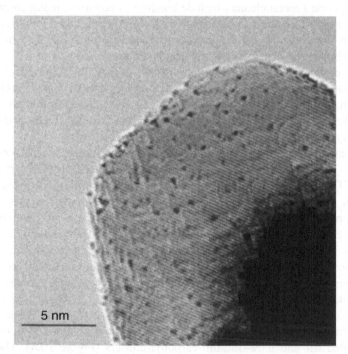

FIGURE 10.5 High-resolution transmission electron micrograph of $[Os_5C(CO)_{14}]^{2-}$ on MgO, prepared as shown in Figure 10.4. This cluster was present with some osmium carbonyl clusters containing three and four Os atoms.[17]

frame of the cluster, but when the decarbonylation takes place at elevated temperatures, migration and aggregation of the metal occur, and structural simplicity is lost.

For example, when $[HIr_4(CO)_{11}]^-$ on MgO was treated in He at 573 K, the CO ligands were removed fully, as shown by infrared and EXAFS spectra, and the Ir_4 tetrahedra remained essentially intact, as shown by EXAFS spectra.[3, 18] Infrared spectra indicated the formation of formate and carbonate on the MgO.[18] When the decarbonylation took place in the presence of H_2, the iridium more readily aggregated into larger clusters. Similar results pertain to the decarbonylation of neutral clusters of Ir and of Rh and to anionic clusters of these metals.[3, 19] The decarbonylation of oxide-supported metal carbonyls yields gaseous products including not just CO, but also CO_2, H_2, and hydrocarbons.[20] The decarbonylation chemistry involves the support surface and breaking of C-O bonds and has been thought to possibly leave carbon on the clusters.

Metal clusters in zeolites. Zeolite NaY-supported $Ir_4(CO)_{12}$ was prepared by direct deposition onto the outside surface of the zeolite crystals and alternatively by reductive carbonylation of $Ir(CO)_2(acac)$ sorbed in its pores.[21] Measurements of infrared and EXAFS spectra during the formation of the entrapped metal carbonyl clusters gave evidence of reaction intermediates, suggested to be $Ir_2(CO)_8$.[22] These samples and the products of their decarbonylation and subsequent recarbonylation were characterized by infrared and EXAFS spectroscopies.[21] The data show that the clusters in the zeolite pores were molecularly dispersed (site isolated) and could be reversibly decarbonylated and recarbonylated, whereas those on the outer surface outside the pores were aggregated under decarbonylation condition (573 K in He) and could not be reversibly recarbonylated. The results demonstrate stabilization of the molecular clusters entrapped in the pores.

Rhodium clusters, $Rh_6(CO)_{16}$, and the decarbonylated clusters approximated as Rh_6, have also been prepared in the cages of zeolite NaY by a ship-in-a-bottle method.[23] Theoretical work shows one of the advantages of these rather simple supported catalysts: Calculations at the density functional level by Vayssilov et al.[24] representing the cluster-zeolite combination show that an Rh_6 cluster in the zeolite (assumed to be bonded at a six-ring, Figure 10.6) with three H ligands in bridging positions is markedly more stable (by 370 kJ mol^{-1} of cluster) than the bare cluster on the zeolite with OH groups. The calculations match the EXAFS data well and also show that the clusters with hydride ligands are much more stable than the clusters with carbon ligands that one might have expected to form from the CO ligands during decarbonylation. Thus, the supported clusters have a strong affinity for hydride ligands formed by reverse spillover, and this result may be general for supported platinum group metals.[24] The theoretical results further demonstrate that spillover is a redox process; in reverse spillover, the support OH groups oxidize the cluster, and the positively charged atoms are at the metal-support interface—another inference that may be general for supported metals. The theoretical and EXAFS results characterizing zeolite-supported Rh_6 raise the question of whether ligand-free clusters or metal complexes are stable on hydroxylated supports and thus whether chemisorption measurements intended to determine the number of bonding sites of supported metal clusters can be done cleanly (which would require initial removal of all the adsorbates by evacuation). Evacuation can remove H_2, at the

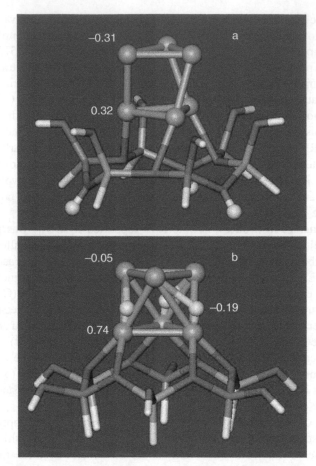

FIGURE 10.6 Models of Rh_6 supported on a zeolite fragment represented by density functional theory.[24] (a, top) Model of Rh_6 supported on a zeolite fragment with three bridging OH groups. (b, bottom) Model of Rh_6 with three hydride ligands supported on a zeolite fragment formed by reverse spillover of hydrogen from a zeolite fragment with three OH groups. Mulliken charges (in e) of the atoms in the supported cluster are shown, indicating that the Rh atoms at the metal-support interface are cationic.

expense of support OH groups, but are the clusters stable during this process, and do their morphologies change?

Supported bimetallic catalysts. Supported bimetallics with extremely high dispersions can be made from molecular bimetallic precursors in which the two metals are bonded to each other. Pt-Ru clusters dispersed on γ-Al_2O_3 were prepared by decarbonylation of molecularly adsorbed $Pt_2Ru_4(CO)_{18}$ by treatment in He or H_2 at temperatures in the range of 573–673 K.[25] EXAFS data show that, after decarbonylation, the Pt-Ru interactions were largely maintained, but the Pt-Ru cluster frame was changed. The average Pt-Pt bond distance apparently increased slightly (from 2.66 to 2.69 Å), and the Ru-Ru distance decreased from 2.83 to 2.64 Å.

The corresponding Pt-Pt and Ru-Ru coordination numbers were found to be 2.0 and 4.0, respectively, indicating that slight agglomeration of the metal took place, and the clusters incorporated, on average, less than three and six Pt and Ru atoms, respectively. These appear to be the smallest supported bimetallic clusters of platinum-group metals yet reported.

When a supported metal on an oxide is prepared from an adsorbed precursor incorporating a noble metal bonded to an oxophilic metal, the result may be small noble metal clusters nested in a cluster of atoms of the oxophilic metal, which is oxidized and anchored to the support through metal-oxygen bonds.[26] The simplest such structure was made from $Re_2Pt(CO)_{12}$; the decarbonylated clusters are modeled as Re_4Pt_2 on the basis of EXAFS data showing that the oxophilic rhenium interacts strongly with the oxygen atoms of the support and also with platinum (Figure 10.7).[26] In general, it appears that when one of the metals in a supported bimetallic cluster is noble and the other oxophilic, the oxophilic metal interacts more strongly with the support than the noble metal; if the bimetallic frame of the precursor is maintained nearly intact, then this metal-support interaction helps keep the noble metal highly dispersed.

Other samples of such nested noble metal clusters on oxides have been made from $Pd_2Mo_2(CO)_6(C_5H_5)_2(PPh_3)_2$,[27] $PtMo_2(CO)_6(C_5H_5)_2(PhCN)_2$,[28] $PtW_2(CO)_6(C_5H_5)_2(PhCN)_2$,[29] $Pt_2W_2(CO)_6(C_5H_5)_2(PPh_3)_2$,[30] and $[Ru_{12}C_2Cu_4Cl_2(CO)_{32}]$ $[PPN]_2$ (Ph is phenyl).[31] Platinum clusters of as few as four atoms each, on average, are indicated by EXAFS data characterizing the Pt-W clusters.[30] The Pt-W samples are quite stable, with the cluster size remaining essentially unchanged even after oxidation-reduction cycles at 673 K.[29, 30] The stability of these samples might be

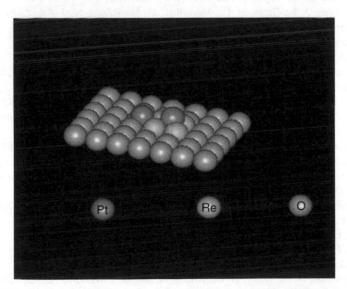

FIGURE 10.7 Simplified model based on EXAFS data of Re_4Pt_2 clusters formed on the surface of γ-Al_2O_3 from $Re_2Pt(CO)_{12}$.[26] The more oxophilic Re atoms are cationic and preferentially bonded to the support.

a significant advantage in catalytic applications, perhaps even justifying the use of organometallic precursors in preparations of technological catalysts.

ACKNOWLEDGMENTS

This research was supported by the U.S. Department of Energy contract number FG02-04ER15513 and by the Petroleum Research Fund, administered by the American Chemical Society.

REFERENCES

1. Basset, J.-M.; Gates, B. C.; Candy, J.-P.; Choplin, A.; Leconte, M.; Quignard, F.; Santini, C. (Eds.) *Surface Organometallic Chemistry: Molecular Approaches to Surface Catalysis*, *NATO ASI series*, Vol. 231, Kluwer: Dordrecht, 1988.
2. Copéret, C.; Chabanas, M.; Saint-Arroman, R. P.; Basset, J.-M. *Angew. Chem. Int. Ed. 42*, 156, 2003.
3. Gates, B. C. *Chem. Rev. 95*, 511, 1995.
4. Cariati, E.; Roberto, D.; Ugo, R.; Lucenti, E. *Chem. Rev. 103*, 3707, 2003.
5. Guzman, J.; Gates, B. C. *Dalton Trans.* 3303, 2003.
6. Hlatky, G. G. *Chem. Rev. 100*, 1347, 2000.
7. Ajjou, J.A.N.; Scott, S. L. *Organometallics 16*, 86, 1997.
8. Ajjou, J.A.N.; Scott, S. L. *J. Am. Chem. Soc. 120*, 415, 1998.
9. Ajjou, J.A.N.; Rice, G. L.; Scott, S. L. *J. Am. Chem. Soc. 120*, 13436, 1998.
10. McKittrick, M. W.; Jones, C. W. *J. Am. Chem. Soc. 126*, 3052, 2004.
11. Goellner, J. F.; Gates, B. C.; Vayssilov, G. N.; Rösch, N. *J. Am. Chem. Soc. 122*, 8056, 2000.
12. Haruta, M. *Catal. Today 36*, 153, 1997.
13. Haruta, M. *Top. Catal. 29*, 95, 2004.
14. Lamb, H. H; Gates, B. C.; Knözinger, H. *Angew. Chem. Int. Ed. Engl. 27*, 1127, 1988.
15. Gates, B. C. *J. Mol. Catal. 86*, 95, 1994.
16. (a) Lamb, H. H.; Fung, A. S.; Tooley, P. A.; Puga, J.; Krause, T. R.; Kelley, M. J.; Gates, B. C. *J. Am. Chem. Soc. 111*, 8367, 1989; (b) Bhirud, V. A.; Panjabi, G.; Salvi, S. N.; Phillips, B. L.; Gates, B. C. *Langmuir 20*, 6173, 2004.
17. Allard, L. F.; Panjabi, G. A.; Salvi, S. N.; Gates, B. C. *Nano Lett. 2*, 381, 2002.
18. Alexeev, O. S.; Kim, D.-W.; Gates, B. C. *J. Mol. Catal. A, 162*, 67, 2000.
19. Gates, B. C. In *Catalysis by Di- and Polynuclear Metal Cluster Complexes*, Adams, R. D.; Cotton, F. A. (Eds.), Wiley-VCH, Weinheim, p. 509, 1998.
20. Smith, A. K.; Theolier, A.; Basset, J.-M.; Ugo, R.; Commereuc, D.; Chauvin, Y. *J. Am. Chem. Soc. 100*, 2590, 1978.
21. Li, F.; Gates, B. C. *J. Phys. Chem. B 107*, 11589, 2003.
22. Li, F.; Gates, B. C. *J. Phys. Chem. B 108*, 11259, 2004.
23. Weber, W. A.; Gates, B. C. *J. Phys. Chem. B 101*, 10423, 1997.
24. Vayssilov, G. N.; Gates, B. C.; Rösch, N. *Angew. Chem. Int. Ed. 42*, 1391, 2003.
25. Alexeev, O. S.; Graham, G. W.; Shelef, M.; Adams, R. D.; Gates, B. C. *J. Phys. Chem. B 106*, 4697, 2002.

26. Fung, A. S.; Kelley, M. J.; Koningsberger, D. C.; Gates, B. C. *J. Am. Chem. Soc.*, *119*, 5877, 1997.
27. Kawi, S.; Alexeev, O.; Shelef, M.; Gates, B. C. *J. Phys. Chem. 99*, 6926, 1995.
28. Alexeev, O.; Kawi, S.; Shelef, M.; Gates, B. C. *J. Phys. Chem. 100*, 253, 1996.
29. Alexeev, O.; Shelef, M.; Gates, B. C. *J. Catal. 164*, 1, 1996.
30. Alexeev, O. S.; Graham, G. W.; Shelef, M.; Gates, B. C. *J. Catal. 190*, 157, 2000.

11 Supported Metal Oxides and the Surface Density Metric

William V. Knowles, Michael O. Nutt, and Michael S. Wong

CONTENTS

ABSTRACT

Supported metal oxides (SMOs) comprise a large class of catalytic materials used in numerous industrial processes. There are many conventional approaches to preparing these materials, ranging from impregnation and equilibrium adsorption to grafting and coprecipitation. Independent of preparation methods, one of the key metrics in characterizing SMOs is surface density, which quantifies the amount of the supported metal oxide relative to the underlying support surface area. Catalytic activity is correlated to the surface density-dependent structure of the supported species. There are different definitions for surface density and different methods for its determination, though, causing some difficulties in reconciling structure-activity results reported by different researchers. Here, a rigorous analysis of the different surface density calculation methods is presented, using tungstated zirconia as an example.

11.1 INTRODUCTION

Many synthetic routes for preparing transition metal oxide catalysts produce a supported metal oxide structure consisting of an active metal oxide phase (the surface oxide) dispersed on a second, high surface area oxide (the support oxide) [1–3]. A key metric in characterizing SMOs is surface density. International Union of Pure and Applied Chemistry (IUPAC) defines surface density as mass per unit area [4]. For supported metal oxides, this is vaguely interpreted as the amount of supported metal oxide active phase per surface area of the underlying oxide support. This broad definition allows considerable latitude in whether total or exposed surface oxide content is considered and whether the surface area is of the uncovered support or final catalyst. Furthermore, absence of standardized methods to measure these parameters introduces additional variability into the determination of surface density.

The surface oxide generally exists in several different molecular and nanoscale structures, depending on the surface density range. Formation of these structures depends on the supported metal oxide content of the catalyst and the support/supported-layer interactions, and also on the synthesis route, support surface area, and calcination conditions [5]. By studying the catalytic properties of an SMO as a function of surface density and by characterizing the surface oxide structures, one can gain an understanding of how the structure relates to catalysis and how to fine-tune the material for improved catalytic performance (e.g., higher activity, higher selectivity).

In this chapter, we briefly review the basic synthesis routes of SMOs, the various surface structures that can arise and their correlations to surface density ranges, and the commonly used characterization methods for SMOs. We discuss the difference between surface saturation and monolayer (ML) coverage, two terms that represent different surface density thresholds but can be confused with one another. We analyze the different surface density calculation methods used, which can lead to numerical discrepancies that complicate investigations into structure-property relationships for SMOs. We discuss the most appropriate calculation method, using tungstated zirconia (WO_x/ZrO_2) prepared through incipient wetness impregnation as the model SMO material.

11.2 SYNTHESIS OF SUPPORTED METAL OXIDE CATALYSTS

Preparation of supported metal oxides for heterogeneous catalysis can be carried out through numerous routes. The following underlying steps are common to all procedures for support metal oxides: preparation of the support precursor, contact of the surface and support precursors, and thermal decomposition of the precursors (through calcination) to their oxide form.

11.2.1 IMPREGNATION

Impregnation entails wetting a solid support material (typically of high surface area) with a liquid solution containing the dissolved surface oxide precursor. Subclassification of impregnation methods depends upon the relationship between impregnating liquid volume (V_{imp}) and support pore volume (V_p) [6]: Capillary, dry, or incipient wetness (IW) is often used to describe the process when $V_{imp} \sim V_p$ whereas equilibrium adsorption, diffusional, or ion exchange is preferred when $V_{imp} > V_p$ (Section 11.2.2). The support may be either a metal oxide or metal oxide precursor, crystalline or amorphous. The impregnating solution may be either aqueous or organic. The dominant driving force for pore filling is capillary pressure (i.e., fluid mechanical), thus rendering the quality of impregnation insensitive to surface interactions between support and surface precursor that might otherwise limit the overall loading by other methods (e.g., equilibrium adsorption) [5]. Based on its simplicity, IM represents the most widely cited synthesis method for supported metal oxides.

Despite its ease of use, IW impregnation is notoriously deficient at achieving uniform surface coverage [7]. Multiple resistances exist to achieving uniform coverage that may lead to undesirable distribution profiles (crusting) of the surface oxide. For example, rising hydraulic pressure at blocked pore ends prevents complete wetting of the surface as pore-filling liquid compresses trapped gas [6]; even after the gas eventually dissolves, different residence times of wetting and adsorption can lead to concentration profiles. Another resistance is the narrowing of pore necks at their entrance created by (1) crystallization of active precursor due to premature solvent evaporation [7] or (2) unfavorable charge balancing between support and surface precursors such as most metal oxides on SiO_2 [6, 7]. It is even conceivable that a sample intended to possess submonolayer surface coverage may, in fact, possess crystalline regions of locally high concentration (i.e., above monolayer coverage), whereas areas of relatively low concentration remain noncrystalline (sub-monolayer coverage).

11.2.2 EQUILIBRIUM ADSORPTION

Impregnation methods that immerse a solid support into an excess of solution ($V_{imp} > V_p$) of dissolved surface oxide precursor for long periods of time are frequently called *equilibrium adsorption* (EA) or *ion exchange* [6]. In contrast to IW impregnation, the dominant driving forces for these methods are the concentration gradient and electrostatic interactions [5, 8]. Almost all supports are oxides that bear a net positive or negative electrical surface charge, which attracts oppositely charged ions in aqueous solution. The charge of these amphoteric oxides is a function of aqueous solution pH and is dictated by the oxide's point-of-zero charge (pzc) [9, 10]. For pH > pzc, terminal hydroxyls on the support deprotonate, leaving the surface with a net negative charge that attracts cationic species. For pH < pzc, the hydroxyls protonate and favor anionic adsorption. For a given oxide support, adjusting the difference between pzc and solution pH allows control of the surface oxide loading [8, 11].

Equilibrium adsorption is generally thought to yield better dispersion of surface metal oxide overlayers than IW impregnation, although this is difficult to measure for metal oxides. By way of comparison, chemisorption reveals much better atomic dispersion of the active species for supported metal catalysts made via EA versus IW impregnation [6]. Nonetheless, despite uniform distribution of active precursor species throughout the support interior, EA does not necessarily provide uniform distribution between interior and exterior regions of the support [7]. Recovery of the solid support from solution leads to surface evaporation of residual solvent, which preferentially deposits noninteracting precursor species on the support's exterior shell, thereby enriching the precursor outer surface concentration. A final wash step should be employed to remove this physisorbed (i.e., weakly bound) active species, but precise control of the wash pH is required to avoid inadvertently expelling the electrostatically bound species. Another disadvantage of EA is that the final loading of active species is rarely known *a priori* and requires subsequent elemental analysis.

11.2.3 DEPOSITION-PRECIPITATION

Deposition-precipitation (DP) shares many similarities with equilibrium adsorption. Both DP and EA methods suspend the solid support in excess solution ($V_{DP} > V_p$) containing dissolved active phase precursor and allow the surface precursor-support interactions (e.g., concentration gradient, electrostatic) to evolve naturally. DP goes one step further than EA, however, by introducing a change to the solution, which triggers deposition and precipitation of the active species. Common triggers include pH, valence state of the precursor, and complexing agent concentration [7]. Similar to EA, < 100% transfer of the active phase from solution to support is typical.

Although straightforward to perform experimentally, detailed mechanisms of the underlying chemical transformations for DP are rarely understood completely and likely differ for many systems [12]. This conclusion is based on the frequent omission of details explaining the competition between nucleation and growth for DP-prepared materials. For instance, after the trigger is applied to initiate deposition, nucleation may occur through (1) the formation of insoluble colloidal nanoparticles in solution and their subsequent surface-adsorption (e.g., electrostatic attraction), or (2) hydrolysis of the soluble active species [13], ligand exchange, and condensation with support hydroxyls to form surface-bound nuclei for the condensation of additional soluble species [7]. Understanding these subtleties is critical to designing better catalysts. The benefits of particle size and distribution of active species afforded by DP are clearly justified for supported metal catalysts, where metal dispersion exerts strong influence on catalytic activity [12, 14]. For supported metal oxide catalysts, there are no clear advantages in preparing SMOs via DP instead of EA.

11.2.4 GRAFTING

Grafting occurs through covalent bond formation between surface functional groups, typically hydroxyl species, of the support and surface active species [15]. Common active species precursors are coordination metal complexes such as metal halides and oxyhalides, metal alkoxides, and organometallics. These species are often moisture-sensitive and require thermal pretreatment of the support to remove physisorbed water. Upon immersing the solid support in an excess of grafting solution ($V_{graft} > V_p$), oxo anchors are formed immediately by condensation between surface and support hydroxyls. A final wash step is often employed to remove physisorbed precursor ligands. The covalent bonds that tether the active species to the support prevent migration (e.g., sintering, agglomeration) during thermal treatment and therefore endow the grafting method with greater control over active species distribution than other catalyst preparation methods. The upper limit to metal oxide loading is set by the hydroxyl surface density (e.g., 4–5 OH/nm^2 for SiO$_2$ [16] and ZrO$_2$ [17]), assuming 100% reaction of these hydroxyl groups with the grafting precursor. This provides a synthesis advantage over impregnation and DP methods, which tend to lead to more heterogeneously covered surfaces. Chemical vapor deposition (CVD) describes the grafting procedure using gas-phase precursors instead of liquid-phase precursors.

11.2.5 COPRECIPITATION

Coprecipitation differs from the other methods significantly. It is a method by which a solid is precipitated from a solution containing soluble precursors of both the support and surface oxides. Nucleation of the solid phase is initiated by mixing the solution with a precipitating agent that either (1) changes the solution pH and leads to precursor condensation to form oxides or hydroxides, or (2) "introduces additional ions into the system by which the solubility product for a certain precipitate is exceeded" [18]. Filtration and washing of counterions from the precipitate yield the final solid. The resulting architecture of the coprecipitated binary framework is more spatially distributed than a strictly supported metal oxide material prepared by the above methods [5]. This distinct structure allows for better interaction between support and active species but also results in partial exclusion of the active species from the surface, rendering it inaccessible for catalysis. Surface density calculations for resulting materials thus overestimate actual values.

Inverse coprecipitation offers an improved alternative to coprecipitation. A limitation for coprecipitation is that the support and surface oxide precursors are unlikely to share similar solubilities (i.e., solubility products). Consequently, dropwise addition of a precipitating agent generates solids dominated by the more insoluble precursor during early stages and rich in the latter precursor at late stages. This gives rise to temporal-spatially inhomogeneous compositions [18]. By contrast, inverse coprecipitation [19] adds the precursor mixture dropwise to an excess of precipitating agent. This approach ensures that a strict ratio of precursors is maintained throughout the course of batchwise addition and leads to better coprecipitate homogeneity [18].

11.2.6 ROLE OF CALCINATION

Calcination (also known as annealing, thermolysis, or pyrolysis) exposes the as-prepared catalyst precursor to high temperatures for the final step in the formation of finished metal oxide catalysts. Although specifically referring only to heat treatment, *calcination* is commonly used to imply all the process variables associated with the furnace: composition of the gas phase atmosphere in contact with the catalyst (e.g., oxidizing, reducing, inert, functionalizing) and the thermal profile (e.g., ramp rates, hold temperatures, and hold times). The source of thermal energy is not considered critical and includes convection (e.g., electric, gas-fired) and microwave ovens. The impact of gas pressure and thermal cooling rate are considered negligible.

There are several purposes for calcination. If the support oxide is formed in a separate step before addition of the surface oxide, calcination may be used to lock in the support's surface area, pore structure, and crystalline phase. The primary use of calcination is to thermally decompose nonoxidic precursors, remove unwanted ligands, and oxidize the support and surface species. Precursor counterions consisting of hydrogen, carbon, or nitrogen often volatilize in the furnace and are swept away, leaving an impurity-free surface; counterions such as alkali and alkaline metals, halides, phosphates, and sulfides mostly remain on the surface and, if not washed, can participate as promoters or poisons in the final catalyst. Proper selection of gas composition permits control of the final oxidation state of the support and surface metal centers. After oxidation (or reduction) of the precursors, calcination provides thermal energy to activate wetting and spreading as the Tammann temperature of the surface oxide is approached [20]. Unfortunately, calcination temperature is not the sole variable affecting wetting; even when the surface oxide has sufficient thermal energy, unfavorable surface free energies between support and surface oxide can lead to poor dispersion [21]. The thermal energy of calcination also controls the crystalline phase and grain size of the support and surface oxides.

Calcination is one of the two main variables used to control surface density; the other is surface oxide concentration. At a constant calcination temperature, increasing the loading of surface oxide directly increases the surface density of that species. Perhaps less intuitively, increasing calcination temperature at a constant surface oxide loading can also increase surface density. Higher temperatures would cause the support to crystallize, and as crystallization progresses, the pore walls of the support cannot withstand the growing internal stresses (leading to pore collapse and the consequent loss of surface area) [22].

11.3 MOLECULAR SURFACE STRUCTURES

The nature of the supported metal oxide species depends upon a number of factors: the preparation method (wet chemical synthesis plus calcination), chemical interactions between the support and surface layers, and surface density (surface oxide weight loading and specific surface area of the support oxide) [5]. Figure 11.1 schematically demonstrates the various dehydrated surface structures commonly observed for a mono-oxo metal oxide: isolated, oligomeric, polymeric, and crystalline species. Several reviews comprehensively catalog the expected surface structures for transition metal mono- and polyoxoanions in four-, five-, and sixfold coordination under nonreaction conditions [3, 23–25].

Except at the extrema of the surface density scale, a distribution of structures is expected at any given surface density. In addition, transition from isolated to polymerized to crystalline species with increasing surface oxide content does not progress stepwise but rather through a continuum; except for the onset of crystallinity, there are no sharp surface density delineations between the various species. Still, the different surface species have been rigorously identified as active sites (or precursor to the active sites) for a number of reactions for some SMO catalyst compositions.

11.3.1 ISOLATED SPECIES

The first structure in Figure 11.1 is that of isolated species in tetrahedral coordination (MO_4), which are typically present at very low concentrations of surface oxide. Isolated species possess only M=O and, presumably, M-O-S bonds, where M represents the metal center of surface oxide, O an oxygen, and S the metal center of the support oxide. It is often difficult to obtain direct spectroscopic evidence of M-O-S bonds [26], yet their existence is consistent with coordination models of

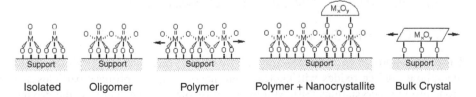

| Isolated | Oligomer | Polymer | Polymer + Nanocrystallite | Bulk Crystal |

FIGURE 11.1 Schematic of surface structures of a generic mono-oxo metal oxide active phase on a metal oxide support. The active phase forms tetragonal structures at low loading (MO_4 coordination), distorted octahedral corner- and edge-sharing structures at intermediate loading (MO_6 coordination), and stoichiometric crystalline structures at high loading (M_xO_y).

surface oxides and infrared (IR) spectroscopy of support hydroxyl titrations [26–29]. Absent are M-O-M bridges, which appear in oligomeric and polymerized surface species.

Only a few examples exist in the literature that positively identify isolated species as the key to catalytic activity, two of which are TiO_x/SiO_2 [30] and ZrO_x/SiO_2 [31] for methanol oxidation (x indicates the surface oxide exists in noncrystalline form). In these cases, the highest turnover frequencies (TOFs) for methanol consumption were observed at the lowest tested concentrations of supported titanate and zirconate, where isolated TiO_4 and ZrO_4 structures were shown to dominate. It was concluded that the number of Ti-O-Si and Zr-O-Si bonds were maximized in monomeric Ti and Zr sites, respectively. Significant TOF decreases concomitant with increasing surface oxide polymerization were rationalized to occur because the number of M-O-S bonds per active M site was reduced.

11.3.2 OLIGOMERIC AND POLYMERIZED SPECIES

In addition to the M=O and M-O-S bonds also present in isolated species, increases in surface oxide concentration leads to condensation of surface oxoanions and the formation of bridging M-O-M bonds characteristic of oligomeric and polymeric species. Figure 11.1 shows this for polyoxoanions in octahedral coordination (MO_6) possessing a combination of corner- and edge-sharing octahedra. Analogous to organic polymer nomenclature, oligomers (e.g., dimers, trimers) are generally considered to be of low molecular weight and incorporate only a limited number of active metal centers; polymers consist of larger domains that are often described as two-dimensional surface layers [2, 3]. Rarely is a distinction between oligomers and polymers drawn, and usually only in relation to reference crystalline materials through fundamental spectroscopic investigations [23, 32]. The formation of polymeric species is sensitive to the support composition. For example, due to unfavorable surface energy interactions, few oxides are known to form oligomeric/polymeric species supported on SiO_2 [2, 3].

A survey of catalysis literature reveals that for the majority of transition metal oxide catalyzed reactions, maximum TOFs most frequently correlate with maxima in polymeric surface species content. A limited subset of examples includes WO_x/ZrO_2 for o-xylene isomerization [33], n-pentane isomerization [29], and 2-butanol dehydration [34]; MoO_x/Al_2O_3 and VO_x/Al_2O_5 for dimethyl ether oxidation [35]; MoO_x/Al_2O_3 for propane oxidative dehydrogenation [36]; and MoO_x supported on TiO_2, ZrO_2, Al_2O_3, and Nb_2O_5 for methanol oxidation [37].

11.3.3 CRYSTALLINE SPECIES

Under the appropriate conditions (i.e., above monolayer coverage), well-dispersed amorphous surface oxides will nucleate into nanocrystalline primary particles and grow into microcrystallites [38]. Minimization of surface free energy is the driving force for the genesis and evolution of crystallite formation (i.e., sintering) for metal oxides [21]. Two common conditions that lead to crystallite growth are (1) supersaturation of the support oxide's capacity to stabilize noncrystalline active oxide (i.e., above monolayer coverage) [2, 26, 39], and (2) unfavorable solid-solid chemical interactions between the surface and support oxides, such as strong electrostatic repulsion in the case of SiO_2 and typical transition metal oxide precursors [2, 5]. These crystalline metal oxide species are characterized by well-defined unit cells exhibiting long-range order [40] amenable to x-ray diffraction investigation [41, 42]. These species have been referred to as nanocrystallites, microcrystallites, or bulk crystallites, terms that are often used interchangeably in supported metal oxide catalysis literature. To be precise, no official size delineation has been found for bulk crystallites, although presumably *nano* refers to < 100 nm and *micro* > 100 nm as per convention.

The oxidative dehydrogenation (ODH) of isobutane to isobutene provides an interesting example where bulk crystalline Cr_2O_3 was demonstrated more active than supported CrO_x/Al_2O_3 catalysts, both with and without nanocrystallites of Cr_2O_3 [43, 44]. The supported materials were found less active but more selective than Cr_2O_3 due to slower dissociative O_2 adsorption and electron transfer from O_2 to O^{2-} (i.e., reoxidation), consistent with common partial oxidation behavior that higher product selectivity is obtained for higher oxygen bond energy [45, 46]. The higher activity of the bulk crystal was assigned to relatively more labile oxygen than that of the supported materials frustrated by Cr-O-Al bonds. Oxygen mobility was higher for Cr_2O_3 than CrO_x/Al_2O_3 due to (1) lower Cr-O bond energy as shown by ~4 times lower activation energy (E_a) for oxygen chemisorption over 250–375°C, (2) roughly twice the oxygen content removal, and (3) peak temperatures (T_p) of oxygen temperature programmed desorption (TPD) and hydrogen TPR lowered by ~40°C.

11.4 DETERMINATION OF SURFACE SATURATION AND MONOLAYER COVERAGE

The primary benchmark in the quantitative investigation between surface density and molecular surface structure is the onset of monolayer coverage of the surface oxide. Two competing definitions for ML coverage have emerged in the literature, and disregard for their subtle differences can cause confusion, particularly when trying to correlate surface density to ML coverage. This chapter adopts the most general definition: Surface saturation (SS) is the experimentally determined onset of crystallinity (which occurs below monolayer coverage), and monolayer coverage is the hypothetical titration of all support hydroxyls. In the second definition, ML coverage (labeled henceforth as ML_2) occurs at the onset of crystallinity.

The analysis of surface coverage places primary importance on the concept of support hydroxyl titration. As has been verified repeatedly for supported metal oxide systems immediately after deposition, precursors of surface oxides (typically salts) interact weakly with the oxide support through hydrogen bonding and electrostatic forces [8, 11, 47] (except in the case of grafting). However, postdeposition thermal treatment leads to formation of a thermally stable, two-dimensional overlayer of surface metal oxide distinctly different in structure than the precursor molecule (under dehydrated conditions) [48]. Numerous infrared studies have shown a loss of support hydroxyl density with an increase in surface oxide loading [8, 27, 28,

49–54] through the formation of M-O-S bridges. ML coverage of the support occurs when 100% of the support hydroxyls are in the form of these M-O-S bridges [28], although in reality steric hindrance and lateral repulsion makes achieving ML coverage difficult [28, 33]. Thus, this leaves a landscape of polyoxoanions of the active oxide separated by patches of exposed support [8], and crystallization of these polyoxoanions can occur at surface densities below ML coverage.

The second definition equates ML coverage with SS, which assumes that crystallization occurs when all exposed hydroxyls have been titrated [26, 39, 55–57]. Hydroxyls remaining after crystallization could be discounted as interior hydroxyls (i.e., located in collapsed pores), which are inaccessible for forming M-O-S bridges [58]. On the other hand, many papers report the presence of support hydroxyls with surface oxide crystals for various compositions (see next section) and indicate that these hydroxyls are located on the support oxide exterior.

There is a relatively wide range of SS and ML coverage values (Table 11.1). In addition to definition differences, there are two other contributing factors: (1) Multiple characterization techniques permit the investigation of surface coverage, SS, and ML coverage. However, these methods do not necessarily lead to the same conclusions, especially if SS and ML coverage are implicitly or explicitly equated. (2) There is a subtle but important difference between surface density values based on the specific surface area of the support-only material and that of the composite SMO material.

TABLE 11.1
Calculation and Experimental Estimates of WO_x/ZrO_2 Surface Density for Surface Saturation (SS) and Monolayer (ML) Coverage of Tungsten Oxide Species

SS	ML	Units of measurement	Method	Chapter section	References
–	7.3	W/(nm² support)	Calculated	11.4.1	[59]
5.5	–	W/(nm² support)	Calculated	11.4.1	[1, 50, 57, 60]
6.4	–	W/(nm² support) [a]	Calculated	11.4.1	[39]
7.8	–	W/(nm² support) [a]	Calculated	11.4.1	[33]
4.0	–	W/(nm² support)	Raman	11.4.2	[26]
~3.0	5.0	W/(nm² support)	ISS	11.4.3	[60]
~2.0	4.6	W/(nm² composite)	CO chemisorption + IR	11.4.4	[50]
~4.5	–	W/(nm² composite)	Raman	11.4.4	[33]
–	~8.0	W/(nm² composite)	UV-vis DRS	11.4.5	[33]
4.5–6.0	–	W/(nm² composite)	Raman	11.5	This chapter
>19.3	–	W/(nm² composite)	Raman + UV-vis DRS	11.6.3	[39]

[a] W/(nm² support) interpreted from the W/nm² units used in the references.

11.4.1 IDEALIZED MODELS OF SMOs

Theoretical models predicting ML coverage idealize the support and surface oxides as vertically stacked, perfectly flat layers. Two different approaches are used, each focusing on a different aspect of the SMO: (1) support layer geometry and (2) surface layer geometry.

Models that focus on support layer geometry assume epitaxial-like growth of the surface oxide without considering unfavorable chemical interactions (e.g., steric hindrance, repulsion) at the surface layer. Based on the (001) projection of ZrO_2 [61], tungstate ML was calculated ~7.3 W/(nm^2 support) [59], in which the WO_6 octahedra was presumably anchored to exposed Zr-OH sites (but calculation details were not given) [62].

Models that concentrate exclusively on the surface layer geometry are support-independent in that they disregard how the surface and support oxides are bound together. One group envisioned surface saturation of the supported metal oxide layer as a two-dimensional close-packing of monomeric sites (Figure 11.2). For WO_3, a close-packing was calculated to be 0.21 g WO_3/(100 m^2), which is equivalent to ~5.5 W/(nm^2 support) [1, 50, 57, 60]. A second group estimated surface saturation of WO_3 at ~6.4 W/nm^2 based on the bulk density of WO_3 [39], and another group determined SS at ~7.8 W/nm^2 "estimated by the density of WO_x species in a two-dimensional plane of corner-shared WO_6 octahedra with W-O bond distances corresponding to those in low-index planes of monoclinic WO_3 crystallites" [33]. In the above examples, the original sources used units of W/nm^2, which we interpret as W/(nm^2 support).

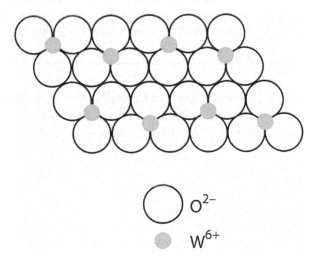

O^{2-}

W^{6+}

FIGURE 11.2 Close-packed monolayer model for WO_3 and other similar transition metal atoms [1].

11.4.2 X-ray Diffraction and Raman Spectroscopy

Many experimental approaches have been used to quantify the surface density at which surface saturation occurs. Investigations relying solely on the onset of crystallization use both x-ray diffraction (XRD) and Raman spectroscopy [26, 39, 57]. Raman spectroscopy is the method of choice, as its detection limits are below the ~4 nm cutoff of crystallite size for XRD, and has been widely applied to many SMOs. Wachs and coworkers demonstrated that surface crystallization occurs at surface densities for specific SMO compositions, as shown in Table 11.2 [26], referring to these values as monolayer coverage (ML_2). Separate Raman investigations of WO_x/ZrO_2 monitoring the WO_3 peak at ~808 cm^{-1} find that WO_3 crystallites form at ~4.0 $W/(nm^2$ support) for materials based on crystalline ZrO_2 [63] and at ~4.5 $W/(nm^2$ composite) for materials based on amorphous zirconium oxyhydroxide [33].

11.4.3 X-ray Photoelectron Spectroscopy and Ion Scattering Spectroscopy

Surface techniques that probe only the uppermost atomic layers of a solid are frequently used to assist in determination of surface saturation. In x-ray photoelectron spectroscopy (XPS) [56, 64], incident x-ray photons eject photoelectrons from surface atoms, and each photoelectron has a characteristic energy, which allows identification of its parent atom. In contrast, ion scattering spectroscopy (ISS) [11, 54, 60, 64, 65] bombards the surface with ions that ricochet with new momentum characteristic of the mass of surface atom with which it collided. Both methods have an advantage over Raman and XRD in that they can measure surface coverage directly as a function of surface loading. Their limitation is that they are spot-size

---

TABLE 11.2
Surface Densities Marking the Onset of Crystallization of the Surface Oxide for Various SMO Compositions, Determined via Raman Spectroscopy of Dehydrated Samples (Units of Atoms/(nm^2 Support)) [26]

Surface oxide cation	Surface saturation or ML_2[a]				
	Al_2O_3	TiO_2	ZrO_2	Nb_2O_5	SiO_2
Re	2.3	2.4	3.3	–	0.54
Cr	4.0	6.6	9.3	–	0.6
Mo	4.6	4.6	4.3	4.6	0.3
W	4.0	4.2	4.0	3.0	0.1
V	7.3	7.9	6.8	8.4	0.7
Nb	4.8	5.8	5.8	–	0.3

[a] These surface density values are called SS and ML_2 coverage values by this chapter and Reference 26, respectively

dependent, and extrapolation of their results to the entire supported phase requires assumption of sample uniformity [60].

In general, these methods monitor the relative intensity of peaks associated with the surface and support metal centers (e.g., W and Zr, respectively). The ratio of W:Zr peak areas is found to increase linearly with W content until SS is reached, above which the W:Zr peak area ratio increases slowly due to WO_3 crystal formation. Complementary XPS and ISS measurements of WO_x/ZrO_2 catalysts prepared through IW showed that SS occurred at ~3.0 W/(nm^2 support) [60]. Additionally, ISS found that ~45% of the ZrO_2 support surface was exposed at SS and that ML coverage was reached at ~5.0 W/(nm^2 support).

11.4.4 INFRARED SPECTROSCOPY AND SELECTIVE CHEMISORPTION

Infrared spectroscopy and site-selective chemisorption have been the primary techniques in determining ML coverage. IR spectroscopy allows one to track the hydroxyl bands of an SMO as a function of surface oxide content; the absence of these bands signifies complete titration of the support hydroxyl groups and, therefore, ML coverage. Complementary data can be obtained by employing a site-selective chemisorption probe molecule and quantitative temperature programmed desorption. Dehydroxylation via heat treatment creates coordinatively unsaturated (cus) support cations (i.e., Lewis acid centers), which can be quantified by site-selective probe molecules. The lack of cus support cations on a dehydroxylated SMO indicates that the support hydroxyls are completely reacted with the surface oxide, marking ML coverage.

Seminal work developing these techniques was performed on MoO_x/Al_2O_3. IR studies conclusively showed that the most basic Al_2O_3 hydroxyls were completely titrated in preference to acidic hydroxyls with increasing MoO_x loading [8, 27, 28, 49, 51, 52, 66–68]. This was bolstered by hydrogen-deuterium (HD) isotopic exchange, which revealed a decrease in surface hydrogen content [69], and pyridine chemisorption, which observed a depletion of the intrinsic Lewis acidity of the alumina support [49]. However, residual alumina hydroxyls were found along with MoO_3 crystals through NMR [70], which indicated complete OH titration was not reached. This data clearly differentiated the onset of surface oxide crystallization from complete support oxide hydroxyl titration; i.e., SS and ML coverage did not occur at the same surface density.

CO chemisorption is an effective probe [8, 28] because CO (a soft Lewis base) adsorbs to coordinatively unsaturated Lewis acid sites, such as the metal centers of dehydroxylated support oxides. Combined IR and CO-TPD on MoO_x/Al_2O_3 confirmed that MoO_x surface saturation occurred prior to ML coverage [28]. Figure 11.3 shows increasing Mo loading reduces the number of surface-exposed Al^{3+}(cus) sites until reaching ~4.2 Mo/nm^2, above which MoO_3 crystals are observed (via XRD) and the number of Al^{3+}(cus) remains constant as Mo concentration is increased. The knee in the data represents MoO_3 SS, indicating ~14 ± 5% residual Al_2O_3 remained surface-exposed. Extrapolation to the hypothetical zero CO adsorption point leads to ML coverage occurring at a surface density of 4.6 Mo/nm^2. Similarly, a CO chemisorption study of WO_x/ZrO_2 samples prepared by EA on zirconium oxyhydroxide indicated

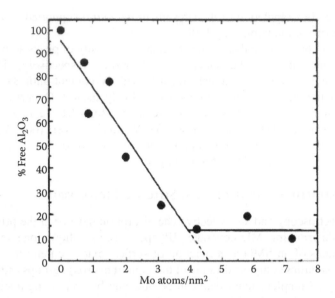

FIGURE 11.3 Percentage of exposed Al_2O_3 surface, determined from CO adsorption (via TPD and FTIR) on MoO_x/Al_2O_3 heat-treated at 1200 K [28].

that the surface saturated at ~2 W/(nm² composite) with ~50% of the zirconia hydroxyls exposed, and ML coverage occurred at ~4.6 W/(nm² composite) [50].

Other probe molecules have been used for chemisorption, but the results may be less reliable. Although initially used to monitor exposed Al-OH, CO_2 chemisorption has been discredited by several groups as a viable technique for monitoring surface oxide coverage on Al_2O_3 because (1) it fails to fully count all alumina hydroxyls, as it preferentially titrates only the most basic Al-OH [71], and (2) the probe molecule experiences steric exclusion from basic Al-OH at high loadings of MoO_x, thereby leading to overestimated surface coverage [28, 51]. Surface coverage estimates by CO_2 chemisorption are also questionable for WO_x/ZrO_2 [54, 64] and other SMO compositions [33, 72]. Chemisorption using benzaldehyde and ammonia was used to measure SS values for WO_x/ZrO_2 [73–75], in which benzaldehyde supposedly titrates the exposed support of a basic metal oxide (e.g., ZrO_2) and ammonia reacts with the surface-bound benzaldehyde to form benzonitrile. Because benzaldehyde is a soft base [76, 77] and would bind to cus Lewis acid sites of the support, complete dehydroxylation of the SMO is required (heat treatment 1200°K [28] for Al_2O_3 and 993 K for ZrO_2 [54]). Also, care must be taken to prevent reduction of the surface oxide, which can contribute to benzaldehyde chemisorption capacity. For example, it was shown that fully oxidized WO_3 adsorbed little benzaldehyde, but $WO_{2.24}$ adsorbed >7 times as much [75]. Failure to recognize the $WO_{2.24}$ contribution would lead to an overestimate of exposed ZrO_2 surface coverage.

11.4.5 UV-Vis Diffuse Reflectance Spectroscopy

If the surface oxide is of a semiconducting composition (e.g., MoO_3, V_2O_5, and WO_3), its electronic band structure can be probed through optical absorption spectroscopy [78]. UV-visible (vis) diffuse reflectance measurement of powders enables experimental determination of the optical absorption gap (E_{opt}), most commonly estimated through Tauc's relation [33, 79–81], although other methods exist [80–82]. A Tauc plot linearly extrapolates $[F(R_\infty)\cdot E]^{\wedge}(1/\eta)$ versus E to zero absorption to calculate $E_{opt,meas}$ as an estimate of E_{opt}, where $F(R_\infty)$ is the Kubelka-Munk function [83, 84]. Choice of η, which depends upon the type of optical transition for crystalline materials [85], is less straightforward for noncrystalline metal oxides [33] and can range from $\eta = 0.5$ [86] to $\eta = 2$ [33, 39].

For strictly crystalline materials, E_{opt} equals the band gap (E_g), the energy separation between valence and conduction band edges (Figure 11.4A) [87]. For amorphous materials, atomic disorder introduces tails to the band edges, which prevents demarcation of a well-defined energy gap (Figure 11.4B); the gap is called the mobility band gap (E_μ sometimes referred to as E_g) [87]. Technically, $E_{opt} \neq E_\mu$ because "no definite correlation has been made between mobility gaps and ... optical gaps" [81]. Nonetheless, E_μ estimates for amorphous SMOs are routinely performed using the Tauc relation (and assuming $E_{opt} = E_\mu$), and in this manner, band gap changes have been correlated with monomer/oligomer/polymer domain size [33, 78], crystallite particle size [88], and coordination number [33, 39, 78, 86]. Precise molecular structure information of the support oxide cannot be derived, though [39].

As an example, Barton et al. calculated $E_{opt,meas}$ ($\eta = 2$) for a series of WO_x/ZrO_2 samples of different WO_x loadings and correlated them to surface densities [33] (Figure 11.5). Optical gap energies were observed to decrease with increasing WO_x loading, suggestive of increasing domain size of polytungstates. They concluded the leveling of the optical absorption energies ≥ 8 $W/(nm^2$ composite) was consistent with the onset of WO_3 formation (see Section 11.6.2 for further discussion).

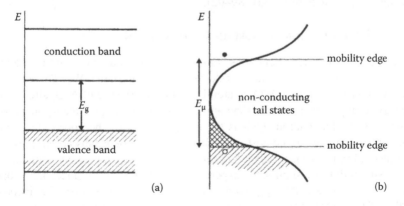

FIGURE 11.4 Energy band diagram for (A) a crystalline semiconductor with band gap, E_g; (B) an amorphous semiconductor with mobility gap, E_μ. Adapted from Reference 87.

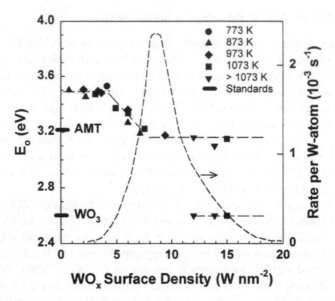

FIGURE 11.5 Optical absorption gap energies and maximum *o*-xylene isomerization rate curve (dashed volcano-shaped curve) of a series of WO_x/ZrO_2 catalysts (calcined at different temperatures) as a function of surface density [33]. E_0 represents $E_{opt,meas}$.

11.5 CHARACTERIZATION OF SURFACE DENSITY

Surface density (labeled as ρ_{surf}) is a calculated value that incorporates both surface oxide loading and specific surface area into a single metric. Surface density is, by nature, an averaged quantity and does not account for nonuniform surface oxide distribution. It does not contain structural information on surface oxide dispersion analogous to metal dispersion in supported metal catalysts (e.g., flat sheets vs. hemispherical islands of metal) [89–92].

11.5.1 DETERMINATION OF SURFACE OXIDE CONTENT

The numerator of surface density is commonly reported as the number of atoms of active metal center of the surface oxide (e.g., tungsten atoms, W). This is obtained by converting the weight of WO_3 to atomic W through the use of Avagadro's number. The various approaches differ in their approach to estimating the incorporated weight of WO_3. Bulk elemental analysis is the most rigorous of the approaches, but it is time-consuming and can be cost-prohibitive if independent labs are employed.

Calculation from mass balance is convenient but depends upon synthetic recipe and is susceptible to significant experimental errors. For instance, mass balance calculations are reasonable approximations for catalyst preparations that incorporate ~100% of the added surface oxide precursor (e.g., incipient wetness impregnation); however, mass balance is not suitable for preparation methods that retain less than 100% of the precursor (e.g., equilibrium adsorption, where a significant concentration

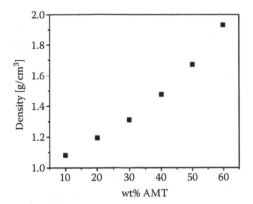

FIGURE 11.6 Density of aqueous $(NH_4)_6H_2W_{12}O_{40} \cdot 5H_2O$ at room temperature as a function of weight percentage.

of the precursor potentially remains in solution after surface adsorption). Also, mass balance calculations assume that there is no loss of the surface oxide, which can happen at high calcination temperatures for certain oxides.

Reliance on mass balance calculations requires extensive knowledge of precursor physical properties. Many transition metal oxide precursors are hydrated salts with variable molecules of water incorporated in their crystal structure, such as ammonium metatungstate $((NH_4)_6H_2W_{12}O_{40} \cdot xH_2O)$. The amount of hydrated water (x) varies, depending on the manufacturer and atmospheric moisture; thermogravimetric analysis should be performed to quantify the water content of the metal salt hydrates for accurate calculations of surface oxide content.

In the impregnation preparation of SMOs, we have found that a potential source of large error is the dissolved precursor solution density. For instance, one may assume that 1 ml of a 60 wt% AMT solution introduces 0.6 g of AMT. However, the density of AMT solutions varies quite significantly, especially for concentrated solutions (Figure 11.6). In actuality, 1 ml of a 60 wt% AMT solution contains

$$1.16\,g\left(=(1\text{ ml solution})\left(\frac{1.93\text{ g solution}}{1\text{ ml solution}}\right)\left(\frac{0.6\text{ g AMT}}{1\text{ g solution}}\right)\right)$$

not 0.6 g, of AMT.

11.5.2 CONSIDERATIONS FOR SURFACE AREA

Surface area values are used to determine surface densities, but literature indicates two types of surface areas are used: the surface area of the support oxide [26, 60] and the surface area of the final composite SMO catalyst [33, 34, 50, 56, 73].

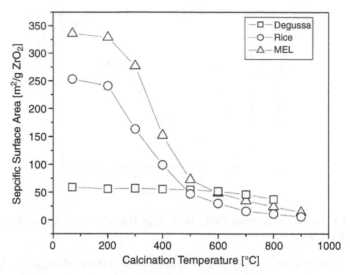

FIGURE 11.7 BET surface areas of crystalline ZrO_2 (Degussa), amorphous $ZrO_x(OH)_{4-2x}$ prepared from $ZrOCl_2 \cdot 8H_2O$ (Rice), and commercial-source amorphous $ZrO_x(OH)_{4-2x}$ (MEL) as a function of calcination temperature.

Nonspecification of the surface area type has led to differences in reported SS ρ_{surf} values. It is reasonable to use the former type only if the support material is known to be structurally stable (e.g., no sintering upon calcination). For example, a crystalline ZrO_2 material (Degussa, VP Zirconoxid PH) maintains SSA up to 600°C ($\sim$55 m²/g) and suffers negligible surface area loss despite crystalline phase transformation (from tetragonal to monoclinic) and grain growth at higher temperatures (Figure 11.7). In contrast, hydrous precipitates like zirconium oxyhydroxide ($ZrO_x(OH)_{4-2x}$) prepared from $ZrOCl_2 \cdot 8H_2O$ using NH_4OH and $ZrO_x(OH)_{4-2x}$ procured from a commercial source (MEL, XZO880/01) show significant sintering (Figure 11.7).

11.5.3 Calculation Methods

From the literature, we identified two different methods for calculating surface density, which we term the *support-normalized method*, (ρ_{surf1}, Equation 11.1), and *composite-normalized method*, (ρ_{surf2}, Equation 11.2):

$$\rho_{surf1}\left[\frac{\text{M atoms}}{\text{nm}^2\,\text{support}}\right] = \frac{(1[\text{g composite}])\left(\frac{\frac{\text{wt\% MO}_x}{100}}{1}\left[\frac{\text{g MO}_x}{\text{g composite}}\right]\right)\left(\frac{1}{Mw}\left[\frac{\text{mol MO}_x}{\text{g MO}_x}\right]\right)\left(\frac{v_{stoich}}{1}\left[\frac{\text{mol M}}{\text{mol MO}_x}\right]\right)\left(\frac{6.022\times10^{23}}{1}\left[\frac{\text{M atoms}}{\text{mol M}}\right]\right)}{(1[\text{g composite}])\left(\frac{1-\frac{\text{wt\% MO}_x}{100}}{1}\left[\frac{\text{g support}}{\text{g composite}}\right]\right)\left(SSA_{\text{virgin support after calcination}}\left[\frac{\text{m}^2\,\text{support}}{\text{g support}}\right]\right)\left(\frac{10^9}{1}\left[\frac{\text{nm}}{\text{m}}\right]\right)^2}$$

(11.1)

$$\rho_{surf2}\left[\frac{g\,MO_x}{g\,composite}\right]=\frac{(1[g\,composite])\left(\frac{\frac{wt\%\,MO_x}{100}\left[\frac{g\,MO_x}{g\,composite}\right]}{}\right)\left(\frac{1}{Mw}\left[\frac{mol\,MO_x}{g\,MO_x}\right]\right)\left(\frac{v_{stoich}}{1}\left[\frac{mol\,M}{mol\,MO_x}\right]\right)\left(\frac{6.022\times10^{23}}{1}\left[\frac{M\,atoms}{mol\,M}\right]\right)}{(1[g\,composite])\left(SSA_{composite\,after\,calcination}\left[\frac{m^2\,support}{g\,support}\right]\right)\left(\frac{10^9}{1}\left[\frac{nm}{m}\right]\right)^2}$$ (11.2)

The numerator of Equation 11.1 and Equation 11.2 includes the molar stoichiometry (v_{stoic}) of the metal atoms in the surface metal oxide (MO_x) and its molecular weight (M_w). These equations differ in calculation of the denominator: Equation 11.1 uses the surface area of the calcined, pure support oxide (no surface oxide), and Equation 11.2 uses the surface area of the calcined composite catalyst. Unless the support oxide shows negligible sintering and the support oxide content is low, ρ_{surf1} and ρ_{surf2} are numerically different.

11.5.4 MOST APPROPRIATE SURFACE DENSITY CALCULATION METHOD

In this section, we discuss the most appropriate calculation method for ρ_{surf} using WO_x/ZrO_2 as the example SMO. We prepared WO_x/ZrO_2 materials via incipient wetness impregnation of crystalline ZrO_2 (Degussa, VP Zirconoxid PH, 59 m²/g, 0.13 cm³/g) and two x-ray amorphous zirconium oxyhydroxide $ZrO_x(OH)_{4-2x}$ powders, one precipitated from $ZrOCl_2 \cdot 8H_2O$ (see appendix for synthesis procedures, 253 m²/g, 0.17 cm³/g) and the other from MEL (336 m²/g, 0.32 cm³/g). The materials were prepared with different WO_x weight loadings and calcined at different temperatures to vary ρ_{surf1} and ρ_{surf2}. We analyzed the materials through XRD and dispersive Raman spectroscopy, specifically tracking the onset of WO_3 crystals to indicate surface saturation. For 94 of 96 samples, there was complete agreement between XRD and FT-Raman. Two samples were found to be x-ray amorphous but crystalline through Raman spectroscopy (Degussa, 11.1 wt% NO_3/ZrO_2, 500°C; Rice, 23.3% wt 9% WO_3/ZrO_2, 600°C.

Figure 11.8 maps ρ_{surf1} and ρ_{surf2} values for the three different sets of WO_x/ZrO_2 samples as a function of calcination temperature. It is clear that, because of the stability of the crystalline ZrO_2 support, ρ_{surf1} values were based on a fairly constant support surface area, and there was little difference compared to ρ_{surf2} values. A fairly well-defined surface density region between samples with noncrystalline WO_x and with crystalline WO_3 surface species can be noted (4.5–6.0 W/(nm² support) or W/(nm² composite), Figure 11.8). In contrast, the ρ_{surf1} values for SMOs prepared from amorphous zirconia increased significantly at calcination temperatures of 600°C and above, which is due to the significant surface area decrease of the calcined, pure support oxide (Figure 11.7) [39, 59]. Not shown in the plots, for example, is a ρ_{surf1} value of 174 W/(nm² support) for ~27 wt% WO_3 (Rice) sample calcined at 900°C.

Additional observations can be made by overlaying the surface density data of Figure 11.8 (Figure 11.9). Figure 11.9A shows no clear demarcations between samples with noncrystalline WO_x and those with crystalline WO_3 surface species, but Figure 11.9B does, supporting the notion that that surface density calculations based on support surface area (Equation 11.1) are less reliable than those based on composite surface area (Equation 11.2). Based on composite surface areas, surface

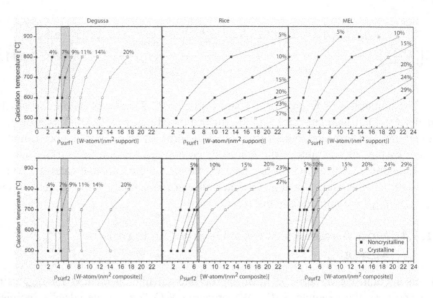

FIGURE 11.8 Comparison of surface density calculations for WO_x/ZrO_2 prepared on three different zirconia supports: crystalline ZrO_2 (Degussa), amorphous $ZrO_x(OH)_{4-2x}$ prepared from $ZrOCl_2 \cdot 8H_2O$ (Rice), and commercial-source amorphous $ZrO_x(OH)_{4-2x}$ (MEL). The lines connect points of constant WO_x loading (wt%). Gray regions roughly mark the surface density boundary range between noncrystalline WO_x species and WO_3 crystals. The left-hand border marks the highest, average surface density for samples with no WO_3 crystals, and the right-hand border marks the lowest, average surface density for samples with WO_3 crystals.

density values for all three sets of WO_x/ZrO_2 samples show great overlap of the noncrystalline/crystalline transition ranges for the Degussa and MEL samples; i.e., 4.5–6.0 W/(nm² composite). As expected, this range is close to that of Rice $ZrO_x(OH)_{4-2x}$, 6.7–7.1 W/(nm² composite). The actual range may be narrower than is currently shown, due to the large sampling intervals. We conclude that surface density based on composite material surface area (ρ_{surf2}, Equation 11.2) is more appropriate than that based on the pure support oxide surface area (ρ_{surf1}, Equation 11.1), because the former has less sensitivity towards the state of the support oxide and calcination temperature.

11.6 COMPARING PUBLISHED SURFACE DENSITY VALUES FOR VARIOUS WO_x/ZrO_2 MATERIALS

We examined published reports of SS and ML coverage values for various WO_x/ZrO_2 catalysts in light of our surface density, SS, and ML coverage discussions (Table 11.1). Because of their recent work on WO_x/ZrO_2, we chose to focus on three research groups: the Wachs Lehigh group, the Iglesia Berkeley group, and the Knözinger München group.

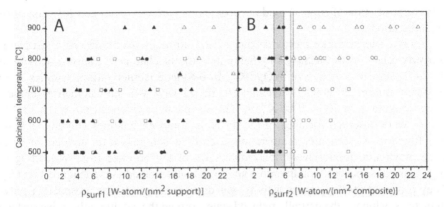

FIGURE 11.9 Overlay of surface density values from Figure 11.8 calculated using (A) support surface area and (B) catalyst surface area. Gray regions roughly mark the surface density boundary range between noncrystalline WO_x species and WO_3 crystals. Symbols are (■) Degussa, (filled) noncrystalline, and (open) crystalline.

11.6.1 WACHS AND COWORKERS

Wachs and coworkers characterized extensively WO_x/ZrO_2 samples prepared via incipient wetness impregnation of crystalline ZrO_2 (Degussa) [3, 63]. Similar to their structural assignments of WO_x/Al_2O_3 [93], they showed that isolated WO_x monomers were tetrahedrally coordinated as WO_4 sites and were found mostly at low surface coverages (<< 4 W/(nm² support)). They concluded that octahedrally coordinated WO_6 sites formed polymerized domains that were found mostly near 4 W/(nm² support) [3]. They also detected the onset of WO_3 crystallization at ~4 W/(nm² support) [63], referring to this surface density value as monolayer coverage.

11.6.2 IGLESIA AND COWORKERS

WO_x/ZrO_2 catalysts were prepared via incipient wetness impregnation of amorphous zirconium oxyhydroxide [33, 62] based on the early work of Hino and Arata [94]. Unlike Wachs and coworkers, this group calculated surface density using the ρ_{surf2} definition. Figure 11.10 summarizes their model of molecular surface structures of WO_x/ZrO_2 as a function of ρ_{surf2} [33, 62]. Based on XANES results for loadings down to 3 W/(nm² composite), they concluded that WO_x species were octahedrally coordinated (WO_6) at all loadings, from isolated to polymerized WO_x [62]. WO_3 crystals were detected at ~4.5 W/(nm² composite) via Raman spectroscopy (but not detectible via XRD), coexisting with polymerized WO_x from 4–8 W/(nm² composite) (from UV-vis spectroscopy, Figure 11.8) [33]. Based on our discussions in this chapter, ~4.5 W/(nm² composite) actually represents SS, consistent with our own analysis of various WO_x/ZrO_2 materials (Section 11.5.4).

They concluded that monolayer coverage began at 8 W/(nm² composite) based on the leveling of the band gap values from UV-vis spectroscopy (Figure 11.8). However, this monolayer coverage appears to differ from the definitions of ML (complete titration of support oxide hydroxyls by surface oxide) and ML_2 (onset of crystallization).

Iglesia and coworkers provide one of the strongest examples of a structure-property relationship for SMOs by being able to correlate acid catalytic activity to the WO_x surface structure of WO_x/ZrO_2. In o-xylene isomerization, samples were prepared at three different W loadings (6, 12, 21 wt% W) and calcination temperatures ranging from 500—1000 K [59]. Three separate o-xylene turnover rate (TOR) curves were observed for each of the different tungsten loadings (Figure 11.11A) as a function of calcination temperature, each of which passed through an intermediate maximum. Impressively, renormalization of the ordinate from W-loading to ρ_{surf2} caused all curves to map onto a single volcano-shaped curve (Figure 11.11B). They proposed that maximum activity was directly related to the presence of polytungstates, which in the partially reduced state, formed the catalytically active surface species. Likewise, they identified a similar volcano-shaped dependence of initial 2-butanol dehydration rates on surface density and concluded that the polymeric WO_x surface species were also responsible for the catalysis (Figure 11.12) [34]. We note,

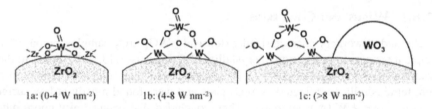

la: (0-4 W nm⁻²)   1b: (4-8 W nm⁻²)   1c: (>8 W nm⁻²)

FIGURE 11.10 WO_x molecular surface structures and coordination as a function of ρ_{surf2} as published by Iglesia and coworkers: (1a) isolated WO_6 octahedra, (1b) polymerized WO_6 octahedra, and (1c) coexisting WO_6 polytungstates and WO_3 crystallites [33, 62].

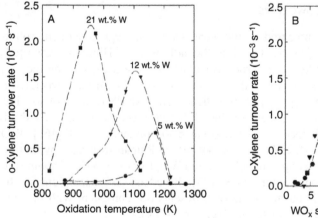
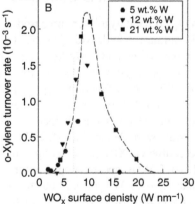

FIGURE 11.11 O-xylene isomerization rates over WO_x/ZrO_2 with different WO_x loadings plotted as a function of (A) calcination temperature and loading, and (B) surface density. Adapted from Reference 59.

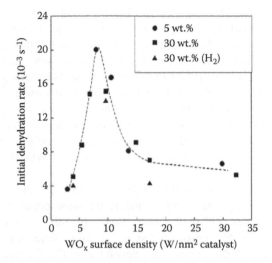

FIGURE 11.12 Initial 2-butanol dehydration rates over WO_x/ZrO_2 [34].

though, that the very small WO_3 crystalline domains (Raman-detectible but XRD-amorphous) coexisting with the polymeric WO_x should not be discounted as active sites.

11.6.3 KNÖZINGER AND COWORKERS

Similar to Iglesia and coworkers, Scheithauer et al. [29, 39] prepared WO_x/ZrO_2 from amorphous zirconium oxyhydroxide, but they used equilibrium adsorption at 383 K instead of incipient wetness impregnation. They considered ML coverage to occur at ~6.4 W/nm^2, although this was not explicitly stated [39] (Section 11.4.1). They used a different metric called WO_x surface coverage (θ_{WOx}, Equation 11.3), in which the surface density values were normalized to the ML coverage value.

$$\theta_{WO_x} = \frac{\rho_{surf2}}{\rho_{surf2-ML}} \quad (11.3)$$

This group found interesting structural information different from the other groups. Similar to Wachs and coworkers, they observed only terminal W=O mono-oxo species, but their dehydrated Raman spectra (bands at 1000 and 1022 cm^{-1}), IR spectra, and ^{18}O-exchange experiments suggested there were two distinct mono-oxo-state species characterized by different bond orders of the W=O bond [39]. They speculated that the surface WO_x network resembled pseudo-heteropolytungstates containing Zr^{4+} cations. CO chemisorption found SS occurred at ~5 $W/(nm^2$ composite) [29].

Scheithauer et al. provided detailed results for two series of WO_x/ZrO_2 materials calcined at different calcination temperatures (923 and 1098 K), but did not compare them against one another [29, 39]. Within each series, they found that the IR bands of the hydroxyl groups on the zirconia surface shifted after CO chemisorption at 77 K, with the magnitude of the shift (Δv_{OH}) a function of WO_x concentration. It is known that interaction between hydroxyl groups (Brønsted acid) and chemisorbed CO (weak base) leads to formation of hydrogen bonds and to a red shift (Δv_{OH}) in hydroxyl band position proportional to the SMO's relative Brønsted acidity [29]. Comparison between Δv_{OH} for WO_x/ZrO_2 (90–170 cm^{-1}) and virgin ZrO_2 (60 cm^{-1}) demonstrated that WO_x increases Brønsted acidity relative to ZrO_2 and that the highest acid strengths (indicated by the largest Δv_{OH}) occurs for samples with the largest WO_x domains (indicated by ρ_{surf} and $E_{opt,meas}$).

These data can be reorganized with the Δv_{OH} data [29, 39] in Table 11.3 as a function of ρ_{surf2} (or θ_{WOx}) to reveal an underlying dependence on ρ_{surf2} not immediately evident in the as-published dataset. Similarly, Table 11.3 shows that UV-vis optical absorption gaps (E_{opt}) estimated as $E_{opt,meas}$ from Tauc plots [78] on the same WO_x/ZrO_2 dataset [39] also exhibit ρ_{surf2} dependence. The dependence of Δv_{OH} and $E_{opt,meas}$ on ρ_{surf2} is another example of how surface density provides a useful basis for analyzing the structure and catalysis of SMOs.

TABLE 11.3
Reorganization of Data From [29, 39] as a Function of ρ_{surf2}

Nomenclature[a]	Surface density[b] [W/(nm² composite)]	θ_{WOx} [–]	$\Delta \tilde{v}_{OH}$ [cm^{-1}]	$E_{opt,meas}$ [eV]
3.6WZ923	1.7	0.27	~90	4.20
5.9WZ923	2.4	0.37	~100	4.15
3.6WZ1098	3.1	0.48	~125	4.15
8.6WZ923	3.2	0.50	~115	4.12
10.5WZ923	3.3	0.52	~125	4.00
5.9WZ1098	3.8	0.60	~160	4.03
13.6WZ923	4.0	0.62	~125	3.94
19.0WZ773	4.1	0.64	~135	
8.6WZ1098	4.9	0.75	~160	3.94
19.0WZ923	5.1	0.80	~160	3.76
10.5WZ1098	5.9	0.92	~170	
13.6WZ1098	8.4	1.31	~170	
23.9WZ923	8.9	1.38	~160	3.69
19.0WZ1098	14.1	2.19	~170	
23.9WZ1098	20.7	3.22	~170	

[a] "xWZy" means a WO_x/ZrO_2 material prepared with x wt% WO_3 and calcined at a temperature of y K.
[b] Calculated from wt% WO_3 and SSA data provided in References 29 and 39.

11.7 SUMMARY

Confusion in SMO literature can arise because there is no generally accepted method for determining surface density. As the metric that characterizes the surface oxide of supported metal oxide catalysts, surface density allows one to consider the various structures of the surface oxide on a common scale, independent of total oxide content, preparation method, calcination treatment, and surface area of the support oxide. Surface saturation and monolayer coverage are important threshold surface density values, at which surface oxide crystals form and at which complete consumption of surface hydroxyl groups of the support oxide occurs, respectively. Inconsistencies in these values come about because of (1) differences in their definitions, (2) difficulties in compatibilizing data from different characterization techniques, and (3) the use of support surface area instead of the overall composite SMO. These inconsistencies can make structural comparison of the same SMO composition, such as WO_x/ZrO_2, difficult across different research groups. Calculated properly, however, the surface density metric provides the most simple and useful basis for understanding the relationship between surface nanostructure and catalytic and surface properties.

11.8 APPENDIX

11.8.1 PREPARATION METHOD FOR $ZrO_x(OH)_{4-2x}$ SUPPORT

Zirconium oxyhydroxide was prepared at room temperature (r.t.) by completely dissolving 313.6 g $ZrOCl_2 \cdot 8H_2O$ (Aldrich, 98%) into 0.7 l deionized water, solution pH ~1. This solution was added dropwise (~1 h) into 600 ml deionized water with 15.8 M NH_4OH added as necessary to maintain batch $10 <= pH <= 11$, resulting in immediate precipitation of a buoyant white powder. After stirring for 24 h at r.t., the precipitate was dried in static air at 70°C, crushed, sieved (<170 mesh), and redispersed in 1.8 l deionized water at pH ~10 for 30 min before separation on a Büchner funnel. Multiple redispersions and filtrations were performed until the supernatant chloride ion concentration reached that of the water background, as verified by $AgNO_3$ titration with K_2CrO_4 indicator. Finally, powder was dried overnight in static air at 70°C before being crushed and sieved (<170 mesh).

11.8.2 GENERAL INCIPIENT WETNESS IMPREGNATION METHOD

Each support master batch (Degussa VP Zirconoxid PH, MEL ZXO880/01, or Rice University's $ZrO_x(OH)_{4-2x}$) was hand-mixed ≥1 h after sieving <170 mesh. For Degussa, five random locations in the sample container were tested by XRD (Rigaku D/Max-2100PC) to verify homogeneity. Prior to incipient wetness impregnation, supports were dried overnight in static air (125°C for Degussa, 70°C for amorphous powders) to remove adventitious moisture. Powders were impregnated with aqueous solutions of ammonium metatungstate (Osram Sylvania) to 95% of pore volume (V_p) as determined by triplicate nitrogen physisorption analyses at

77 K ($V_p \sim P/P_0 = 0.984$, Micromeritics ASAP2010) and mixed ~30 min by hand to ensure uniform solution dispersion. Average V_p were 0.13 cm^3/g (Degussa), 0.32 cm^3/g (MEL 880), and 0.17 cm^3/g (Rice). After drying overnight at 70°C in static air, powders were crushed and sieved (< 170 mesh). For samples undergoing single impregnation, calcination occurred at the desired temperature (3.2°C/min ramp) in static air for 3 h (natural cooling rate). Samples requiring repeat impregnation were calcined to 350°C for 3 h (3.2°C/min ramp), after which V_p was remeasured once via nitrogen physisorption prior to crushing, sieving (< 170 mesh), drying overnight at 70°C in static air, and second impregnation. Elemental analyses (ICP-MS) on select samples were performed (Lehigh Testing Laboratories, Inc., New Castle, DE) to confirm mass balance methodology for calculating wt% WO$_3$ loadings. All measured values matched within 3% of the expected content (e.g., 7.4% measured vs. 7.2% calculated, 24.9% measured vs. 24.4% calculated).

ACKNOWLEDGMENTS

Special thanks to Prof. I.E. Wachs for many helpful discussions. Material samples were obtained from Degussa (VP Zirconoxid PH), MEL (ZXO 880/01), and Osram Sylvania (ammonium metatungstate).

REFERENCES

1. Y.-C. Xie and Y.-Q. Tang, "Spontaneous monolayer dispersion of oxides and salts onto surfaces of supports: Applications to heterogeneous catalysis," in D.D. Eley, H. Pines and P.B. Weisz (Eds.), *Advances in catalysis*, Academic Press: San Diego, pp. 1–43, 1990.
2. G. Mestl and H. Knözinger, "Vibrational spectroscopies," in G. Ertl, H. Knözinger and J. Weitkamp (Eds.), *Handbook of heterogeneous catalysis*, Vol. 2, Wiley-VCH: Weinheim, pp. 539–574, 1997.
3. I.E. Wachs, "Molecular structures of surface metal oxide species: Nature of catalytic active sites in mixed metal oxides," in *Metal oxides: Chemistry and applications*, Taylor & Francis Group, LLC: Boca Raton, FL, pp. 1–30, 2006.
4. International Union of Pure and Applied. Chemistry, "Surface density," in *IUPAC compendium of chemical terminology*, 2nd ed., Blackwells: Oxford, U.K., 1997.
5. M.S. Wong, "Nanostructured supported metal oxides," in J.L.G. Fierro (Ed.) *Metal oxides: Chemistry and applications*, Taylor & Francis Group, LLC: Boca Raton, FL, pp. 31–54, 2006.
6. M. Che, O. Clause and C. Marcilly, "Impregnation and ion exchange," in J. Weitkamp (Ed.) *Handbook of heterogeneous catalysis*, Vol. 1, Wiley-VCH: Weinheim, pp. 191–207, 1997.
7. J.W. Geus and A.J. van Dillen, "Preparation of supported catalysts by deposition-precipitation," in G. Ertl, H. Knözinger and J. Weitkamp (Eds.), *Handbook of heterogeneous catalysis*, Vol. 1, Wiley-VCH: Weinheim, pp. 240–257, 1997.
8. W.K. Hall, "The genesis and properties of molybdena-alumina and related catalyst systems," in *Chem. Uses Molybdenum, Proc. Int. Conf., 4th*, 224–233, 1982.

9. G.A. Parks, "The isoelectric points of solid oxides, solid hydroxides, and aqueous hydroxo complex systems," *Chem. Rev.*, *65(2)*, 177–198, 1965.

10. H.H. Kung, *Transition metal oxides: Surface chemistry and catalysts*, Elsevier: Amsterdam, 1989.

11. V. Sazo, L. Gonzalez, J. Goldwasser, M. Houalla and D.M. Hercules, "Determination of the surface coverage of Mo/Al_2O_3 catalysts," *Surf. Interface Anal.*, *23(6)*, 367–373, 1995.

12. F. Moreau, G.C. Bond and A.O. Taylor, "Gold on titania catalysts for the oxidation of carbon monoxide: Control of pH during preparation with various gold contents," *J. Catal.*, *231(1)*, 105–114, 2005.

13. C.F. Baes, *Hydrolysis of cations*, Wiley: New York, 1976.

14. M. Haruta, S. Tsubota, T. Kobayashi, H. Kageyama, M.J. Genet and B. Delmon, "Low-temperature oxidation of CO over gold supported on TiO_2, α-Fe_2O_3, and Co_3O_4," *J. Catal.*, *144(1)*, 175–192, 1993.

15. C. Louis and M. Che, "Anchoring and grafting of coordination metal complexes onto oxide surfaces," in G. Ertl, H. Knözinger and J. Weitkamp (Eds.), *Handbook of heterogeneous catalysis*, Vol. 1, Wiley-VCH: Weinheim, pp. 207–216, 1997.

16. R.K. Iler, *The chemistry of silica: Solubility, polymerization, colloid and surface properties, and biochemistry*, Wiley: New York, p. 633, 1979.

17. E. Mamontov, "Dynamics of surface water in ZrO_2 studied by quasielastic neutron scattering," *J. Chem. Phys.*, *121(18)*, 9087–9097, 2004.

18. F. Schüth and K. Unger, "Precipitation and coprecipitation," in G. Ertl, H. Knözinger and J. Weitkamp (Eds.), *Handbook of heterogeneous catalysis*, Vol. 1, Wiley-VCH: Weinheim, pp. 72–86, 1997.

19. F.E. Wagner, S. Galvagno, C. Milone, A.M. Visco, L. Stievano and S. Calogero, "Mössbauer characterisation of gold/iron oxide catalysts," *J. Chem. Soc. Faraday Trans.*, *93(18)*, 3403–3409, 1997.

20. H. Knözinger and E. Taglauer, "Spreading and wetting," in G. Ertl, H. Knözinger and J. Weitkamp (Eds.), *Handbook of heterogeneous catalysis*, Vol. 1, Wiley-VCH: Weinheim, pp. 216–231, 1997.

21. S.H. Overbury, P.A. Bertrand and G.A. Somorjai, "The surface composition of binary systems. Prediction of surface phase diagrams of solid solutions," *Chem. Rev.*, *75(5)*, 547–560, 1975.

22. U. Ciesla, S. Schacht, G.D. Stucky, K.K. Unger and F. Schüth, "Formation of a porous zirconium oxo phosphate with a high surface area by a surfactant-assisted synthesis," *Angew. Chem.-Int. Edit. Engl.*, *35(5)*, 541–543, 1996.

23. G. Busca, "Differentiation of mono-oxo and polyoxo and of monomeric and polymeric vanadate, molybdate and tungstate species in metal oxide catalysts by IR and Raman spectroscopy," *J. Raman Spectrosc.*, *33(5)*, 348–358, 2002.

24. L. Dixit, D.L. Gerrard and H.J. Bowley, "Laser Raman spectra of transition metal oxides and catalysts," *Appl. Spectrosc. Rev.*, *22(2–3)*, 189–249, 1986.

25. J.M. Stencel, *Raman spectroscopy for catalysis*, Van Nostrand Reinhold: New York, p. 220, 1989.

26. I.E. Wachs, "Raman and IR studies of surface metal oxide species on oxide supports: Supported metal oxide catalysts," *Catal. Today*, *27(3–4)*, 437–455, 1996.

27. N. Topsøe, "Infrared study of sulfided Co-Mo-Al_2O_3 catalysts—The nature of surface hydroxyl-groups," *J. Catal.*, *64(1)*, 235–237, 1980.

28. A.L. Diaz and M.E. Bussell, "An infrared spectroscopy and temperature-programmed desorption study of CO on MoO_3/Al_2O_3 catalysts: Quantitation of the molybdena overlayer," *J. Phys. Chem.*, *97(2)*, 470–477, 1993.

29. M. Scheithauer, T.K. Cheung, R.E. Jentoft, R.K. Grasselli, B.C. Gates and H. Knöz-inger, "Characterization of WO_x/ZrO_2 by vibrational spectroscopy and n-pentane isomerization catalysis," *J. Catal., 180(1)*, 1–13, 1998.

30. X.T. Gao, S.R. Bare, J.L.G. Fierro, M.A. Banares and I.E. Wachs, "Preparation and *in situ* spectroscopic characterization of molecularly dispersed titanium oxide on silica," *J. Phys. Chem. B, 102(29)*, 5653–5666, 1998.

31. X.T. Gao, J.L.G. Fierro and I.E. Wachs, "Structural characteristics and catalytic properties of highly dispersed ZrO_2/SiO_2 and $V_2O_5/ZrO_2/SiO_2$ catalysts," *Langmuir, 15(9)*, 3169–3178, 1999.

32. J.A. Horsley, I.E. Wachs, J.M. Brown, G.H. Via and F.D. Hardcastle, "Structure of surface tungsten oxide species in the WO_3/Al_2O_3 supported oxide system from x-ray absorption near-edge spectroscopy and Raman spectroscopy," *J. Phys. Chem., 91(15)*, 4014–4020, 1987.

33. D.G. Barton, M. Shtein, R.D. Wilson, S.L. Soled and E. Iglesia, "Structure and electronic properties of solid acids based on tungsten oxide nanostructures," *J. Phys. Chem. B, 103(4)*, 630–640, 1999.

34. C.D. Baertsch, K.T. Komala, Y.H. Chua and E. Iglesia, "Genesis of Brønsted acid sites during dehydration of 2-butanol on tungsten oxide catalysts," *J. Catal., 205(1)*, 44–57, 2002.

35. H.C. Liu, P. Cheung and E. Iglesia, "Effects of Al_2O_3 support modifications on MoO_x and VO_x catalysts for dimethyl ether oxidation to formaldehyde," *Phys. Chem. Chem. Phys., 5(17)*, 3795–3800, 2003.

36. K. Chen, S. Xie, A.T. Bell and E. Iglesia, "Structure and properties of oxidative dehydrogenation catalysts based on MoO_3/Al_2O_3," *J. Catal., 198(2)*, 232–242, 2001.

37. H.C. Hu and I.E. Wachs, "Catalytic properties of supported molybdenum oxide catalysts: *In situ* Raman and methanol oxidation studies," *J. Phys. Chem., 99(27)*, 10911–10922, 1995.

38. D.D. Eberl, V.A. Drits and J. Srodon, "Deducing growth mechanisms for minerals from the shapes of crystal size distributions," *Amer. J. Sci., 298*, 499–533, 1998.

39. M. Scheithauer, R.K. Grasselli and H. Knözinger, "Genesis and structure of WO_x/ZrO_2 solid acid catalysts," *Langmuir, 14(11)*, 3019–3029, 1998.

40. B.D. Cullity and S.R. Stock, *Elements of x-ray diffraction*, Prentice Hall: Upper Saddle River, NJ, p. 664, 2001.

41. A.F. Wells, *Structural inorganic chemistry*, Oxford University: New York, p. 1382, 1984.

42. F.A. Cotton and G. Wilkinson, *Advanced inorganic chemistry*, Wiley: New York, p. 1455, 1988.

43. B. Grzybowska, J. Sloczynski, R. Grabowski, K. Wcislo, A. Kozlowska, J. Stoch and J. Zielinski, "Chromium oxide alumina catalysts in oxidative dehydrogenation of isobutane," *J. Catal., 178(2)*, 687–700, 1998.

44. J. Sloczynski, B. Grzybowska, R. Grabowski, A. Kozlowska and K. Wcislo, "Oxygen adsorption and catalytic performance in oxidative dehydrogenation of isobutane on chromium oxide-based catalysts," *Phys. Chem. Chem. Phys., 1(2)*, 333–339, 1999.

45. A. Bielanski and J. Haber, *Oxygen in catalysis*, Dekker: New York, 1991.

46. H.H. Kung, "Oxidative dehydrogenation of light (C_2 to C_4) alkanes," *Adv. Catal., 40*, 1–38, 1994.

47. L. Wang and W.K. Hall, "The preparation and genesis of molybdena-alumina and related catalyst systems," *J. Catal., 77(1)*, 232–241, 1982.

48. H.C. Hu, I.E. Wachs and S.R. Bare, "Surface structures of supported molybdenum oxide catalysts: Characterization by Raman and Mo L$_3$-edge XANES," *J. Phys. Chem.*, *99(27)*, 10897–10910, 1995.

49. K. Segawa and W.K. Hall, "Site selective chemisorptions on molybdena-alumina catalysts," *J. Catal.*, *77(1)*, 221–231, 1982.

50. G. Ferraris, S. De Rossi, D. Gazzoli, I. Pettiti, M. Valigi, G. Magnacca and C. Morterra, "WO$_x$/ZrO$_2$ catalysts. Part 3. Surface coverage as investigated by low temperature CO adsorption: FT-IR and volumetric studies," *Appl. Catal. A-Gen.*, *240(1–2)*, 119–128, 2003.

51. L. Gonzalez, J.L. Galavis, C. Scott, M.J.P. Zurita and J. Goldwasser, "A study of the CO$_2$ chemisorption on supported molybdena-alumina catalysts," *J. Catal.*, *144(2)*, 636–640, 1993.

52. N.Y. Topsøe and H. Topsøe, "FTIR studies of Mo/Al$_2$O$_3$-based catalysts. 1. Morphology and structure of calcined and sulfided catalysts," *J. Catal.*, *139(2)*, 631–640, 1993.

53. M. Cornac, A. Janin and J.C. Lavalley, "Application of FTIR spectroscopy to the study of sulfidation of molybdenum catalysts supported on alumina or silica (4000–400 cm^{-1} range)," *Infrared Phys.*, *24(2–3)*, 143–150, 1984.

54. K. Stoppek-Langner, J. Goldwasser, M. Houalla and D.M. Hercules, "Infrared and carbon dioxide chemisorption study of Mo/ZrO$_2$ catalysts," *Catal. Lett.*, *32(3–4)*, 263–271, 1995.

55. R.A. Boyse and E.I. Ko, "Crystallization behavior of tungstate on zirconia and its relationship to acidic properties. 1. Effect of preparation parameters," *J. Catal.*, *171(1)*, 191–207, 1997.

56. M. Valigi, D. Gazzoli, I. Pettiti, G. Mattei, S. Colonna, S. De Rossi and G. Ferraris, "WO$_x$/ZrO$_2$ catalysts. Part 1. Preparation, bulk and surface characterization," *Appl. Catal. A-Gen.*, *231(1–2)*, 159–172, 2002.

57. B. Zhao, X. Xu, J. Gao, Q. Fu and Y. Tang, "Structure characterization of WO$_3$/ZrO$_2$ catalysts by Raman spectroscopy," *J. Raman Spectrosc.*, *27(7)*, 549–554, 1996.

58. V.M. Mastikhin, A.V. Nosov, V.V. Terskikh, K.I. Zamaraev and I.E. Wachs, "^{1}H MAS NMR studies of alumina-supported metal oxide catalysts," *J. Phys. Chem.*, *98(51)*, 13621–13624, 1994.

59. D.G. Barton, S.L. Soled, G.D. Meitzner, G.A. Fuentes and E. Iglesia, "Structural and catalytic characterization of solid acids based on zirconia modified by tungsten oxide," *J. Catal.*, *181(1)*, 57–72, 1999.

60. N. Vaidyanathan, M. Houalla and D.M. Hercules, "Surface coverage of WO$_3$/ZrO$_2$ catalysts measured by ion scattering spectroscopy and low temperature CO adsorption," *Surf. Interface Anal.*, *26(6)*, 415–419, 1998.

61. J.D. McCullough and K.N. Trueblood, "The crystal structure of baddeleyite," *Acta Cryst*, *12*, 507–511, 1959.

62. E. Iglesia, D.G. Barton, S.L. Soled, S. Miseo, J.E. Baumgartner, W.E. Gates, G.A. Fuentes and G.D. Meitzner, "Selective isomerization of alkanes on supported tungsten oxide acids," *101(Pt. A, 11th International Congress on Catalysis--40th Anniversary, 1996, Pt. A)*, 533–542, 1996.

63. D.S. Kim, M. Ostromecki and I.E. Wachs, "Surface structures of supported tungsten oxide catalysts under dehydrated conditions," *J. Mol. Catal. A-Chem.*, *106(1–2)*, 93–102, 1996.

64. N. Vaidyanathan, D.M. Hercules and M. Houalla, "Surface characterization of WO$_3$/ZrO$_2$ catalysts," *Anal. Bioanal. Chem.*, *373(7)*, 547–554, 2002.

65. M.A. Eberhardt, M. Houalla and D.M. Hercules, "Ion scattering and electron spectroscopic study of the surface coverage of V/Al$_2$O$_3$ catalysts," *Surf. Interface Anal.*, *20(9)*, 766–770, 1993.

66. T. Fransen, O. Vandermeer and P. Mars, "Investigation of surface-structure and activity of molybdenum oxide-containing catalysts. 1. Infrared study of surface-structure of molybdena-alumina catalysts," *J. Catal.*, *42(1)*, 79–86, 1976.

67. P. Ratnasamy and H. Knözinger, "Infrared and optical spectroscopic study of Co-Mo-Al$_2$O$_3$ catalysts," *J. Catal.*, *54(2)*, 155–165, 1978.

68. Y. Okamoto and T. Imanaka, "Interaction chemistry between molybdena and alumina: Infrared studies of surface hydroxyl-groups and adsorbed carbon dioxide on aluminas modified with molybdate, sulfate, or fluorine anions," *J. Phys. Chem.*, *92(25)*, 7102–7112, 1988.

69. W.S. Millman, M. Crespin, A.C. Cirillo, Jr., S. Abdo and W.K. Hall, "Studies of the hydrogen held by solids. XXII. The surface chemistry of reduced molybdena-alumina catalysts," *J. Catal.*, *60(3)*, 404–416, 1979.

70. A.C. Cirillo, Jr., F.R. Dollish and W.K. Hall, "Studies of the hydrogen held by solids. XXVI. Proton resonance from alumina and molybdena-alumina catalysts," *J. Catal.*, *62(2)*, 379–88, 1980.

71. M. Bensitel, O. Saur, J.C. Lavalley and G. Mabilon, "Acidity of zirconium oxide and sulfated ZrO$_2$ samples," *Mater. Chem. Phys.*, *17(3)*, 249–258, 1987.

72. F.M. Mulcahy, K.D. Kozminski, J.M. Slike, F. Ciccone, S.J. Scierka, M.A. Eberhardt, M. Houalla and D.M. Hercules, "Chemisorption of carbon dioxide on alumina-supported catalysts," *J. Catal.*, *139(2)*, 688–690, 1993.

73. N. Naito, N. Katada and M. Niwa, "Tungsten oxide monolayer loaded on zirconia: Determination of acidity generated on the monolayer," *J. Phys. Chem. B*, *103(34)*, 7206–7213, 1999.

74. M. Niwa, S. Inagaki and Y. Murakami, "Alumina: Sites and mechanism for benzaldehyde and ammonia reaction," *J. Phys. Chem.*, *89(12)*, 2550–2555, 1985.

75. M. Niwa, K. Suzuki, M. Kishida and Y. Murakami, "Benzaldehyde-ammonia titration method for discrimination between surfaces of metal oxide catalysts," *Appl. Catal.*, *67(2)*, 297–305, 1991.

76. J.A. Lercher, C. Gründling and G. Eder-Mirth, "Infrared studies of the surface acidity of oxides and zeolites using adsorbed probe molecules," *Catal. Today*, *27(3–4)*, 353–376, 1996.

77. H. Knözinger, "Infrared spectroscopy for the characterization of surface acidity and basicity," in G. Ertl, H. Knözinger and J. Weitkamp (Eds.), *Handbook of heterogeneous catalysis*, Vol. 2, Wiley-VCH: Weinheim, pp. 707–732, 1997.

78. R.S. Weber, "Effect of local structure on the UV-visible absorption edges of molybdenum bride clusters and supported molybdenum oxides," *J. Catal.*, *151(2)*, 470–474, 1995.

79. J. Tauc, R. Grigorovici and A. Vancu, "Optical properties and electronic structure of amorphous germanium," *Physica Status Solidi*, *15(2)*, 627–637, 1966.

80. W.B. Jackson, S.M. Kelso, C.C. Tsai, J.W. Allen and S.J. Oh, "Energy-dependence of the optical matrix element in hydrogenated amorphous and crystalline silicon," *Phys. Rev. B*, *31(8)*, 5187–5198, 1985.

81. C.R. Wronski, S. Lee, M. Hicks and S. Kumar, "Internal photoemission of holes and the mobility gap of hydrogenated amorphous-silicon," *Phys. Rev. Lett.*, *63(13)*, 1420–1423, 1989.

82. K. Chew, Rusli, S.F. Yoon, J. Ahn, Q. Zhang, V. Ligatchev, E.J. Teo, T. Osipowicz and F. Watt, "Gap state distribution in amorphous hydrogenated silicon carbide films deduced from photothermal deflection spectroscopy," *J. Appl. Phys.*, *91(7)*, 4319–4325, 2002.

83. I.Y. Ravich, B.A Efimova and I.A. Smirnov, *Semiconducting lead chalcogenides*, Plenum Press: New York, pp. 52–53, 1970.

84. W.N. Delgass, G.L. Haller, R. Kellerman and J.H. Lunsford, *Spectroscopy in heterogeneous catalysis*, Academic Press: New York, 1979.

85. R.A. Smith, *Semiconductors*, Cambridge University Press: Cambridge, U.K., 1978.

86. X.T. Gao, M.A. Banares and I.E. Wachs, "Ethane and *n*-butane oxidation over supported vanadium oxide catalysts: An *in situ* UV-visible diffuse reflectance spectroscopic investigation," *J. Catal.*, *188(2)*, 325–331, 1999.

87. H.P. Myers, *Introductory solid state physics*, Taylor & Francis: London, pp. 276, 295, 1997.

88. L.E. Brus, "Electron-electron and electron-hole interactions in small semiconductor crystallites—The size dependence of the lowest excited electronic state," *J. Chem. Phys.*, *80(9)*, 4403–4409, 1984.

89. P.H. Emmett and S. Brunauer, "The use of low temperature van der Waals adsorption isotherms in determining the surface area of iron synthetic ammonia catalysts," *J. Am. Chem. Soc.*, *59(8)*, 1553–1564, 1937.

90. L. Spenadel and M. Boudart, "Dispersion of platinum on supported catalysts," *J. Phys. Chem.*, *64(2)*, 204–207, 1960.

91. H.L. Gruber, "Chemisorption studies on supported platinum," *J. Phys. Chem.*, *66(1)*, 48–54, 1962.

92. T.J. Osinga, B.G. Linsen and W.P. Vanbeek, "Determination of specific copper surface area in catalysts," *J. Catal.*, *7(3)*, 277–279, 1967.

93. M.A. Vuurman and I.E. Wachs, "*In situ* Raman-spectroscopy of alumina-supported metal-oxide catalysts," *J. Phys. Chem.*, *96(12)*, 5008–5016, 1992.

94. M. Hino and K. Arata, "Synthesis of solid superacid of tungsten oxide supported on zirconia and its catalytic action for reactions of butane and pentane," *J. Chem. Soc.-Chem. Commun.*, *(18)*, 1259–1260, 1988.

12 Solid-State Ion-Exchange of Zeolites

Geoffrey L. Price

CONTENTS

12.1 INTRODUCTION

One of the most important properties of zeolites is their ability to undergo ion-exchange. Zeolitic frameworks are inherently anionic by virtue of incorporation of trivalent cations (commonly Al^{3+}) for quadrivalent (commonly Si^{4+}) cations, which are central to the tetrahedral building blocks comprising most zeolites. The anionic framework charge is satisfied by cations, and these cations are often accessible to ion-exchange through the zeolitic pore system. The ability of zeolites to undergo ion-exchange leads directly to their utility in applications such as water softening, but the property also allows chemical modifications to the zeolites, which lead to their utility as catalysts. Protonic forms of zeolites, routinely derived from ion-exchange of alkali-metal-containing zeolites, are the most common type of zeolite catalysts, but transition metal and other metal forms are also well known.

Aqueous ion-exchange is the most common cation-exchange process applied to zeolites. In a batch process, a soluble salt of the desired ingoing cation is prepared, and the zeolite is slurried in the solution. The slurry is often heated to speed the ion-exchange process, and the temperature is thus limited to the boiling point of the solution, though 40–80°C is common. Occasionally, the process may be performed in a pressure vessel, which allows a further increase in the operating temperature. As the ion-exchange process proceeds, ingoing cations replace outgoing (zeolitic) cations, and outgoing cations appear in the aqueous solution. As the number of outgoing cations in the aqueous solution increases, an equilibrium distribution between cations in aqueous solution and cations in the zeolite is eventually established, and close approach to equilibrium can take anywhere from a few minutes to several hours or more.

But because of the reversible nature of the exchange process, a single batch exchange will not accomplish complete exchange, and multiple exchanges are necessary if complete exchange is required. To accomplish this, the aqueous solution is filtered away from the zeolite, and a fresh solution of the ingoing cation is prepared and contacted with the zeolite. By using a large excess of the ingoing cation in the aqueous solution, the number of exchanges to accomplish the required degree of exchange can be minimized, but this leads either to wasted ingoing cations or to a problem of separating and recycling ingoing cations. On a laboratory scale, these difficulties are not severe, but on a large scale, environmental and economic concerns are especially evident, complicating the overall process. A countercurrent-type process, where the zeolite phase goes one direction and the aqueous exchange solution goes the other, can be applied to help improve efficiency.

Solid-state ion-exchange (SSIE) occurs when the ingoing cations are in the solid state, then are contacted with the zeolite, usually in some type of grinding process, and are finally driven into the zeolite under the influence of a driving force. In the simplest case, the driving force is a high temperature, but processes are known that take place at room temperature when moisture is present. Reducing or oxidizing agents may also be required. In most cases, the zeolite is in its proton form prior to the SSIE process, and in that case, the product of the process is volatile and leaves the zeolite. Such processes can generally be considered irreversible if the volatile product is swept away into a purge gas or removed via vacuum. Cases are known where the zeolite is not in the proton form and ingoing cations compete with cations already present in the zeolite for framework sites, and extraframework cations and anions wind up populating the zeolitic pores. This phenomena was termed *occlusion* in an early paper on the subject of SSIE by Rabo et al. [1] and a later review article by Rabo [2].

In this chapter, we focus on the important aspects of preparing catalytic materials via SSIE in keeping with the overall purpose of the book. For a rather complete review of the papers covering the specific subject area, a work from a few years ago by Karge and Beyer [3] is important and highly recommended.

12.2 ADVANTAGES OF SOLID-STATE ION-EXCHANGE OVER AQUEOUS ION-EXCHANGE

Though aqueous ion-exchange is a common method used to replace cations in zeolites, there are virtually no advantages to the method other than familiarity. Some may argue that an advantage of preparing materials by aqueous exchange is that they do not require high-temperature heating to facilitate exchange, but the high-temperature heating process required in the SSIE case would simply compare with calcination, which is generally required for the materials prepared by aqueous ion-exchange. Solid-state techniques, however:

- May yield metal loadings higher than conventional techniques. This is clear for the case where solid-state techniques yield cations in the 1+ oxidation state while corresponding aqueous cations are not stable. Cations in the 1+ oxidation state can be used to achieve loadings of 1 Me/framework-Al in the zeolite.
- Offer chemistry at temperatures widely different from aqueous conditions. Open containers utilizing aqueous techniques are limited to about 100°C, though closed vessels could be used to 200°C or so routinely with water. Solid-state techniques would most often be limited to the upper temperature at which the zeolite is stable, which is often at least 600°C. There is a wealth of metal chemistry available between 200 and 600°C or higher.
- Eliminate solvation shells, which are present in aqueous solution and which have the effect of greatly enhancing the effective size of the cation. This effect can greatly reduce the ability of a cation to enter a zeolitic pore, thwarting aqueous ion-exchange particularly in smaller pore systems.
- Can eliminate zeolitic proton acidity. Acidity may cause unwanted side reactions such as coking and polymerization, and aid in the removal of framework-Al from the zeolite when the zeolite is exposed to water. Acidity is virtually impossible to reduce to insignificant levels when aqueous solutions are used.
- Do not result in a spent liquor that must be dealt with in an environmentally friendly manner. Solid-state techniques may generate an off-gas to deal with, but the off-gas is water if metal oxides or hydroxides are used as the metal source, and in any event, the off-gas is in a more concentrated state than in the aqueous case.
- Probably require less postexchange processing than their aqueous exchange counterparts. As a minimum, the materials prepared by aqueous exchange must be filtered, then dried. They may further require one or more washing and filtering sequences, thus generating more wastewater to deal with.

Criticism that has sometimes been directed at solid-state techniques purports that simpler aqueous techniques (which are not simpler at all—perhaps the correct phrasing would be *more familiar*) can often be used to generate the same material as a solid-state procedure. Though it is true that similar products can often be made, this criticism does not take into account the possible processing benefits that can be derived from the smaller processing volumes, smaller volumes of byproducts, and simpler postprocessing, as detailed above. Therefore, solid-state techniques may be preferable even in the case where aqueous exchange can be used to prepare the same material.

There are many further considerations and comparisons that could be made; e.g., solid-state processes require the mixing of the two solid phases, but aqueous techniques require preparation of the exchange solution and filtration of the product from the spent liquor. Thus, there are reasons to consider solid-state techniques from a perspective of improving the practical aspects of processing in addition to the possible advantage of preparing materials that may be impossible to prepare through aqueous techniques.

A question that then arises is that if SSIE techniques offer so many advantages, why have they not taken over as the method of choice for ion-exchange of zeolites? The answer is that there are many variables that affect the efficiency of solid-state processes, and the product of the ion-exchange process may not be fully predictable without experimentation. Even the simplest of variables, temperature and time required for a given process, are not predictable for solid-state techniques or well known. In contrast, the chemistry of aqueous solutions is well known, and therefore, the process of aqueous ion-exchange is generally predictable, though it is the author's belief that many researchers make assumptions regarding the extent of aqueous ion-exchange processes that are not justified. Nonetheless, when literature references are consulted, it is generally plausible to construct an aqueous ion-exchange process that will give a product of high enough quality to be useful. The sparsity of specific examples of SSIE makes a similar construction tenuous at best, so that experimentation and evaluation of the SSIE product are still required. The inefficiencies associated with aqueous ion-exchange are not severe enough to overcome the uncertainties of the solid-state processes for the laboratory scale. One goal the author has set for this chapter is to help narrow this knowledge gap.

SSIE requires two primary elements. The first element is an intimate physical mixture, the second is a driving force to "push" the ingoing metal into the zeolite. We will discuss each of these factors.

12.3 THE PHYSICAL MIXTURE

Almost all studies utilizing SSIE to prepare catalysts use a mortar and pestle to make the physical mixture of the ingoing cation compound with the zeolite. This is not an efficient method for making these mixtures and certainly could not be implemented in industrial practice. A more efficient and reproducible method is highly recommended.

Let us begin by considering evidence that the quality of the mechanical mixture is an important factor in the preparation of metal-containing zeolites by SSIE. Ball-milling is the method we have used routinely in our lab because it is relatively easy to set up and efficient in operation. Our ball-mill is a stainless-steel cylindrical chamber that is 3 in. deep, 5 in. inner diameter, and rotated at 86 rpm. We load approximately 20 g of catalytic material in the ball-mill along with thirty 1/4-in.-diameter, twenty 3/8-in.diameter, and fifteen 1/2-in.-diameter stainless-steel balls. Using different-sized balls helps keep the landing spot of each ball random as the mill is rotated. In general, we grind the two phases for 3 or 4 h in this device.

To show the effect of ball-milling time on the efficiency of an SSIE process, we used the particular ball-mill described above to prepare Ga_2O_3/H-MFI[1]

---

[1] MFI is the International Zeolite Association's structure code for Mobil's ZSM-5 framework. We generally prefer the MFI terminology, but in this chapter, we have tried to use whatever term the original authors used in their writings. The reader should note that the two terminologies apply to materials that have the same framework connectivities.

mixtures, then performed the SSIE process in a microbalance. In this particular case, we were actually performing the reductive solid-state ion-exchange (RSSIE) process, whereby a reducing agent such as H_2 gas is required for the SSIE process to proceed. As we have reported [4], the reaction can be written stoichiometrically as

$$Ga_2O_3 + 2H\text{-}Z + 2H_2 \rightarrow 2Ga\text{-}Z + 3H_2O,$$

where H-Z is the proton form of the zeolite.

Results of these experiments [5, 6] show that depending on the Ga_2O_3 loading, after about 3 to 4 h of ball-milling time, no improvement in the overall rate of the ion-exchange process is seen. More ball-milling time is required for low loadings of Ga_2O_3, whereas less time is required for high loadings.

More qualitative details of the RSSIE process can be seen in Figure 12.1, which depicts the weight loss of $Ga_2O_3/H\text{-}MFI$ mechanical mixtures determined by microbalance. In each case, the samples were dried in pure He at 550°C, then H_2 was switched in as noted. The sample ball-milled for only 0.75 h shows a much slower overall rate than the same formulation ball-milled for 6 h. This experiment shows the utility of very intimate mixtures for enhancing the rate of the RSSIE process.

Also shown in Figure 12.1 is a sample taken from the material ball-milled for 0.75 h, then placed in a vibrating mill—also known as a *wiggle bug*—which is usually used for preparing KBr mixtures for use in IR investigations. In this device,

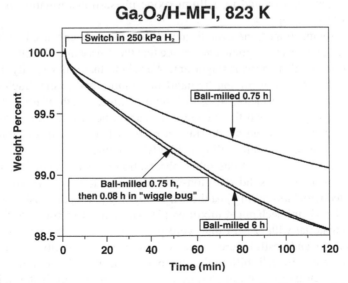

FIGURE 12.1 Effect of mechanical mixing on the reduction of gallium containing MFI.

a sample is placed in a small ampoule along with a stainless-steel ball. The ampoule assembly is then placed in a vibrating sample holder and vibrated for a period of time, usually 3–5 min. Note that this process is also effective for making a very intimate mixture, and its only drawback is that it can only prepare samples as large as 100 mg or so.

Consider also some of the kinetic details that can be seen in Figure 12.1. After a very short upset caused by the gas switch, about 0.1% weight is lost very quickly (less than 1 min). We believe this weight loss corresponds to reduction of a small amount of $(GaO)^+$, which is ion-exchanged into the zeolite during the drying phase, though there is no direct evidence to support this postulate. Next, we note a region that is nearly linear over the next 60–80 min. We believe this region corresponds to areas where there is direct contact between the zeolite and the ingoing (Ga_2O_3) phase as the result of the initial grinding process. Enhanced grinding yields more contact area between the two phases up to a point, and therefore a higher rate ensues. Once these direct contact regions are gone, a slower region of weight loss begins that decays exponentially; these regions are not fully completed in Figure 12.1. The mechanism of the SSIE process that is occurring in this region is unclear, but one factor that might be considered is filling of the anionic sites near the pore mouths with ingoing cations. The rate of such a process might become dependent upon cations moving deeper and deeper into the zeolite, and thus, the exponential decay of the rate could result. However, this cannot fully explain the behavior, as all three samples should begin to experience these diffusional limitations at the same gallium loading that would correspond to the same weight loss, and this is not the case. Therefore, another mechanism probably related to gallium oxide phase transformation is affecting the rate in this region.

As already mentioned, the overall rate of the process is enhanced by increasing the ball-milling time up to a point, and we see that the degree of ball-milling, however we might quantify that entity, is important. We submit that it is virtually impossible to quantify the process of mixing by hand in a mortar and pestle. Such a process will be dependent upon factors such as length of time, length of each stroke, and pressure on each stroke, and these conditions cannot be quantified and reproduced from human hand to human hand. Thus, a reproducible method for preparing the physical mixture such as ball-milling is highly recommended.

We should also point out one other observation we have made in our studies. In the case of reductive solid-state ion-exchange of Ga_2O_3 with H-MFI, we have experimented with ball-milling just the Ga_2O_3 ingoing phase, then using this micro-crystalline Ga_2O_3 in mechanical mixtures [7]. Mechanical mixtures of this micro-crystalline material with H-MFI that were prepared even by an inefficient method (using a mortar and pestle) were found to be reduced as rapidly as the mixtures that were prepared by ball-milling the two phases together. Thus, in this particular case, the difficult part of forming a quality mechanical mixture is the reduction in the crystallite size of the Ga_2O_3 phase. Ball-milling the two phases together appar-ently is effective because the ball-milling process reduces the particle size of the Ga_2O_3 phase.

Finally, we have also been wary of the possibility that mechanical mixtures may be susceptible to spontaneous separation due to normal building vibrations whenever there are significant differences in the densities of the solid phases. To ensure that this does not happen, we pelletize all of our mechanical mixtures for storage or complete the SSIE process immediately after removal from the milling device.

We conclude this section by suggesting that some experimentation is required to ensure that the initial mechanical mixture is of high quality. We cannot predict *a priori* what will constitute an effective mixture or how long it might require in a ball-mill or other device to properly prepare. In each instance, we should try varying the length of the milling process, and we should use some type of reproducible method for carrying out the process so that we can be sure of reproducing the same result in future experiments.

12.4 DRIVING FORCE FOR THE SOLID-STATE ION-EXCHANGE PROCESS

In order for the SSIE process to proceed, there must be some type of driving force making the cations in the ingoing phase mobile enough to break them apart from the solid matrix where they begin, such that they can travel into the zeolite. In this chapter, we break these processes down into three fundamental types: (1) moisture induced, (2) high-temperature induced, and (3) reaction induced.

12.4.1 MOISTURE-INDUCED SSIE

Moisture-induced examples are often not useful for the preparation of catalytic materials and have especially been observed on alkali metal systems, though some other systems are notable. Historically, these types of SSIE processes have often been referred to as low-temperature exchange reactions or contact-induced exchanges. In these processes, moisture that is adsorbed by the zeolite exists naturally in humid air, or is part of the matrix of the ingoing solid phase that partially dissolves the ingoing phase and apparently acts as a conduit for the ion-exchange process, thus making the process quite similar to aqueous exchange processes. This type of exchange is limited to ingoing phases that are water soluble. Some of the best examples of moisture-induced exchange were the earliest reports of SSIE processes, including a very early paper by Ataman and Mark [8] regarding the exchange of an ammonium-zeolite with potassium bromide, which was being used as a pelletizing agent for the production of wafers for infrared investigation. More recent papers by Lazar et al. [9, 10] on the formation of Fe-containing zeolite catalysts report that upon mixing $FeCl_2$ with NH_4-Y at room temperature, Fe hydroxides are formed in the zeolite. The process was followed closely by Mossbauer spectroscopy, so that the authors were able to conclude unequivocally that water participated in the ion-exchange process, and because water is known to participate, the only remaining possible advantages of this procedure over aqueous exchange processes relate to processing improvements.

12.4.2 High-Temperature-Induced SSIE

High-temperature-induced examples of SSIE are the most common type reported in the open literature. We will break these kinds of SSIE into two different types: those that create a volatile product, and those that do not.

To the best of our knowledge, all cases of SSIE that produce a volatile product begin with the proton form of the zeolite. In that case, the volatile product will be water if the ingoing phase is a metal oxide, or a hydrogen halide if a halide salt is used, which is common practice. In a few instances, the ammonium form of the zeolite has been used, but this breaks down into NH_3 and the proton form of the zeolite on heating, and thus, the proton form of the zeolite actually undergoes exchange.

A recent paper by Kinger et al. [11] is a good example of the type of SSIE process where a volatile product is formed. In this investigation, H-MFI, H-MOR, and H-BEA were loaded with Ni via several methods including SSIE. $NiCl_2 \cdot 6H_2O$ was mechanically mixed with the proton form of the zeolite in a mortar and pestle. The mixtures were calcined in vacuum or a flowing helium stream up to a final temperature of 500°C, yielding a hydrogen chloride volatile product. The authors especially noted that high loadings of Ni were possible using this method, further citing that a quantitative reaction of the zeolite with the Ni salt occurs. Compared to aqueous ion-exchange, the SSIE process induced fewer structural defects in the zeolite, and the authors stated that the liquid-phase process was sterically constrained, resulting in lower Ni loadings.

When this type of SSIE occurs, zeolitic protons are consumed, and analytical methods that detect zeolitic protons are especially useful in determining the extent of the solid-state reaction. FTIR in the -OH region is useful for this purpose and has been exploited recently by El-Malki et al. [12, 13], Wang et al. [14, 15], and Kinger et al. [11]. Recent papers show that many metals are amenable to SSIE between H-zeolites and chlorides such as $CoCl_2$/HZSM-5 [13–15], $EuCl_3/NH_4$-Y [16], CuCl/MCM-41 [17], CuCl/HY [18], $FeCl_2.4H_2O$/H-ZSM-5 [19], $FeCl_2$/H-MFI [12], and $CsCl/NH_4$-Y[20].

In the case where there is no volatile product of the SSIE process, an occluded salt or oxide resides in the zeolitic pores along with the zeolitic charge compensating cations after the SSIE process. Some of the earliest reports explaining this effect date back to reports by Rabo et al. [1, 2]. Basically, salt occlusion results when an excess of cations populate a zeolitic pore along with extra anions to provide charge compensation for the excess cations. Take, for example, an SSIE between NaY and $CuCl_2$. Rabo [2] reported that when such a mixture is treated at 550°C for 48 h, 1.12 cation equivalents per framework-Al resulted, so that both Cu and Na cations coexist in the zeolite, along with some extra Cl anions to work with the anionic zeolite framework for charge compensation. Note that because these cations coexist and are physically near each other, it is not clear which cations satisfy the zeolitic anion vacancies and which are associated with the extra framework anions. Instead, all the cationic charge is shared among all the anionic charge distributed around the pore. Rabo has noted that these occluded salts increase the stability of the system by providing shielding for cations in the zeolitic pores. This can be explained by considering that the population of cations within a zeolitic pore must face each other and be relatively close to one another, yet the cations tend to repel each other because they are the same charge. However, if an occluded salt molecule is present, the anion associated with the occluded cation can sit toward the center of the pore and shield the cations from one another. This has the effect of stabilizing the system.

From the point of view of catalysis, occluded salts might be detrimental by restricting pore space. Few examples of this type of exchange exist in the catalysis literature, but one recent example is known. Thoret et al. [21] have investigated the occlusion of $MnCl_2.xH_2O$ into NaY and LaNaY by heat treatment at 313 to 1173 K. Various Mn-containing phases including Na_4MnO_4 and $Na_2Mn_5O_{10}$ were detected. The results were compared with previous studies involving solid-state reactions with oxides, but unfortunately, catalysis characterization of the materials was not reported.

12.4.3 Reaction-Induced SSIE

The acronym *RSSIE* was first used by Price et al. in early work on the Ga_2O_3/H-MFI system [22] and Kanazirev and Price CuO/H-MFI [23] system for the conversion of light paraffins to aromatics. The gallium-containing system made use of a reactant (hydrogen) as a reducing agent, whereas the copper-containing system relied on spontaneous reduction at high temperature in the absence of oxygen, and both fall into the general category of reaction-induced forms of SSIE.

One of the earliest reports of this type of SSIE goes back to CrO_3/H-ZSM-5 (Kucherov et al. [24]), although the authors did not emphasize that the process involved both reaction and SSIE. Thermal treatment of CrO_3/H-ZSM-5 mechanical mixtures resulted in the appearance of an ESR signal attributable to Cr^V ions. This signal was thought to originate from $(CrO_2)^+$ cations in the zeolite. Thus, the solid-state process involves both reduction and ion-exchange. A similar reaction was reported by Price et al. [25] for mechanical mixtures of CuO/H-MFI, whereby Cu^+ zeolitic cations were formed spontaneously by high-temperature treatment in the absence of oxygen.

The two most widely reported systems using RSSIE are Ga_2O_3/H-zeolites and In_2O_3/H-zeolites. Recent reports on Ga_2O_3/H-zeolite include Mihalyi et al. [26], who have reported on incorporating Ga in H-Y zeolites in a fashion similar to earlier reports on Ga_2O_3/H-MFI; Fuchsova et al. [27], who compared aqueous ion-exchange and RSSIE for preparing Ga/MFI and Ga/BEA catalysts; and Raichle et al. [28], who investigated Ga/ZSM-5 prepared by RSSIE for the conversion of cycloalkanes into a steam cracker feedstock.

A number of recent reports have appeared on In-containing zeolites. Mavrodinova et al. [29] prepared In-containing beta zeolite by RSSIE and studied the resulting catalysts for *m*-xylene transformation. In this study, conditions very similar to the analogous Ga_2O_3/H-MFI system were used, except that lower temperatures were required for the reduction process in hydrogen. Dimitrova et al. [30] successfully prepared In-containing [Al]-beta and [B]-beta using RSSIE. They reported that In^+ cations, incorporated into zeolitic cation positions upon the RSSIE process using hydrogen, in some cases were converted to $(InO)^+$ species via reaction with silanol hydroxyls. Schmidt et al. [31] found both small intra-zeolitic and large extra-zeolitic

In-oxide clusters along with a preponderance of atomically dispersed In cations in
H-MOR prepared by RSSIE. The resulting catalyst is useful for low-temperature
selective catalytic reduction (SCR) of NO_x. Schutze et al. [32] compared In-containing
H-MFI and H-MOR prepared by aqueous ion-exchange, RSSIE, precipitation, and
combinations of these techniques for the methane SCR of NO_x. They further investigated Ce promotion of the catalysts. Sowade et al. [33] used $InCl_3$ as the source
of In for SCR of NO_x catalysts they prepared and found activities similar to catalysts
prepared by RSSIE from In_2O_3/H-zeolites. In light of the report by Mihalyi and
Beyer [34], where In_2O_3/H-ZSM-5 was found to undergo autoreduction to In^+ cations
incorporated as zeolitic cations upon thermal treatment under vacuum, the possibility
that $InCl_3$ undergoes a similar autoreduction cannot be overlooked. In such a situation,
the overall products of the two methods might be very similar.

Two recent works have also appeared whereby CCl_4 has been used as a reactive
component for SSIE. Lazar et al. [35] used air saturated with CCl_4 during thermal
treatment of mechanical mixtures of $FeCl_2$/NH_4-zeolites and Fe_2O_3/NH_4-zeolites and
found that Fe-Cl species are formed and move into the zeolite freely. Kucherov and
Slinkin [36] came to similar conclusions in the case of CCl_4 used to promote the
transfer of MoO_3 into several different H-zeolites. Mo^V could be stabilized in the
zeolites using this procedure.

12.5 CONCLUSIONS AND RECOMMENDATIONS

There are a number of advantages that can be gleaned from the use of solid-state
ion-exchange for the preparation of catalytic materials. Many good examples exist
in the open literature, so that there are suitable examples as a starting point for
researchers. However, the science of solid-state ion-exchange is not currently well
developed enough to fully allow prediction of the outcome of preparation, so that
experimentation with preparation conditions is required. A technique that is highly
recommended is the use of a reproducible method for preparing the physical mixture.
Other conditions that require experimentation include the temperature and time for
the exchange process to occur, and other possibilities include experimentation with
reacting agents that may be used to promote the exchange reactions.

REFERENCES

1. Rabo, J. A.; Poutsma, M. L. and Skeels, G. W.; "New zeolite-salt adducts and
their catalytic properties," *Proc. 5th Int. Congr. Catal.*, Hightower (Ed.), 98–1353
(1973).
2. Rabo, J. A., *Zeolite Chemistry and Catalysis, ACS Monograph Series* 171, 332 (1976).
3. Karge, H. G. and Beyer, H. K. "Solid-state ion exchange in microporous and mesoporous materials," *Molecular Sieves* 3(Post-Synthesis Modification I, 2002), 43–201
(2002).
4. Price, G. L. and Kanazirev, V., "Ga_2O_3/HZSM-5 propane aromatization catalysts:
Formation of active centers via solid-state reaction," *Journal of Catalysis* 126,
267–278 (1990).

5. Price, G. L.; Kanazirev, V. and Dooley, K.M., "Gallium-containing zeolite catalysts," U.S. Patent # 5,149,679 (1992).

6. Price, G. L. and Kanazirev, V., "The oxidation state of Ga in Ga/ZSM-5 light paraffin aromatization catalysts," *Journal of Molecular Catalysis* 66, 115–120 (1991).

7. Price, G. L.; Kanazirev, V.; Dooley, K.M. and Hart, V.I., "On the mechanism of propane dehydrocyclization over cation-containing, proton-poor MFI zeolite," *Journal of Catalysis* 173, 17–27 (1998).

8. Ataman, O. Y.; and Mark, H. B., Jr., "Ion exchange between ammonium zeolite and the supporting matrix in potassium bromide pellets," *Analytical Letters* 9(12), 1135–1141 (1976).

9. Lazar, K.; Pal-Borbely, G.; Beyer, H. K. and Karge, H. G., "Solid-state ion exchange in zeolites. Part 5. NH$_4$-Y-iron(II) chloride." *Journal of the Chemical Society, Faraday Transactions* 90(9), 1329–1334 (1994).

10. Lazar, K.; Pal-Borbely; G.; Beyer, H. K. and Karge, H. G., "Catalysts by solid-state ion exchange: Iron in zeolite," *Studies in Surface Science and Catalysis* 91(Preparation of Catalysts VI): 551–559 (1995).

11. Kinger, G.; Lugstein, A.; Swagera, R.; Ebel, M.; Jentys, A. and Vinek, H. "Comparison of impregnation, liquid- and solid-state ion exchange procedures for the incorporation of nickel in HMFI, HMOR and HBEA zeolites. Activity and selectivity in n-nonane hydroconversion," *Microporous and Mesoporous Materials* 39(1–2), 307–317 (2000).

12. El-Malki, El-M.; van Santen, R. A. and Sachtler, W. M. H., "Active sites in Fe/MFI catalysts for NO$_x$ reduction and oscillating N$_2$O decomposition," *Journal of Catalysis* 196(2), 212–223 (2000).

13. El-Malki, El-M.; Werst, D.; Doan, P. E. and Sachtler, W. M. H., "Coordination of Co^{2+} cations inside cavities of zeolite MFI with lattice oxygen and adsorbed ligands." *Journal of Physical Chemistry B* 104(25), 5924–5931 (2000).

14. Wang, X.; Chen, H.-Y. and Sachtler, W. M. H., "Catalytic reduction of NO$_x$ by hydrocarbons over Co/ZSM-5 catalysts prepared by different methods," *Applied Catalysis, B: Environmental* 26(4), L227-L239 (2000).

15. Wang, X.; Chen, H. and Sachtler, W. M. H., "Selective reduction of NO$_x$ with hydrocarbons over Co/MFI prepared by sublimation of CoBr$_2$ and other methods," *Applied Catalysis, B: Environmental* 29(1), 47–60 (2001).

16. Nassar, E. J. and Serra, O. A., "Solid state reaction between europium III chloride and Y-zeolites," *Materials Chemistry and Physics* 74(1), 19–22 (2002).

17. Li, Z.; Xie, K. and Slade, R. C. T., "High selective catalyst CuCl/MCM-41 for oxidative carbonylation of methanol to dimethyl carbonate," *Applied Catalysis, A: General* 205(1,2), 85–92 (2001).

18. Li, Z.; K. Xie and Slade, R.C.T., "Studies of the interaction between CuCl and HY zeolite for preparing heterogeneous Cu^I catalyst," *Applied Catalysis, A: General* 209(1,2), 107–115 (2001).

19. Long, R. Q. and Yang, R. T., "Fe-ZSM-5 for selective catalytic reduction of NO with NH$_3$: A comparative study of different preparation techniques," *Catalysis Letters* 74(3–4), 201–205 (2001).

20. Concepcion-Heydorn, P.; Jia, C.; Herein, D.; Pfander, N.; Karge, H. G. and Jentoft, F. C., "Structural and catalytic properties of sodium and cesium exchanged X and Y zeolites, and germanium-substituted X zeolite," *Journal of Molecular Catalysis A: Chemical* 162(1–2), 227–246 (2000).

21. Thoret, J.; Man, P. P.; Ngokoli-Kekele, P. and Fraissard, J., "Solid-state modification of Y zeolites (NaY or LaNaY) by $MnCl_2 xH_2O$: Comparison with V_2O_5, MoO_3 and Sb_2O_3," *Microporous and Mesoporous Materials*, 49(1–3), 45–56 (2001).
22. Price, Geoffrey L.; Kanazirev, Vladislav I. and Dooley, Kerry M., "Characterization of [Ga]MFI via thermal analysis,", *Zeolites*, 15(8), 725–731 (1995).
23. Kanazirev, Vladislav I. and Price, Geoffrey L., "Propane conversion on Cu-MFI zeolites," *Journal of Molecular Catalysis A: Chemical*, 96(2), 145–154 (1995).
24. Kucherov, A. V.; Slinkin, A. A.; Beyer, G. K. and Borbely, G., "Zeolites H-[Ga]ZSM-5 and H-ZSM-5: A comparative study of the introduction of transition-metal cations by a solid-state reaction," *Journal of the Chemical Society, Faraday Transactions 1: Physical Chemistry in Condensed Phases* 85(9), 2737–2747 (1989).
25. Price, G. L.; Kanazirev, V. and Church, D. F., "Formation of Cu-MFI NO decomposition catalyst via reductive solid-state ion exchange," *Journal of Physical Chemistry*, 99(3), 864–868 (1995).
26. Mihalyi, R. M., Beyer, H. K. and Keindl, M., "Incorporation of Ga ions into Y zeolites by reductive solid-state ion exchange," *Studies in Surface Science and Catalysis* 135(Zeolites and Mesoporous Materials at the Dawn of the 21st Century), 1561–1568 (2001).
27. Fuchsova, J.; Bulanek, R.; Novoveska, K.; Cermak, D. and Lochar, V., "On the incorporation of gallium ions into a cationic sites of pentasil ring zeolites and their acid-base, redox and catalytic properties," *Scientific Papers of the University of Pardubice, Series A: Faculty of Chemical Technology* 8, 137–150 (2002).
28. Raichle, A.; Moser, S.; Traa, Y.; Hunger, M. and Weitkamp, J., "Gallium-containing zeolites as valuable catalysts for the conversion of cycloalkanes into a premium synthetic steam-cracker feedstock," *Catalysis Communications* 2(1), 23–29 (2001).
29. Mavrodinova, V. P.; Popova, M. D.; Neinska, Y. G. and Minchev, C. I., "Influence of the Lewis acidity of indium-modified beta zeolite in the *m*-xylene transformation," *Applied Catalysis, A: General*, 210(1,2), 397–408 (2001).
30. Dimitrova, R., Neinska,Y., Mihalyi, M. R., Tsoncheva, T. and Spassova, M., "Catalytic activity of boron-beta zeolite modified with indium in the epoxidation of cinnamyl alcohol," *Reaction Kinetics and Catalysis Letters* 74(2), 353–362 (2001).
31. Schmidt, C.; Sowade, T.; Schuetze, F.-W.; Richter, M.; Berndt, H. and Gruenert, W., "A comparison of different preparation methods of indium-modified zeolites as catalysts for the selective reduction of NO_x," *Studies in Surface Science and Catalysis* 135(Zeolites and Mesoporous Materials at the Dawn of the 21st Century), 4973–4980 (2001).
32. Schutze, F.-W.; Berndt, H.; Richter, M.; Lucke, B.; Schmidt, C.; Sowade, T. and Grunert, W., "Investigation of indium loaded zeolites and additionally promoted catalysts for the selective catalytic reduction of NO_x by methane," *Studies in Surface Science and Catalysis*, 135(Zeolites and Mesoporous Materials at the Dawn of the 21st Century), 1517–1524 (2001).
33. Sowade, T.; Schmidt, C.; Schutze, F.-W.; Berndt, H. and Grunert, W., "Relations between structure and catalytic activity of Ce-In-ZSM-5 catalysts for the selective reduction of NO by methane. I. The In-ZSM-5 system," *Journal of Catalysis* 214(1), 100–112 (2003).
34. Mihalyi, M. R. and Beyer, H. K., "Direct evidence for the incorporation of univalent indium into high-silica zeolite, H-ZSM-5, by thermal auto-reductive solid-state ion exchange," *Chemical Communications (Cambridge, United Kingdom)*(21), 2242–2243 (2001).

35. Lazar, K.; Micheaud, N.; Mihalyi, M. R., Pal-Borbely, G. and Beyer, H. K., "Attempts to exchange iron into H-Y and H-ZSM-5 zeolites by *in situ* formed chloride-containing mobile species," *Reaction Kinetics and Catalysis Letters* 74(2), 289–298 (2001).
36. Kucherov, A. V. and Slinkin, A. A., "Zeolite modification by *in situ* formed reactive gas-phase species. Preparation and properties of Mo-containing zeolites," *Studies in Surface Science and Catalysis* 118(Preparation of Catalysts VII), 567–576 (1998).

13 Strong Electrostatic Adsorption of Metals onto Catalyst Supports

John R. Regalbuto

CONTENTS

13.1 INTRODUCTION

Among the simplest, least expensive, and most prevalent methods to prepare supported metal catalysts begins with the process known as impregnation, whereby a high surface area oxide or carbon support is contacted with a liquid solution containing dissolved metal ions or coordination complexes such as platinum hexachloride $[PtCl_6]^{-2}$ (derived from chloroplatinic acid, CPA), or platinum tetraammine $[(NH_3)_4Pt]^{+2}$ (PTA). After impregnation, wet slurries of support and metal precursor are dried and then heated in various oxidizing or reducing environments in order to remove the ligands and to reduce the metal to its active elemental state. This sequence of steps in catalyst synthesis is illustrated in Figure 13.1.

It is very often the goal of the synthesis to create high metal surface area or, in other words, small metal crystallites anchored onto the support. The efficiency of

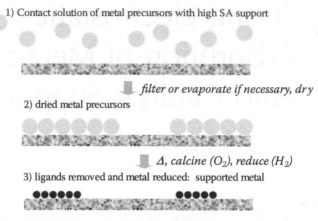

1) Contact solution of metal precursors with high SA support

filter or evaporate if necessary, dry

2) dried metal precursors

Δ, *calcine (O$_2$), reduce (H$_2$)*

3) ligands removed and metal reduced: supported metal

FIGURE 13.1 Stages of catalyst preparation by aqueous impregnation.

metal utilization is commonly defined as *dispersion*, the fraction of metal atoms at the surface of a metal particle (and thus available to interact with adsorbing reaction intermediates) divided by the total number of metal atoms. Metal dispersion and crystallite size are inversely proportional; nanoparticles about 1 nm in diameter or smaller have dispersions of 100%, that is, every metal atom in the catalyst is available for reaction. If high dispersion is desired, an average particle size of 1 nm, or *nanoparticles* in the truest sense of the word, is a fitting benchmark.

Unlike supported metal oxides, which can be prepared with high dispersion in many cases simply by calcining mixtures of bulk oxide and support, it appears that well-dispersed metals are most easily produced from well-dispersed metal precursors. Once deposited in this fashion, an appropriately mild reduction treatment will preserve the high dispersion of the precursor in the reduced metal particles. This chapter will demonstrate that a simple method based on electrostatic adsorption, employing common, relatively cheap precursors, is widely applicable to synthesize highly dispersed metals on simple supports, metals highly dispersed onto a supported promoter, and highly dispersed bimetallic catalysts.

13.2 TYPES OF IMPREGNATIONS

The most common types of impregnations can be classified as in Table 13.1 and will be distinguished from the method referred to here as *strong electrostatic adsorption* (SEA). The first, dry impregnation, is procedurally the simplest. The thick paste formed from contacting the support with just the amount of liquid needed to fill the pore volume contains a precise metal loading and does not need to be filtered. The drawbacks from this method arise when metal precursors do not interact strongly with the support surface. Metal complexes that remain in solution can migrate significantly during drying [1, Chapter 16 of this text]. Chapter 15 of this text describes methods that overcome this limitation.

Impregnations can be considered wet whenever an amount of solution in excess of the support pore volume is employed. Wet impregnation (WI) is as simple as contacting the solution for a certain time and then recovering the solid by filtration.

TABLE 13.1
Types of Catalyst Impregnation

Type of impregnation	Distinguishing characteristics	Advantages/disadvantages
Dry impregnation (DI)*	Use sufficient metal solution to fill pore volume of catalyst support; adjust metal concentration for desired weight loading	A: Simplest to employ; no filtering; metal content is fixed D: Strong precursor-support interaction is not guaranteed
Wet impregnation (WI)	Amount of solution in excess of pore volume of support; precursors that interact weakly with support are washed/filtered away	A: Mixing is improved D: Filtering required; metal loading must be measured; metal wasted if it does not strongly interact with support
Strong electrostatic adsorption (SEA)**	Excess solution; pH held at optimal value for strong precursor interaction with support	A: Strong; monolayer adsorption of metal precursor D: Optimal pH must be determined and achieved during preparation
Deposition-precipitation (DP)	Excess solution; pH slowly and homogeneously increased to precipitate precursor at support surface	A: High metal loadings are most easily achieved D: pH must be altered during preparation
Ion exchange (IE)	Exchange of cationic metal precursors with counterions in zeolite framework	A: Strong interaction of precursor with zeolite is assured D: Limited to cationic precursors
Reactive adsorption	Driving force exists for reaction of precursor with support	A: Strong interaction of precursor with support is assured D: Limited in applicability to certain systems

* Also known as *pore filling* and incipient wetness.
** Also known as *ion adsorption*.

The amount of metal retained by the solid must be determined by analysis of either the solid or liquid. The extent of metal retained by the support is a function of the precursor-support interaction, which may be chemical or physical (electrostatic) in nature. As WI is defined here, no attention is paid to controlling the impregnation conditions (varying pH, for example) to optimize the interaction.

One way strong interactions can be created is via the electrostatic adsorption mechanism illustrated in Figure 13.2. An oxide surface contains terminal hydroxyl groups that protonate or deprotonate, depending on the acidity of the impregnating solution. The pH at which the hydroxyl groups are neutral is termed the *point of zero charge* (PZC). Below this pH, the hydroxyl groups protonate and become positively charged, and the surface can adsorb anionic metal complexes such as CPA. Above the PZC, the hydroxyl groups deprotonate and become negatively charged, and cations such as PTA can be strongly adsorbed.

As depicted in Figure 13.2b, the maximum density of adsorbed CPA corresponds to a close packed monolayer of complexes that retain one hydration sheath [1], which corresponds to a surface density of about 1.6 μmoles/m^2, or 1 complex per nm^2.

a)

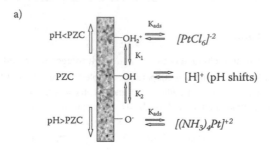

b)

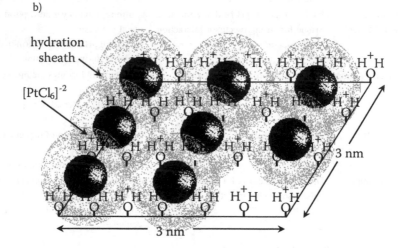

FIGURE 13.2 An electrostatic adsorption mechanism; (a) surface charging, metal adsorption, and proton transfer, (b) monolayer coverage of Pt anions with hydration sheath.

PTA cations, on the other hand, appear to retain two hydration sheaths [1–3] and adsorb at an inherently lower surface density of about 0.86 μmoles/m², or about 1 complex per 2 nm². The historical development of the electrostatic mechanism of adsorption in the field of catalysis is given in a separate review [4].

A typical SEA system is PTA cations adsorbed at high pH over silica, which has a PZC of 4 and charges strongly and negatively at high pH [2]. This preparation has been historically referred to as *ion exchange* (IE) [5–9] based on the overall exchange of protons from the silanol groups at the support surface with Pt cations. A distinction between SEA and IE may be reasonable, however, based on recent fundamental studies of metal adsorption.

Typical results from such a study are shown in Figure 13.3. Here, adsorption of PTA as a function of pH at constant metal concentration is compared over an amorphous silica (Figure 13.3a) and an Na-Mordenite (Figure 13.3b) [10]. The pH dependence of adsorption over silica is highly concentration dependent and is well

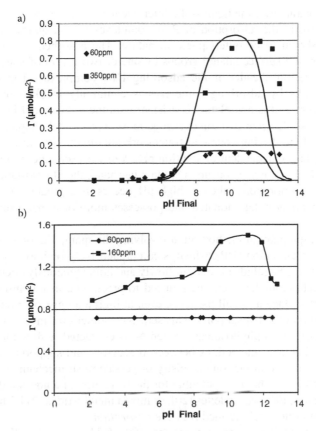

FIGURE 13.3 PTA uptake onto (a) amorphous silica, (b) Na-Mordenite. Adapted from Reference 10.

described by the revised physical adsorption (RPA) model [2, 11, 12], which is based solely on the coulombic interaction of ionic metal complexes and oppositely charged surfaces in the manner of Figure 13.2. Adsorption increases above silica's PZC of 4 as the hydroxyl groups become progressively deprotonated and the negative surface charge builds. At the highest pH, high ionic strength effectively screens the surface charge from the metal complex, and the adsorption equilibrium constant drastically decreases [2, 12]. Over the mordenite, metal uptake is complete and independent of pH as long as the metal concentration is below the exchange capacity of the zeolite. This behavior is most certainly ion exchange of PTA for Na^+ counterions. When the metal concentration is increased above the exchange capacity of the zeolite, it is interesting to note that the shape of the uptake curve bears resemblance to that over silica. In Reference 10, the overexchange of Pt and Cu ammines in low aluminum zeolites was explained by the dual adsorption mechanism of ion exchange at the framework aluminum sites, and electrostatic adsorption over internal silanol groups.

The physical nature of Pt adsorption on silica has been further confirmed by a cutting-edge combination of x-ray reflectivity and resonant anomalous x-ray scattering to characterize the liquid-solid interface in the most direct manner possible [13]. In this work, the adsorption of PTA complexes over a quartz single crystal surface was shown to be outer-sphere in nature. The adsorbed Pt complexes are separated from the surface silanol groups by one or two layers of water, presumed to be the hydration sheaths of the adsorbing PTA complexes. The retention of hydration sheaths by adsorbing complexes is commonly invoked in the colloid science literature and is completely consistent with the RPA model [14].

In sum, it would seem that ion exchange most appropriately applies to a microscopic electrostatic driving course at framework aluminum sites in zeolites, which is very strong and independent of solution pH. A direct exchange of a metal cation and an H$^+$ or Na$^+$ counterion occurs. On the other hand, the electrostatic interaction of cations with amorphous silica at high pH occurs at a greater distance over a surface that is already deprotonated and possesses more of a macroscopic surface charge.

Another impregnation method often involving the manipulation of solution pH is deposition-precipitation (DP), which is comprehensively reviewed in Chapter 14 of this book. The general idea is that by slowly and homogeneously altering solution conditions, metal complexes can be induced to precipitate in a controlled fashion on the support surface as small particles, even at high loadings over oxide [15, 16] as well as carbon [17, 18] supports. This sequence sometimes involves reaction, as in the formation of Ni phyllosilicates when Ni is contacted with amorphous silica [19]. In fact, chemical interactions between precursors and supports can be understood in terms of a coordination chemistry or geochemical mechanism, the driving force for which is the chemical potential for the formation of a new solid phase such as phyllosilicates [19], hydrotalcites [20], or mixed metal oxides [21]. These impregnations can be collectively termed *reactive adsorption*.

From this point forward, the components of the SEA approach will be discussed. These include a consideration of proton transfer between the liquid solution and the support surface, the determination of the PZC of a particular material, the pH dependence of metal adsorption, the correlation of SEA to high metal dispersion, and the extension of the SEA method to mixed oxide surfaces for the synthesis of promoted and bimetallic catalysts.

13.3 OXIDE SURFACE CHARGING AND pH BUFFERING

To achieve strong electrostatic adsorption over a catalyst support, attention must be paid to achieving the optimal (equilibrium) value of pH. This is not so straightforward, due to the coupling of the protonation-deprotonation chemistry of the surface hydroxyl groups with the bulk solution pH (see Figure 13.2).

The dramatic effect of catalyst support surfaces on solution pH was first demonstrated in the method called *mass titration* [22], a depiction of which is shown in Figure 13.4a. Starting at a pH value above the oxide's PZC, the pH lowers as successive pinches of oxide are placed into solution and protons are released by the oxide. If a sufficient amount of oxide is placed into solution, the pH arrives at the PZC of the oxide, at which point no driving force for proton transfer exists.

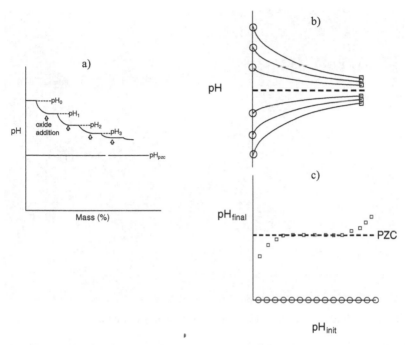

FIGURE 13.4 Proton transfer and pH shifts; (a) the mass titration demonstration of Reference 22, (b, c) the transformation of mass titration to EpHL.

At this condition, the surface is essentially uncharged, as the number of OH groups on the oxide surface far outnumbers the quantity of OH originally present in solution.

The shifts in bulk pH that occur when aqueous solutions are contacted by various amounts of oxides have been quantified [23]. A critical parameter in these systems is the oxide surface area per volume of solution. This parameter, with units of m^2/liter, was termed the *surface loading* (SL) and is illustrated in Figure 13.5 for an oxide support with specific surface area of 105 m^2/g. Low surface loadings give thin slurries; high surface loadings give thick slurries. For any particular oxide support, impregnation to incipient wetness (or dry impregnation) in which the pore volume of the support is just filled with aqueous solution represents the highest tenable value of surface loading and is typically on the order of several hundreds of thousands of m^2/l. When comparing oxides with different surface areas, the mass of oxide can be adjusted so as to achieve the same SL.

The quantitative pH shift model [23] combined (1) a proton balance between the surface and bulk liquid with (2) the protonation-deprotonation chemistry of the oxide surface (single amphoteric site), and (3) a surface charge-surface potential relationship assumed for an electric double layer. Given the mass and surface area of oxide, the oxide's PZC, its protonation-deprotonation constants K_1 and K_2 (Figure 13.2), and the hydroxyl density, these three equations are solved simultaneously and give the surface charge, surface potential, and final solution pH. The mass titration experiment of Figure 13.4 can be quantitatively simulated, but perhaps the most powerful simulation is a comprehensive prediction of final pH versus initial pH, as a function of

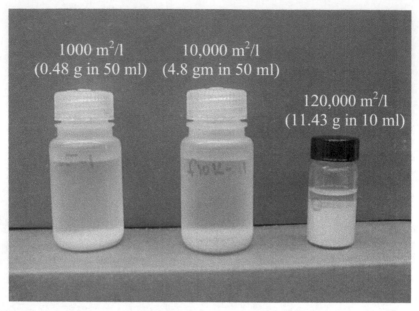

FIGURE 13.5 Bottles containing different surface loadings of alumina.

oxide surface loading. This relationship, for parameters representative of alumina [24], is shown in Figure 13.6a and is comprised of a number of key features.

First, the effect of surface loading is immediately apparent. Low surface loadings can be employed in the laboratory to minimize pH shifts. At high surface loadings, oxides exhibit a dramatic effect on pH. In fact, this plot predicts that the final pH of dry impregnations is almost always at the PZC of the oxide, unless the starting solutions are extremely acidic or basic [23]. Put another way, in dry impregnation, the hydroxyl groups on the oxide surface far outnumber the protons or hydroxide ions initially in solution, and the surface never becomes significantly charged. Strong electrostatic interactions with ionic metal complexes can only occur when this buffering effect of the oxide is overcome.

The second main feature of the pH shift plot of Figure 13.6a is the wide plateau of final pH seen at the higher surface loadings. That is, starting from a wide range of initial pH, the final pH is always the same and is, in fact, the oxide PZC. This suggests that oxide PZC can be measured simply with a pH probe, by measuring the final pH of a series of oxide-solution slurries at high surface loading [23]. This method was called *EpHL*, the measurement of equilibrium pH at high loading. The relationship of EpHL to mass titration is seen in Figure 13.4b and Figure 13.4c. If multiple initial pH values are employed with mass titration, and high mass loadings are used, the final pH values that result (squares in Figure 13.4b) corresponding to the initial pH (circles) will cluster about the PZC. This is clearly seen when final pH is plotted versus initial pH, as in Figure 13.4c. The width of the pH plateau is a function of oxide surface loading; the higher the SL, the wider the plateau. For common high surface area supports, a wide plateau can be achieved in slurries that are thin enough (for example, in Figure 13.5 at 10,000 or 120,000 m²/l)

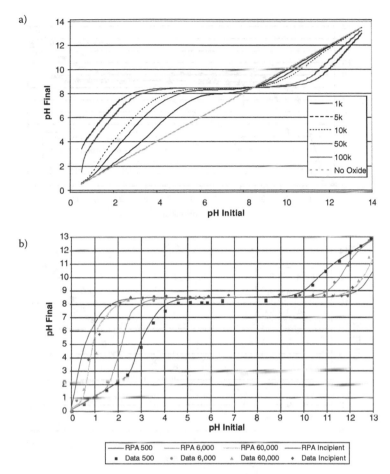

FIGURE 13.6 pH shifts over alumina; (a) theory, with SL as parameter, (b) experimental, fit with RPA model.

that the pH can be measured with a standard pH electrode. This represents a simple, accurate way to measure support PZC [23]. Related PZC measurements have been developed based on mass titration [24, 25].

In a recent treatment of surface charging, pH shift data generated at different surface loadings are fit to the model so as to obtain the best values of K_1 and K_2 [26]. Representative experimental and model results are shown for alumina in Figure 13.6b. Having obtained the oxide charging parameters in the absence of metal adsorption, the parameters can be used with no adjustment in the RPA model to simulate metal uptake, which is described in the next section.

Before proceeding to that section, some final comments can be made on the utility of final pH-initial pH plots. In Figure 13.7a, this plot is compared for the same surface loading (60,000 m²/l) of γ-alumina and XC-72 carbon black. Both materials have similar PZCs. All unoxidized carbons typically have PZCs in this range [27, 28].

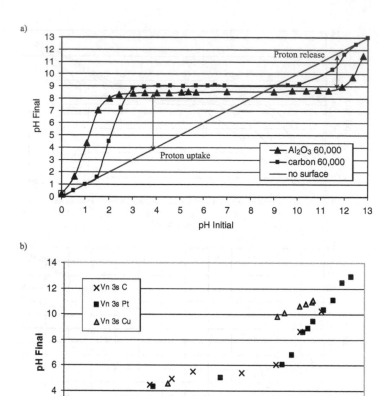

FIGURE 13.7 Uses of pH shift data; (a) a comparison of proton transfer over alumina and carbon, (b) a comparison of Pt and Cu adsorption.

First of all, it is noticed that the greater breadth of the pH plateau for alumina at the same SL implies a higher density of proton-exchanging sites. The nature of those sites is also hinted at in the plots. The amphoteric nature of the OH groups on the alumina surface can be easily surmised: The deviation of the curve from the diagonal below the PZC signifies proton uptake by the surface, and the deviation above the PZC connotes proton release by the surface. The proton equilibrium of the carbon behaves much more in a single direction: adsorbing protons at pH below its PZC, but releasing little above it. This is consistent with literature citing the protonation of the pi bonds of aromatic rings at carbon surfaces [29]. This implies that unoxidized carbons with high PZCs should be able to adsorb anions but not cations, which is addressed in the next section.

A second final pH-initial pH plot, in Figure 13.7b, compares the pH shifts over silica of metal free solutions and those containing Pt and Cu ammines [10]. In the former case, the metal-containing pH shifts are identical to the metal free control experiment, which can be interpreted by the stability of the PTA complex and the independence of adsorption and proton transfer. The pH shifts during copper ammine adsorption are significantly different from the control experiment and are thought to arise as the copper ammine complex hydrolyzes to the bridged $Cu_2(OH)_2$ dimer at the lower pH near the silica surface [10].

13.4 METAL ADSORPTION SURVEYS

The overarching hypothesis of the SEA approach is that a correlation exists between strong adsorption of the metal precursor and high dispersion of the reduced metal. The steps of the SEA approach for any particular metal/support system are then (a) the measurement of support PZC (which determines which charge of metal ion and which pH range to employ), (b) uptake-pH surveys to determine the pH of strongest interaction, and (c) tuning the reduction treatment to preserve high dispersion. These steps are summarized in Figure 13.8 for the case of Pt/silica.

13.4.1 CATION ADSORPTION OVER SILICA

The first step in the SEA approach is to determine the PZC of the support. Low PZC materials accrue a strong negative charge at high pH and can strongly adsorb cations, whereas high PZC materials charge positively at low pH and strongly adsorb anions. Mid-PZC materials might strongly adsorb either in the respective pH range.

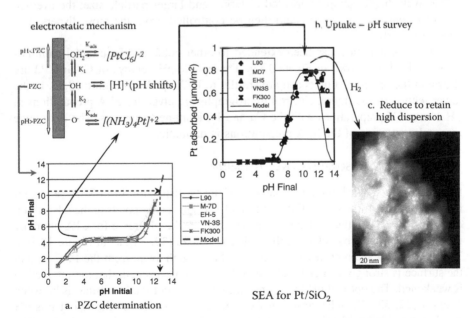

FIGURE 13.8 Pt/silica catalyst synthesis by strong electrostatic adsorption (SEA); (a) electrostatic mechanism, (b) PTA uptake-pH survey to locate optimal pH over silica, (c) reduction of sample prepared at optimum conditions yields 1 nm metal particles.

PZC is easily determined as mentioned in the section above by measuring final pH versus initial pH at high surface loading. The pH shifts for a number of silicas are shown in Figure 13.8a [2] and are all about 4. Correspondingly, a cationic metal complex, PTA, will be chosen as the catalyst precursor, and the pH region above pH 4 will be surveyed to determine the pH of strongest interaction.

Uptake-pH surveys were conducted over a series of five silicas [2]. Differences in the surface areas were corrected by using different masses, such that all experiments employed the same surface loading of silica. Because all have nearly the same PZC and the surface loading was identical, the uptake curves for all were very similar and could be modeled with a single curve [2]. The optimal pH is seen to be about 10.5, at which point silica dissolution is not too severe [2]. This final pH can be attained in practice by adjusting the pH after the solution is contacted with the support, or by anticipating the pH shift that will occur due to oxide buffering. The latter method works, provided that metal adsorption is independent of proton transfer, as indicated earlier for the PTA/silica system in the discussion of Figure 13.7b. The correct initial pH is made by consulting the final pH-initial pH plot; in Figure 13.8a, the final pH of 10.5 is traced back to an initial pH of 12.5. The discrepancy between the initial and final pH values is a function of the SL employed; the higher the SL, the more the pH will shift.

Adsorbing at this optimal pH and reducing directly at 200°C (with no calcination) led to 1 nm Pt particles, as seen in the Z contrast electron micrograph of Figure 13.8, which was confirmed by CO chemisorption and EXAFS [30]. Calcination at successively higher temperatures led to larger and larger particle size; the average size of Pt particles on silica could then be controlled, from the most dispersed to very large particles.

The SEA approach has been extended to other noble and base metal ammines on silica [31], as shown in Figure 13.9. The uptake-pH surveys of Cu and Pd are shown in the uppermost plots and are similar to PTA uptake of Figure 13.8b. In the lower section of the figure, electron micrographs are given for SEA preparations at pH 11 and DI preparations with the Cu and Pd ammine precursors. The difference in metal dispersion of the SEA preparations is dramatic.

13.4.2 ANION ADSORPTION OVER ALUMINA

In complementary fashion, a comparison of the Pt dispersion from SEA versus DI has also been performed with CPA/alumina [32]. This corresponds to the adsorption of anions over a positively charged surface. The uptake-pH curve for CPA on many types of alumina is characteristically volcano shaped [32, 33], as in Figure 13.10a. The optimal pH again corresponds to a pH far enough away from the PZC so that the surface is strongly charged, but not so high in ionic strength that the interaction is weakened. The optimal pH for strongest Pt adsorption for all aluminas is between 3 and 4 [32, 33]. Particles resulting from an SEA synthesis performed at this pH are shown in Figure 13.10b, and a comparison of CO chemisorption results for SEA and DI preparations at various surface loadings is shown in Figure 13.10c [32]. In every case, the SEA preparation yields a higher dispersion than DI and gives values near 100% even at relatively high Pt loadings. The DI preparations were not so poor,

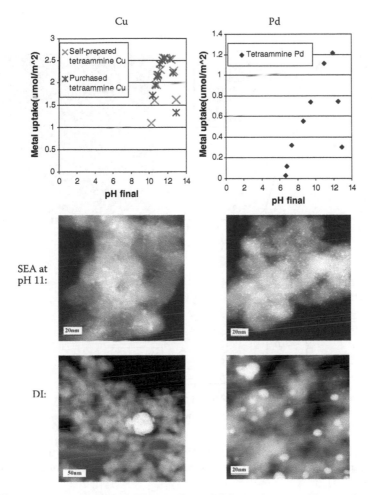

FIGURE 13.9 The extension of SEA to Cu and Pd ammines on silica, and an electron microscopy comparison of SEA to DI for each.

as the acidity of the chloroplatinic acid (H_2PtCl_6) is sufficient to substantially charge the alumina surface [32].

13.4.3 ANION AND CATION ADSORPTION OVER CARBON

The PZC of carbons can be irreversibly changed by oxidizing the surface; in a series of papers, it has been demonstrated that Pt anion and cation uptake can be manipulated on this basis [27, 28, 34, 35]. Working with carbons of unknown surface oxidation (and PZC), the complete SEA approach illustrated in Figure 13.8 can be employed. First, the PZC is determined. Second, uptake-pH surveys are conducted to determine the pH of strongest interaction, and third, a sample synthesized at the SEA condition is reduced in a way that retains high dispersion.

In the upper set of uptake versus pH curves in Figure 13.11, a set of three activated carbons (Norit SX2, SX4, and SXU) and three carbon blacks (Ensaco 250,

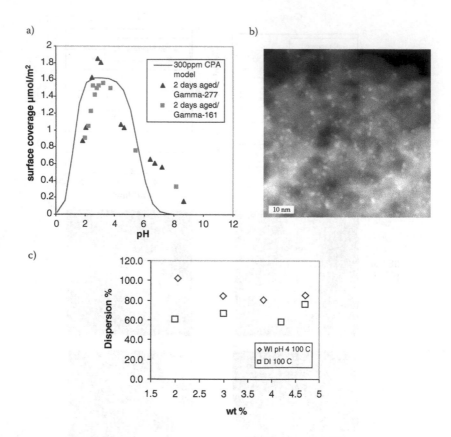

FIGURE 13.10 The extension of SEA to Pt(IV) chloride on alumina; (a) uptake-pH survey, (b) micrograph of nanoparticles, (c) a CO chemisorption comparison of SEA versus DI dispersions.

Ensaco 350, and Vulcan XC-72) with PZCs near 9 was surveyed for CPA adsorption (left-hand side), whereas a set of two graphitic (Asbury and TImrex) and two activated carbons (Norit S51 and Darco KB-B) with PZCs near 4 was monitored for PTA uptake (right-hand side) [28]. The different surface area of the carbons was normalized by utilizing different masses of each to achieve the same SL for each carbon in the low and high pH ranges. When the differences in the surface area are properly accounted for, the uptake of CPA at low pH follows the same trend for all carbons. A sharp maximum in uptake is seen at pH 2.9. The uptake volcano is much narrower than over alumina (Figure 13.10a), which is consistent with the lower density of protonated sites over the carbon support seen in Figure 13.7a.

In the high pH range (Figure 13.10, right side), monolayer adsorption is nearly attained for the two graphites, which possess relatively low surface area and large pore size. Over the two high surface area carbons (CA and KB), however, uptake is about half that predicted by the RPA model. The discrepancy was explained [28]

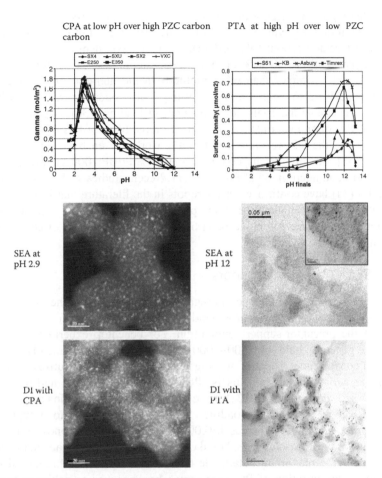

FIGURE 13.11 The extension of SEA to Pt chlorides and ammines on carbon, and an electron microscopy comparison of SEA to DI for each.

by steric exclusion of the large Pt ammine complexes, believed to retain two hydration sheaths [5, 13] from the smallest micropores of the high surface area activated carbon.

The consequences of SEA and DI preparations of Pt anions and cations over carbon are shown in the lower portion of Figure 13.11. On the right-hand side, a loading of 9 wt% Pt PTA was applied to high surface area graphite (TIMREX) at pH 12 [28]. For the sake of comparison, dry impregnation was conducted with a neutral pH solution of PTA (and a final pH near the PZC of the carbon, as usually occurs in DI). Both samples were directly reduced at 200°C following drying at 100°C. The DI preparation results in large Pt particles (lower right-hand micrograph). On the other hand, using SEA yields predominantly 1–1.5 nm particles almost as well dispersed as the Pt/SiO$_2$ sample of Figure 13.4c.

On the left-hand side, micrographs are shown for SEA conducted over an unoxidized BP2000 carbon black, which exhibits CPA uptake completely consistent with the other high PZC carbons of Figure 13.11 [35]. The adsorbed CPA precursors were reduced in flowing hydrogen at 200°C. One advantage of utilizing this support is its high surface area (about 1500 m²/gm); at a monolayer density of 1.6 μmol/m², it was possible to adsorb 30 wt% Pt in a single, simple adsorption step [35]. This procedure is much simpler than other methods in the literature for synthesizing Pt/carbon fuel cell electrocatalysts. The resulting Pt dispersion is as good as or better than the best available materials, for equivalent Pt loadings [36]. A patent for the synthesis of Pt/carbon materials has been filed on the basis of this method [37].

Surprisingly, the DI preparation with CPA worked even better than the SEA preparation [38]. The high Pt dispersion of a 30 wt% Pt/BP 2000 sample is seen in the lower left-hand micrograph of Figure 13.11. Recent studies with *in situ* XANES and EXAFS [34] have confirmed earlier repots in the literature that Pt(+4) in CPA is reduced by carbon when applied at low pH [39]. This is a special case of reactive adsorption. A provisional patent has been filed pertaining to the use of DI with CPA on carbon [38].

13.4.4 SEA IN WI AND DI MODES

A final note can be made in this section regarding the generality of the SEA method. *Strong electrostatic adsorption* corresponds to the pH value which maximizes the strength of the precursor-support interaction. For benchtop experiments, surface loadings are generally held low (500–2000 m²/l) in order to minimize pH shifts for the sake of convenience. There is nothing inherently incorrect in utilizing SEA at much higher surface loadings. This is illustrated in the pH shift and adsorption calculations shown in Figure 13.12, for the uptake of CPA by alumina as a function of surface loading. The lowest loading, 500 m²/l, corresponds to a typical SL employed in the laboratory, whereas 150,000 m²/l roughly corresponds to the SL at DI conditions of typical γ-aluminas. Uptake curves at all SLs show the characteristic volcano shape. The maximum obtainable uptake decreases at the highest SLs due to the higher concentration of Pt, which must be employed to achieve the same surface density. The higher ionic strength of the CPA solutions suppresses the adsorption equilibrium constant of the Pt complex, which is not quite compensated by the higher Pt concentration. For a 200 m²/gm alumina, a surface density of 1 μmol/m² corresponds to a Pt loading of about 4 wt%, which is much higher than

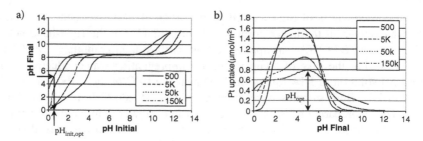

FIGURE 13.12 Simulation of SEA at high surface loadings; (a) pH shift simulation, (b) uptake-pH simulation.

that normally utilized in industrial catalysts apart from fuel cell electrocatalysts. In theory, then, it would be possible to employ dry impregnation with a solution acidified to the correct value so as to achieve SEA. From Figure 13.12a, the correct initial pH for a dry impregnation at 150,000 m^2/l would be about 0.5. This mode might be thought of as *charge enhanced dry impregnation* (CEDI).

13.5 EXTENSION OF THE SEA APPROACH TO PROMOTED AND BIMETALLIC CATALYSTS

The electrostatic control of metal complex adsorption might also be achieved at the nanoscale over surfaces containing two oxides: The SEA method can be extended to provide a simple, scientific method to prepare a wide range of bimetallic catalysts and promoted catalysts. The idea is illustrated in Figure 13.13 in the simulation of surface potential versus pH for a surface consisting of an oxidized carbon support, with a PZC of 4, which supports particles of cobalt oxide, which has a PZC of about 9. At a pH of 6, the cobalt oxide phase will be protonated and positively charged, whereas the carbon surface will be deprotonated and negatively charged. Hexachloroplatinate anions should then be adsorbed selectively onto the cobalt oxide particles. Subsequent reduction in H_2 will be used to form the bimetallic PtCo particles. If the composite surface was instead low PZC niobia (as supported promoter) and high PZC alumina, PTA could be used for selective adsorption onto the niobia phase.

13.5.1 EXTENSION OF SEA TO PROMOTED CATALYSTS

Though proposed years ago [39, 40], the use of electrostatic interactions to direct the adsorption of a metal onto one of two oxides has not been successfully executed to date. In an attempt to deposit Co^{+2} onto negatively charged WO_3 (low PZC oxide)

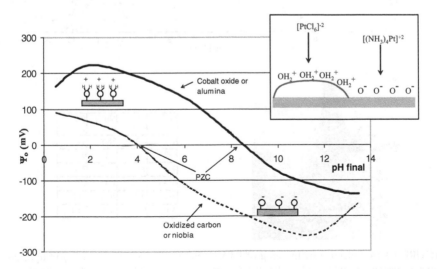

FIGURE 13.13 Simulation of surface potential of mixed oxide (or carbon) surface with different PZCs.

and away from positively charged alumina at pH 6, more Co deposited on the alumina than the tungsta [39]. Co particle size on either phase was seen by TEM to be 60 to 80 nm; clearly no strong interaction occurred, probably due to an excess of Co and perhaps precipitation of the Co species. Using Pd^{+2} instead of Co^{+2}, even larger, undirected Pd clusters appeared [40]. Precipitation is again the likely cause.

A clear demonstration of selective partitioning of PTA over niobia/alumina has now been demonstrated by a comparison of SEA studies of the individual oxides to a bulk physical mixture, and STEM characterization of the impregnated mixture [41]. In Figure 13.14a, experimental adsorption data is given for Pt on pure niobia (diamonds), pure alumina (triangles), and a physical mixture of alumina and niobia (squares). With a low PZC of 2.5, niobia adsorbs cationic PTA over a very wide pH range, whereas the same area of alumina, with a PZC of 9, adsorbs very little and over a very narrow basic pH range. The uptake on the individual oxides can be readily simulated using the RPA model [41] and, with no adjustment of parameters, can be extended to a model of the mixture of oxides. The result of this simulation predicts that PTA is almost always adsorbed onto niobia, and the simulation agrees reasonably well with the experimental adsorption data.

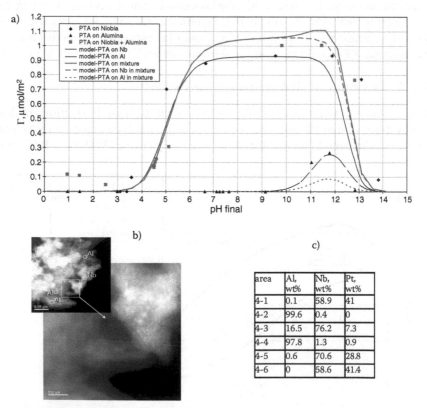

FIGURE 13.14 Selective partitioning of PTA in niobia-alumina mixtures; (a) Pt uptake-pH surveys over single and mixed oxides, (b) Z-contrast imaging of Pt adsorbed over the alumina-niobia mixture, c) elemental analysis of various regions in (b).

The partitioning of PTA onto the niobia fraction of the physical mixture impregnated at a pH of 7 has been born out by STEM imaging and energy dispersive x-ray spectroscopy (EDXS). Alumina is seen in the lower left-hand fraction of the zoomed-in STEM image (Figure 13.6b); niobia covered by Pt is seen in the upper right-hand portion of the figure. Analyzing dozens of images and thousands of Pt particles [41], Z-contrast imaging and EDXS (Figure 13.6c) reveal the complete absence of Pt on alumina, and the ubiquitous presence of Pt on niobia. The average size of reduced Pt particles is about 1.1 nm, close to 100% dispersion.

13.5.2 EXTENSION OF SEA TO BIMETALLIC CATALYSTS

Although there are some groups producing well-defined model bimetallic catalysts with precursors such as organometallic clusters or dendrimers (see chapters 9 and 10), it is more typical that high surface area materials made from common precursors are prepared by sequential or simultaneous impregnation or coprecipitation, and so the creation of intimate contact between the two metals is largely left to chance. A notable exception is the ingenious method of surface redox preparations [42–44], in which a reduced metal surface that is bare in some cases, or covered with adsorbed hydrogen in others, has sufficient chemical potential to reduce an aqueous-phase cation of the second metal. The redox reaction occurs only at the surface of the initially deposited metal, creating bimetallic particles. One disadvantage of this strategy is that preparations are limited to a pH range in which metals exist as bare ions such as Pd^{+2}, Cu^{+2}, and Co^{+2}. It was indicated in the previous section that these metal species readily hydrolyze and precipitate over a large range of pH. A more practical disadvantage of this method is that the bare or H-covered surface of the first metal must be maintained in that state as it is placed back into aqueous solution. This means that the solution must be sparged with nitrogen to purge away dissolved oxygen, and that the entire process must be done in a glovebox. There may certainly be cases where this painstaking, multistep process pays dividends in the quality of the bimetallic catalyst produced.

The SEA method can be extended to the synthesis of bimetallics and may represent a simpler, more versatile alternative to surface redox reactions. The syntheses of bimetallics are the same as described in the previous section, only that the adsorbing supported oxide, like the CO_3O_4 depicted in Figure 13.13, is itself reducible and, after reduction, forms a bimetallic particle in intimate contact with the second metal precursor that had adsorbed directly onto it. This process can be conducted at ambient conditions, with an intermediate calcination in air to create the first metal oxide from a deposited or adsorbed precursor. The first metal might itself be deposited by SEA in well-dispersed form by precursors such as cationic cobalt hexa-ammine on silica. Thus, there is the potential to create homogeneous bimetallic particles with very high dispersion, using simple methods with common metal precursors.

Preliminary work has been conducted with the Pt/Co/silica and Pt/Co/carbon systems [45]. In the former case, CPA was impregnated at pH 3 onto Co_3O_4 (PZC 9) and at the same pH onto a physical mixture of Co_3O_4 and silica of equivalent exposed areas. TPR experiments comparing CPA on silica, pure Co_3O_4, CPA-impregnated

Co_3O_4, and a CPA-impregnated physical mixture of Co_3O_4 and silica are shown in Figure 13.15a. The results show that the reduction of silica-supported CPA begins at 50°C and is complete by 200°C. Reduction of pure Co_3O_4 begins at 200°C and ends at 360°C; the reduction profile is consistent with stepwise reduction of Co^{+3} to Co^{+2}, followed by Co^{+2} to metal. The reduction of the CPA-impregnated Co_3O_4 occurs at almost 100° lower than the Pt-free cobalt oxide. Finally, the TPR pattern of the CPA-impregnated physical mixture of silica and cobalt oxide is very similar to that of the CPA-impregnated Co_3O_4 pattern; this provides indirect evidence that the CPA adsorbed selectively onto the Co_3O_4.

Direct evidence of the interaction of Pt with the cobalt oxide phase in the physical mixture has been obtained with EXAFS [45]. In the Fourier-transformed spectra of Figure 13.15b, Pt foil is given for reference and is fit by the typical first shell bond distance of 2.77 Å. The EXAFS spectrum of the reduced CPA/(Co_3O_4+SiO_2) sample

a)

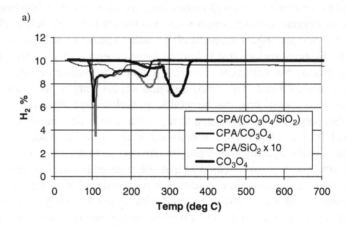

b)

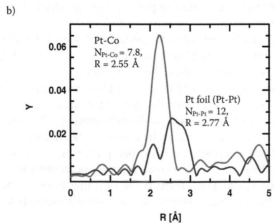

FIGURE 13.15 Analysis of Pt/Co bimetallics; (a) TPR patterns of CPA/SiO_2, Co_3O_4, CPA/Co_3O_4, and CPA/(Co_3O_4 + SiO_2), (b) Fourier-transformed EXAFS spectra of Pt foil and Pt/(Co_3O_4 + SiO_2).

shows no Pt-Pt bonds, but only Pt-Co bonds at 2.55 Å, with eight nearest neighbors of Co. Virtually all the Pt in this sample is alloyed.

As well, EXAFS has been used to confirm that several series of carbon-supported PtCo catalysts prepared by SEA are completely alloyed. The development of these alloys for fuel cell electrocatalysts, along with carbon supported Pt/Mo and Pt/Ru, is presently under way using SEA.

13.6 SUMMARY

A simple electrostatic model of metal adsorption onto oxide surfaces has great utility for the preparation of typical supported metal catalysts. Owing to the chemistry of the surface hydroxide groups, the oxide surfaces become protonated and positively charged at pH values below their PZC, and capable of strongly adsorbing metal anions. Above the oxide PZC, the surface is deprotonated and negatively charged and capable of strongly adsorbing cations. In either case, however, the oxide buffering effect must be overcome. This is accomplished in practice by measuring and controlling the final pH of the slurry solution. It is hypothesized that once the metal precursor has been strongly adsorbed, it can be reduced to its active elemental state at conditions that retain its high dispersion.

The strong electrostatic adsorption approach can be applied to a novel system in three steps: (1) measure the PZC of the oxide (or carbon), and choose a metal cation for low PZC materials and an anion for high PZC materials, (2) perform an uptake-pH survey to determine the pH of strongest interaction in the appropriate pH regime (high pH for low PZC and vice versa), and (3) tune the calcination/reduction steps to maintain high dispersion. Highly dispersed Pt materials have been prepared in this way over silica, alumina, and carbon. Other oxides can be employed similarly. Bimetallics should be effectively synthesized by adsorbing a second metal complex selectively onto a precursor oxide phase of the first metal and then reducing the intimately contacted metals. Over promoter/support surfaces, pH can be used to achieve selective adsorption of the metal complex onto the promoter and not the support.

REFERENCES

1. Santhanam, N., Conforti, T.A., Spieker, W.A. and Regalbuto, J.R., Catal. Tod. 21, 141, 1994.
2. Schreier, M., and Regalbuto, J.R., J. Catal. 225, 190, 2004.
3. Spieker, W., Regalbuto, J., Rende, D., Bricker, M., and Chen, Q., Stud. Surf. Sci. Catal. 130, 203, 2000.
4. Regalbuto, J.R., A Scientific Method to Prepare Supported Metal Catalysts, in Richards, R. (Ed.), *Surface and Nanomolecular Catalysis*, Chapter 6, Boca Raton, FL: Taylor and Francis/CRC Press, in press.
5. Benesi, H.A., et al., J. Catal. 10, 328, 1968.
6. Gonzalez, R.D., and Zou, W., Catal. Lett. 15, 443, 1992.

7. Bond, G.C., and Wells, P.B., Appl. Catal. 18, 225, 1985.
8. Arai, M., Guo, S.L., and Niahiyama, Y., Appl. Catal. 77, 141, 1991.
9. Goguet, A., et al., J. Catal. 209, 135, 2002.
10. Schreier, M., et al., Nanotech. 16, S582-S591, 2005.
11. Hao, X., Spieker, W.A., and Regalbuto, J.R., J. Coll. Interf. Sci. 267, 259, 2003.
12. Spieker, W.A., and Regalbuto, J.R., Chem. Eng. Sci. 56, 2365, 2000.
13. Park, C., Fenter, P., Sturchio, N., and Regalbuto, J.R., Phys. Rev. Lett. 94, 076104, 2005.
14. Agashe, K. and Regalbuto, J.R., J. Coll. Interf. Sci. 185, 174, 1997.
15. Moreau, F., Bond, G.C., and Taylor, A.O., J. Catal. 231, 105, 2005.
16. Zanella, R., Delannoy, L., and Louis, C., Appl. Catal. A., in press, 2005.
17. Bitter, J.H., et al., Catal. Lett. 89, 139, 2003.
18. Toebes, M.L., et al., J. Phys. Chem. B 108, 11611, 2004.
19. Burattin, P., Che, M., and Louis, C., J. Phys. Chem. B 102, 6171, 1998.
20. Paulhiac, P., and Clause, O., J. Am. Chem. Soc. 117, 11471, 1995.
21. Lambert, J.F., J. Phys. Chem.
22. Noh, J.S., and Schwarz, J.A., J. Coll. Interf. Sci. 130, 157, 1989.
23. Park, J. and Regalbuto, J.R., J. of Coll. Interf. Sci., 175, 239, 1995.
24. Bournikas, K.; Vakros, J.; Kordulis, C.; and Lycourghiotis, A. Potentiometric Mass Titration: Experimental and Theoritical Establishment of a New Technique for Determining the point of a New Technique for Determining the point of Zero Charge (PZC) of metal (Hydr)oxides. *Journal of Physical Chemistry B*, 2003.
25. Bournikas, K., and Lycourghiotis, A., J. Coll. Interf. Sci. 28x, 4100, 2005.
26. Schreier, M., Timmons, M., Feltes, T., and Regalbuto, J.R., manuscript in preparation.
27. Hao, X., Quach, L., Korah, J., and Regalbuto, J.R., J. Molec. Catal. 219, 97, 2004.
28. Hao, X., and Regalbuto, J.R., manuscript in preparation.
29. van Dam and van Bekkum.
30. Miller, J.T., Schreier, M., Kropf, A.J., and Regalbuto, J.R., J. Catal. 225, 203, 2004.
31. Jiao, L., and Regalbuto, J.R., manuscript in preparation.
32. Liu, J., and Regalbuto, J.R., manuscript in preparation.
33. Regalbuto, J.R., Navada, A., Shadid, S., Bricker, M.L., and Chen, Q., J. Catal. 184, 335, 1999.
34. Hao, X., Miller, J.T., Kropf, A.J., and Regalbuto, J.R., manuscript in preparation.
35. Castorano, M., Robles, J., and Regalbuto, J.R., manuscript in preparation.
36. Gasteiger, H.A., Kocha, S.S., Sompalli, B., and Wagner, F.T., Appl. Catal. B: Env. 56, 9, 2005.
37. Hao, X., and Regalbuto, J.R., U.S. Patent 03/28586, filed September 9, 2003.
38. Regalbuto, J.R., U.S. provisional patent filed November 11, 2005.
39. Zhang, R., Schwarz, J.A., Datye, A., and Baltrus, J.P., J. Catal. 135, 200, 1992.
40. Schwarz, J.A., Ugbor, C.T., and Zhang, R., J. Catal. 138, 200, 1992.
41. Zha, Y., and Regalbuto, J.R., manuscript in preparation.
42. Montasseir, J.C., et al., J. Molec. Catal. 70, 65, 1991.
43. Melendrez, R., et al., J. Molec. Catal. A., Chem. 157, 143, 2000.
44. Kerkeni, S., Lamy-Pitara, E., and Barbier, J., Catal. Today 75, 35, 2002.
45. D'Souza, L., and Regalbuto, J.R., manuscript in preparation.

14 Deposition-Precipitation Synthesis of Supported Metal Catalysts

Catherine Louis

CONTENTS

14.1 INTRODUCTION

The method of deposition-precipitation has been known for a long time in the industry (Table 14.1), but it has been more recently extensively studied from a

TABLE 14.1
History of Deposition-Precipitation, from Reference 10

Year	Assignee/author	Description	Reference
1943	IG Farben	DP of (hydr)oxides, sulfides, selenides	11
1967	Stamicarbon-JWG	DP with urea; powder supports	12
1970	Unilever	DP with urea from concentrated solutions	13
1970	Stamicarbon-JWG	DP with reduction reaction; powder supports	14
1977	Geus and colleagues	Basic studies on DP, thermodynamic mechanism; powder supports	1, 2
1988	Shell-de Jong	DP on preshaped supports	15
1998	Louis and colleagues	Basic studies on DP, molecular mechanism; powder supports	4–7

fundamental point of view to understand the underlying chemical phenomena, mainly by Geus and colleagues (1–3), then by Burattin and colleagues (4–7). Several reviews were published on this preparation method (8, 9). The purpose of this new review is to summarize what has been published since the publication of the last one in 1997 (9). We will see that sonication (Section 14.5) and evaporation of ammine (Section 14.6) can be considered as other alternatives to the deposition-precipitation. In contrast, deposition-precipitation at a fixed pH (Section 14.3.b) can be considered as a distortion of the pristine principle of deposition-precipitation.

14.2 PRINCIPLE OF DEPOSITION-PRECIPITATION

The method of deposition-precipitation is a modification of the precipitation methods in solution. It consists of the conversion of a highly soluble metal precursor into another substance of lower solubility, which specifically precipitates onto a support and not in solution. The conversion into the low soluble compound (above the solubility or saturation curve S in Figure 14.1), and then into the precipitate (above the supersolubility or supersaturation curve SS in Figure 14.1), is usually achieved by raising the pH of the solution. This can also be done by decreasing the pH, or by changing the valence state of the metal precursor through electrochemical reactions (16) or by using a reducing agent (17), or by changing the concentration of a complexing agent. The principle of these alternative methods can be found in the Geus reviews (8, 9). Except for the two procedures referenced above, the others belong to patents that can be found in References 8 and 9.

To perform the precipitation exclusively on the surface of the support, two conditions must be fulfilled:

1. Interaction between the soluble metal precursor and the surface of the support is required. In such a case, the supersolubility curve SS in Figure 14.1 is shifted towards lower concentrations in the presence of the support (curve

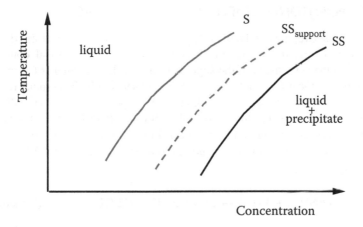

FIGURE 14.1 Schematic phase diagram for a precipitate in equilibrium with its solution and in the presence of the solid support; (S) solubility curve; ($SS_{support}$) supersolubility curve in the presence of the support (precipitation on the support); (SS) supersolubility curve in liquid (precipitation in liquid).

$SS_{support}$ in Figure 14.1). As a consequence, when the concentration of the precursor increases, this supersolubility curve is encountered before the other.

2. The concentration of the precursor must be maintained between the concentrations of the solubility (curve S) and the supersolubility curves (curve SS) to avoid the precipitation in solution (Figure 14.1).

The key point for a successful deposition-precipitation is, therefore, the gradual addition of the precipitating agent to avoid local rise of concentration above the supersolubility curve (curve SS in Figure 14.1), which would cause a rapid nucleation of the precipitate in solution. According to Hermans and Geus (2), the interaction with the support decreases the nucleation barrier as long as nucleation at the surface of the support can proceed at a concentration between the solubility and the supersolubility curve. So, at concentrations between those of the supersolubility curves, $SS_{support}$ and SS, the compound exclusively precipitates onto the support surface. It may be added that according to Geus (8), the difference in concentrations between the two supersolubility curves, $SS_{support}$ and SS, is related to the bond strength between the precipitate and the support surface.

In principle, this method of preparation enables the deposition of a controlled amount of metal precursor up to high loading, and the interaction between the metal precursor and the support leads to the formation of highly dispersed active phase after thermal treatment.

In practice, the support is suspended into the solution containing the soluble precursor. The suspension is thoroughly stirred, then the precipitating agent is added. After a given time of gradual and controlled addition of the precipitating agent, the solid sample is gathered, washed, dried, and activated.

14.3   DEPOSITION-PRECIPITATION WITH A BASE

As mentioned above, raising the pH is the most common way to perform deposition-precipitation. To avoid precipitation in solution, and to have local concentration differences in the suspension of the support minimized, it is better if the mixing and the generation of the precipitant can be carried out separately. It is crucial to maintain the concentration continuously below that of the supersolubility curve SS.

At the laboratory scale, this is possible with urea ($CO(NH_2)_2$). This is a delay-base, which permits mixing and basification in two separate steps: mixing at room temperature, and basification when the mixture is heated above 60°C, leading to urea hydrolysis:

$$CO(NH_2)_2 + 3\ H_2O \rightarrow CO_2 + 2\ NH_4^+ + 2\ OH^- \qquad \text{(Equation 14.1)}$$

and to a gradual rise in pH (Figure 14.2a). Cyanate of an alkali metal or sodium nitrite can also be used when high metal loadings are required. Indeed, their decomposition does not lead to the formation of ammonium as in the case of urea, which may dissolve the precipitate. The drawback with these cyanates as precipitating agents is that oxygen must be avoided.

For larger scale preparation, urea cannot be used because of the large volumes of wasted water containing nitrogen compound. Injection of a basic solution through a tube ending below the surface of the suspension can be an alternative, but it must be accompanied by a thorough stirring to avoid local high concentration and precipitation in solution.

The deposition-precipitation method using a base has been applied for the preparation of various catalysts. Upon raising the pH of the solution, the precipitation of a hydroxide onto the support is expected. In fact, it was shown in several cases (Table 14.2) that mixed compounds such as phyllosilicates for silica support or hydrotalcite for alumina support formed, involving support dissolution and neoformation of a mixed compound with a layered structure.

14.3.1   EXAMPLE OF Ni/SiO$_2$

The phenomenon of neoformation of a mixed compound has been extensively studied in the case of the Ni/SiO$_2$ system, which forms nickel phyllosilicate of 1:1 type, also referred to as nickel hydrosilicate. 1:1 nickel phyllosilicate exhibits a stacked structure, each layer consisting of a brucite-type sheet containing Ni(II) in octahedral coordination and a sheet containing linked tetrahedral SiO$_4$ units (Figure 14.3a).

In practice, the support was suspended into the solution containing nickel nitrate and urea at room temperature (2, 4). Nitric acid was also added to better follow the changes in pH. The atomic ratio Ni:Si was equal to 1.1 in the suspension. The suspension was stirred while heating at 90°C, and the deposition-precipitation started upon decomposition of urea (Equation 14.1), leading to a gradual rise in pH (Figure 14.2d and Figure 14.2e). Due to the interaction between nickel and the support, the pH of the suspension is below that of the solution without support (Figure 14.2b).

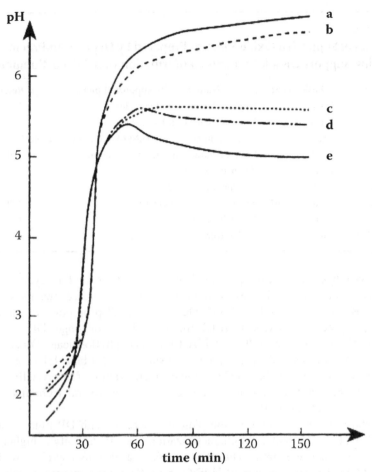

FIGURE 14.2 pH curves versus time (up to 150 min) of urea hydrolysis at 90°C in the presence of (a) urea and nitric acid; (b) urea, nitric acid, silica-400 m²/g; (c) urea, nitric acid, nickel nitrate; (d) urea, nitric acid, nickel nitrate, silica-50 m²/g; (e) urea, nitric acid, nickel nitrate, silica-400 m²/g; (urea: 0.42 M, nitric acid: 0.02 M, nickel nitrate: 0.14 M, silica: 7.6 g L⁻¹); from Reference 5.

After a given time of deposition-precipitation (DP time), the solid sample was gathered, washed, dried, and reduced.

A detailed study (4) showed that the nature of the supported nickel phase depends in fact on the DP time and on the silica surface area; two porous silicas were studied, one of ~ 50 m²/g (Figure 14.2d) and one of ~ 400 m²/g (Figure 14.2c). In both cases, the Ni loading gradually increased with the DP time and reached a plateau when all the nickel was consumed (Figure 14.4). On the silica of low surface area and after DP times ≤4 h, the supported phase is mainly a turbostratic nickel hydroxide, α-Ni (OH)₂, i.e., a nickel hydroxide with a disordered stacking of brucitic layers of octahedral Ni(II) (Figure 14.3b). The layers are separated by intercalated nitrate and isocyanate anions. For longer DP times (< 16 h), the supported Ni(II) phase becomes

TABLE 14.2
Examples of Supported Oxide Samples Prepared by Deposition-Precipitation, Involving Support Dissolution and Neoformation of a Mixed Compound

System	Basic agent	Nature of the supported phase	Reference
Ni/SiO$_2$	Urea or NaOH	Ni phyllosilicate + Ni hydroxide	2, 4, 5, 9
Ni/Al$_2$O$_3$	NaOH	Ni hydrotalcite + Ni hydroxide	8, 18
Cu/SiO$_2$	Urea	Basic Cu nitrate (Cu$_2$(OH)$_3$NO$_3$) or Cu phyllosilicate	9
Cu/SiO$_2$	NaOH	Basic Cu nitrate, then Cu phyllosilicate	8, 9
Zn/SiO$_2$	Urea	Zn phyllosilicate	19
Zn/Al$_2$O$_3$	NaOH	Probably Zn hydrotalcite	9
Fe^{II}/SiO$_2$	Urea or NaOH	Fe phyllosilicate + FeOOH	8, 9
Co/SiO$_2$	Urea	Co phyllosilicate	20
Cu-Zn/Al$_2$O$_3$	Urea	Cu-Zn hydrotalcite	21

an ill-crystallized 1:1 nickel phyllosilicate with nitrate and isocyanate anions entrapped in the structure. In the case of the silica of high surface area, the supported Ni(II) phase is a 1:1 nickel phyllosilicate with a small proportion of turbostratic nickel hydroxide even at very short DP times ($\leq$2.5 h). For longer DP times (< 16 h), the Ni(II) phase is also an ill-crystallized 1:1 nickel phyllosilicate. For still longer DP times ($\geq$16 h), both silicas are totally consumed, and a bulk 1:1 nickel phyllosilicate is obtained. It may be noted that the characterization of ill-crystallized Ni(II) phases was not an easy task and required the use of several techniques: FTIR, EXAFS, TPR, and XRD (4)

Crystallinity and stacking of the platelets increase with DP time, as attested by XRD, FTIR, TPR, and TEM characterization (4). Crystallinity is higher for the silica of low surface area. The BET surface areas also evolve with time (Figure 14.5). The surface area of Ni/SiO$_2$-50 m^2/g always increases with the DP time (Figure 14.5a). The increase may be related to the increasing amount of layered compounds formed, first as α-Ni (OH)$_2$ up to 4 h of DP, then as 1:1 nickel phyllosilicate, which both develop new surface area because of their disordered stacking. In the case of Ni/SiO$_2$ – 400 m^2/g, the surface area increases up to ~4 h of DP because of the development of thin platelets of 1:1 nickel phyllosilicates with random orientation (Figure 14.5b). Then, it decreases because of the consumption of the network of high surface area silica. The increasing crystallinity of nickel phyllosilicates with DP time probably also contributes to the decrease of the surface area.

Almost the same Ni phases are obtained whether silica is porous or not. One difference is that for short DP times, the Ni(II) phase on nonporous silica of low surface area is already a mixture of 1:1 nickel phyllosilicate and nickel hydroxide (5). In addition, as observed by TEM, the Ni(II) phase is better crystallized and the surface of contact between the support and the Ni(II) phase is larger with nonporous silica (Figure 14.6b) than with porous ones (Figure 14.6a), because of the more regular shape of the silica particles.

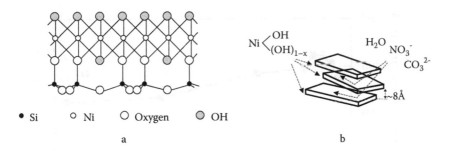

FIGURE 14.3 Scheme of (a) a layer of 1:1 nickel phyllosilicate (projection on the *bc* plane); (b) α-nickel hydroxide (turbostratic hydroxide), from Reference 4.

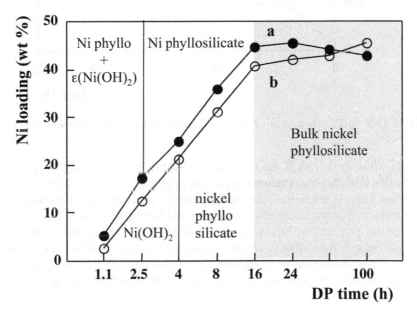

FIGURE 14.4 Influence of the DP time on the Ni loading and on the nature of the Ni phase; (a) silica-400 m²/g; (b) silica-50 m²/g.

The conditions of deposition-precipitation used to prepare Ni/SiO_2 samples (type of silica; time, i.e., Ni loading; nature of the Ni(II) phase) have a strong influence on the size of nickel metal particles obtained after temperature programmed reduction up to 900°C (6). Table 14.3 reports the results obtained with the samples with the lowest loadings (DP times ≤4 h), i.e., at a stage where silica is not fully consumed. It shows that the average metal particle size varies between 27 and 79 Å. For a given silica, the nickel particle size does not depend on the Ni loading, but the metal particles are smaller and the size distribution is narrower when the supported phase is a 1:1 nickel phyllosilicate (Ni/SiO_2-400 m²/g⁻¹) rather than a nickel hydroxide

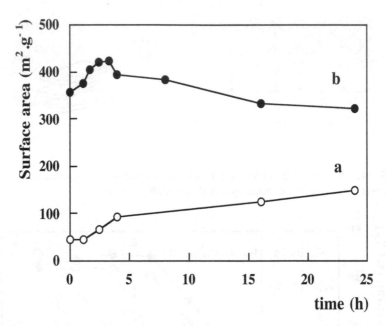

FIGURE 14.5 BET surface area of (a) Ni/silica-50 m²/g; (b) Ni/silica-400 m²/g; from Reference 4.

(Ni/SiO$_2$-50 m²/g^{-1}). This is probably due to the fact that 1:1 nickel phyllosilicate is reducible at higher temperature than nickel hydroxide, which induces less sintering. When silica is nonporous, the nickel particles are even smaller because of the better surface of contact between the support and the Ni(II) phase.

Catalysts prepared by this method exhibit advantages compared to more conventional ones: It is possible to obtain high nickel loadings (> 20 wt %) depending on the time of deposition-precipitation; after reduction, small metal particles are obtained with rather narrow size distribution (6, 7); these particles are highly resistant toward sintering.

14.3.2 MOLECULAR APPROACH OF THE MECHANISM OF DEPOSITION-PRECIPITATION

From the results summarized above, a mechanism of deposition-precipitation of Ni(II) on silica has been proposed (5). It complements the kinetic and thermodynamic approach proposed by Geus (8), which explains that the maximum of pH as obtained in Figure 14.2d and Figure 14.2e, results from a phenomenon of nucleation and growth of the Ni(II) precipitate onto the support. The sudden nucleation induces a drop in pH because of the OH⁻ consumption, after which the growth of the nickel phyllosilicate proceeds at a lower pH-level because of the establishment of a dynamic equilibrium between the formation of OH⁻ ions in solution and their consumption for the deposition-precipitation of the Ni(II) phase.

The mechanism of deposition-precipitation can be also explained at a molecular level, as follows (Figure 14.7): At the beginning of the basification of the solution,

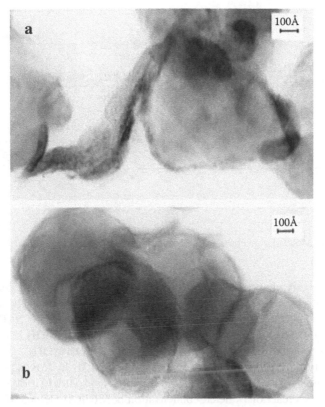

FIGURE 14.6 TEM micrographs of Ni/silica-50 m²/g; (a) porous silica (DP time: 2.5 h); (b) nonporous silica (DP time: 4 h); from Reference 4.

TABLE 14.3
Characteristics of Supported Ni Samples Prepared by Deposition-Precipitation with Urea (0.14 M of Nickel Nitrate, 0.42 M of Urea, 7.6 g/l^{-1} of Silica) from Reference 6

Samples	Ni loading (wt %)	Ni(II) phase	TPR peak (°C)	Metal particles size distr d (Å)	
Ni/SiO$_2$-400 m²/g (70 min)	5.2	ε(Ni(OH)$_2$) + Ni phyllosillicate	380 (sh) + 460	10–120	53
" (2.5 h)	17.2	Ni phyllosillicate	510	10–90	47
" (4 h)	25.1	Ni phyllosillicate	545	10–100	50
Ni/SiO$_2$-50 m²/g (70 min)	2.8	Ni(OH)$_2$	380	20–280	78
" (2.5 h)	12.5	Ni(OH)$_2$ + ε(Ni phyllosillicate)	380, 450 (sh)	20–280	79
" (4 h)	21.3	Ni(OH)$_2$ + ε(Ni phyllosillicate)	380, 450–500 (sh)	10–240	77
Ni/SiO$_2$-400 m²/g n.p.* (2.5 h)	17.2	1:1 Ni phyllosillicate	530	10–90	37
Ni/SiO$_2$-50 m²/g n.p.* (2.5 h)	14.7	Ni(OH)$_2$ + Ni phyllosillicate	410, 450–500 (sh)	10–70	27

* n.p.: Nonporous silica.

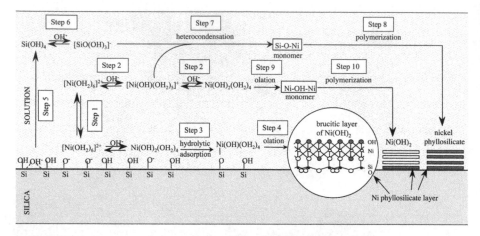

FIGURE 14.7 Molecular mechanism of deposition-precipitation of nickel on silica with urea; from Reference 5.

silica is negatively charged and exhibits cation adsorption capacity because pH is higher than its point of zero charge (PZC ~2). In consequence, the first step of deposition-precipitation is probably an electrostatic adsorption of the Ni(II) hexa-aqua complexes on the silica support (step 1 in Figure 14.7). The Ni(II) hexa-aqua complexes are in equilibrium with $[Ni(OH)(OH_2)_5]^+$ and $Ni(OH)_2(OH_2)_4$ (step 2; for the sake of brevity, H_2O produced is not reported). The Ni(II) hydroxoaqua complexes close to the silica surface can react with the silanol groups via a hydrolytic adsorption (step 3). This is a reaction of heterocondensation between Si-OH and Ni-OH, which leads to the formation of Si-O-Ni bonds. Moreover, the Ni(II) hydroxoaqua complexes can also react with each other via olation reaction (step 4):

$$2\ Ni(OH_2)_4(OH)_2 \longrightarrow (HO)_2(H_2O)_3Ni\text{-}OH\text{-}Ni(OH_2)_4(OH) + H_2O$$

and form a brucitic layer of octahedral Ni(II) bonded to the silica surface. This brucitic layer bonded to silica can also be considered as a sheet of 1:1 nickel phyllosilicate (Figure 14.3a) at the support-solution interface. Meanwhile, silica starts dissolving and silicic acid starts to be released in solution (steps 5 and 6 in Figure 14.7). At this point, there are two possibilities according to the amount of released $Si(OH)_4$, which depends on the silica surface area and the DP time:

- For silica of high surface area (400 m²/g), the dissolution is kinetically favored because of the large support-solution interface. Hence, the amount of silicic species is large enough to react with the nickel complexes in solution via a heterocondensation reaction (step 7):

$$Si(OH)_4 + Ni(OH_2)_4(OH)_2 \longrightarrow (HO)_3Si\text{-}O\text{-}Ni(OH_2)_4(OH) + H_2O$$

and forms Si-O-Ni monomers. Further polymerization forms layers of 1:1 nickel phyllosilicate (step 8 in Figure 14.7), which grows on the brucitic layer of Ni(II) bonded to silica, and leads to the formation of supported 1:1 nickel phyllosilicates.

- The solubility of silica of low surface area (50 m^2/g) is weaker and its rate of dissolution is slower because of the smaller support solution interface. Therefore, the amount of Si(OH)$_4$ is low at the beginning of the deposition-precipitation ($\leq$4 h). So, Si-O-Ni heterocondensation is a minor reaction, and olation between the Ni(OH)$_2$(OH$_2$)$_4$ complexes is the main reaction (step 9 in Figure 14.7). Further polymerization forms brucitic layers of Ni(II) (step 10), which grow on the brucitic layer of Ni(II) bonded to silica, and lead to the formation of supported nickel hydroxide.

It can be noted that the brucitic layer of Ni(II) bonded to silica acts as nuclei for the growth of supported 1:1 nickel phyllosilicate or supported nickel hydroxide. The heterocondensation reaction is faster than the olation one, but it is limited by the concentration and diffusion in solution of silicic acid arising from silica dissolution, which itself depends on the silica surface area, i.e., on the extent of support-solution interface.

This mechanism enables us (i) to explain why nickel hydroxide or nickel phyllosilicate are obtained, depending on the silica surface area and on the DP time, and (ii) to interpret the changes in the pH curves, the nature of the Ni(II) phase as a function of several parameters of preparation: the concentrations of urea and nickel nitrate, and the silica loading (5).

14.4 RECENT STUDIES USING DEPOSITION-PRECIPITATION BY DIRECT ADDITION OF A BASE

14.4.1 PREPARATION OF Ni/SILICIOUS SUPPORT

The method of deposition-precipitation by injection of sodium carbonate was used for the deposition of nickel on silicious diatomaceous earth containing 1 wt % of alumina (Celite FC, 45 m^2/g) (22). At the end of the slow addition of NaHCO$_3$ performed at different temperatures (25, 70, and 90°C) up to pH 7.6 (probably the pH value of the plateau as in Figure 14.2d), the solids were aged in the solution for various times at the same temperature. The same high nickel loading (~50 wt %) was obtained whatever the aging time, indicating that the precipitation had occurred during the rise of pH. The rate of injection of sodium carbonate was not reported. During aging time and particularly at 70 and 90°C, the surface area increased, nickel reducibility decreased, and nickel surface area after reduction increased. The Ni(II) phase was not characterized, but the authors evoked the formation of a nickel silicate. In another paper, the same authors (23) developed an empirical model to determine the degree of reduction and the metallic area of Ni/Celite FC from reduction parameters such as final reduction temperature, heating rate, flow rate, and hydrogen content.

330   Catalyst Preparation: Science and Engineering

14.4.2 PREPARATION OF SUPPORTED NOBLE METALS

The procedure of preparation by deposition-precipitation by direct addition of a base has been altered with the preparation of supported noble metals. The corresponding papers still refer to deposition-precipitation, but the procedure of preparation is different. The base is added either slowly, dropwise, or it is unspecified, so as to reach a final pH. The support can be added before or after pH adjustment. Then, the suspension ages for a given time at this pH. In other word, it turns out that the preparation is performed at a fixed pH. The supported phase is assumed to be a hydroxide of the corresponding metal (Table 14.4), which would exclusively precipitate on the oxide surface, but no characterization has been performed. Moreover, most often the experimental conditions do not allow us to know whether all the metal complex in solution is deposited onto the support. Such kind of modified procedure has also been used for the preparation of oxide supported on oxide, such as Cu on CeO_2 ($Cu_{0.1}Ce_{0.9}O_x$), with the goal of modifying the redox properties of ceria (with $NaHCO_3$ at pH 8.5 for 2 h, probably at room temperature) (24).

The most famous papers using this type of method for the preparation of supported noble metal catalysts are those of Tsubota (31, 32) with the preparation of gold catalysts. They are famous because this is one of the few methods of preparation of gold catalysts that led to the formation of small gold particles (2–3 nm) and to the discovery of the catalytic reactivity of gold.

Au/TiO_2 catalysts were prepared at various pH by deposition-precipitation using NaOH as a base (31, 32). Typical procedure of preparation was as follows: After adding the support to an aqueous solution of $HAuCl_4$, the pH of the suspension was raised to a fixed value by adding sodium hydroxide or carbonate, after which it was heated at 70 or 80°C with stirring for 1 h. After thorough washing with water to remove as much of the sodium and chlorine as possible, the solid

TABLE 14.4
Supported Metal Samples Prepared by Deposition-Precipitation with NaOH

System	Basic agent	pH	Time (h)	T (°C)	Assumed nature of the supported phase	Metal loading (wt %)	Deposition yield*	Reference
Pd/ZrO₂	NaOH	10.0	1	n.r.	Pd hydroxide	2.00	n.r.	25
Pd/CeO₂	Na₂CO₃	10.0	1	70	Pd hydroxide	3.00	n.r.	26, 27, 28
Ir/TiO₂	NaOH	3.0	1	25	Ir hydroxide	0.57	32%	29
		5.0				0.64	36%	
		7.0				1.14	63%	
		8.0				1.48	82%	
		10.0				0.85	47%	
Pd/ZrO₂	Na₂CO₃	10.5	1	n.r.	n.r.	1.00	n.r.	30

* Percentage of the metal precursor, which is deposited onto the support.

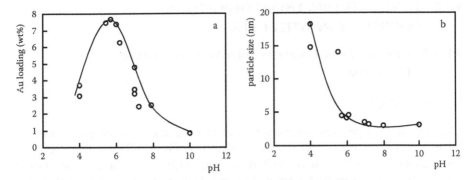

FIGURE 14.8 Deposition-precipitation of gold on titania with NaOH for various pH; (a) Au loading; (b) gold particle size; from Reference 32.

was dried under vacuum at 100°C, and then thermally treated to obtain gold metal particles. With a nominal amount of 13 wt % of Au in solution and within a pH range between 7 and 10, this method permitted the deposition of no more than 3 wt % of Au, but gold particles were small (~3 nm) (Figure 14.8). Higher Au loading of 8 wt % could be achieved at lower pH (~5.5), but much larger particles were obtained (~10 nm). Hence, the amount of gold deposited on TiO_2 was always lower than the amount of gold contained in solution; i.e., the yield of DP was lower than 100%. The conditions required for a successful preparation is a pH around 7–8 for titania (pH above the PZC of titania), because it corresponds to the best compromise between the gold loading (Figure 14.8a) and the gold particle size (Figure 14.8b).

So, this method of preparation does not correspond to the procedure of deposition-precipitation with a gradual rise of pH, and neither does the mechanism because gold hydroxide cannot precipitate under the conditions of pH and gold concentration used for these preparations (33, 34). The chemical mechanism occurring during this preparation at pH 7–8 has been studied (33, 34), and it can be summarized as follows: At pH 7–8, both the titania surface and the gold complex $[Au(OH)_3Cl]^-$ ($[AuCl_4]^-$ is hydrolyzed as pH increases) are negatively charged, so they cannot electrostatically interact, but they could form a surface complex by reaction with the surface OH of titania still present at this pH:

$$TiOH + [AuCl(OH)_3]^- \rightarrow Ti\text{-}[O\text{-}Au(OH)_3]^- + H^+ + Cl^- \qquad \text{(Equation 14.2)}$$

The evolution of the metal loading with the solution pH (Figure 14.8a) with a maximum at pH 6, close to the point of zero charge of titania (PZC ~ 6), is consistent with Equation 14.2. The decreasing amount of chloride in the catalyst, due to the hydrolysis of the Au complex when pH increases, is also consistent with the decreasing gold particle size obtained after thermal treatment (Figure 14.8b) because chlorides favor mobility of gold and particle aggregation. The mechanism of gold deposition is, therefore, closer to grafting reaction than to deposition-precipitation.

14.5 RECENT STUDIES USING DEPOSITION-PRECIPITATION WITH UREA

14.5.1 PREPARATION OF CATALYSTS SUPPORTED ON CARBON NANOFIBERS

14.5.1.1 Ni/CNF

The method of deposition-precipitation can also be used for the preparation of catalysts based on inert supports, like carbon nanofibers (CNFs). This has been demonstrated first in the case of the Ni/CNF system (35). To make the CNF surface reactive, the CNFs were first submitted to an oxidative pretreatment by reflux with acidic solution, HNO_3 (36), or a mixture of HNO_3 and H_2SO_4 (35). This treatment induced the formation of polar oxygen-containing surface groups resulting from the hydrolysis of anhydric carboxyl groups. The consequence is a better wettability of the aqueous precursor solutions and a decrease of the PZC from a pH of ~ 5 for the untreated fibers to a pH of ~ 2–3.

Using the method of deposition-precipitation with urea, 45 wt % of Ni could be deposited on fishbone carbon nanofibers (35). The Ni-support interaction is weaker than in the case of the Ni/SiO_2 system, as indicated by the higher nickel reducibility (~300°C instead of ~600°C), but the nickel metal particles are rather small (9 nm) and well distributed on the surface (Figure 14.9a). In contrast with Ni/SiO_2, the deposited species would not be a mixed compound, but a nickel hydroxide, which would nucleate and grow on the oxygen-containing surface groups

14.5.1.2 Pt/CNF

The method has also been used for the preparation of CNF-supported platinum catalysts (36). A suspension of CNF in water acidified with nitric acid to pH 3 was heated to 90°C under inert atmosphere. Then urea and $Pt(NH_3)_4(NO_3)_2$ (nominal Pt loading of 5 wt %) were added. After the pH had reached a constant value, ~6 after 6–18 h under stirring at 90°C, the solid was filtered, washed with water, dried at 80°C in a nitrogen flow, and reduced in H_2 at 200°C for 1 h. The authors found that not all Pt was deposited (3.9 instead of 5 wt %), indicating that the mechanism of platinum deposition is not deposition-precipitation, as for nickel. Moreover, they found a linear correlation between the number of acid sites of CNF and the amount of platinum that is deposited. For the highest loaded catalyst (3.9 wt %), platinum/adsorption site ratio was equal to 0.5. It was concluded that the mechanism of deposition is a mechanism of cation adsorption, with two acidic oxygen-containing groups of CNF per Pt as anchoring sites. However, this method results in significantly higher platinum loadings (3.9 wt %) than usual cation adsorption performed at pH ~ 6 and at 20 or 90°C (1.4 and 1.8 wt %, respectively). The difference is attributed to the creation of additional adsorption sites due to the hydrolysis of anhydric carboxyl groups in the acidic conditions at which HDP is started. After reduction at 200°C, small uniform platinum particles (1–2 nm) were obtained homogeneously distributed over the CNF (Figure 14.9b).

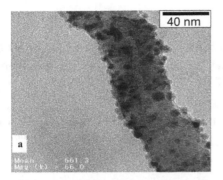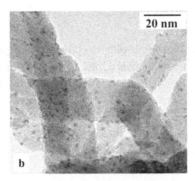

FIGURE 14.9 TEM micrographs of (a) Ni/CNF (10 wt %) after reduction at 500°C; from Reference 35; (b) Pt/CNF (5 wt %) after reduction at 200°C; from Reference 36.

CNF-supported ruthenium catalysts were also prepared by the procedure of deposition-precipitation using $RuNO(NO_3)_3(H_2O)_2$ as precursor (36). In this case, the 5 wt % of Ru in solution could be deposited, and small ruthenium particles (1.1–2.2 nm) were obtained after reduction. Hydrolysis of the Ru precursor in solution leads to the cationic complex $[RuNO(NO_3)_2(H_2O)_3]^+$. The authors calculated that the ruthenium/adsorption site ratio was equal to 1.2, i.e., close to the ratio expected for one-to-one adsorption. Implicitly, they concluded to a mechanism of cation adsorption. However, they did not attempt to increase the Ru loading to check if there is really a limitation in the Ru loading due to the limited adsorption capacity of support.

For both types of sample, ruthenium and platinum particles are thermally stable because their size is not modified by an extra thermal treatment under nitrogen at 500°C.

14.5.2 PREPARATION OF SUPPORTED GOLD CATALYSTS

The method of deposition-precipitation with urea was also applied for the preparation of gold catalysts supported on oxides such as titania, alumina, and ceria (34, 37). As mentioned in Section 14.3.b, the challenges in the preparation of supported gold catalysts are to obtain small gold particles (≤5 nm) and to be able to control the gold loading. The initial mixture contained the oxide support in suspension in a solution containing $HAuCl_4$ and urea. The initial pH was ~2, and the mixture was heated at 80°C for a given time before thorough washing with water and then thermal treatment. Table 14.5 shows that all the gold (8 wt %) was deposited onto titania, alumina, and ceria within the first hour, while the pH of the suspension was still acidic (~ 3). Thereafter, samples age while the pH continues rising, reaching a plateau at pH ~ 7 after 4 h. After calcination in air at 300°C performed to obtain gold metal particles, the particle sizes were found to decrease as the time of deposition-precipitation was longer (Table 14.5).

The mechanism of this preparation has been investigated (34). Gold that is deposited is not $Au(OH)_3$, because according to gold chemistry, it cannot form under the conditions of concentration and pH used for the preparation. As judged by

TABLE 14.5
Au/oxide Samples Prepared by Deposition-Precipitation with Urea at 80°C with a Nominal Gold Loading of 8 wt %, Followed by Thermal Treatment at 300°C, from References 34 and 37

Oxide support	PZC	DP time (h)	Final pH	Au loading (wt %)	Cl loading (wt %)	Average particle size (nm)
TiO_2 (45 m²/g)	~ 6.0	1	3.0	7.8	0.041	5.6
		2	6.3	6.5	0.122	5.2
		4	7.0	7.7	< 0.030	2.7
		16	7.3	6.8	< 0.030	2.5
		90	7.8	7.4	< 0.030	2.4
CeO_2 (256 m²/g)	~ 6.0	1	4.3	7.9	—	8.1
		16	6.6	8.2	—	< 5.0*
$\gamma\text{-}Al_2O_3$ (100 m²/g)	~ 7.5	1	4.3	6.9	—	6.9
		16	7.1	7.2	—	2.3
SiO_2 (250 m²/g)	~ 2.0	1	5.2	2.9	—	> 20.0
		16	7.0	3.7	—	> 20.0

* Estimated by XRD (poor contrast between gold particles and CeO_2 by TEM).

comparison of EXAFS spectra, the same gold compound was obtained by precipitation of the same solution heated at 80°C in the absence of a support. The precipitate was orange in color, and its chemical composition was $AuN_{2.2}O_{1.2}C_{0.9}H_{4.2}Cl_{0.1}$. The precipitate was amorphous, but its formation was found to arise from reaction between gold complexes in solution and the products of hydrolysis of urea. Indeed, there is no precipitation of the orange compound as long as the suspension is not heated, i.e., as long as urea is not decomposed, as attested by Raman spectroscopy (34). The mechanism proposed is the following (34): Some anionic gold species present in the solution at pH between ~ 2 (initial pH) and ~ 3 (pH of precipitation) adsorb on the support and act as nucleation sites for the precipitation of the orange compound. This interpretation is based on the fact that for the silica support, which has a PZC close to 2, all gold is not deposited and gold particles are much larger (Table 14.5).

The decrease in the gold particle size with the time of deposition-precipitation (Table 14.5) was attributed to a phenomenon of peptization (redispersion), as observed, for instance, when nanocolloids of titania, formed by hydrolysis of alkoxides, agglomerate rapidly to produce large precipitates, but can be slowly redispersed through the action of nitric acid.

This method is applicable to the same supports as those used for deposition-precipitation with other bases (Section 14.3.b), i.e., to those that have a PZC greater than 3–4. It also leads to small gold particles, but it has the advantage that all the gold in solution is deposited onto the support, so the Au loading can be easily controlled. Moreover, it is easy to prepare a set of samples with the same gold loading but with different particle sizes, playing with the DP time, and not with the treatment temperature.

It may be noted that the fast deposition of a gold compound containing nitrogen instead of the slow precipitation of an hydroxide, and the decrease of gold particle size versus the DP time, indicate that the mechanism of gold deposition during deposition-precipitation with urea is different from that occurring during the preparation of Ni/SiO$_2$ system by the same method (Section 14.2).

14.5.3 MISCELLANEOUS

14.5.3.1 Preparation of Ni/H-Beta Zeolite

The method of deposition-precipitation with urea has been recently applied to the preparation of Ni in H-beta zeolite (38), which is a three-dimensional large pore zeolite (7.6 × 6.4 Å and 5.5 × 5.5 Å). The Ni/Hβ samples were prepared for different times (1 to 4 h). The Ni loading increased with the DP time in a highly reproducible way, from 5.5 to 22 wt %. The deposition-precipitation of nickel seems to proceed in the same way as on silica of low surface area because various characterization studies (BET, XRD, TPR, FTIR, and TEM) indicate that, as in the case of silica of low surface area (Table 14.3, Figure 14.4), for short DP times (≤ 2 h), nickel hydroxide is the main Ni(II) phase deposited on Hβ zeolite, whereas for longer DP times (3 and 4 h), it is a mixture of nickel hydroxide and 1:1 nickel phyllosilicate. The Ni(II) phase is mainly deposited on the external surface of the zeolite, as indicated by the IR bands of OH groups. After reduction at 450°C, the samples contain well-dispersed Ni metal particles with an average size from 3 to 4.5 nm depending on the DP time, but also mainly located on the external surface.

14.5.4 PREPARATION OF Y$_2$O$_3$-Eu^{3+}/SILICA

Deposition-precipitation with urea has been also used to prepare Eu^{3+} doped-Y$_2$O$_3$/silica materials (39), which find application not in catalysis, but in the field of phosphor emitting in fluorescent lamps and cathode ray tubes. The method has been applied for depositing yttria (9.7 wt %) doped with Eu (0.30 wt %) from a mixture of nitrates heated at 80°C in the presence of urea and silica for 90 min. After thermal treatment at 900°C, Y$_2$O$_3$ particles are 12 nm large, coated by a silicate, and after thermal treatment at 1000°C, the α-Y$_2$Si$_2$O$_7$ phase forms. It exhibits interesting luminescent properties, and the authors conclude that this method of preparation is more promising than the sol-gel method.

14.6 DEPOSITION-PRECIPITATION BY SONOLYSIS

Sonication of a solution containing metal precursors can also generate precipitation. For instance, the sonication of a solution of molybdenum hexacarbonyl (Mo(CO)$_6$)) leads to the precipitation of a molybdenum oxide under air (40–42) and of molybdenum carbide under argon atmosphere (41). In the presence of a silica support, sonication of Mo(CO)$_6$ solution in decalin leads to molybdenum carbide (Mo$_2$C)-silica material under argon atmosphere and to molybdenum blue oxide (Mo$_2$O$_5$.2H$_2$O)-silica under air at room temperature (40, 41). Therefore, sonication can be considered as an alternative method of deposition-precipitation. A more recent paper called this method *ultrasonically controlled deposition-precipitation* (42). The mechanism of deposition of Mo oxide from $Mo(CO)_6$ precursor in decane on nonporous silica microspheres (Stober's silica) (40, 41) and in decalin on silica-based mesoporous material (Al-MCM-41) (42) was investigated using various techniques such as NMR, XPS, and FTIR. Sonication was carried out under ambient air at room temperature for periods of up to 4 h. The solid product was separated by centrifugation, thoroughly washed with dry pentane, and dried in a vacuum at room temperature.

The sonolysis of $Mo(CO)_6$ in decalin containing dissolved oxygen leads to its decomposition-oxidation reaction into a hydrated Mo oxide:

$$\text{decalin}$$

$$2Mo(CO)_6 + 9.5O_2 \rightarrow Mo_2O_5\ 2H_2O + 12CO_2$$

In this reaction, decalin acts as a source of protons for the formation of water molecules in solvated Mo oxide. The amorphous silica undergoes structural reorganization with breakage of strained Si-O-Si bonds. This favors the formation of Si-O-Mo interfacial linkage (Figure 14.10) and, as a consequence, strong inter-action between the molybdenum oxide and the silica surface, and a well-dispersed and uniform coating. The interaction increases the Mo oxide deposition rate at the support's surface by an order of magnitude, compared with that of bulk precipitation. Thus, ultrasound-induced cavitation appears to play a dual role in the decomposition of molybdenum hexacarbonyl and in the activation of the silica surface for the adhesion of the resulting Mo species (Figure 14.10). This method leads to a complete monolayer coverage of Al-MCM-41 surface with Mo, which corresponds to ~ 45 wt % MoO_3 loading.

14.7 DEPOSITION-PRECIPITATION BY AMMONIA EVAPORATION

It had been mentioned by Geus (8) that deposition-precipitation by evaporation of solvent is another way to gradually increase the concentration of a metal precursor in solution, but he also reported that this leads to an inhomogeneous distribution of the active material. However, ammonia evaporation of solutions containing ammine or ammine carbonate metal complexes has been reported as a way to gradually decrease the pH of the solution; to induce precipitation of hydroxy carbonate compound on alumina support; and to prepare Co/Al_2O_3 (43, 44), Ni/Al_2O_3 (45, 46), Cu/Al_2O_3 catalysts (47). The preparation method was very broadly described in the patent references. It consists of a suspension of alumina

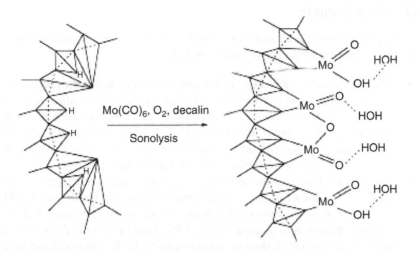

$Mo(CO)_6$, O_2, decalin

Sonolysis

FIGURE 14.10 Scheme of deposition-precipitation by sonolysis of $Mo(CO)_6$ at the Al-MCM-41 surface; from Reference 42.

powder in an aqueous solution containing ammonia and appropriate amount of nickel ammine, or cobalt or copper ammine carbonate complex. The initial pH of the solution is high (9 to 11, depending on the metal), and the solution is heated (>70°C). Upon heating, ammonia is evaporated, so pH drops, which leads to precipitation and possible aging of the precipitate, depending on the time in solution. At these pH levels, interaction between metal cations and the negatively charged alumina surface is made possible. These adsorbed cations probably act as nucleation sites for the deposition of hydroxy carbonate compounds formed during ammonia evaporation:

$$evaporation$$

$$[Ni(or~Co)(NH_3)_6]^{2+} \rightarrow Ni(or~Co)(OH)_{2-x}(CO_3)_x~on~alumina$$

As for deposition-precipitation with urea, this method allows the deposition of high metal loadings (>40 wt %) in a highly dispersed state and in a single step. For 22 wt % Co/Al_2O_3, the average size of Co^0 particles was 3 nm, for 34 wt % Co/Al_2O_3, 5 nm (43), and for 20 wt % Ni/Al_2O_3, the largest Ni^0 particles were smaller than 10 nm (45). An advantage of this method compared to deposition-precipitation with urea seems to be that the reaction with the support, and therefore the formation of mixed compounds (such as hydrotalcite for the present study), is minimized because the solution pH decreases during preparation, of course providing that the aging time is rather short. As a consequence, the loss of metal in the mixed compound is minimized as well, and the metal reducibility is higher. Moreover, the metal particle size is stable upon high-temperature reduction (for instance, up to 800°C for Co/Al_2O_3).

14.8 CONCLUSION

Since the publication of the last review on deposition-precipitation in 1997 (9), there have been several significant new results published:

- The studies on the molecular mechanisms occurring during the preparation of the Ni/SiO_2 system, which has been one of the most studied systems.
- The use of supports different from oxides, i.e., carbon nanofibers. In the case of supported nickel, the chemical mechanism also corresponds to a mechanism of deposition-precipitation, whereas in the case of Ru and Pt, i.e., noble metals, a mechanism of cation adsorption prevails.
- This method has been also applied to the development of supported gold catalysts. In the case of the so-called deposition-precipitation with NaOH, the procedure of preparation has been, in fact, completely modified and rather corresponds to preparations at a fixed pH in excess of solution, in other words, to a procedure of ion adsorption. In the case of deposition-precipitation with urea, there is precipitation of a gold compound, not gradual as in classical deposition-precipitation, but sudden and fast at low pH, followed by a period of maturation during the increase of pH, which induces a redispersion of the supported phase.
- Finally, new alternative methods of deposition-precipitation have been published involving sonolysis techniques and decrease in pH by ammonia evaporation.

ACKNOWLEDGMENT

Thanks to Prof. Krijn de Jong for his reading of and comments on this review. Thanks also to Dr. Lok for his information on deposition-precipitation by ammonia evaporation.

REFERENCES

1. J. A. van Dillen, J. W. Geus, L. A Hermans, J. van der Meijden, in *Proceedings of the 6th International Congress on Catalysis*, London, 1976; G. C. Bond, P. B. Wells, F.C. Tompkins (Eds.), The Chemical Society: London, p 677, 1977.
2. L. A. Hermans, J. W. Geus, Stud. Surf. Sci. Catal., 4, 113, 1979.
3. C. J. G. van der Grift, P.A. Elberse, A. Mulder, J. W. Geus, Appl. Catal., 59, 275, 1990.
4. P. Burattin, M. Che, C. Louis, J. Phys. Chem. B, 101, 7060, 1997.
5. P. Burattin, M. Che, C. Louis, J. Phys. Chem. B, 102, 2722, 1997.
6. P. Burattin, M. Che, C. Louis, J. Phys. Chem. B, 103, 6171, 1997.
7. P. Burattin, M. Che, C. Louis, J. Phys. Chem. B,104, 10482, 2000.
8. J. W. Geus, Stud. Surf. Sci. Catal., 16, 1, 1983.
9. J. W. Geus, A. J. van Dillen, in *Handbook on Heterogeneous Catalysis*, G. Ertl, H. Knözinger, J. Weitkamp (Eds.), VCH, Weinheim, Vol. 1, p.240, 1997.
10. K.P. de Jong, Stud. Surf. Sci. Catal., 63, 19, 1991.
11. German Patent 740,634 to IG Farben, 1943.
12. Netherlands Patent Application 67,05259 to Stamicarbon, 1967.
13. US Patent 3,668,148 to Lever Brothers Company, 1970.

14. Netherlands Patent Application 68,16777 to Stamicarbon, 1970.
15. European Patent Specification 258,942 to S.I.R.M.-B.V., 1988.
16. P. C. M. van Stiphout, H. Donker, C. R. Bayense, J. W. Geus, Stud. Surf. Sci. Catal., 31, 55, 1987.
17. K. P. de Jong, J. W. Geus, Appl. Catal., 4, 41, 1982.
18. P.K. de Bokx, W.B.A. Wasserberg, J.W. Geus, J. Catal., 104, 86, 1987.
19. C. Chouillet, F. Villain, M. Kermarec, H. Lauron-Pernot, C. Louis, J. Phys. Chem. B, 107, 3565, 2003.
20. R. Trujillano, J. Grimoult, C. Louis, J.-F. Lambert, Stud. Surf. Sci. Catal., 130B, 1055, 2000.
21. S. Catillon, thesis, Paris, Université Pierre et Marie Curie, 2004.
22. V. M. M. Salim, D. V. Cesar, M. Schmal, M. A. I. Duarte, R. Frety, Stud. Surf. Sci. Catal., 91, 1017, 1995.
23. D.L. Bhering, M. Pele, J. C. Pinto, V. M. M. Salim, Appl. Catal. A, 234, 55, 2002.
24. W. Shan, Z. Feng, Z. Li, J. Zhang, W. Shen, C. Li, J. Catal., 228, 206, 2004.
25. Y. Matsumura, M. Okumura, Y. Usami, K. Kagawa, H. Yamashita, M. Anpo, M. Haruta, Catal. Letters, 44, 189, 1997.
26. W.-J. Shen, Y. Matsumura, J. Mol. Catal. A, 153, 165, 2000.
27. W.-J. Shen, Y. Ichihashi, M. Okumura, Y. Matsumura, Catal. Letters, 64, 23, 2000.
28. W.-J. Shen, Y. Ichihashi, H. Ando, M. Okumura, M. Haruta, Y. Matsumura, Appl. Catal. A, 217, 165, 2000.
29. M. Okumura, N. Masuyama, E. Konishi, S. Ichikawa, T. Akita, J. Catal., 208, 485, 1997.
30. R. Gopinah, N. Lingaiah, N. Seshu Babu, I. Surynarayana, P.S. Sai Prasad, A. Obuchi, J. Mol. Catal. A, 223, 289, 2004.
31. S. Tsubota, M. Haruta, T. Kobayashi, A. Ueda, Y. Nakahara, Stud. Surf. Sci. Catal., 72, 695, 1991.
32. S. Tsubota, D. A. H. Cunningham, Y. Bando, M. Haruta, Stud. Surf. Sci. Catal., 91, 227, 1995.
33. F. Moreau, G. C. Bond, A. O. Taylor, J. Catal., 231, 105, 2005.
34. R. Zanella, L. Delannoy, C. Louis, Appl. Catal. A, 291, 62, 2005.
35. J. H. Bitter, M. K. van der Lee, A. G. T. Slotboom, A. J. van Dillen, K. P. de Jong, Catal. Letters, 89, 139, 2003.
36. M. L. Toebes, M. K. van der Lee, L. M. Tang, M. H. Huis, T . Veld, J. H. Bitter, A. J. van Dillen, K. P. de Jong, J. Phys. Chem. B, 108, 11611, 2004.
37. R. Zanella, S. Giorgio, C. R. Henry, C. Louis, J. Phys. Chem. B, 106, 7634, 2002.
38. R. Nares, J. Ramírez, A. Gutiérrez-Alejandre, C. Louis, T. Klimova, J. Phys. Chem. B, 106, 13287, 2002.
39. (a) C. Cannas, M. Casu, A. Mainas, A. Musinu, G. Piccaluga, Compos. Sci. Technol., 63, 1175; (b) C. Cannas, M. Casu, M. Mainas, A. Musinu, G. Piccaluga, S. Polizzi, A. Speghini, M. Bettinelli, J. Mater. Sci., 13, 3072, 2003.
40. N. Arul Dhas, A. Gedanken, J. Phys. Chem. B, 101, 9495, 1997.
41. N. Arul Dhas, A. Gedanken, Chem. Mater., 9, 3144, 1997.
42. M. V. Landau, L. Vradman, M. Herskowitz, Y. Koltypin, A. Gedanken, J. Catal. 201, 22, 2002.
43. C. M. Lok, Stud. Surf. Sci. Catal. 147, 283, 2004.
44. C. M. Lok, G. Gray, G. J. Kelly, WO Patent Application No. 01/87480, 2001.
45. C. M. Lok, S. V. Norval, abstract, 13th ICC, Paris, July 2004.
46. C. M. Lok, PCT Patent Application WO No. 00/47320, 2000.
47. C. M. Lok, U.S. Patent Application No. 6,703,342, 2004.

15 Production of Supported Catalysts by Impregnation and (Viscous) Drying

John W. Geus

CONTENTS

15.1 SUMMARY

After a discussion of the demands solid catalysts have to meet, the distribution of the active component within support bodies adapted for different situations is discussed. Next, impregnation of support bodies with a solution of a precursor of the active component(s) is addressed. Adsorption of the active precursor on the surface of the support is distinguished from reaction of the (surface of the) support with the impregnated solution.

Drying of impregnated support bodies is most extensively discussed. A general treatment of drying of porous materials is followed by a rough classification of the processes proceeding during drying of materials having pores of widths varying from 2 mm to about 20 nm. A survey of the literature on the drying of impregnated support bodies precedes a discussion of experiments on impregnation and drying of model supports and real supports performed in the author's laboratory. In experiments with a range of solids, it appeared that most of the moisture is removed during the period of constant drying rate, in which the transport of liquid water to the external edge of the support bodies can readily keep up with the evaporation of water at the edge. The result is that a substantial fraction of the dissolved precursor is deposited on the external edge of the support bodies.

It has been established that drying support bodies impregnated with solutions of active precursors, the viscosity of which rises during drying, can provide the generally desired uniform distribution of the active component throughout the support bodies. The reason for the uniform distribution has been extensively investigated. The viscosity is certainly important, but also the interaction of the liquid with the surface of the support, as was shown by experiments with the flat supports provided by silicon wafers. The use of solutions of citrate or gluconate salts of the active component or addition of compounds such as hydroxy-ethyl-cellulose thus can provide a uniform distribution of the active precursor.

15.2 INTRODUCTION

Solid catalysts have to meet a considerable number of (conflicting) demands, such as a high mechanical strength, a high attrition resistance, a large pore volume, a high thermal stability, an elevated active surface area per unit volume, and a good accessibility of the active area for reactants, which often implies a large external surface area and a high bulk density. It is usually not possible to achieve an acceptable compromise with the catalytically active component alone. Therefore, most solid catalysts contain two components, viz., the active component and a support. The support provides the shape and size of the catalyst bodies and thus determines the external surface area, the mechanical strength, and the porous structure. Generally, the catalytically active component has to be applied finely divided on the internal surface of the support. The desired distribution of the active component within the catalyst bodies, however, depends upon the conditions under which the catalyst is to be employed.

To ensure good accessibility of the active sites for the reactants, small catalyst bodies are most favorable. As indicated by Thiele's modulus [1], the length of the pores of a catalyst affects the apparent rate of the reaction much more than the diameter of the pores. However, the size of the catalyst bodies determines the pressure drop over the catalyst bed. Usually, the installation in which the catalyst is to be employed cannot operate at a high pressure drop. Furthermore, with very small catalyst bodies, the pressure drop can be so high that channeling in the catalyst bed occurs or the catalyst bodies are even blown out of the reactor. With fixed catalyst beds, the size of the catalyst bodies, therefore, can generally not be much smaller than about 5 mm. With pores of a length characteristic with catalyst bodies of at least about 5 mm, a relatively slow catalytic reaction does not exhibit transport limitations. Then, a uniform distribution of the active component throughout the support body is most attractive. However, the overall rate of fast catalytic reactions will also be determined by the rate of transport to the active sites. In that case, it is more attractive to deposit the catalytically active particles at the external edge of the catalyst bodies.

Also, when a subsequent catalytic reaction can lead to an undesired product and thus lower the selectivity, it is desired to apply the active component within the external edge of the catalyst bodies. Catalyst bodies having catalytically active particles in the external surface, on the other hand, are liable to lose active material by attrition. It is, therefore, more advantageous to deposit the active component in the subsurface layer, leading to an egg-white distribution. Also, when the reactants

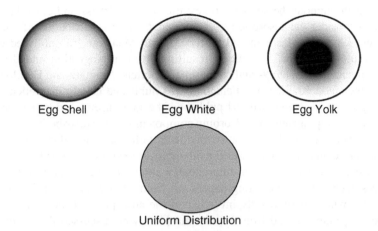

FIGURE 15.1 Different distribution of catalytically active component within support bodies.

to be processed contain some components poisoning the active sites, an egg-white distribution may be favorable. The poisoning components can then be captured within the layer at the external edge. Figure 15.1 represents different distributions of the active component(s) throughout catalyst bodies.

Catalysts to be employed technically can be developed most readily by application of the active component(s) by impregnation and drying. Commercially, a large range of support bodies is available of widely different shapes and sizes, of specific surface areas and porous structures, and, most importantly, of the required mechanical strength. After selection of the desired support, application of the active component(s) with the desired distribution into the support bodies is the remaining task, which can be readily executed by impregnation and drying. Another attractive feature of impregnation and drying is that no wastewater is produced and that loss of active component(s) with wastewater is not possible. Especially with precious metals, the negligible risk of losing active component(s) is highly advantageous.

This chapter deals with the production of supported catalysts by impregnation and drying. Many interesting papers have been published about the preparation of supported catalysts by impregnation and drying. The treatment of impregnation had been generally more satisfactory than that of drying. We will first discuss theoretical and experimental results on impregnation. Then, the literature involving drying will be reviewed. We will consider not only drying of impregnated catalyst bodies, but also some literature about drying of beds of materials, such as sand, and of hygroscopic porous solids. The last part of this chapter describes experimental results on catalyst preparation by impregnation and drying obtained in the author's laboratory.

15.3 ADSORPTION ON THE SURFACE OF THE SUPPORT

Adsorption of a precursor of the active component from the impregnating solution onto the surface of the support can determine the final distribution of the active component throughout the support bodies after drying. Weisz and coworkers were the first to consider the adsorption of the precursor on the distribution of the active component [2, 3, 4]. Maatman and Prater [5], and Maatman [6] published experimental results about the distribution resulting from impregnation and drying of platinum hexachloride into χ- and α-alumina. Kheifets, Neimark and Fenelonov [7] and later Neimark, Kheifets and Fenelonov [8] published mostly theoretical results on impregnation. Lee and Aris [9] published the most useful and extensive paper on the effect of impregnation of adsorbing components on the distribution throughout the support body. The authors exhaustively dealt with experimental results published in the literature and developed useful mathematical expressions. For literature published earlier on adsorption from impregnated solutions, we therefore refer to Lee and Aris's paper. Other interesting papers are from Komiyama, Merrill and Harnsberger [10], who experimentally investigated the adsorption of nickel from nickel chloride on γ-alumina, and from Komiyama [11], who discussed the results more extensively.

When the active precursor is applied by adsorption on the surface of the support, the loading that can be achieved is small. Adsorption of the active precursor is, therefore, often employed with precious metals, the loadings of which are generally low. Two different procedures can be employed with impregnation of support bodies. The first procedure involves addition of the impregnating solution to the dried or the evacuated support bodies. Most effective is application of a volume of the impregnating solution that is equal to the pore volume of the support bodies. This procedure is known as *incipient wetness impregnation* or *dry impregnation* [12]. The impregnation can be performed rapidly, because the uptake of the liquid by the support bodies proceeds fast. The dissolved active precursor is transported by convection into the pores of the support. An analogous procedure is dipping impregnation [12]. The support bodies are immersed in a solution of the active precursor and subsequently drained. When a liquid volume larger than the pore volume contacts the support bodies, the dissolved precursor present in the liquid phase out of the support bodies must diffuse into the pores of the support, because flow does not proceed.

Diffusion through liquids is a slow process. A calculation based on the Fourier number indicates that it can readily take about 10 h to assume the same concentration throughout the solution and the pore system of the support body. The other impregnation procedure aims at the establishment of an egg-shell distribution of the active precursor. Now, the support bodies preferably present in a fixed bed are filled with pure water, and a solution containing the active precursor is passed through the bed with the support bodies. Adsorption of the active precursor now proceeds almost exclusively at the external edge of the support bodies. Extending the period of time that the support bodies contact the solution containing the active precursor leads to a slow rise in thickness of the layer of the active component.

Thus far, literature has not distinguished adsorption on and chemical interaction with the support. Important is the impregnation of chloroplatinum acid into γ-alumina supports. Van den Berg and Rijnten [13] carried out a potentiometric titration of hexachloro platinum acid. The larger part of the alkali added was consumed at a pH level of 2, a level at which alumina is attacked. Vermeulen, Geus, Stol and de Bruyn

[14] extensively investigated the precipitation of alumina. Raising the pH of an aluminum nitrate solution homogeneously, the authors did not observe the solution to produce a visible precipitate, whereas consumption of hydroxyl ions is apparent from the fact that the pH level stayed at about 3. When the addition of hydroxyl ions was continued, hydrated aluminum oxide precipitated within a narrow pH range. Later, Vogels, using ^{27}Al NMR, showed that the aluminum(III) hydrolyzes gradually at the pH level of 3. The hydrolyzed aluminum(III) reacts to a precipitate at a pH level of about 5.5 [15]. After drying, the precipitate exhibits a faint x-ray diffraction pattern of bayerite (α-Al(OH)$_3$). Adding a solution of a pH level below 3 (H$_2$PtCl$_6$ displays a pH of 2, according to van den Berg and Rijnten [13]) to an alumina suspension will consequently lead to dissolution of alumina, resulting in some dissolved aluminum chloride. Consequently, impregnation of an H$_2$PtCl$_6$ solution into γ-alumina will lead to a rapid reaction with the alumina.

Spieker and Regalbuto [16] mention as one of the three adsorption mechanisms the complexation of the hexachloroplatinate ions with the dissolved oxide. The reaction will proceed where the acid solution contacts the alumina and, thus, immobilizes the platinum. The other two adsorption mechanisms involve chemical interactions between the metal complexes of the precursor and the surface of the solid support, and pure electrostatic interaction of charged complexes of the precursor with oppositely charged sites on the surface of the support. Results obtained by Roth and Reichard [17] demonstrate the immobilization by local neutralization of the hexachloroplatinum acid by reaction with the support. When an aqueous ammoniacal solution of platinum diamino dinitrite solution is impregnated, reaction with the surface of the support does not proceed, and a more uniform platinum profile results.

Komiyama and colleagues [10, 11] impregnated nickel chloride solutions into γ-alumina. The reactivity of (the surface of) γ-alumina is high. Consequently, reaction of the surface layer of the alumina with nickel to a hydrotalcite, viz., an analogous compound as takovite, Ni$_3$Al$_2$(CO$_3$)5H$_2$O, can proceed. The reaction will lead to immobilization of the nickel. Especially at higher temperatures, reaction of the alumina surface is likely. When reaction to a hydrotalcite has proceeded, subsequent calcination leads to a spinel, NiAl$_2$O$_4$, which is apparent from the fact that a significant fraction of the nickel can only be reduced at temperatures above about 450°C. Recently, Lekhai, Glasser, and Khinast [18] discussed the influence of pH and ionic strength on the metal profile of impregnation catalysts based on adsorption equations introduced by Agashe and Regalbuto [19] and by Regalbuto et al. [20, 21]. With nickel, Lekhai et al. sometimes extrapolate their method too far. Because nickel(II) has already completely precipitated at a pH level slightly below 7, consideration of diffusion of nickel(II) within a solution of pH 11 does not seem to be realistic.

Incipient wetness impregnation will lead to an egg-shell distribution when an acid platinum hexachloride solution that is not highly concentrated is impregnated into γ-alumina. As dealt with above, an egg-white distribution of platinum is often preferred. Maatman [6], van den Berg and Rijnten [13], and Michalko [22] describe an egg-white distribution that can be achieved by addition of acids to the impregnating platinum hexachloride solution. In addition to inorganic acids, such as, nitric acid, organic dibasic acids (such as oxalic acid, tartraric acid, and citric acid), and aromatic acids with hydroxyl groups (such as salicylic acid) are employed to achieve an egg-white distribution. The acid reacts more readily with the alumina surface than the platinum hexachloride, which is consequently immobilized in more interior parts of the support bodies.

With active precursors that do not significantly adsorb on the surface of the support, pore-volume (incipient wetness) impregnation leads to a uniform distribution of the solution of the precursor throughout the pore system of the support. It is highly relevant to investigate whether subsequent drying of the impregnated support results in a nonuniform distribution of the active component.

15.4 DRYING OF IMPREGNATED SUPPORTS

As mentioned above, application of an impregnated precursor by adsorption on the surface of the support only leads to relatively low loadings of the support. Impregnation of a precursor not adsorbing on the surface of the support can result in higher loadings, provided the pore volume is sufficiently large and the solubility of the precursor is sufficiently elevated. After impregnation, the distribution of a dissolved precursor not adsorbing on the surface of the support is uniform. To achieve a uniform distribution of the active component in the final catalyst, drying may not change the distribution of the precursor. To investigate whether drying can affect the distribution of the precursor, it is interesting to consider the literature dealing with drying of porous materials. Transport of moisture in porous materials can proceed by movement of liquid to the external surface, by flow of water vapor through the pores of the material, and by movement within adsorbed water layers. It is apparent that only movement of liquid within the support bodies or within adsorbed water layers can affect the distribution the active precursor.

As discussed in Coulson and Richardson [23], who consider only flow of liquid water and of water vapor, the course of the drying rate as a function of time displays three different ranges, which are represented by curve A in Figure 15.2. In the first range, the drying rate is constant; this period is known as the constant rate period. The transport of vapor from the external surface of the body to be dried into the gas

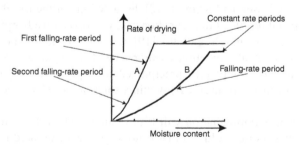

FIGURE 15.2 Rates of drying. Curve A is characteristic for materials such as sand; curve B is characteristic for materials such as soap and catalyst supports. Curve A initially exhibits a constant rate and, subsequently, a linear drop in drying rate (the first falling-rate period), and finally, a more slowly falling rate (the second falling-rate period).

flow determines the drying rate, together with the heat flow to the drying body. Usually, the drying solid assumes the wet bulb temperature, provided the pores of the drying material are not so narrow that the equilibrium pressure of water vapor is depressed. Within the porous material, capillary forces are responsible for the rapid transport of liquid water to the external surface. When the transport of the liquid to the external surface of the body can no longer keep up all over the external surface with the rate of evaporation, areas of the surface become dry. The temperature of the dried areas can rise. Now, the rate of drying is determined by evaporation from a decreasing area at the external surface and possibly also by flow of vapor through the empty pores underneath the dried areas.

The surface of the dry areas increases during the second period of the drying process. Consequently, the rate of drying is dropping during the second range, the first falling rate period. The effect of dried patches at the external surface on the rate of evaporation is, however, debated. It has been stated that the transport to the gas flow, controlled by the thickness of the (schematic) laminary boundary layer, remains to dominate also when a limited fraction of the external surface is dry. Van Brakel, therefore, maintains that the end of the constant rate period reflects the almost-complete recession of the evaporation front into the porous solid [24]. However, others have stated that the first falling rate period precedes the penetration of the evaporation front into the porous solid. The other authors believe that the flow of liquid to the external surface is too small to keep the wetting of the external surface at the level exhibited during the constant rate period.

Due to the much lower density of water vapor, transport of water vapor is much slower than that of liquid water. Ceaglske and Hougen [25] have demonstrated the fact that transport of water vapor through a porous solid proceeds much more slowly than the transport of liquid water. The authors placed a layer of coarse sand onto a bed of fine sand and studied the drying of the fine sand bed after the coarse sand layer had been dried. The moment the coarse sand layer became dry corresponded with a steep drop in the drying rate. Though the drop in the rate of evaporation might be attributed to a lower heat transfer to the sand layer still wet, it is likely that the transport of water vapor through the dry sand layer proceeds much more slowly than the uptake of water vapor from the interstices between the coarse sand particles at the external surface, as long as the interstices are filled with liquid water.

As long as the regions filled with the liquid are present as a continuous network up to the external surface, the system is said to be in the *funicular* state. When the continuous network of liquid elements breaks up, the system is said to be in the *pendular* state. Now, the rate is completely determined by transport of vapor through the emptied pores of the solid. Because the fraction of the pores filled with liquid shrinks to the center of the impregnated body, the rate of evaporation drops more rapidly.

Movements of liquid water within porous solids have been experimentally observed. Haines studied the movement of liquid water within porous media during drying [26, 27]. He observed rapid movements of water elements in the porous system (Haines jumps). The capillary pressure transports water to narrow pores at the surface. Larger volumes surrounded by narrow pores can be rapidly emptied, and smaller volumes can be filled when the water is removed out of the narrow

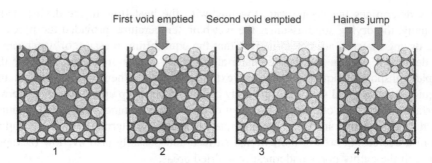

FIGURE 15.3 Haines jump schematically represented as occurring during drying of porous solid. Going from 3 to 4 in the above drawings, the void at the left-hand side is filled, and in the center, a larger void is emptied. The liquid redistributes rapidly.

pores surrounding the larger volumes. Figure 15.3 shows an example of a Haines jump. When the pores are neither wide nor very narrow and the viscosity of the liquid is not high, transport of the liquid to the external surface of the porous body is usually sufficiently rapid to keep up with the rate of evaporation from the external surface for a considerable period of time. Usually, the constant rate period involves about 90% of the water present in the pore volume of the porous material.

The results of Newitt, Nagar and Papadopoulus [28] are good examples. The authors employed three beds of sand of different sizes, viz., about 215 μm, 68 μm, and 12 μm. They impregnated the sand beds with either water or salt solutions of different concentrations and measured the rate of drying and established the distribution of the salt within the dried sand beds. A large amount of water was evaporating rapidly during the constant rate period, in which water can freely move to the external surface. Most of the salt turned out to be deposited within the top layer of the sand beds. Separate measurement of the concentration of the liquid at the top of the sand beds indicated a high concentration, which demonstrates that back-diffusion due to the concentration gradient is small as compared with the forward flow. The authors also found another important feature, viz., choking of the narrow pores at the surface of the bed of very fine sand, which was apparent from the relative moisture content at the end of the constant rate period. The results of Newitt, Nagar and Papadopoulus unambiguously demonstrate the migration of a considerable amount of a dissolved compound with the liquid to the external edge of a porous body during drying.

Coulson and Richardson [23] arrived at an important classification of porous solids. With particles of a size of 0.2 to 2.0 mm and, hence, pores of the same dimensions, the capillary and gravitational forces are dominating the transport of moisture in the porous material, which results in most of the moisture to be removed during the constant rate period. With particles and pores of 2 to 20 μm, flow by capillary forces is dominating the transport of the moisture, whereas the friction forces are not yet affecting the movement of liquid elements significantly. With materials having pores of radii between 2 μm and 2.0 mm, most of the moisture will evaporate during the period that a constant rate of evaporation is displayed, the constant rate period. With particles and pores less than 2 μm, the capillary and friction forces are controlling the flow of liquid elements. When the viscosity is not low, the evaporation front now may penetrate into the porous material before most of the moisture has evaporated.

Coulson and Richardson's [23] discussion of the transport of liquid elements in porous solids during drying is very important to predict the distribution of active precursors after drying. When a large amount of the liquid is removed during drying in the constant rate period, most of the dissolved precursor is deposited at the external surface of the support bodies. During the first decreasing rate period, the rate of transport of liquid to the external surface is dropping, so more of the dissolved precursor will be deposited within the support. Deposition at the center of the support bodies will result during the second falling rate period, when the moisture is transported as water vapor out of the porous support. Two items not discussed by Coulson and Richardson are the presence of liquid water remaining after the pores at the external surface have been partially emptied, due to narrow pores present at the contact areas of the sand particles, and transport of liquid in a layer adsorbed on the (internal) surface of the porous body. Spreading of a liquid solution over a hygroscopic surface is usually exhibited. The generally employed alumina and silica supports are drying agents and, thus, have a hygroscopic surface. Especially when the dissolved active precursor is deliquescent, spreading of the solution of the active component is likely.

Kheifets, Neimark and Fenelonov published a series of three papers dealing with catalyst preparation by impregnation and drying [7, 8, 29]. The authors consider two limiting regions of drying, viz., rapid and slow drying. With rapid drying, the constant rate period is negligible. This implies that the vapor is removed thus fast from the external edge of the support bodies and the heat is supplied so rapidly to the support bodies that the capillary flow ceases almost immediately after initiation of the drying process. Consequently, the front of the liquid recedes into the support bodies. Because the authors assume that the concentration of the liquid rises to the level of the supersaturation only after the front of the liquid has arrived fairly far within the support bodies, the active precursor will be deposited in the center of the support bodies. With the limit of slow drying, the rate of capillary flow, film flow, and recondensation is higher than the rate of removal of the vapor at the external edge of the support bodies. The authors assume that, also with slow drying, the evaporation front moves into the impregnated porous solid before crystallization of the impregnated active precursor takes place. They maintain that with slow drying, the boundary between the gas and the liquid phase passes through pores of the same radii, which are presumably uniformly distributed throughout the support bodies. Consequently, the active precursor will be deposited uniformly within the support bodies. Neimark, Kheifets and Fenelonov [8] derived a complex number, which is not dimensionless as mentioned by the authors, to predict whether slow or rapid drying prevails. In the complex, the Nusselt number, which pertains to heat transfer, apparently is the Sherwood number, which pertains to mass transfer. To calculate the complex, the pore size distribution of the solid and the drying conditions have to be known in detail, which is often difficult to achieve.

Presumably, the classification of Coulson and Richardson [23] is more useful to predict the distribution of the active precursor within the support bodies resulting from drying of a solution that is initially uniformly distributed throughout the support. During drying of a support with pores of a size of 2 µm to 2 mm, the liquid will rapidly migrate to the external surface, which will result in deposition of a considerable fraction of the dissolved material at the external edge of the support bodies. An important item with supports having wide pores is that the dissolved species may be deposited as small crystallites within the pore mouths. The small crystallites enclose narrow pores that exhibit a high capillary suction. Emptying of the pores at the mouths will, therefore, not proceed readily, which brings about that most of the dissolved material will be deposited at the external edge. However, the deposition of crystallites within the pore mouths at the external surface can also severely lower the rate of evaporation. As demonstrated by the experiments of Newitt, Nagar and Papadopoulus [28] with 12 µm sand, the pore mouths can be plugged by the deposited precursor, which will strongly decrease the rate of evaporation. Sand of 218 and 68 µm did not display plugging of the pore mouths. Another possibility is that crystallization of the dissolved material takes place on the external edge of the bodies, leading to a crust of dry, finely divided solid material that also brings about a very low drying rate.

With wide pores, it may be that the rate of drying can be thus elevated that the evaporation front migrates into the porous solid. With pores of 2 to 20 µm, the capillary forces are dominating. The thermal energy has to be supplied very rapidly, and the vapor has to be transported extremely fast from the external edge of the impregnated support bodies, to arrive at conditions where flow due to capillary forces cannot keep up with the removal of vapor from the external surface. With pores smaller than 2 µm, the friction due to the viscosity of the liquid becomes of the same order of magnitude as the capillary forces. Now, the viscosity of the impregnated solution is highly important. When the viscosity of the liquid is low, transport to the external surface will still prevail, unless the pores are very narrow. However, with very narrow pores, the water vapor pressure is diminished according to Kelvin's law, which will lead to a low rate of evaporation. The lower vapor pressure brings about that the liquid transport to the external surface can keep up more readily with the evaporation. Because the high capillary pressure of narrow pores furthermore lowers the pressure within the liquid inside the porous system, vapor bubbles arise easily in the center of the support bodies. The highly porous supports usually facilitate the formation of vapor bubbles. As long as the rate of evaporation has not been quantitatively established and the pore size distribution is not known in detail, it is very difficult to specify whether the front of evaporation has penetrated into the porous solid before the onset of crystallization of the dissolved material.

In the literature, a discrepancy about the effect of rapid drying on the distribution of the active precursor within support bodies is often mentioned. Whereas Neimark, Kheifets and Fenelonov [8] maintain that rapid drying will lead the active precursor to be deposited near the center of the catalyst body and not at the external edge, van den Berg and Rijnten [13] presented results indicating that fast drying brings about preferential deposition at the external edge. Though Maatman and Prater [5] speculated about redistribution of active precursors into small pores and deeper penetration into the interior of the support body during the drying step, they did not provide experimental evidence for a more uniform distribution being established during drying. Komiyama, Merrill and Harnsberger [10], on the other hand, mentioned that rapid drying leads to deposition of nickel(II) chloride impregnated in γ-alumina, preferably near the center of the support bodies.

It is well possible that in the experiments of van den Berg and Rijnten [13], transport of the liquid by capillary forces can keep up readily with the rate of evaporation, which results in an egg-shell or pellicular distribution. The authors performed the impregnation of γ-alumina with a copper(II) chloride solution, which migrated to the external edge during rapid drying. However, it is also important to take precipitation of basic copper chloride into consideration. Van der Meijden [30] extensively studied the precipitation of copper(II) from solutions of copper chloride, copper nitrate, copper sulfate, and copper perchlorate by raising the pH homogeneously either by the hydrolysis of urea or by injection of sodium hydroxide. Figure 15.4 represents the results he measured for copper(II) chloride. He observed precipitation of a basic copper chloride, $Cu_2(OH)_3Cl$ at a pH level of only 3.2. With the hydrolysis of urea, nucleation of the basic chloride is suppressed, which results in the formation of well-shaped bipyramides of the size of 50 μm. The basic copper chloride crystallites were much smaller with the injection of sodium hydroxide into a solution of copper(II) chloride kept at room temperature.

Van der Meijden [30] also demonstrated that nucleation of basic copper chloride is not promoted at the surface of silica or alumina, and that the basic copper chloride is thus not deposited onto the surface of the support. It can be expected that the pH of a copper(II) chloride solution rises in volume elements where the liquid contacts the alumina. The local growth of the pH level will lead to nucleation of

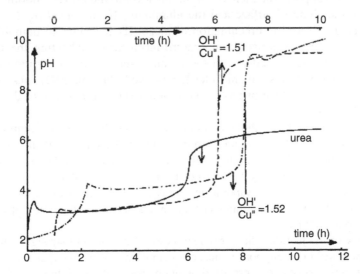

FIGURE 15.4 Precipitation of basic copper(II) chloride, $Cu_2(OH)_3Cl$, from homogeneous solution by hydrolysis of urea and injection of a sodium hydroxide solution below the level of the intensively agitated liquid. With the precipitation with urea, the copper concentration was 0.176 mol/l; with the precipitations with sodium hydroxide, the copper concentrations were 0.191 and 0.019 mol/l. The copper concentration of 0.019 mol/l brought about precipitation at pH level of about 4.1.

basic copper chloride. When the drying proceeds rapidly, the nuclei cannot grow, and small particles of basic copper chloride are transported together with the remaining dissolved copper chloride to the external edge of the support bodies. Slow drying enables the basic copper chloride nuclei to grow and to flocculate, which will severely impede transport with the liquid to the external surface of the support bodies.

Komiyama, Merrill and Harnsberger [10] investigated impregnation of γ-alumina with nickel(II) chloride; they observed that drying with a heating rate of 600°C/h resulted in no segregation of a nickel species at the external surface of the support, whereas a heating rate of 100°C/h brought about a substantial segregation at the surface. To prevent hydrolysis of nickel(II), the authors in some cases added nitric acid to the impregnated nickel(II) chloride solution. De Bokx, Wassenberg and Geus [31] showed that the first layers of nickel(II) deposited onto γ-alumina from a homogeneous solution reacts with the nickel surface to the above compound takovite, $Ni_6Al_2(OH)_{16}CO_3.5H_2O$, with a hydrotalcite structure. Titulaer [32] investigated the precipitation of takovite. He found that at a pH level below about 7, nitrate ions were taken up into the interlayers, preferably over carbonate ions. It is also likely that chloride ions are accommodated, preferentially in the interlayers at low pH levels.

A highly important difference between nickel(II) and copper(II) is that nickel precipitates at higher pH levels, where the interaction with alumina and silica support is much stronger. Accordingly, precipitation from a homogeneous solution of nickel(II) usually leads to a very uniform distribution of nickel(II) over the surface of the support. The increase in the pH level of the nickel chloride solution contacting the alumina surface and the aluminum(III) dissolved by the addition of nitric acid will lead to precipitation of nickel(II) and, thus, to immobilization of the nickel. At the faster rising temperature during the rapid heating, the dissolved nickel(II) reacts more rapidly with the alumina surface. After addition of nitric acid to prevent hydrolysis of nickel(II), dissolved aluminum(III) ions are more abundantly present, which brings about a more extensive reaction of the impregnated nickel ions.

Because the reaction with the alumina surface and the dissolved aluminum ions immobilizes the nickel, a more uniform distribution of the nickel(II) throughout the γ-alumina bodies results. Accordingly, the lack of segregation at the external surface of the γ-alumina bodies may be due not to the rate of capillary movement being unable to keep up with the evaporation rate at the surface, but to a more extensive reaction of nickel with the alumina surface and, hence, immobilization of nickel(II) at higher temperatures.

In view of the fact that the porous structure of the support strongly determines the transport within the support bodies, correct interpretation of experimental results involving active precursor crystallizing upon evaporation of the liquid and not precipitation due to a change in pH level or interaction with the support calls for a detailed knowledge of the porous structure of the support and the rate of evaporation. When the system contains many very narrow pores, the friction during the transport of liquid through the pores is thus high that the evaporation is more rapid and the front of the liquid moves at an early stage of the drying process into the support bodies. Whereas with pores smaller than 2 μm the friction is dominating, capillary forces are dominating with pores between about 2 and 20 μm. When capillary forces dominate, transport of liquid to the external surface can proceed very rapidly.

The general conclusion is that capillary forces lead generally to a rapid transport of the liquid to the external surface of the support bodies. When the mouths of the pores in the external surface of the supports are fairly narrow, the pressure of the liquid within the supports can be very low due to the capillary suction. Accordingly, vapor bubbles can readily arise within the support bodies. Nucleation of vapor bubbles on the finely divided support particles usually proceeds smoothly. Transport of most of the liquid to the external surface of the support bodies, therefore, can proceed readily. As a result, most of the active precursor is deposited at the external edge of the support bodies, which is generally observed experimentally.

It is questionable whether the evaporation occurs so slowly that the liquid front is established at pores of the same radius throughout the support bodies, unless the support contains almost exclusively very narrow pores. It is more likely that the liquid is rapidly transported to pores of the critical radius situated at the external edge from which the evaporation proceeds. Only with very narrow pores can the conditions of slow evaporation and, hence, a uniform distribution be achieved. With a support containing regions of very narrow pores, the active precursor can be deposited, preferably within the narrow pores. Kotter has presented some useful time constants to compare the rate of evaporation, the rate of transport of liquid within pores by capillary forces, and the rate of diffusion in the liquid within the pores [33]. This author also mentioned that organic salts of transition metals with ammonia and amines react to complexes that are soluble in water. At increasing content of amines, the viscosity of such solutions grows, which is attractive, because transport by capillary forces to the external edge of support bodies is suppressed when the viscosity of the liquid is high. Kotter and Riekert [34] also raised the viscosity of the impregnating solution by addition of hydroxy-ethyl-cellulose to the impregnating solution. It appeared that a much more uniform distribution of the active precursor results, due to the slower transport of the liquid by capillary forces.

15.5 EXPERIMENTS ON IMPREGNATION AND DRYING OF HIGHLY POROUS SUPPORTS

Above, we mentioned the scarcity of data about the rate of evaporation of liquid from impregnated support bodies of a well-known porous structure. To confirm the above results of Ceaglske and Hougen [25], Knijff [35] performed an experiment with a bed of glass spheres of a diameter between 1 and 53 μm and a depth of 13 mm. On top of the bed, a layer of larger glass spheres of a diameter of 230 to 320 μm was placed. A flow of air was passed over the bed at a linear velocity of about 5 m/s; the rate of evaporation was measured by periodically determining the weight of the bed. Knijff observed that after assuming the wet bulb temperature, the water present within the large glass spheres evaporated at an almost-constant rate. Subsequently, the rate of evaporation dropped steeply from a level of about 35 mg/min to a level of only about 2 mg/min, while the temperature of the bed rose from the wet bulb temperature 283 to 291 K. The results demonstrate that a layer of 2 mm of a wide-porous material severely decreases the rate of evaporation, in agreement with the results of Ceaglske and Hougen.

Knijff[35] made some important observations about the migration of liquids in porous materials. He mentioned that catalyst supports generally consist of stackings of more or less spherical particles or of acicular or platelike particles. The pores within such stackings of small particles almost never resemble the tortuous capillaries of a uniform cross-sectional area considered in theoretical treatments of the transport of liquids inside catalyst supports. At the contact areas of the small particles, narrow voids are present that retain the liquid by capillary forces after the large pore adjacent to the narrow voids has been emptied. Accordingly, the liquid flow to the external surface of a support body can continue after the liquid completely filling the wide pore adjoining the narrow voids has been evaporated. The special character of the pores inside catalyst support bodies (and, presumably, also in sand beds) brings about that approximately 90% of the moisture in the support can evaporate from liquid water that has been transported to the external surface. Knijff also mentioned that due to the suction of the narrow pore mouths at the external surface of the support bodies, vapor bubbles can readily arise within the liquid filling the support bodies. Schematically, Knijff arrived at the sequence of liquid arrangements in drying support bodies filled with pure water represented in Figure 15.5.

It is highly important that Knijff [35] also measured drying curves for a range of supports, the surface area, pore volume (the porosity), and the pore size distribution of which has been accurately determined. He passed a nitrogen flow through a bed of support bodies impregnated with pure water and measured continuously the water content of the gas flow out of the bed of support bodies. The support bodies were installed in a cylindrical wire basket of a length of 8 cm and diameter of 2 cm positioned coaxially in a thermostatted double wall horizontal glass tube, which was rotated at a rate of about 1 turn per second.

Figure 15.6 shows the results for rings of α-alumina and for sol-gel spheres of silica at 323 K. The pore width of the two supports is very different; the pore radii of the α-alumina (1 to 10 μm) are much larger than that of the sol-gel spheres (12 nm). From Figure 15.6, it can be seen that after heating the support bodies to the wet bulb temperature and after removal of the adhering water, the water evaporates almost completely within the constant drying rate period. Figure 15.6 indicates that the capillary flow of water to the external surface of the support bodies can keep up with the evaporation until the water has almost completely been removed. The slower drop from the rate of evaporation of the adhering water to that of the constant rate period exhibited by the sol-gel spheres is due to the fact that the evaporation of adhering water present in between contacting spheres overlaps with evaporation out of the pores of the support. Though Coulson and Richardson predict that the friction will dominate with pores of a diameter less than 2 μm [23], the sol-gel spheres having a pore radius of only 25 nm still exhibit a sufficiently rapid capillary flow. It is apparent that the capillary flow to the

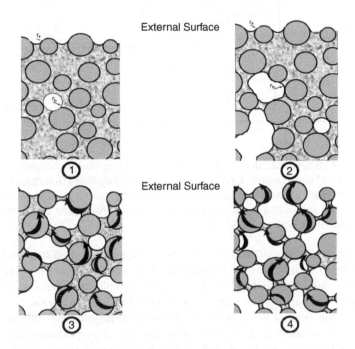

FIGURE 15.5 Schematic representation of arrangements of liquid within drying support bodies initially filled with water. First, formation of gas bubbles within the liquid due to the suction of the narrow pore mouths at the external surface (the two pictures at the top). Next, emptying of the pores, which leads to transport of liquid along narrow voids at the contact areas of the elementary particles of the support to take over the transport of liquid to the external surface of the support body (pictures at the bottom).

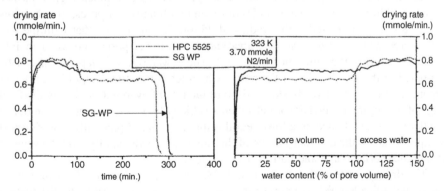

FIGURE 15.6 Left-hand side: drying rate as a function of drying time; right-hand side: drying rate as a function of water content. Drying at 323 K. HPC 5525 sieve fraction 2 to 4 mm of α-alumina rings (Norton), pore volume 0.40 ml/g, surface area 0.27 m^2/g, porosity 0.62, pore radii 1 to 10 mm. SG WP shell silica sol-gel spheres 2 to 3 mm, pore volume 0.85, surface area 70 m^2/g, porosity 0.66, pore radius 25 nm.

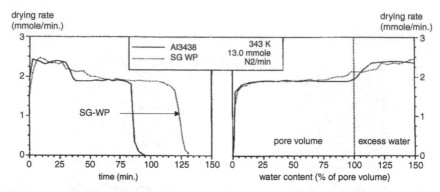

FIGURE 15.7 Left-hand side: drying rate as a function of drying time; right-hand side: drying rate as a function of water content. Drying at 343 K. Al 3438 Engelhard α-alumina cylindrical bodies 3×3 mm, pore volume 0.20 ml/g, surface area 1.0 m²/g, porosity 0.45, pore radii 0.1 to 10.0 μm. SG WP shell silica sol-gel spheres 2 to 3 mm, pore volume 0.85, surface area 70 m²/g, porosity 0.66, pore radius 25 nm.

external edge of the support bodies will lead to deposition of an egg-shell of the active precursor with drying of a solution of an active precursor. Figure 15.7 compares the rate of drying at 343 K of the same sol-gel spheres with that of another α-alumina support having pore radii of 0.1 to 10 μm. Also, the last support exhibited a rapid flow of liquid water to the external edge of the support bodies. The conclusion is that it is difficult to raise the rate of evaporation thus far that the capillary flow cannot supply sufficiently liquid water to the external edge, without raising the temperature to the boiling point of the liquid inside the support bodies with α-alumina bodies and with the silica sol-gel spheres having pores of a radius of 12 nm.

Figure 15.8 compares the drying rate of sol-gel spheres at 343 K. The drying of sol-gel spheres SG WP has already been represented in Figure 15.6 and Figure 15.7. The drying rate is compared with that of sol-gel spheres of a much higher surface area, viz., 230 m²/g as compared to 70 m²/g, and with narrower pores, viz., a radius of 7 to 8 nm as compared with a radius of 25 nm. It can be seen that with the sol-gel spheres with the higher surface area and the narrower pores, the water mainly evaporates during the constant rate period. The lower rate of evaporation can be attributed to the smaller size of the pores, which leads to a lower equilibrium water vapor pressure. The relatively slower drop of the rate of evaporation during the final stage of the drying of the NP sol-gel spheres is likely to be due to evaporation out of smaller pores and, hence, to a lower equilibrium pressure. Apparently, the transport of liquid water to the external surface of the sphere proceeds rapidly with the spheres of a higher surface area and narrower pores, too.

The drying curve of the alumina Al 4723, compared in Figure 15.9 with that of the sol-gel silica SG-WP, is interesting. This γ-alumina has a very wide distribution of pore radii, viz., from 7 to 1000 nm, whereas 70% of the pore volume is present within narrow pores. The drying period of constant rate involves 30% of the pore volume, whereas two periods with falling rates are subsequently exhibited. Apparently, the narrow pores retain the water, whereas the liquid front retracts relatively

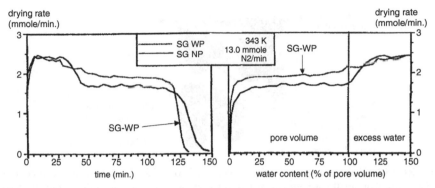

FIGURE 15.8 Left-hand side: drying rate as a function of drying time; right-hand side: drying rate as a function of water content. Drying at 343 K. SG WP shell silica sol-gel spheres 2 to 3 mm, pore volume 0.85, surface area 70 m²/g, porosity 0.66, pore radius 25 nm. SG NP shell silica sol-gel spheres 2 to 3 mm, pore volume 0.88 ml/g, surface area 230 m²/g, porosity 0.67, pore radius 7 to 8 nm.

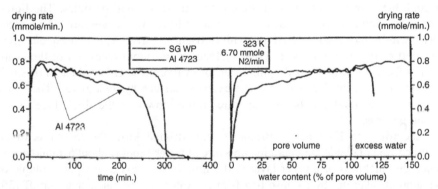

FIGURE 15.9 Left-hand side: drying rate as a function of drying time; right-hand side: drying rate as a function of water content. Drying at 323 K. Al 4723 Engelhard γ-alumina spheres 3 mm. Pore volume 0.52 ml/g, surface area 200 m²/g, porosity 0.66, pore radii 7 to 1000 nm. SG WP shell silica sol-gel spheres 2 to 3 mm, pore volume 0.85, surface area 70 m²/g, porosity 0.66, pore radius 25 nm.

rapidly from the external edge through the wide pores. The dropping rate of evaporation indicates that the evaporation proceeds from increasingly narrow pores. Apparently, the γ-alumina contains clusters of very small particles enclosing pores of about 4 to 10 nm, whereas the clusters enclose pores the diameter of which is 2 orders of magnitude larger, viz., about 1000 nm. The interpretation is that liquid water is evaporating from narrow pores ending at the external surface, whereas evaporation of water present in narrow pores within the bodies displays the final dropping evaporation rate. This alumina support apparently better fulfills the conditions required for the slow evaporation as described by Neimark, Kheifets and Fenelonov [8].

Figure 15.10 schematically represents the structure of the γ-alumina Al 4723. A significant fraction of the clusters of very small particles is situated at the

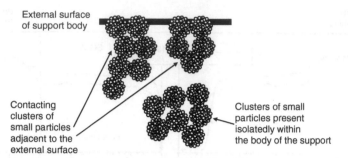

FIGURE 15.10 Explanation of the drying curve of Al 4723. The liquid flow can continue along clusters of small particles connected to the external surface, though the rate of evaporation falls due to the narrower pores. Clusters present without a connection via other clusters of small particles to the external surface dry very slowly due to the slow transport of water vapor through the porous body.

external edge of the support bodies. The liquid can migrate to the external surface through the narrow capillaries of the clusters of very small particles. The liquid evaporates slowly from the narrow pores due to lower vapor pressure. The narrow pores within the clusters lead to a high friction force. The consequent slow transport of the liquid to the external surface brings about recession of the evaporation front, resulting in a dropping rate of evaporation. Other clusters of small particles are present, isolated from the external surface by wide pores. The water present in the isolated clusters evaporates more slowly in the final stage of the drying process.

According to Hagymassy, Brunauer and Mikhail [36], the thickness of an adsorbed water layer at the saturation pressure of water vapor is about 6 monolayers. When the thickness of a water layer is 0.3 nm, the fraction of the pore volume of the support taken up by an adsorbed film of water can be calculated. Knijff [35] established that the fraction of the pore volume filled with 6 monolayers of water agrees well with the moisture content, at which the drying rate displays the first, slight drop below the level of the constant rate period. The rate of evaporation that is maintained during that period indicates that transport of the water film proceeds more slowly when the thickness of the water layer decreases below about 6 monolayers. However, the drying experiments clearly indicate that liquid water can be transported not only by capillary pressure, but also as a film adsorbed on the surface of oxidic supports. A prerequisite for this mechanism of transport may be that the surface of the support is hydrophilic.

Knijff [35] also measured drying curves for a support, part of the surface of which is hydrophobic, viz., activated carbon. Figure 15.11 represents the drying curve measured for activated carbon. The oxidized carbon, A Ox, is more hydrophilic than the carbon that has not been oxidized (A Wa). The drying curves exhibit two plateaus and two periods where the drying rate is dropping more rapidly. The activated carbon bodies thus exhibited, after a constant rate drying period, a decreasing rate of evaporation over an extended range of moisture contents. Knijff argued that activated carbon has both hydrophilic and hydrophobic surfaces. Investigation

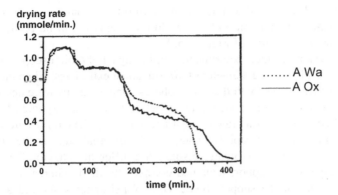

FIGURE 15.11 Course of drying rate as a function of time of activated carbon. A Wa surface area (nitrogen) 1690 m²/g, 1.56 ml/g pore volume. A Ox surface area (nitrogen) 1690 m²/g, 1.47 ml/g pore volume. Both supports contain pores from 4 to 5000 nm.

of activated carbons in the transmission electron microscope supports the presence of surfaces of a different nature. Beside very finely divided carbon particles, graphitic plates also show up in the micrographs. For the carbon A Wa, the total pore volume is 1.56 ml/g, the macropore volume is 0.61 ml/g, and the mesopore volume is 0.32 ml/g, which leaves 0.63 ml/g for the micropores. After oxidation, the pore volume of A Ox is 1.47 ml/g, the macropore volume is 0.59 ml/g, and the mesopore volume is 0.32 ml/g, which leaves 0.56 ml/g for the micropores.

Knijff [35] established the amount of water evaporated during the constant rate period and found 0.57 g/g for the A Wa carbon and 0.56 g/g for the A Ox carbon, quantities that agree very well with the macropore volume as determined from mercury porosimetry. Knijff assumed that the water evaporated during the final period, in which the drying rate falls slowly, is present in micropores, which he refers to as *bound water*. The drying rate curve suggests an amount of bound water of 0.65 g/g for A Wa and 0.58 g/g for A Ox, which agrees reasonably with micropore volumes of 0.63 ml/g and 0.56 ml/g, respectively. The low final rate of evaporation is attributed to the high interaction of water with the oxidized surfaces of the activated carbon. The amount of water evaporating during the steep drop in the rate of evaporation corresponds nicely to the volume of the mesopores. Apparently, the evaporation front moves rapidly into the carbon bodies during the evaporation of water present within the mesopores. It is likely that the presence of hydrophobic surfaces in the carbon bodies promotes the penetration of the evaporation front into the carbon bodies. The fact that interaction of water with the oxidized carbon surface is larger is apparent from the lower rate of evaporation during the final stage of the oxidized carbon A Ox in Figure 15.11.

Knijff [35] also performed drying experiments with α-alumina bodies and sol-gel silica spheres impregnated with solutions of 1.0 molar $K_3Fe(CN)_6$ and 5.0 molar $CuCl_2$. Both dissolved species were deposited on or into the external surface of the support bodies. The deposition of the dissolved material led to a severe drop in the rate of evaporation, due to either plugging of the pore mouths or the deposition of a crust of crystallites onto the edge of the support bodies. The γ-alumina Al 4723, on the other hand, exhibited a much more uniform distribution of the deposited solutes. The better distribution within the alumina support can be predicted from the rate of evaporation curve of Figure 15.9.

Van den Brink performed experiments employing a light microscope on a model support of glass spheres of a diameter of 40 μm, silica extrudates, silica tablets, and silica sol-gel spheres [37]. With the glass spheres, Haines jumps were apparent that led to the evaporation front penetrating in between the glass spheres before the moisture had completely evaporated. Apparently, the capillary forces are not sufficient to transport the water rapidly enough to the external edge at the end of the drying process. The silica extrudates displayed a different behavior. The pores at the external edge of the support remained completely filled, whereas dry volumes arose in the center of the support body. The sol-gel spheres showed a different behavior, though some spheres also became dry in the center and the external edge remained wet. Most of the sol-gel spheres, however, first developed a dry external edge, after which smoothly dry channels developed within the support. Apparently, a uniform pore size as present within the sol-gel spheres leads to this behavior. Van den Brink's results are very relevant, because they demonstrate that the porous structure of the support has to be known in detail to explain the phenomena during drying of impregnated support bodies. Depending upon the porous structure, very different phenomena can be observed.

Employing glass spheres of 40 μm, van den Brink investigated the deposition of ammonium iron(II) sulfate upon drying [37]. He observed the crystallization of small crystallites at the external edge of the body of glass spheres. The narrow pores enclosed by the small crystallites took up the liquid, which resulted in a bundle of crystallites at the external edge. Also, the highly porous supports exhibited the formation of an egg-shell of the active precursor, when a crystallizing active precursor, such as ammonium iron(II) sulfate, was employed. Some larger voids in the support bodies were apparently enclosed by very narrow pores, which brought about crystallization of some large crystallites within the support bodies. Liquid water cannot escape through the very narrow pores. Consequently, the water can only evaporate, leaving behind relatively large crystallites.

When van den Brink employed dissolution of iron(III) nitrate and iron(III) chloride, the results were different [37]. Hydrated iron oxide had deposited at the contact areas of the glass spheres. Van der Giessen [38] and, later on, de Bruyn and coworkers [39, 40] showed that already at a pH level of 2, iron(III) has reacted to small particles of hydrated iron oxide. Van der Giessen mentioned a particle size of 3 nm and de Bruyn and coworkers, a size of 4 nm. Heating the solutions of iron(III), which are acid due to hydrolysis of the iron(III), leads to loss of acid and, hence, to a rise in pH value, which brings about flocculation of the hydrated iron oxide particles. It has been shown that the hydrated iron oxide particles do not attach to either glass or to silica or alumina supports, possibly due to both having a positive electrostatic charge. Consequently, clusters of hydrated iron oxide particles are generated in between the glass spheres that cannot migrate with the liquid to the external edge. The clusters can, however, migrate with the remaining liquid to nearby contact areas of the glass spheres.

Van den Brink obtained analogous results with the real silica supports [37]. Iron(II) ammonium sulfate was deposited as an egg-shell at the external edge of the support bodies, whereas the hydrolyzing and flocculating iron(III) species remained within the support bodies. However, the desired distribution of very small iron oxide particles was not achieved; calcination resulted in large iron oxide particles distributed uniformly throughout the support. Calcination led to the loss of water of the flocculated hydrated iron(III) oxide that sinters to large iron(III) oxide particles. Van den Brink prepared catalysts by impregnating silica extrudates prepared from a Degussa Ox 50 support. The pore volume was 0.8 ml/g, the mean pore radius was 35 nm, and the BET surface area was 44 m^2/g. After impregnation, the extrudates were dried in a rotating reactor for 2 h at room temperature in a flow of about 15 l/min of air. The temperature was next raised to 60°C for 3 h, and to 120°C for another 3 h. The final calcination was performed at 600°C for 3 h.

Investigation of the calcined support bodies in the light microscope showed that impregnation with iron(III) chloride and iron(III) nitrate had resulted in formation of uniformly red-colored bodies, due to the intensive color of iron(III) oxide. The same uniform distribution of clusters of hydrated iron oxide particles was observed when the drying process was studied in the light microscope, where the rate of evaporation was different. It is very interesting that the hydrolyzing iron(III) salts exhibited a different distribution with the tabletted Ox 50 support. The tablets displayed, after impregnation with iron(III) nitrate or iron(III) chloride, a pale circle in the center of the extrudate cross section and a dark-colored cylinder at the external edge. It is well known that tabletting leads to fairly narrow pores at the external surface of the bodies. The narrow pores provide a high capillary suction force without contributing much to the friction, in view of their short length. However, the rate of evaporation from narrow pore mouths is relatively low, due to the lower saturation pressure. Apparently, the loss of acid components from the liquid is slowed down such that a significant fraction of the tiny hydrated iron oxide particles are transported with the liquid to the external edge of the tablets before the flocculating has proceeded.

Terörde [41] confirmed van den Brink's results. With potassium iron(III) cyanide, a compound that crystallizes well, he also obtained an egg-shell distribution, whereas from iron nitrate and iron chloride solutions, flocculated hydrated iron oxide clusters were deposited uniformly throughout the support bodies. Subsequent calcination in Terörde's experiments also resulted in large iron(III) oxide particles. Above in the discussion of data from the literature, we emphasized the importance of chemical reactions in the liquid phase of the impregnated species, during the rise in temperature during drying. The results of van den Brink unambiguously indicated the importance of the hydrolysis and flocculation of iron(III) during drying.

Drying of a liquid drop on a solid surface is an interesting limiting case of the drying of an impregnated support. Transport by capillary forces cannot proceed with the drying of drops. Some interesting papers have been published about the stains left by drying drops [42, 43, 44]. Most well known are rings left by dried coffee drops. The authors show that the meniscus of the liquid is pinned at the edge of the initial drop, where small solid particles present in the liquid are being deposited. However, the meniscus of a liquid not containing solid particles was also found to be pinned; the cause of the pinning has not been disclosed. The authors stated that the outward flow of liquid is due to the tendency of the interface to minimize its surface energy; they developed an involved theory to account for the outward flow of liquid. The outward flow was experimentally demonstrated from the movement of tiny polystyrene microspheres. Deegan and coworkers [42] have observed the outward flow of liquid with drying drops for a large range of liquids and substrate, provided the surface was partially wet by the liquid [44]. The results on the drying of drops are important for the drying of impregnated supports. The experiments show that transport of liquid can also proceed within water layers not completely filling the pores. It is not certain whether the menisci of the liquid are also pinned at the external surfaces of support bodies, and it is also uncertain whether the drying phenomena are also displayed on curved surfaces.

That the deposition of small solid particles is important in determining the deposition of crystallizing compounds is apparent from Figure 15.12, which represents a micrograph of a dried drop of copper nitrate. It can be seen that the copper nitrate crystallites have been deposited at the edge of the original drop. An obvious explanation, not taking into account the above work on drying drops, is that the tiny copper nitrate crystallites, initially crystallized at the edge of the drop, enclose narrow pores that take up the liquid by capillary forces. As a result, the copper nitrate crystals have been deposited almost completely at the edge of the drop. Analogously, Boon [45] observed that drying of a capillary glass tube filled with a copper nitrate solution resulted in deposition of copper nitrate crystallites at the external edge of the capillary end and not within the capillary. Drying of supports with wide pores, such as α-alumina, therefore is likely to result in deposition of the active precursor at the external edge of the pores, as indicated in Figure 15.13.

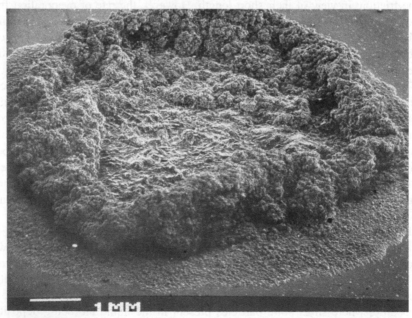

FIGURE 15.12 Drop of solution of copper(II) nitrate dried on microscope glass slide. Note preferential build-up of crystallites at the rim of the drop.

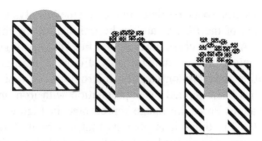

FIGURE 15.13 Deposition of small crystallites at the external edge of pores in supports having wide pores, such as a-alumina, as observed experimentally by Boon [45].

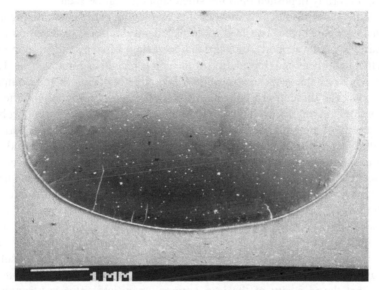

FIGURE 15.14 Drop of solution of copper(II) citrate dried on microscope glass slide. Note absence of large crystallites; no transport to the rim of the drop.

Figure 15.14 represents a dried drop of water containing copper(II) citrate. Apparently, no visible crystallites have formed from the citrate, and migration of the dissolved species to the edge of the drop has not proceeded. Apparently, the crystallization proceeds badly with the citrate salt, and thus no large crystallites can be seen. The result shown in Figure 15.14 suggests that the desired uniform distribution of the active precursor throughout the support bodies can be obtained by using salts of badly crystallizing anions of the active precursor.

Accordingly, van den Brink [37] demonstrated that impregnation with a solution of iron(III) ammonium citrate and iron(III) gluconate resulted after drying in a uniform distribution of the active precursor within the support body. Transmission electron microscopy of the subsequently calcined catalysts displayed the presence of very small iron oxide particles, homogeneously dispersed over the silica. With impregnation of iron(III) ammonium EDTA, the pH of the impregnating solution is important. A solution of a pH of 5.3 produced an egg-shell distribution, and a solution of pH 7.1, a faint egg-shell. However, solutions of a pH of above 8.5 led to uniform distributions and tiny iron oxide particles in the calcined catalysts. The effect of the pH on the deposition of the EDTA salt can be rationalized by the fact that at low pH levels, iron EDTA rapidly crystallizes as $NH_4FeEDTA$. At higher pH levels, the iron is present in anions that crystallize badly. Figure 15.15a and Figure 15.15b shows high-resolution transmission electron micrographs of a silica support, onto which iron oxide has been applied by impregnation with iron ammonium citrate, drying, and calcination. From the image represented in Figure 15.15a taken at a magnification of 210 kx, times it is obvious that a very uniform distribution of very small particles can be achieved with badly crystallizing salts. Figure 15.15b represents an image taken at a higher magnification; the presence of extremely small particles, uniformly deposited onto the silica support, can be seen.

From the above results, it is obvious that employing badly crystallizing salts to impregnate support bodies can lead to an excellently uniform distribution of very small particles on the support. An additional advantage of using salts of complexing acids is that hydrolysis and flocculation of the active precursor cannot occur. Terörde [41] set out to establish the reason of the favorable behavior of the complexing organic salts. The difficult crystallization cannot be the only reason why the active precursor does not migrate with the liquid to the external edge of the support bodies. If the viscosity of the liquid or the interaction with the surface of the support would not be different with the salt of the organic complexing acids, the liquid would migrate to the external edge of the support bodies, and the active precursor would be deposited as an egg-shell. When the viscosity of the liquid is operative in impeding the flow of the liquid, it is most effective when the viscosity during impregnation is low, to enable the impregnating liquid to penetrate readily into narrow pores of the support. During drying, a high viscosity is desired. A constituent of the impregnating solution that raises the viscosity when water is removed is, therefore, attractive. Terörde measured the viscosity as a function of the extent of drying for solutions of iron ammonium citrate and iron nitrate. Figure 15.16 represents his results. It is apparent that the viscosity of the iron ammonium nitrate solution strongly rises during drying, in contrast to the solution of iron(III) nitrate. The steep rise in the viscosity certainly will assist in restricting the flow of the liquid to the external edge of the support bodies.

The viscosity can also affect the growth of crystallites. Above, we discussed the results of Kotter [33] and Kotter and Riekert [34], who raised the viscosity of impregnating solutions by addition of hydroxy-ethyl cellulose. Terörde [41] studied in a light microscope the crystallization of potassium iron(III) hexacyanide with 1 wt.% and 2 wt.% hydroxy-ethyl cellulose. Without hydroxy-ethyl cellulose, large crystallites at the edge of the drop resulted, whereas smaller crystallites more uniformly distributed over the area of the drop were observed from the solution with hydroxy-ethyl cellulose. Impregnating support bodies of α-alumina with a solution of potassium iron(III) hexacyanide led to an egg-shell distribution, but impregnation with the same solution containing either 1 or 2 wt.% of hydroxy-ethyl cellulose resulted in a uniform distribution. Figure 15.17 shows the results of the impregnation; the effect of the hydroxy-ethyl cellulose is striking.

The viscosity of the impregnated solution thus affects the movement of the liquid within the pores as well as the crystallization of the dissolved active precursor.

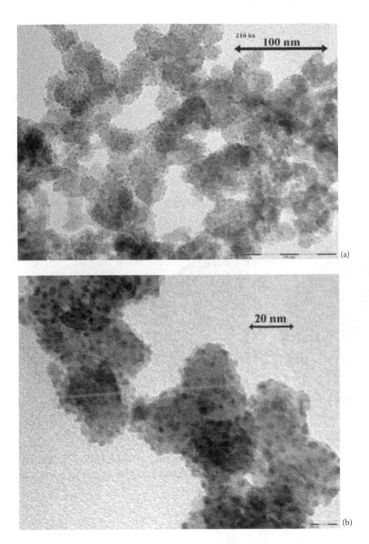

FIGURE 15.15 Transmission electron micrograph of iron(III) oxide deposited on silica. The catalyst has been prepared by impregnation of silica with ammonium iron(III) citrate, drying, and calcination. (a) The very uniform distribution of tiny iron oxide crystallites is apparent. (b) At a more elevated magnification, the small size of the iron oxide particles is evident.

However, the interaction with the surface of the support can also influence the distribution of the support. Above, we dealt with the work of Knijff [35], who observed a layer of a thickness of about six water molecules to be adsorbed on hydrophilic surfaces at the saturation pressure of water, which pressure will be present within the partly emptied pore system of the support. Figure 15.18 represents the structure of citrate and gluconate ions. It is likely that the complexing organic molecules containing one or more hydroxyl groups will affect the thickness and the mobility of the adsorbed layer of water molecules via hydrogen bonding. Also,

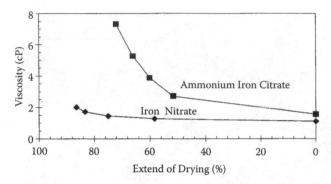

FIGURE 15.16 Course of viscosity during drying of solutions of ammonium iron(III) citrate and iron(III) nitrate.

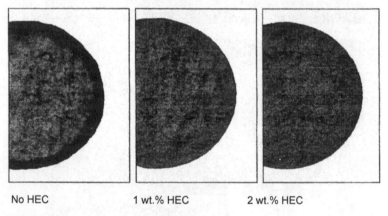

No HEC   1 wt.% HEC   2 wt.% HEC

FIGURE 15.17 α-alumina impregnated with solution of potassium hexacyanoferrate and subsequently dried. From the left- to the right-hand side, the solutions contained 1 wt.% and 2 wt.% hydroxy-ethyl cellulose (light microscopy).

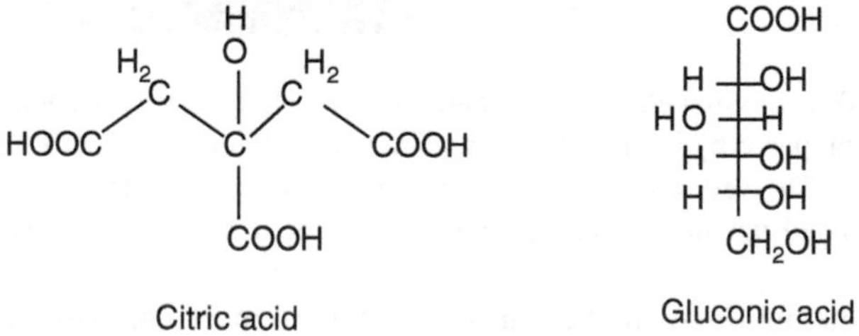

FIGURE 15.18 Structure of citrate and gluconate ions.

reactions between different complexing molecules may proceed, leading to dimers or oligomers. It is, therefore, interesting to investigate the effect of the interaction of the constituents of the impregnating solutions with the surface of the support. Model experiments with the atomic force microscope on the surface of silicon wafers turned out to be very informative.

Ten Grotenhuis et al. [46] investigated the distribution of copper nitrate crystallites deposited from cyclohexane on the surface of silicon wafers. The solution of copper nitrate in cyclohexane was applied by spin coating on the surface of the wafer. Figure 15.19 shows the result. It can be seen that clusters of tiny crystallites have been deposited on the surface of the wafer. The size of the crystallites is apparent from the high magnification. The natural surface of silicon wafers is hydrophobic. The silicon oxide at the surface of the wafer is not hydroxylated; hydrogen bonding with water is not possible. Treatment with an aqueous mixture of ammonia and hydrogen peroxide leads to a hydrophilic surface.

Figure 15.20 represents the distribution and the size of the copper nitrate crystallites deposited on the hydrophilic silicon surface. On the hydrophilic surface, tiny $Cu(NO_3)_2.6H_2O$ crystallites have been deposited uniformly. The effect of the

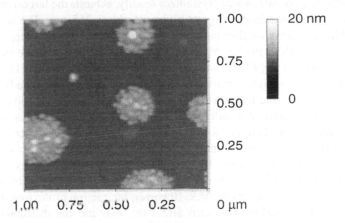

FIGURE 15.19 Copper nitrate ($Cu(NO_3)_2.6H_2O$) crystallites deposited from a copper nitrate solution in cyclohexane on the natural surface of a silicon wafer [44].

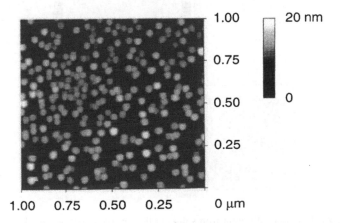

FIGURE 15.20 $Cu(NO_3)_2.6H_2O$ crystallites deposited from a copper nitrate solution in cyclohexane on the hydrophilic surface of a silicon wafer [44]. Treatment with an aqueous solution of ammonia and hydrogen peroxide has resulted in a hydrophilic surface on the silicon wafer.

interaction of the solution with the surface is evident. On the surface of the hydrophobic silicon wafer, the solution of $Cu(NO_3)_2 \cdot 6H_2O$ prefers to wet the surface of the initially deposited copper nitrate crystallites. Consequently, the solution is taken up in between the initially deposited small crystallites, where evaporation of the water leads to more small crystallites. Figure 15.19 shows the resulting clusters of tiny crystallites. With a hydrophilic surface, the copper nitrate solution is not selectively attracted by the initially crystallized small copper nitrate crystallites, and crystallization will take place uniformly over the surface.

Terörde performed experiments with different iron salts on hydrophilic silicon surfaces [41]. Figure 15.21 represents his results. The size of the deposited particles is apparent from the height of the particles rather than from the apparent lateral extension, due to the size of the tip of the atomic force microscope. Mohr's salt, iron(II) ammonium sulfate, which crystallizes readily, exhibits the largest crystallites. The height of the crystallites suggests a size of about 215 nm. The hydrolyzing iron(III) chloride shows smaller particles, viz., of about 15 nm, whereas iron(III) nitrate displays smaller particles, viz., of about 7 nm. The reason is the flocculation of the initially present tiny (3 to 4 nm) particles of hydrated iron oxide. The volatility of hydrogen chloride is higher than that of nitric acid. Consequently, the flocculation of the tiny particles in the liquid will proceed faster with the $FeCl_3$ solution, which will lead to larger clusters. With nitric acid, resulting from the hydrolysis of iron(III), the volatilization proceeds more slowly, and the flocculation starts after more water has evaporated, which results in smaller clusters. With ammonium iron citrate, note the higher magnification in Figure 15.21; an almost-continuous film results. Formation of the continuous film may be due to the rise in viscosity during drying, which impedes formation of larger crystallites. However, interaction of the iron complexes with the hydrophilic surface may also affect the size and the distribution of the crystallites.

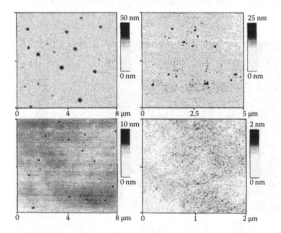

FIGURE 15.21 Drying of different iron salts applied by spin-coating on silicon wafers, the surface of which is hydrophilic due to treatment with an aqueous H_2O_2-NH_3 solution. Top left: iron(II) ammonium sulfate; top right: iron(III) chloride. Bottom left: iron nitrate; bottom right: iron ammonium citrate.

To establish the interaction of the impregnating solutions with the surface of silica supports, Terörde measured the amount of water required to completely remove different iron salts applied onto the surface of silicagel in a column chromatograph [41]. It appeared that an elution volume of water of 5.0 ml was required to remove the iron ammonium citrate, and of 6.2 ml to remove the Mohr's salt. Iron(III) chloride called for an elution volume of 6.3 ml, and iron(III) nitrate for a volume of 7.7 ml. Apparently, the iron species in the initially impregnated solution of iron ammonim citrate does not interact strongly with the silica surface. The relatively small interaction of the iron(III) complex of citric acid with the surface of silica indicates that the presence of the citric acid affects the adsorbed layer of water molecules that remains within the partially dried impregnated support. A thicker, less mobile layer of water molecules, citric acid anions, and iron(III) ions remains than without citric acid, where a layer of about six fairly mobile water molecules is present.

To demonstrate the effect of the interaction with the silica surface, Terörde measured the distribution resulting from drying silica supports impregnated with ammonium iron(III) citrate with different amounts of hydroxyl groups at the surface [41]. A low concentration of hydroxyl groups was achieved by calcining the support at 750°C and performing the impregnation immediately after the support has been cooled to room temperature in an exsiccator. Another silica support was treated with the aqueous solution of ammonia and hydrogen peroxide to obtain the maximum number of hydroxyl groups. A final silica support was denoted as a standard silica support; the support was stored under atmospheric conditions for several months after having been calcined at 750°C. Figure 15.22 represents the results. It can be seen that the liquid is transported to the external edge with the support containing almost no hydroxyl groups at the surface. A substantial fraction of the liquid and all of the dissolved ammonium iron citrate complex remains localized within the support, with the support treated with the ammonium and hydrogen peroxide. The standard support displays some migration of the complex to the external edge; however, transmission electron microscopy performed on samples taken from the

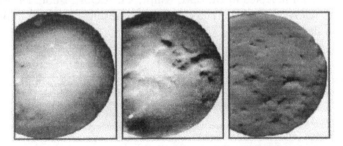

FIGURE 15.22 The distribution of iron oxide of silica supports impregnated with iron(III) ammonium citrate, dried, and calcined. The support at the left-hand side has been previously calcined at 750°C and was impregnated immediately after being cooled to room temperature. The support at the right-hand side was treated with an aqueous ammonia and hydrogen peroxide solution to raise the number of surface hydroxyl groups. The support at the center had previously been equilibrated under atmospheric conditions for some months (light microscopy).

edge and from the center did not exhibit a large difference in density of iron oxide crystallites. The experiments with the different silica support, therefore, demonstrate unambiguously the effect of the interaction of the solution in a later stage of the drying process on the resulting distribution of the iron oxide crystallites. As mentioned above, we feel that the thickness and the mobility of the water film remaining on the surface of the support in the final stage of the drying process is responsible for the uniform distribution of the active precursor.

15.6 CONCLUSIONS

The experimental data discussed above have demonstrated that a number of often complicated chemical phenomena can severely affect the distributions of the active component(s) resulting from impregnation and drying of preshaped support bodies. Though the production of supported catalysts by impregnation and drying seems to be fairly easy, the chemistry of the impregnated active precursors have to be considered carefully. Also, the effect of the porous structure of the supports on the course of the drying rate and, consequently, on the distribution of the active precursor within the dried support bodies appeared to be appreciable. Accordingly, it is worthwhile to establish drying rate curves on well-characterized support bodies. The nature of the surface of the support, hydrophilic or hydrophobic, also turned out to be important.

Thus far, treatment of impregnation and drying has been dominated by strictly physical considerations of transport and mathematical descriptions of the transport. The above-presented experimental results hopefully have also shown the importance of actual transport of liquid elements in drying support bodies and the proceeding chemical processes.

ACKNOWLEDGMENT

The author is much indebted to Dr. L.M. Knijff, Dr. A.Q.M. Boon, Dr. P.J. van den Brink and Dr. R.J.A.M. Terörde. The experimental results discussed in this chapter resulted from their Ph.D. work.

REFERENCES

1. C.H. Satterfield, *Mass Transfer in Heterogeneous Catalysis* M.I.T. Press, Cambridge, MA, p.131, 1970.
2. P.B. Weisz, Trans. *Faraday Soc.* 63, 1801, 1967.
3. P.B. Weisz and J.S. Hicks, Trans. *Faraday Soc.* 63, 1807, 1967.
4. P.B. Weisz and H. Zollinger, Trans. *Faraday Soc.* 63, 1815, 1967.
5. R.W. Maatman and C.D. Prater, *Ind. Eng. Chem.* 49, 253, 1957.
6. R.W. Maatman, *Ind. Eng. Chem.* 51, 913, 1959.
7. L.I. Kheifets, A.V. Neimark and V.B. Fenelonov, Kinet. i Kataliz 20, 760, Translation, p. 626, 1979.

8. A.V. Neimark, L.I. Kheifets and V.B. Fenelonov, *Ind. Eng. Chem. Prod. Res. Dev.* 20, 439, 1981.

9. S.-Y. Lee and R. Aris, *Catal. Rev.-Sci. Eng.* 27, 207, 1985.

10. M. Komiyama, R.P. Merrill and H.F. Harnsberger, *J.Catal.* 63, 35, 1980.

11. M. Komiyama, *Catal. Rev.-Sci. Eng.* 27, 341, 1985.

12. G. Berrebi and Ph. Bernusset, in *Preparation of Catalysts, Scientific Bases for the Preparation of Heterogeneous Catalysts*, B. Delmon, P.A. Jacobs and G. Poncelet, (Eds.), Elsevier Scientific Publishing Company, Amsterdam, p. 25, 1976.

13. G.H. Van den Berg and H.Th. Rijnten, in *Preparation of Catalysts II, Scientific Bases for the Preparation of Heterogeneous Catalysts*, B. Delmon, P. Grange, P. Jacobs and G. Poncelet, (Eds.), Elsevier Scientific Publishing Company, Amsterdam, p. 265, 1979.

14. A.C. Vermeulen, J.W. Geus, R.J. Stoll and Ph.L. de Bruyn, *J. Coll. Interf. Sci.* 51, 449, 1975.

15. R.J.M.J. Vogels, Thesis, Utrecht University, Chapter 7, Preparation of Pillaring Agents using the Hydrolysis of Urea, 1996.

16. W.A. Spieker and J.R. Regalbuto, *Chem. Eng. Sci.* 56, 3491, 2001.

17. J.F. Roth and T.E. Reichard, *J. Res. Inst. Catal.* Hokkaido Univ. 20, 85, 1972.

18. A. Lekhai, B.J. Glasser and J.G. Khinast, *Chem. Eng. Sci.* 59, 1063, 2004.

19. K.B. Agashe and J.R. Regalbuto, *J. Coll. Interf. Sci.* 185, 174, 1997.

20. J.R. Regalbuto, A. Navada, S. Shahid, M.L. Bricker and Q. Chen, *J. Catal.* 184, 335, 1999.

21. J.R. Regalbuto, K. Agashe, A. Navada, M.L. Bricker and Q. Chen, in *Preparation of Catalysts VII,* B. Delmon, P.A. Jacobs, R. Maggi, J.A. Martens, P. Grange and G. Poncelet (Eds.), Elsevier Scientific Publishing Company, Amsterdam, p. 147, 1998.

22. E. Michalko, U.S. Patent 3,259,589, July 5, 1966.

23. J.M. Coulson and J.F. Richardson, *Chemical Engineering Volume 2,* Third Edition, Pergamon Press, Oxford, Chapter 16, p. 710, 1978.

24. J. van Brakel, in *Advances in Drying,* Vol. I, A.S. Mujumdar (Ed.), Hemisphere Publishing Corp., New York, p. 217, 1980.

25. N.H. Ceaglske and O.A. Hougen, Trans. *Am. Inst. Chem. Eng.* 33, 283, 1937.

26. W.B. Haines, *J. Agric. Science* 17, 264, 1927.

27. W.B. Haines, *J. Agric. Science* 20, 97, 1930.

28. D.M. Newitt and P. Na. Nagar, and A.L. Papadopoulus, *Trans. Inst. Chem. Engrs.* 38, 273, 1960.

29. A.V. Neimark, V.B. Fenelonov and L.I. Kheifets, *Reaction Kinetics and Catal.* Letters 5, 67, 1976.

30. J. van der Meijden, Thesis, Utrecht University, Chapter 3, 1981.

31. P.K. de Bokx, W.B.A. Wassenberg and J.W. Geus, *J. Catal.* 104, 86, 1987.

32. M. Titulaer, Thesis, Utrecht University, Chapter 10, 2003.

33. M. Kotter, *Chem. Ing. Techn.* 55, 170, 1983.

34. M. Kotter and L. Riekert, in *Preparation of Catalysts II, Scientific Bases for the Preparation of Heterogeneous Catalysts*, B. Delmon, P. Grange, P. Jacobs and G. Poncelet (Eds.), Elsevier Scientific Publishing Company, Amsterdam, p. 51, 1979.

35. L.M. Knijff, Thesis, Utrecht University, Chapters 3 and 4, 1993.

36. J. Hagymassy, Jr., S. Brunauer and R. Sh. Mikhail, *J. Coll. Interface Sci.* 29, 485, 1969.

37. P.F. van den Brink, Thesis, *Utrecht University,* Chapter 3, 1992.

38. A.A. van der Giessen, *J. Inorg. Nucl. Chem.* 28, 2155, 1966.

39. J. Dousma and P.L. de Bruyn, *J. Int. Coll. Sci.* 56, 527, 1976; *Ibid.* 64, 154, 1978; *Ibid.* 72, 314, 1979.

40. J.H.A. van der Woude and P.L. de Bruyn, *Colloids and Surfaces* 8, 55, 1983; J.H.A. van der Woude, P. Verhees and P.L. de Bruyn, *Colloids and Surfaces* 8, 79, 1983; J.H.A. van der Woude, P.L. de Bruyn and J. Pieters, *Colloids and Surfaces* 9, 173, 1984.

41. R.J.A.M. Terörde, Thesis, Utrecht University, Chapter 2, 1996.

42. R.D. Deegan, O. Bajakin, T.F. Dupont, G. Huber, S.R. Nagel and T.A. *Witten Nature* 389, 827, Oct. 23, 1997.

43. R.D. Deegan, *Phys. Rev. B,* 81, 475, 2000.

44. R.D. Deegan, O. Bajakin, T.F. Dupon, G. Huber, S.R. Nagel and T.A. Witten, *Phys. Rev. B,* 62, 756, 2000.

45. A.Q.M. Boon, Thesis, Utrecht University, Chapter 2, 1990.

46. E. ten Grotenhuis, J.C. van Miltenburg, J.P. van der Eerden, R. van Wijk, O.L.J. Gijzeman and J.W. Geus, *Catal. Lett.* 28, 109, 1994.

Part III

Catalyst Finishing

16 Drying of Supported Catalysts

Azzeddine Lekhal, Benjamin J. Glasser, and Johannes G. Khinast

CONTENTS

16.1 INTRODUCTION

Catalysts are used in a wide variety of applications, ranging from catalytic converters and fuel cells in cars to the production of the newest drugs. More than 90% of all chemical and environmental processes use catalysts in at least one step, and there is a trend to replace older stoichiometric processes by clean, catalytic ones. The current worldwide market for catalysts is about $10 billion annually [1], and although over the last 100 years there has been significant progress in the understanding of catalysis and catalytic processes, many aspects of catalyst manufacturing are still poorly understood. Thus, there is a lack of a scientific basis for catalyst manufacture, and the process design often relies on empiricism and trial and error. Clearly, advances and improvements of many key catalyst manufacturing steps could potentially have a large impact on the entire $500 billion chemical industry [2]. The Rutgers Catalyst Manufacturing Science and Engineering Consortium, consisting of several catalyst companies and academic members, addresses exactly these issues: There is a focused effort to get a better scientific basis for catalyst manufacture. This chapter addresses one aspect of catalyst manufacturing: the drying of impregnation catalysts.

Catalysts are compounds that accelerate chemical reactions, ideally without being consumed themselves. Different types of catalysts exist; i.e., there are heterogeneous and homogeneous catalysts, as well as biocatalysts, which combine features of both. In this review, only solid, heterogeneous catalysts are addressed. A good heterogeneous catalyst is required to have high activity, selectivity, and stability. *Activity* is measured as moles of product formed per unit time and mass of catalyst. *Selectivity* is most generally defined as the amount of desired product divided by the total amount of products, both desired and undesired, produced in a certain amount of time. Catalyst *stability* describes a wide variety of properties, such as resistance to poisoning or sintering, mechanical strength, and thermal stability, or the ability to be regenerated after deactivation occurs (typically between 1 s and 5 years). All of these fundamental characteristics are affected by the preparation technique and by the physical and chemical properties of the chosen system.

16.1.1 SUPPORTED CATALYSTS

This chapter reviews research studies on the drying of supported impregnation catalysts. The motivation for investigating drying of impregnation catalysts is the fact that the drying step can have a strong influence on the final catalyst properties and quality. As described in detail below, these catalysts are produced by impregnation of a porous, catalytically inert support material with a solution of the active catalyst, which is adsorbed onto the support surface. The reason for using support materials is that in many cases the active catalytic material, i.e., the *precursor*, is an expensive metal (e.g., Pt, Pd, Rh, Au), which needs to be used sparingly. This can be achieved by using supported catalysts, where the active metal or the metal oxide is dispersed on a porous solid support with a high surface area with a predefined pore structure and particle shape. Hence, the weight fraction of the active catalyst can be low. The support particle not only offers a high surface area, but it also acts as a thermal stabilizer, and it provides easy access of the reactive gas. It may also have bifunctional activity. The preparation of supported catalysts clearly poses a significant challenge for the manufacturing step; i.e., the challenge is how to distribute a microscopic amount of metal such that a large surface area is maintained, and a stable catalyst with high activity and selectivity is obtained.

For identical composition and overall loading, the performance of supported catalysts critically depends on the distribution of the metal within the porous support. Thus, in the past few years, many studies dealing with the preparation of supported catalysts have been concerned with the control of the metal profile within the solid support. Reviews on this subject are given by Gavriilidis et al. [3] and Morbidelli et al. [1]. In many cases, a uniform metal distribution is desirable. However, the advantage of nonuniform distributions has been demonstrated by several authors for some heterogeneous reactions [1]. The choice of the optimal metal profile is determined by the required activity, selectivity, and other characteristics of the chemical reaction, such as intrinsic kinetics and mass transfer limitations. Catalyst stability is also an important factor. The four main categories of catalyst profiles, namely *uniform, egg-shell, egg-white,* and *egg-yolk*, are shown in Figure 16.1.

Several authors have reported that under isothermal conditions, the egg-yolk catalysts have the best activity at low values of Thiele's modulus (no mass transfer limitation) [1]. Clearly, egg-shell catalysts, where the catalyst layer is concentrated close to the external pellet surface, are advantageous in the case of fast reactions with strong diffusional restrictions. Additionally, egg-shell catalysts give higher selectivities for reactions, where diffusion is the limiting step [1]. When the external surface of the catalyst is exposed to poisoning or intense abrasion and attrition, as seen in fluidized or moving beds, the egg-white and egg-yolk distributions have a longer lifetime because the active layer is not impacted [4]. Uniform metal profiles

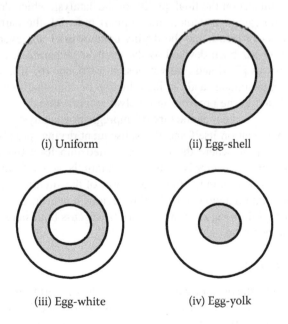

(i) Uniform   (ii) Egg-shell

(iii) Egg-white   (iv) Egg-yolk

FIGURE 16.1 Main types of metal distribution in supported catalysts.

are advantageous for reactions that are kinetically controlled. Intermediate profiles, i.e., decreasing egg-shell, egg-white, and egg-yolk, can also be optimal for certain applications. Under nonisothermal conditions and in the case of multiple reactions, optimization of the catalyst profile becomes increasingly difficult, and it is not surprising that in industrial practice, catalyst profiles are often chosen based on trial and error.

Depending on the environment under which the catalyst will operate, different types of solid support materials are used. The most common supports in industry are solid oxide supports. Typically, these materials include alumina (Al_2O_3), silica (SiO_2), and titania (TiO_2). Oxide surfaces are generally covered with hydroxyl groups, which are usually represented as S-OH, where S stands for aluminum, silica, or titanium. When contacted with water (during the impregnation step), these groups can behave as Brønsted acids or bases by exchanging a proton with the liquid solution, which gives them the ability to interact with several catalytic precursors. However, under aggressive conditions, oxide supports may dissolve (high pH) or be chemically attacked (low pH). Thus, as an alternative, carbon-supported catalysts, which are more stable under these conditions, can be used [5]. In addition, the use of carbon supports allows straightforward recovery of the active metal, as the support can be easily burned off. A detailed description of the different support materials, their advantage and disadvantages, and the preparation techniques are given in Ertl et al. [6].

Clearly, the impregnation step in the manufacturing process of supported cata-lysts has a strong impact on the final quality of the catalyst, which depends strongly on the specific impregnation method, the support material, the particle properties, the nature of the active metal, and the drying conditions. Thus, over the last years, increasing attention has been devoted to the catalyst preparation, as evidenced by several review articles [7–9] and catalyst-design textbooks [6, 10]. Major advances in characterization techniques and the introduction of combinatorial methodologies have also contributed to the progress in catalyst manufacture.

As described above, the manufacture of impregnation catalysts includes two key steps, i.e., the impregnation itself and the subsequent drying, possibly followed by calcination or reduction, where the metal is turned into its active form. In some cases, pretreatment in the reactor is required to obtain the active form of the catalyst, e.g., reduction by H_2 or CO. In the literature, most of the attention has been devoted exclusively to the impregnation step. In this review, however, drying is the focus. Nevertheless, in the following sections, we briefly discuss both steps because drying is coupled to impregnation.

16.1.2 IMPREGNATION

During impregnation, the precursor is deposited on the support from a liquid solution, which in most cases is water-based. If the support surface is hydrophobic or if hydrolysis of the support must be avoided, a nonaqueous solution is used [11]. Typically, the support is immersed in a solution that contains the inert precursor as a metal salt. In the case of *capillary impregnation*, the support is initially dry, whereas during *diffusional impregnation,* the support is initially filled with the liquid solvent prior to the impregnation [6]. In both cases, air can be trapped in the pore volume, which may hinder impregnation. As the liquid penetrates the support due to capillary pressure, the entrapped air is compressed. The liquid penetration stops when the capillary pressure equals the pressure of the entrapped gas. For this reason, the air is often evacuated before impregnation or the particles are exposed to a soluble gas. If the pore radius is very small, the capillary pressure is much larger than the pressure of the entrapped air, and air removal will not lead to any appreciable differences in the results. At the end of the impregnation step, the distribution of the precursor in the support is typically a function of the metal/surface interaction, the pore structure, and the diffusivity of the dissolved precursor in the impregnating solution. In some cases, the deposition can be performed from a gas phase, such as in the chemical vapor deposition (CVD) process.

The desired metal profile can be obtained either by impregnation of one component, or by successive or competitive impregnation of two or more components. The second method is known as *multicomponent impregnation*. When only two components are used during the process, it is called *coimpregnation*. Extensive studies have been performed on the impregnation of support catalysts, and several excellent reviews and textbooks have been published in this area, such as Ertl et al. [6] and Campanati et al. [9].

16.1.3 DRYING

Drying, which follows impregnation, is usually carried out at temperatures between 50 and 250°C, causing the evaporation of the liquid solvent. The liquid solution is transported by capillary flow and diffusion towards the external surface of the support [12–14], and the precursor may be redistributed by adsorption/desorption phenomena [4, 11]. Drying is a three-stage process, consisting of a *preheating period*, a *constant rate period*, and a *falling rate period* (Figure 16.2). In the preheating period, the support is heated up by the drying medium. The rate of liquid vaporization at the support's surface and the rate of drying increase as the temperature increases. During the constant-rate period, vapor is removed from the support's surface. The capillary

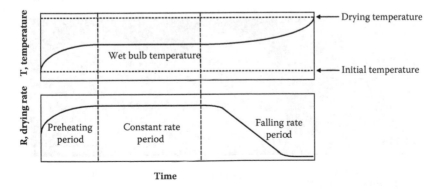

FIGURE 16.2 Variation of drying rate and temperature of a wet porous material during drying.

flow is sufficient to keep the surface saturated [13]. In this stage, the rate of drying depends on the rate of heat transfer to the evaporating surface. The temperature of the wet support is constant and is known as the *wet-bulb temperature*. During the first two stages, the metal dissolved in the liquid phase is transported by liquid convection and diffusion. The falling rate period begins when the moisture transport inside the support is no longer sufficient to keep the surface saturated. As a consequence, the drying rate decreases and dry patches appear near the surface. The temperature of the support begins to rise. A drying front develops, and vaporization takes place inside the solid. In this case, convective flow of the vapor dominates the moisture transport, and convective transport of the metal in the liquid phase is negligible. The evolution of the drying front has been experimentally observed by Hollewand and Gladden [15] and Koptyug et al. [14] by using nuclear magnetic resonance (NMR).

16.2 PHYSICAL EFFECTS DURING DRYING

Typically, a specific metal profile is obtained during impregnation, followed by a drying step, which is intended to not modify the active metal profile [1]. However, this is rarely the case, as the metal(s) can be redistributed during the drying stage [11, 16–18]. This metal migration after the impregnation stage can have significant consequences, as an undesired metal profile can lead to reduced activity and selectivity, potentially requiring costly recycling (if the catalyst is a noble metal), storage (if the active agent is a toxic compound such as chromium), or disposal of an entire catalyst batch, which can translate into many thousands of dollars of loss per barrel. Furthermore, the low selectivity of a catalyst not meeting specifications may lead to the production of waste, requiring separation or workup of desired and undesired species.

An example of the effects of drying on the metal distribution in impregnation catalysts is shown in Figure 16.3 for a Pt/alumina catalyst. It is evident from this illustration that the egg-shell distribution, obtained at the end of the impregnation stage (Figure 16.3a), can either be preserved (Figure 16.3b) or altered (Figure 16.3c) as the drying conditions are changed.

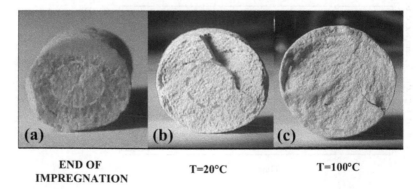

(a)	(b)	(c)
END OF IMPREGNATION	T=20°C	T=100°C

FIGURE 16.3 Example on the effect of drying conditions: Pt/alumina catalyst.

The redistribution of the active metal in the support is governed by several physical and chemical processes. The most important ones, which determine the final metal profile, are (1) adsorption of the active metal onto the support, (2) transport of the solvent, (3) transport of various dissolved species in the solvent, (4) heat transport, and (5) mass and heat transfer at the surface of the support particle (Figure 16.4a). In the following sections, these effects are described in detail.

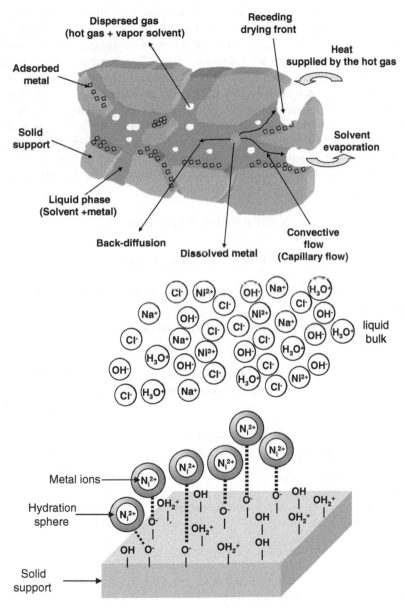

FIGURE 16.4 (a) Transport mechanisms during drying of supported catalysts; (b) schematic of metal adsorption on an oxide support.

16.2.1 ADSORPTION

The deposition kinetics and the adsorption equilibrium of the precursor on the support are a strong function of the chemical and physical interactions between the precursor(s) and the support. Many mechanisms and models have been proposed to account for experimentally observed phenomena. Examples of different adsorption models and catalytic systems are summarized in Table 16.1.

The degree of coverage of the porous support with the catalytically active component is an important factor, and it depends on the pH of the impregnating solution, the point of zero charge (PZC) of the support surface, and the nature of the metal ions/complexes. In the literature, the PZC is sometimes referred to as *isoelectric point* (IEP). Numerous experimental studies have shown that acids and salts can affect the catalyst adsorption during the impregnation and drying stages. The effect of pH on metal adsorption is due to the surface charging of the solid support, usually a porous oxide support such as alumina (Al_2O_3) or silica (SiO_2). The processes at the solid support/liquid solution interface are schematically depicted in Figure 16.4b. When contacted with aqueous solutions, the hydroxyl groups at the surfaces behave as Brønsted acids or bases according to the following surface ionization reactions:

$$S\text{-}OH + H_2O \rightarrow S\text{-}O^- + H_3O^+$$

$$S\text{-}OH + H_3O^+ \rightarrow S\text{-}OH_2^+ + H_2O$$

TABLE 16.1
Adsorption Models for Different Catalytic Systems

Catalyst system	Adsorption model	References
Pt/alumina	Revised Physical Adsorption Model (RPA)	Hao *et al.* [23]
Co/SiO$_2$	RPA	Agashe and Regalbuto [22] and Hao *et al.* [23]
Sn/γ-alumina	Triple Layer Model (TLM)	Karakonstantis *et al.* [60]
Pt/ γ-alumina, Pt// γ-alumina	TLM	Mang *et al.* [61]
Cd/carbon and Zn/carbon	TLM	Aurora *et al.* [62]
Cd/carbon and Zn/carbon	TLM	Marzal *et al.* [63]
Cu/carbon and Ni/carbon	TLM	Seco *et al.* [64]
Vanadia/Titania	Double Layer Model (DLM)	Georgiadou *et al.* [65]
Cu/carbon, Ni/carbon, and Pb/carbon	DLM	Faur-Brasquet *et al.* [66]
As/α-alumina	Constant Capacitance Model (CCM)	Halter and Pfeifer [67]
Cu/silica	CCM	Vlasova [68]
Review paper	DLM, CCM, and TLM	Blesa *et al.* [69]
Pt/alumina	Stern Model (SM)	Olsbye *et al.* [70]

The resulting surface charge, which arises from an excess of either anions or cations, is a function of the pH. At a critical value of the pH, the amount of negative charges (S-O$^-$) exactly balances the amount of positive charges (SOH$_2^+$), leading to a zero net charge. This value is characteristic for an oxide support and is known as the point of zero charge or the isoelectric point. Therefore, when the oxide support comes into contact with a liquid solution with a pH below its PZC, the surface is positively charged and a larger number of anions are adsorbed to balance the positive charges. Conversely, in aqueous solutions with a pH larger than the PZC, it is negatively charged and adsorbs cations preferentially. Some supports, like alumina, are amphoteric and can adsorb anions and cations. In multicomponent impregnation, the coimpregnant can affect the deposition of the metal precursor by changing the ionic strength of the liquid solution (e.g., by adding NaCl, NaNO$_3$, CaCl$_2$), by changing the pH of the system (HCl, HNO$_3$, NaOH), or by competition with the precursor for the adsorption sites [19].

According to Spieker and Regalbuto [20], the following three metal deposition mechanisms can be identified:

1. Coordination chemistry, in which the adsorption occurs after the complexation of the metal with the dissolved oxide. This mechanism was proposed for the adsorption of hexachloroplatinate anions on alumina support [21]. This generally happens under very aggressive conditions obtained at very low pH.
2. Chemical adsorption mechanisms that are based on chemical interactions between the metal complexes and the solid support. In this case, all the ionic species present in the liquid solution compete for the active sites at the solid surface. This type of adsorption is usually described by surface ionization models, such as the triple-layer model and its extended versions (e.g., four-layer model) [1].
3. Physical adsorption mechanisms, where it is assumed that the interaction between metal ions or complexes and the solid support is purely physical. It is assumed that the surface charges are balanced by an equal number of charged metal complexes in a region adjacent to the solid surface. In this mechanism, it is also assumed that only the metal ions can be adsorbed onto the solid support. An excellent model of this type is the revised physical adsorption (RPA) model developed by Regalbuto's group [20, 22, 23]. It has been experimentally verified for several metal/support systems.

In the past, the metal adsorption during impregnation and drying has been modeled using a simple Langmuir adsorption isotherm [11, 19, 24]. Although this approach is valid only within a limited range of conditions (concentration, pH), Langmuir isotherms are frequently used, because they are easy to implement and require estimation of only a few parameters. In the Langmuir model, the rate of adsorption R_i is given by

$$R_i = k_{ads}\, c_{l,i}\, (c_{sat} - c_{s,i}) - k_{des}\, c_{s,i} \qquad (16.1),$$

where k_{ads} and k_{des} are the adsorption and desorption rate constants, and c_{sat} is the maximum concentration at a given adsorption site. If the adsorption is not the limiting step, the deposited metal concentration is in equilibrium with the dissolved precursor in the liquid phase, and the adsorption/desorption process can be characterized by the equilibrium constant K_{eq}:

$$K_{eq} = \frac{k_{ads}}{k_{des}} \tag{16.2}$$

Surface ionization models provide a more detailed description of the processes at the surface, but more parameters need to be adjusted in order to fit experimental data, which can limit theory generality. The RPA model proposed by Regalbuto's group [22] employs a Langmuir isotherm adsorption:

$$c_{s,M} = S_{ox} \Gamma_{max} \frac{K_{eq} c_{l,M}}{1 + K_{eq} c_{l,M}} \tag{16.3},$$

where the adsorption constant K_{eq} is calculated from the Gibbs free energy of adsorption ΔG_{ads}. ΔG_{ads} is negative when positively charged metal ions or metal complexes adsorb onto a negatively charged solid surface, and positive in the opposite case. Γ_{max} is the maximum adsorption density of the adsorbed metal, and S_{ox} is the surface area of the solid support. Due to the detailed consideration of the ionic strength and the pH of the liquid phase, the RPA model predicts a drop in the adsorption constant at extreme pH levels, which is observed in several metal/support systems, with good accuracy without any adjustable parameter. More details on the RPA model can be found in Agashe and Regalbuto [22]. Recently, the RPA model has been used by Lekhal et al. [18] to characterize the impact of pH and ionic strength of the liquid solution on the final metal profile.

It is important to note that when the liquid solution becomes supersaturated, metal precipitation can also take place. Little is known about metal precipitation during drying, and to the best of our knowledge, none of the models existing in the open literature have discussed its effect on the metal deposition and distribution. Thus, more research is needed in this area.

16.2.2 Transport Effects in the Support Particle

During drying, part of the active metal is adsorbed on the solid support surface, while another part remains in the liquid solution filling the pores. It is predominantly the dissolved metal in the liquid phase that is subject to migration during the drying stage. However, redissolution and readsorption also may take place.

Migration of the metal in the liquid phase is controlled by the convective flow of the solvent, which transports the dissolved metal towards the external surface and the back-diffusion, which acts in the opposite direction (Figure 16.4a). Figure 16.5 shows the variation of the metal profile during drying, as a

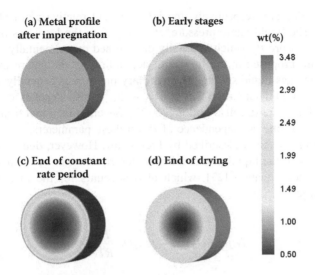

(a) Metal profile
after impregnation

(b) Early stages

wt(%)

3.48

2.99

2.49

(c) End of constant
rate period

(d) End of drying

1.99

1.49

1.00

0.50

FIGURE 16.5 Variation of the metal profile during drying in case of moderate adsorption strength, from Lekhal et al. [16]. The metal concentration wt (%) at a given point is expressed in grams of total metal (metal dissolved in the liquid solution plus metal adsorbed on the solid support) per gram of solid support.

result of these two phenomena. The convective flow is a result of the capillary forces, which develop as the liquid solvent vaporizes. Across each pore, a meniscus forms, forcing the liquid solvent to flow from large pores to small ones at the external surface (capillary flow is inversely proportional to pore size). Therefore, at the early stage of drying, the metal concentrates at the external surface of the support (Figure 16.5b). The metal concentration keeps building up at the external surface and reaches a maximum at the end of the constant rate period (Figure 16.5c). The convective flow decreases (in some cases, it can totally cease) during the falling rate period, as the moisture content approaches zero and diffusion becomes the controlling transport mechanism. During this last drying stage, the concentration of the metal at the external surface decreases as back-diffusion transports most of the catalyst back to the particle center (Figure 16.5d). If the liquid phase becomes discontinuous before or at the end of the constant rate period, the back-diffusion effect is suppressed and steep egg-shell metal profiles are obtained.

In porous media, the transport of the liquid phase due to capillary forces is best represented by Darcy's law, i.e., the flux is proportional to the pressure gradient of the liquid:

$$N_{1,\text{solvent}} = -c_{1,i} \frac{KK_{1,\text{eff}}}{\eta_1} \nabla P_1 \qquad (16.4),$$

where K and $K_{l,eff}$ are, respectively, the intrinsic and relative permeability of the liquid phase. The liquid-phase pressure, P_l, is equal to the local gas pressure, less the capillary pressure P_c, which is usually determined experimentally [25]. Several correlations are proposed in the literature, depending on the nature of the porous material and the gas-liquid system. The capillary pressure is generally a decreasing function of the liquid saturation $S_l = \varepsilon_l / (\varepsilon_g + \varepsilon_l)$. It also depends on the surface tension, γ, and the pore size distribution [25, 26]. Several functional forms have been considered to describe the dependence of P_c on these parameters.

Metal diffusion can be described by Fick's law. However, due to the presence of ionic species in the liquid phase, a better description can be obtained by using the Nernst-Planck equation [27], which also accounts for the migration due to electrical charges:

$$N_{l,i} = -D_{l,i}\nabla c_{l,i} - c_{l,i}z_i D_{l,i}\frac{F}{RT}\nabla \phi \qquad (16.5),$$

where $\nabla \phi$ is the electrostatic potential gradient, which is a function of the number of charges and the gradient of concentration of the charged particles. It can be determined using the no-current equation [16]. The net flux of the metal species in the liquid phase is equal to the sum of the capillary flow (Equation 16.4) and diffusion (Equation 16.5).

16.2.3 HEAT TRANSPORT

The heat supplied by the hot drying medium can be transported in the wet porous support by convection, conduction, radiation, and condensation/evaporation. Radiation is usually neglected, but may be a factor for special drying units. Convective heat transport in the liquid phase is a result of the capillary flow. In the gas phase, the drying medium and the solvent vapor are transported from the support center to the external surface at the beginning of drying and flow in the opposite direction during the falling rate period. Heat conduction occurs in the liquid and solid phase, while the gas thermal conductivity is very small. Therefore, the thermal conductivity of the porous support is a strong function of the volume fraction of gas in the system (during drying, the volume fraction of gas increases with time). Evaporation and condensation of the liquid solvent also contribute to the heat effects in the particle. In fact, evaporation is the largest heat sink in a particle. It occurs in the hot spots of the support, whereas condensation, which is accompanied by a heat release, simultaneously takes place in colder zones.

16.2.4 MASS AND HEAT TRANSFER AT THE EXTERNAL SURFACE

Drying is mainly controlled by the heat and mass transfer rates at the external surface. Mass and heat transfer control the drying process during the constant rate period, where enough liquid solvent is constantly flowing towards the external surface. While heat is transferred from the drying medium to the solid support, the evaporating solvent is removed from the particles. The mass and heat transfer rates can be adjusted by changing the drying temperature of the hot gas in the bulk, the space velocity of the dryer, and the composition of the drying medium. High mass and heat transfer rates are obtained for high drying temperatures and gas velocities, and for low relative humidity of the drying medium.

16.3 IMPACT OF DRYING ON THE METAL PROFILE

From the above discussion of the physical effects influencing the redistribution of the active metal, it is clear that this redistribution is highly dependent on the transport mechanisms. Strong convective flow due to capillary forces (high temperatures) causes the transport of the metal towards the external region, whereas metal diffusion acts in the opposite direction. Nevertheless, both effects have only marginal impact in the case of strong metal adsorption, because the profile does not change significantly during drying. Several metal(s)/support systems have been studied experimentally in the literature to confirm these trends. The most important studies are reviewed in this section, followed by a section reviewing catalyst-drying models.

16.3.1 EXPERIMENTAL STUDIES

16.3.1.1 Single-Component Systems

Maatman and Prater [28, 29] were among the first authors to show that metals can redistribute during drying. By placing wet pellets in a closed container at 100% relative humidity, they found that an egg-shell profile was transformed into a uniform distribution after 3 h. They concluded that under these conditions, diffusion caused a smoothening of the profile. Similar observations were made by Chen et al. [30] during the preparation of $Cu(NO_3)_2/\alpha\text{-}Al_2O_3$ and $Cr(NO_3)_3/\gamma\text{-}Al_2O_3$. However, it was observed that the copper region widened more than the chromium one, due to a stronger metal adsorption in the latter case.

For strongly adsorbing metal precursors, the impregnation-drying process leads in most of the cases to egg-shell profiles, unless the impregnation time is large enough to reach a uniform distribution. For such metal-support systems, the drying has a only a limited impact on metal distribution [1, 4, 11, 16, 18] and the final profile is controlled by the impregnation parameters (time, concentration of the precursor, and pH of the liquid solution). This fact was experimentally confirmed by Santhanam et al. [31], who investigated the effect of adsorption strength for several Pt and Pd systems.

When studying the effect of drying on metal distribution of $NiCl_2$ on $\gamma\text{-}Al_2O_3$, Komiyama et al. [32] found that a uniform profile obtained after impregnation can be transformed into a decreasing egg-shell profile. Similar behavior was observed by Van der Berg and Rijnten [33] for $CuCl_2/\gamma\text{-}Al_2O_3$. They also showed that the metal accumulation at the external surface increased with the drying rate, which was later also reported for other metal-support systems; i.e., Uemura et al. [34] for Ni/alumina, Goula et al. [35] for $Mo/\gamma\text{-}Al_2O_3$, Li et al. [36] for $Ni/\gamma\text{-}Al_2O_3$, Yasuaki et al. [37] for Mo/Al_2O_3, and Vergunst et al. [38] for a monolithic catalyst impregnated with a nickel precursor. The same behavior was later confirmed by modeling studies, e.g., by Lee and Aris [11] and Lekhal et al. [16, 18].

In most of the studies cited above, the drying rate, which is one of the controlling factors, was adjusted through the drying temperature. Another alternative is to vary the capillary flow of the liquid solution. This can be done by changing the viscosity of the liquid solution or by varying the pore size distribution of the porous support. When they studied CuO on α-alumina catalysts, Kotter and Riekert [39, 40] found that the metal distribution changes from egg-shell to uniform when the viscosity of the liquid solvent was increased.

Although there exist several experimental studies focusing on different precursor/support systems, as described above, many effects have not been explored experimentally, as of yet. For example, the deposition rate of the metal is controlled by the pH and the ionic strength of the liquid solution. For positively charged metal ions (for example, Ni^{2+} or Ba^{2+}), the adsorption constant is high when the solid surface is negatively charged and low when it is positively charged. For oxide supports, negative charges appear at the support surface when pH is above the PZC, and positive charges when it is below it. Although several modeling studies were published in the past few decade (e.g., Regalbuto's group[22]), to the best of our knowledge, there are no experimental investigations addressing this issue.

16.3.1.2 Coimpregnation Systems

In coimpregnation systems, the catalyst is impregnated with a solution containing two different metal species. The purpose of the second component can be to obtain a metal distribution with a maximum concentration at a set distance from the external surface, i.e., egg-white or egg-yolk catalysts. Under most conditions, this is not attainable with only one metal precursor. (To the best of our knowledge, only Hariott [41] reports an accumulation of metal inside the particle for a one-metal system.) For two components, the metal with the higher adsorption rate is deposited close to the external surface, whereas the metal with the lower adsorption strength, usually the active ingredient, travels into the support due to a lack of adsorption sites. Hepburn et al. [42] found that the profiles established during this impregnation can be maintained at very high drying temperatures (around 500°C). They argued that the liquid solution evaporates very quickly, and the adsorbed metals do not have time to redistribute. Under milder drying conditions, diffusion becomes more important, leading to a blurring of the catalyst layer. Competition between convective flow due to capillary forces and diffusion can lead to the splitting of the catalyst layer and the formation of a bimodal distribution, as observed by Becker and Nuttall [43] and Hepburn et al. [42]. At ambient conditions (around 20°C), metal diffusion totally controls the drying stage, leading to uniform distribution of the metal [42].

In summary, only a few experimental studies exist on coimpregnation systems, and more work needs to be carried out in this area.

16.3.1.3   Measurement of the Metal Distribution

The most common technique to determine metal profiles in a cross section of the support is electron microprobe analysis (EPMA). In this technique, x-rays are genererated after the sample is targeted with a beam of electrons from a tungsten wire at approximately 2700 K. The x-rays are detected by a spectrometer, and identification of the wavelengths at which the x-rays are emitted allows qualitative analysis. X-ray intensities provide a quantitative analysis at the targeted area. More details on the EPMA technique can be found in Reed's textbook [44]. Typical EPMA data are shown in Figure 16.6 for an Ni/Al_2O_3 system dried at 20, 60, and 100°C. Other analytical techniques include light transmission [32, 45–47], spectroscopic analysis [35, 48, 49], and staining [19].

16.3.2   Modeling Approaches

Neimark et al. [4] were among the first to theoretically study the phenomenon of metal migration during drying. They stated that there exist two limiting regimes: *fast drying,* where the vapor removal at the external surface is faster than the capillary flow, and *slow drying,* where the opposite is true. In order to characterize the drying regime, they introduced a dimensionless number α_1, which compares the capillary flow rate to the vapor flow rate. High values of α_1 correspond to fast drying regime, whereas low values are for slow drying regimes. In the fast drying regime, the constant rate period is suppressed, and a drying front moves very quickly towards the center of the support, which does not give time for the metal to redistribute. This regime is attainable at very high temperature, around 500°C [1]. In the slow drying regime, the metal migration is controlled by the capillary flow and diffusion in the liquid phase. When diffusion is strong in comparison to the convective flow, as can be the case at ambient temperature, the metal profile is smoothed out.

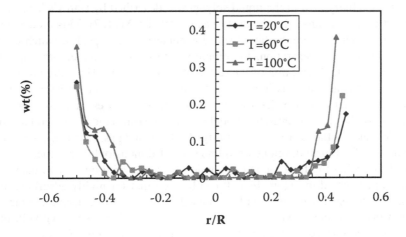

FIGURE 16.6 Example of metal profiles obtained by the EPMA technique for Ni/Al_2O_3 catalysts dried at different temperatures.

Later, Lee and Aris [11] formulated a more descriptive drying model, where the metal migration was assumed to take place only during the constant rate period by capillary flow and back-diffusion. This model can capture only slow drying conditions, where a drying front moves from the center of the support to the external surface. The capillary flow rate was described by the Washburn equation, and it is assumed that metal deposition onto the solid support takes place by adsorption only. Metal precipitation was not considered. This assumption is valid as long as the concentration of the dissolved metal does not exceed the saturation limit. The effect of different drying conditions, such as drying temperature or heat and mass transfer at the external surface, was not reported in this study. The model is accurate enough to qualitatively capture most of the important features that were reported in experimental studies, such as the accumulation of metal at the external surface at intermediate drying conditions and the formation of uniform metal profiles at low drying temperatures.

Uemura et al. [34] proposed a drying model that accounted for capillary flow and metal diffusion. In this model, it was assumed that the macropores act as reservoir tanks supplying liquid solution to the micropores, which are located at the external surface. Unlike Lee and Aris [11], Uemura et al. computed the capillary flow rate from the overall drying rate during the constant rate period. The effect of drying during the falling rate period was ignored by the authors. Li et al. [36] developed a correlation where the metal dissolved in the liquid solution can move from the support center to the external surface only. Metal migration in the opposite direction, due to metal diffusion, was neglected. The model involves an adjustable parameter that needs to be determined by fitting the mathematical expression with the experimental data, which considerably reduces its capabilities. Such a model does not provide any information on the controlling parameters during drying, or on how process parameters affect the final metal profile.

A more sophisticated model was recently proposed by Lekhal et al. [16–18], where the capillary flow of the liquid solvent was described by Darcy's law, and the transport in the gas phase by the dusty gas model (DGM) [27]. This model can be applied to systems where the liquid solution contains ionic species, which is often the case in catalyst preparation. In comparison to previous models, this approach allows an accurate description of the different drying stages. Thus, the rate of each transport mechanism and its impact on the final profile can be individually monitored. The model can also describe different drying regimes, ranging from slow drying to fast drying. However, the model requires the estimation of several transport properties, such as the relative permeability of solvent and gas, based on experiments. Lekhal et al. [16] have developed a framework of dimensionless numbers that can be used to predict the final metal distribution for cases with one active metal. The results are presented in Figure 16.7. It can be seen that for highly adsorbing metal precursors, egg-shell catalysts can be obtained for intermediate drying regimes and uniform ones for slow drying regimes (Figure 16.7a). Decreasing egg-yolk metal profiles can be formed for weakly adsorbing precursors (Figure 16.7b).

In the model, it is assumed that the particle consists of three phases: (1) the gas phase, containing the drying medium, usually air and the solvent vapor, (2) the liquid phase, i.e., the liquid solvent (water) and the metal (in the form of a salt), and (3)

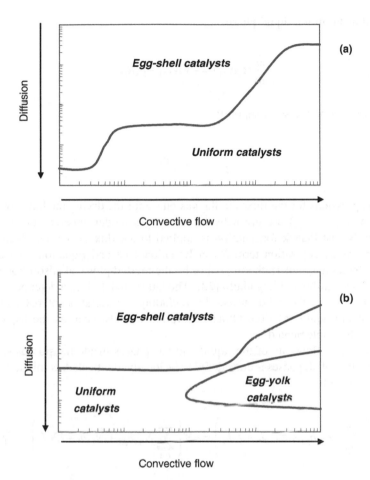

FIGURE 16.7 Metal profiles at the end of drying as a function of the convective flow, diffusion, and adsorption rates in case of (a) strong and (b) weak metal adsorption.

the solid phase, i.e., the solid support and the deposited metal. Due to a low flow rate of the gas and liquid phases, it is assumed that the temperature at a given point is the same within the three phases. The continuum transport equations for the different species are

Drying medium in the gas phase:

$$\frac{\partial}{\partial t}(\varepsilon_g c_{g,a}) = -\nabla N_{g,a} \tag{16.6}$$

Solvent in the gas and liquid phase:

$$\frac{\partial}{\partial t}(\varepsilon_l c_{l,s}) = -\nabla(N_{l,s} + N_{g,v}) \tag{16.7}$$

Dissolved metal in the liquid phase:

$$\frac{\partial}{\partial t}(\varepsilon_l c_{l,M}) = -\nabla N_{l,M} - \rho_s R_M \qquad (16.8)$$

Adsorbed metal on the solid support:

$$\frac{\partial}{\partial t}(c_{s,M}) = R_M \qquad (16.9)$$

The expressions for the fluxes in the gas phase for the drying medium ($N_{g,a}$) and the solvent vapor ($N_{g,v}$) are given by the dusty gas model, whereas in the liquid phase, the Nernst-Planck formulation is applied to the flux of the dissolved metal ($N_{l,M}$). The metal deposition term R_M in Equation 16.8 and Equation 16.9 can be described by a Langmuir isotherm [16] or by the revised physical adsorption model developed by Agashe and Regalbuto [22]. The latter model gives a detailed description of the state of the solid surface by including the variation of pH and ionic strength of the liquid solution. Other adsorption models, such as the triple-layer model, can be implemented.

Due to a low flow rate of the liquid and gas phases inside the porous support, the temperature of all phases is considered to be locally identical. A combined energy balance is given by

$$\frac{\partial}{\partial t}\left(\sum_{i=1}^{2}\varepsilon_g c_{g,i} h_{g,i} + \varepsilon_l c_{l,s} h_l + \rho_s h_s\right) = -\nabla\left(\sum_{i=1}^{2} N_{g,i} h_{g,i} + N_{l,s} h_l - \lambda\nabla T\right) \qquad (16.10)$$

The energy balance accounts for the heat transfer by convection in the gas and liquid phase and conduction.

16.3.2.1   Single-Component Systems

In this section, the effect of several parameters, such as temperature, permeability, and pH, will be analyzed using a drying model where the metal adsorption is described by the RPA model [23]. All the simulations were performed with an initially uniform metal profile in the liquid and solid phase unless stated otherwise. The effect of other parameters, such as diffusion and adsorption constants, can be found in Lekhal et al. [16, 18].

Effect of temperature: The temperature is one of the parameters that can be adjusted easily. If the drying temperature is raised, the heat and mass transfer rates at the external surface increase, leading to higher evaporation and capillary flow rates. As a consequence, the concentration of the dissolved metal will increase near the external surface, which leads to an increase of the surface coverage near the external surface, as shown in Figure 16.8. These results are for a moderately

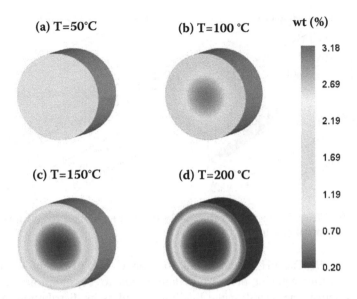

FIGURE 16.8 Effect of the drying temperature on the final profile for moderately adsorbing metals. The metal concentration wt (%) at a given point is expressed in grams of total metal (metal dissolved in the liquid solution plus metal adsorbed on the solid support) per gram of solid support.

adsorbing precursor. For higher adsorption strengths, the model predicts a marginal impact of drying, which is in agreement with reported results in the literature. The model also shows that an increase of the metal concentration at the external surface can be achieved, if the flow rate of the drying medium (higher mass transfer rate) outside the porous support is increased.

Effect of permeability: Another parameter of interest during catalyst preparation is the permeability of the support, which is a function of the pore-size distribution, pore shape, connectivity, and tortuosity. Different types of pores can exist within the support, i.e., *micropores* with diameters below 2 nm, *mesopores* with diameters between 2 and 50 nm, and *macropores* with diameters above 50 nm. Micropores are usually cracks within the primary particles of the support, whereas mesopores and macropores represent the space between these particles. Usually, it is this space that is available to the metal deposition. Solid supports with high permeability have larger pores, whereas the ones with very small pores have low permeability. Higher permeability facilitates the transport in the liquid phase, which leads to an increased metal concentration at the external surface compared to cases with lower permeability (Figure 16.9a). For very low permeabilities, the convective flow of the liquid solvent is weak, leading to metal concentration near the metal center. For high adsorption strength, the impact of permeability is less pronounced (Figure 16.9b). The final metal profile is only slightly different from the one achieved during the impregnation stage.

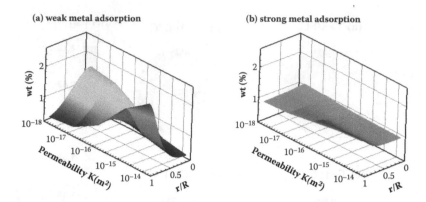

FIGURE 16.9 Effect of the support permeability on the final profile for (a) weak and (b) strong adsorbing metal precursors.

Effect of pH: The pH of the liquid solution is one of the most important parameters during catalyst preparation and needs to be monitored carefully, because it controls the overall metal loading as well as its distribution. As previously stated, when a solid oxide support is put into contact with a liquid solution, it will develop positive charges at its surface if the pH is below its PZC, and negative ones if it is above it. Figure 16.10 shows the effect of pH on the final metal profile for a cationic metal. It is evident that when the pH is below the PZC, the metal is weakly attracted to the solid support. As a result, the metal cations remain in solution until the end of drying, when their deposition occurs. Under these conditions, the final metal profile is entirely controlled by the drying conditions. When the pH is above the PZC, a strong interaction exists between the metal ions and the support. In such a case, the drying has a limited effect on the metal distribution. Obviously, the opposite effects will be obtained for negatively charged species.

Effect of support size: The size of the catalytic support varies from very small particles (~10 μm in slurry reactors and ~100 μm in fluidized beds) to large ones (~1 cm in packed beds, 10–100 cm in monoliths). When the size of the support is increased, the effect of back-diffusion is reduced, leading to the accumulation of metal at the external surface. This effect is illustrated in Figure 16.11, which shows that the final metal profile changes from egg-shell to nearly uniform when the size of the porous support is decreased.

16.3.2.2 Two-Component Systems

The results presented in this section assume that the active metal (catalyst layer) has a weaker adsorption strength than the second metal (coimpregnant). The equations are described in detail by Lekhal et al. [17].

Transient behavior: Initially, the active metal forming the catalyst layer and the coimpregnant concentrate near the external surface, where the combined effect of the capillary flow and solvent evaporation is more important than the back-diffusion

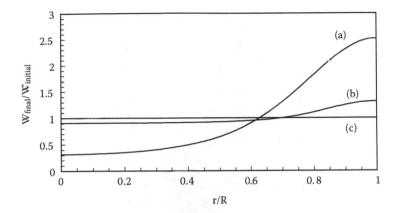

FIGURE 16.10 Effect of liquid pH on the final metal profile: (a) metal profile after impregnation, (b) pH > PZC, (c) pH < PZC. The y-axis is the ratio of the final to initial metal profile.

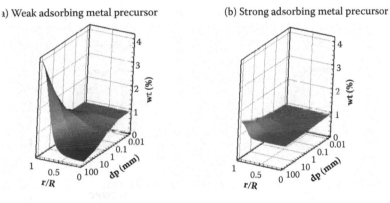

FIGURE 16.11 Effect of support size (d_p) on the final metal profile for (a) weak and (b) strong adsorbing metal precursors.

(Figure 16.12). As time advances, more coimpregnant is adsorbed near the external surface, forcing the active metal to diffuse inside, where deposition is possible. As a consequence, a peak in the catalyst layer is formed. The intensity of the peak is more pronounced for higher drying rates, mainly because the fraction of coimpregnant near the surface is higher. At the end of drying, under the effect of diffusion, this maximum flattens out. A nearly flat metal profile is obtained for slow drying conditions. It is clear from this illustration that, as for the single-component system, metal redistribution occurs as a result of capillary flow, diffusion, and solvent evaporation. The relative intensity of adsorption of both components is another significant parameter.

Effect of temperature: The effect of temperature on the final metal profile for a two-component system is shown in Figure 16.13. An increase of the drying temperature generates high solvent evaporation rates at the external surface of the support, which results in significant convective flow in the liquid phase, causing an increase

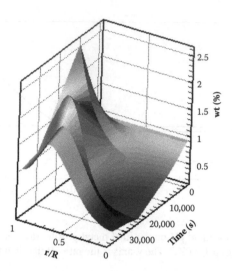

FIGURE 16.12 Variation of the metal profile during drying for coimpregnation systems.

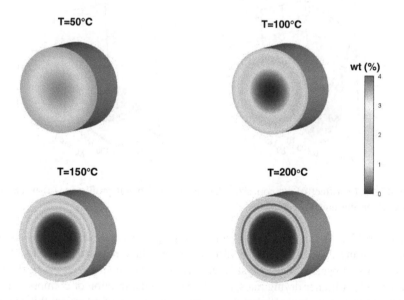

FIGURE 16.13 Variation of the final metal profile with drying temperature in coimpregnation systems; the active metal and the coimpregnant are moderately and strongly adsorbed to the solid support, respectively.

in the concentration of the coimpregnant and metal in this region. At high temperatures, convective flow effects are large, and therefore, most of the coimpregnant adsorbs in a thin layer at the outer region, freeing more adsorption sites inside the support, which allows the active metal to adsorb just behind the coimpregnant layer. At very high temperature (200°C), diffusion effects become negligible, and the metal concentrates in a narrow layer (red color in Figure 16.13d). At low temperatures, metal back-diffusion controls the flow in the liquid phase. The metal layer becomes more diffuse, and the final profile is nearly uniform.

Effect of diffusion: Figure 16.14 shows the effect of the diffusion coefficient of the coimpregnant on the catalyst layer, which can be controlled to an extent by controlling the viscosity of the impregnation solution. The final catalyst layer profile changes from egg-shell to egg-white when the diffusion coefficient of the coimpregnant is varied from 10^{-9} m^2/s to 10^{-12} m^2/s. A decrease in the coimpregnant diffusion coefficient leads to an increase of the fraction of coimpregnant that adsorbs at the surface, leaving fewer available adsorbing sites for the active metal. Thus, the active metal in the liquid solution is transported back towards the support center, where it adsorbs due to a higher number of vacant sites. As the diffusion coefficient of the coimpregnant decreases, the coimpregnant deposits in an increasingly narrow layer closer to the external surface (Figure 16.14). As this occurs, the catalyst layer narrows, and it moves close to the external surface. This trend is also obtained when the adsorption constants of both components are equal.

From the above discussion, is has become clear not only that drying changes the metal profile, but that drying may also be used to achieve a desired metal distribution, if the drying parameters are controlled tightly.

16.4 OTHER DRYING EFFECTS

Recent studies on catalyst preparation focused on the impact of drying on the nature of the metal phase on the solid support (amorphous or crystalline) and on the size of the metal particles. For example, for Cu/SiO$_2$ catalysts, Toupance et al. [50] found that the drying conditions may have a strong influence on the size of the metal particles. They showed that for this particular case, chemical bonding between the metal and the support may develop during drying.

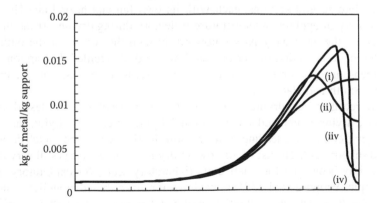

FIGURE 16.14 Effect of the coimpregnant diffusion coefficient on the catalyst layer during drying. From (i) to (iv), the diffusion coefficient of the coimpregnant is decreased, while the metal diffusion coefficient is kept constant.

Using different characterization techniques, such as x-ray diffraction and absorption spectroscopies, Chouillet et al. [51] concluded that in Zn/SiO_2 supported catalysts, the formation of ZnO (the active form of the catalyst) strongly depends on the drying temperature. The active form of the catalyst was obtained only when the catalyst was dried at intermediate temperatures (between 90 and 150°C). Slow and fast drying did not allow the formation of the active form.

More examples of the effect of drying in cases where one precursor is impregnated can be found in the literature, e.g., Ni/MgO [52] and Ni/silica [53]; Co/SiO_2 [54], and Mo/SnO_2 [55]. Similar work has also been carried out for supported catalysts where more than one metal is impregnated. These include systems where a chemical promoter is added to the system [56].

It is important to note that none of these studies were aimed at simultaneously understanding the impact of drying on metal distribution (macroscale effects) and metal structure (microscale effects). This should be the subject of future studies.

16.5 DRYING TECHNOLOGIES

Impregnation catalysts are dried in units, where solvent removal is possible without mixing of the solid support in order to avoid any damage, such as attrition or abrasion, that may cause a loss of the active metal. Typically, tray and screen-conveyor dryers are preferred. The use of rotary dryers is limited due to a high rate of attrition.

Tray dryers are batch units that consist of a rectangular chamber of sheet metal, containing trucks that support racks. Heated air is circulated by a fan, and part of the moist air is continuously vented through an exhaust duct. The tray dryers are useful for small production scale. They can also be operated under vacuum. Tray dryers are, however, very expensive to operate, because they require a considerable amount of time for loading and unloading. This makes screen-conveyor dryers, which operate in a continuous mode, very attractive. The wet supports are slowly carried on a traveling metal screen through a long drying chamber, which is usually separated into several sections, each with its own fan and heated air. This offers the possibility of applying a temperature trajectory during drying. At the inlet end of the dryer, the air usually passes upward through the screen; at the outlet end, where the material is dry, the air passes downward. A detailed description of tray and screen-conveyor dryers is available in several textbooks, for example, in McCabe, Smith and Harriott [57].

More recently, new drying technology has been developed and tested, such as microwave drying. Compared to conventional drying, microwave drying can provide rapid drying conditions, leading to a uniform distribution of the metal throughout the solid support [58]. The use of microwave drying at large scale is still very limited. It has mainly been used for small units (laboratory scale) for exploratory studies. For example, Vergunst et al. [38] showed that for nickel-based monolithic catalysts, it is possible to prevent the accumulation of metal at the external surface, confirming the hypothesis that uniform evaporation of the liquid solvent may be obtained (as heat transport does not occur via conduction). During the preparation of $Pd-Fe/Al_2O_3$ coimpregnation catalysts, Berry et al. [59] found that microwave drying influences the size of metal particles and changes the nature of the palladium species. The variation of metal distribution was not reported.

Future work in the area of microwaves drying should focus on getting a better understanding of the effect of different process parameters on the catalyst properties (metal distribution and dispersion) as well as on the scale-up of this process.

16.6 CONCLUSION

Very often, it is believed that in the production of impregnation catalysts, the impregnation stage itself determines the catalyst distribution. However, several research studies discussed in this chapter clearly show that drying can drastically affect catalyst properties at different length scales, e.g., metal distribution (large-scale effects) and metal structure on the solid support (small-scale effects). These studies suggest that changes in metal distribution during drying occur as a consequence of metal migration in the liquid phase, which is controlled by the convective flow of the solvent (due to capillary forces), molecular diffusion, and adsorption of the metal. Strong metal adsorption usually minimizes the effect of drying.

Several experimental studies on one-component impregnation catalysts for weakly or moderately adsorbing metal precursors showed that egg-shell catalysts are obtained when high convective flow rates are achieved. This can be achieved by increasing the drying temperature (process conditions), by decreasing the viscosity of the liquid solvent, or by increasing the intrinsic permeability of the solid support. At low convective flow rates, the diffusion becomes important, and more uniform metal profiles are obtained. pH also has significant impact on the final profile. For coimpregnation catalysts (two-component systems), it was found that the final profile of each component depends on the same transport mechanisms, i.e., convective flow, diffusion, and adsorption. However, the relative importance of diffusion and adsorption determines the location of each metal layer within the support. High adsorption rates allow the metal to adsorb near the outer region, whereas low adsorption leads to the deposition of the metal closer to the particle center. Thin metal layers are obtained for high convective flows (fast drying conditions), whereas diffuse ones are achieved when diffusion effects become important (slow drying conditions).

Characterization of the metal properties at a very small scale using novel experimental techniques revealed that drying also affects the size of metal particles, their structure, and the way they are deposited onto the solid support. More recently, new drying methods, such as microwave drying, were tested and their effect on catalyst properties was analyzed. Even though only a few studies exist, very promising results were obtained. However, more thorough investigations are still needed in this area, not only to examine drying effect on catalyst properties but also to be able to apply this technique for large-scale production.

In summary, the results presented here provide insight into how to use drying alone to achieve a desired profile. This brings to light the possibilities for smart drying, i.e., drying technology that enables the design of a desired metal profile in the support. However, further work coupled with experiments is needed to demonstrate the feasibility of the suggested strategies.

ACKNOWLEDGMENT

The support of Akzo Nobel, Chevron Texaco, Engelhard, Grace Davison, Haldor Tapsoe, and UOP, which are full members of the Rutgers Catalyst Manufacturing Science and Engineering Consortium, is gratefully acknowledged.

REFERENCES

1. Morbidelli, M., A. Gavriilidis, and A. Varma, *Catalyst design: Optimal distribution of catalyst in pellets, reactors and membranes.* Cambridge, U.K: Cambridge University Press, 2001.
2. NRC, *Catalytic process technology.* Washington, DC: National Research Council, 2000.
3. Gavriilidis, A., A. Varma, and M. Morbidelli, Optimal distribution of catalysts pellets. *Catal. Rev.-Sci. Eng.,* 35: pp. 399–456, 1993.
4. Neimark, A.V., L.I. Kheifez, and V.B. Fenelonov, Theory of preparation of supported catalysts. *Ind. Eng. Chem. Prod. Res. Dev.,* 20: pp. 439–450, 1981.
5. Auer, E., et al., Carbon as supports for industrial precious metal catalysts. *Applied Catalysis A: General,* 173: pp. 259–271, 1998.
6. Ertl, G., H. Knozinger, and J. Weitkamp, *Preparation of solid catalysts.* Weinheim: Wiley-VCH, 1999.
7. Schwarz, J.A., Method for preparation of catalytic materials. *Chem. Rev.,* 95: pp. 477–510, 1995.
8. Perego, C. and P. Villa, Catalysts preparation methods. *Catalysis Today,* 34: pp. 281–305, 1997.
9. Campanati, M., G. Fornasari, and A. Vaccari, Fundamentals in the preparation of heterogeneous catalysts. *Catalysis Today,* 77: pp. 299–314, 2003.
10. Twigg, M.V., *Catalysis handbook.* London: Manson Publishing, 1996.
11. Lee, S.-Y. and R. Aris, The distribution of active ingredients in supported catalysts prepared by impregnation. *Catal. Rev.-Sci. Eng.,* 27: pp. 207–340, 1985.
12. Hollewand, M.P. and L.F. Gladden, Probing the structure of porous pellets: An NMR study of drying. *Magnetic Resonance Imaging,* 12: pp. 291–294, 1994.
13. Kowalski, S.J., Toward a thermodynamics and mechanics of drying processes. *Chem. Eng. Sci.,* 55: pp. 1289–1304, 2000.
14. Koptyug, I.V., et al., A quantitative NMR imaging study of mass transport in porous solids during drying. *Chem. Eng. Sci.,* 55: pp. 1559–1571, 2000.
15. Hollewand, M.P. and L.F. Gladden, Visualization of phases in catalyst pellets and pellet mass transfer processes using magnetic resonance imaging. *Trans IChemE,* 70A: pp. 183–185, 1992.
16. Lekhal, A., B.J. Glasser, and J.G. Khinast, Impact of drying on the catalyst profile in supported impregnation catalysts. *Chem. Eng. Sci.,* 56: pp. 4473–4487, 2001.
17. Lekhal, A., J.G. Khinast, and B.J. Glasser, Predicting the effect of drying on supported coimpregnation catalysts. *Ind. Eng. Chem. Res.,* 40: pp. 3989–3999, 2001.
18. Lekhal, A., B.J. Glasser, and J.G. Khinast, Influence of pH and ionic strength on the metal profile of impregnation catalysts. *Chem. Eng. Sci.,* 59: pp. 1063–1077, 2004.
19. Papageogiou, P., et al., Preparation of Pt/g-Al$_2$O$_3$ pellets with internal step distribution of catalyst: Experiments and theory. *J. Catal,* 158: pp. 439–451, 1996.

20. Spieker, W.A. and J.R. Regalbuto, A fundamental model of platinum impregnation onto alumina. *Chem. Eng. Sci.*, 56: pp. 3491–3504, 2001.

21. Xidong, W., Y. Yongnian, and Z. Jiayu, Influence of soluble aluminium on the state and surface properties of platinum in a series of reduced platinum/alumina catalysts. *Applied Catalysis*, 40: pp. 291–313, 1988.

22. Agashe, K.B. and J.R. Regalbuto, A revised physical theory for adsorption of metal complexes at oxide surfaces. *J. Colloid. Int. Sci*, 185: pp. 174–189, 1997.

23. Hao, X., W.A. Spieker, and J.R. Regalbuto, A further simplification of the revised physical adsorption (RPA) model. *J. Colloid. Int. Sci*, 267: pp. 259–264, 2003.

24. Melo, F., J. Cervello, and E. Hermana, Impregnation of porous supports: Theoretical study of the impregnation of one or two species. *Chem. Eng. Sci.*, 35: pp. 2165–2174, 1980.

25. Dullien, F.A., *Porous media: Fluid transport and pore structure.* Second ed. San Diego: Academic Press, 1992.

26. Scheidegger, A.E., *The physics of flow through porous media.* Toronto: University of Toronto Press, 1974.

27. Krishna, R. and J.A. Wesselingh, The Stefan-Maxwell approach to mass transfer. *Chem. Eng. Sci.*, 52: pp. 861–911, 1997.

28. Maatman, R.W. and C.D. Prater, Adsorption and exclusion in impregnation of porous catalytic supports. *Ind. Eng. Chem. Res.*, 49: pp. 253–257, 1957.

29. Maatman, R.W. and C.D. Prater, How to make a more effective platinum-alumina catalyst. *Ind. Eng. Chem. Res.*, 51: pp. 913–914, 1959.

30. Chen, H.-C., G.C. Gillies, and R.B. Anderson, Impregnating chromium and copper in alumina. *J. Catal*, 62: pp. 367–373, 1980.

31. Santhanam, N., et al., Nature of metal catalyst precursors adsorbed onto oxide supports. *Catal. Today*, 21: pp. 141–156, 1994.

32. Komiyama, M., R.P. Merill, and H.F. Harnsberger, Concentration profiles in impregnation of porous catalysts: Nickel on alumina. *J. Catal*, 63: pp. 35–52, 1980.

33. Van der Berg, G.H.v.d. and H.T. Rijnten. *The impregnation and drying step in catalyst manufacturing, in Preparation of catalysts II. Proceedings of the second international symposium on the scientific bases for the preparation of heterogeneous catalysts.* Amsterdam: Elsevier, 1979.

34. Uemura, Y., Y. Hatate, and A. Ikari, Formation of nickel concentration profile in nickel/alumina catalyst during post-impregnation drying. *J. Chem. Eng.* Japan, 20: pp. 117–123, 1987.

35. Goula, M.A., C. Kordulis, and A. Lycourghiotis, Influence of impregnation parameters on the axial Mo/g-alumina profiles studied using a novel simple technique. *J. Catal*, 133: pp. 486–497, 1992.

36. Li, W.D., et al., Theoretical prediction and experimental validation of the egg-shell distribution of Ni for supported Ni/Al$_2$O$_3$ catalysts. *Chem. Eng. Sci.*, 49: pp. 4889–4895, 1994.

37. Yasuaki Okamoto, S.U., Y. Arima, K. Nakai, T. Takahashi, K. Uchikawa, K. Inamura, Y. Akai, O. Chiyoda, N. Katada, T. Shishido, H. Hattori, S. Hasegawa, H. Yoshida, K. Segawa, N. Koizumi, M. Yamada, A. Nishijima, T. Kabe, A. Ishihara, T. Isoda, I. Moshida, H. Matsumoto, M. Niwa, T. Uchijima, A study on the preparation of supported metal oxide catalysts using JRC-reference catalysts. I. Preparation of a molybdena-alumina catalyst. Part 3. Drying process. *Applied Catalysis A: General*, 170: pp. 343–357, 1998.

38. Vergunst, T., F. Kapteijn, and J.A. Moulijn, Monolithic catalysts: Non-uniform active phase distribution by impregnation. *Applied Catalysis A: General*, 213: pp. 179–187, 2001.

39. Kotter, M. and L. Riekert. *The influence of impregnation, drying and activation on the activity and distribution of CuO on a-alumina,* in *Preparation of catalysts II. Proceedings of the second international symposium on the scientific bases for the preparation of heterogeneous catalysts.* Amsterdam: Elsevier, 1979.

40. Kotter, M. and L. Riekert, Impregnation type catalysts with nonuniform distribution of the active component on inert carriers. *Chem. Eng. Fund.*, 2: pp. 31–38, 1983.

41. Hariott, P., *Diffusion effects in the preparation of impregnation catalysts.* J. Catal, 14: pp. 43–48, 1969.

42. Hepburn, J.S., H.G. Stenger, and C.E. Lyman, Effect of drying on the preparation of HF co-impregnated rhodium/Al_2O_3 catalysts. *Applied Catalysis*, 55: pp. 287–299, 1989.

43. Becker, E.R. and T.A. Nuttall. *Controlled catalyst distribution on supports by co-impregnation,* in *Preparation of catalysts II. Proceedings of the second international symposium on the scientific bases for the preparation of heterogeneous catalysts.* Amsterdam: Elsevier, 1979.

44. Reed, S.J.B., *Electron microprobe analysis.* Second edition. Cambridge, U.K.: Cambridge University Press, 1993.

45. Heise, M.S. and J.A. Schwarz, Preparation of metal distributions within catalysts supports I. Effects of pH on catalytic metal profiles. *J. Colloid. Int. Sci*, 107: pp. 237–243, 1985.

46. Heise, M.S. and J.A. Schwarz, Preparation of metal distributions within catalyst supports II. Effect of ionic strength on catalytic metal profiles. *J. Colloid. Int. Sci*, 113: pp. 55–61, 1986.

47. Heise, M.S. and J.A. Schwarz, Preparation of metal distributions within catalyst supports III. Single component modeling of pH, ionic strength and concentration effects. *J. Colloid. Int. Sci*, 123: pp. 51–58, 1988.

48. Blanco, M.N., et al., Influence of the operative conditions on the preparation of pelleted Mo/Al_2O_3 catalysts. *Applied Catalysis A: General*, 33: pp. 231–244, 1987.

49. Goula, M.A., et al., Development of molybdena catalysts supported on g-alumina extrudates with four different Mo profiles: Preparation, characterisation and catalytic properties. *J. Catal*, 137: pp. 285–305, 1992.

50. Toupance, T., M. Kermarec, and C. Louis, Metal particle size in silica-supported copper catalysts. Influence of the conditions of preparation and thermal pretreatments. *J. Phys. Chem. B.*, 104: pp. 965–972, 2000.

51. Chouillet, C., et al., Relevance of the drying step in the preparation by impregnation of Zn/SiO_2 supported catalysts. *J. Phys. Chem. B.*, 107: pp. 3565–3575, 2003.

52. Malet, P., et al., Influence of drying temperature on properties of Ni-MgO catalysts. *Solid State Ionics*, 95: pp. 137–142, 1997.

53. Gonzalez-Marcos, M.P., et al., Effect of thermal treatments on surface chemical distribution and catalyst activity in Nickel on silica systems. *J. Mol. Catal*. A: Chem., 120: pp. 185–196, 1997.

54. Steen, E.V., et al., TPR study on the preparation of impregnated Co/SiO_2 catalysts. *J. Catal*, 162: pp. 220–229, 1996.

55. Daturi, M. and L.G. Appel, Infrared spectroscopic studies of surface properties of Mo/SnO2 catalyst. *J. Catal*, 209: pp. 427–432, 2002.

56. Haddad, G.J. and J.G. Goodwin, The impact of aqueous impregnation on the properties of prereduced vs. precalcinated Co/SiO2. *J. Catal*, 157: pp. 25–34, 1995.

57. McCabe, W.L., J.C. Smith, and P. Harriott, *Unit operations of chemical engineering*. Chemical and Petroleum Engineering Series. New York: McGraw-Hill, 1993.

58. Bond, G., R.B. Moyes, and D.A. Whan, Recent applications of microwave heating in catalysis. *Catal. Today*, 17(3): pp. 427–437, 1993.

59. Berry, F.J., et al., Microwave heating during catalyst preparation: Influence on the hydrodechlorination activity of alumina-supported palladium-iron bimetallic catalysts. *Applied Catalysis A: General*, 204(2): pp. 191–201, 2000.

60. Karakonstantis, L., K. Bourikas, and A. Lycourghiotis, Tungsten-oxo species deposited on alumina. *J. Catal*, 162: pp. 295–305, 1996.

61. Mang, T., B. Breitscheidel, and P. Polanek, Adsorption of platinum complexes on silica and alumina: Preparation of non-uniform metal distributions within support pellets. *Applied Catalysis A: General*, 106(2), 1993.

62. Aurora, S., et al., Study of the adsorption of Cd and Zn onto an activated carbon: Influence of pH, cation concentration, and adsorbent concentration. *Separ. Sci. Tech.*, 34: pp. 1577–1593, 1999.

63. Marzal, P., et al., Cadmium and zinc adsorption onto activated carbon: influence of temperature, pH and metal/carbon ratio. *J. Chem. Tech. Biotechnol*, 66: pp. 279–285, 1996.

64. Seco, A., et al., Absorption of heavy metals from aqueous solutions onto activated carbon in single Cu and Ni systems and in binary Cu Ni, Cu-Cd and Cu-Zn systems. *J. Chem Tech. Biotech.*, 68(1): pp. 23–30, 1997.

65. Georgiadou, I., et al., Preparation and characterization of various titanias (anatase) used as supports for vanadia-supported catalysts. *Coll. Surf. A: Phys. Eng. Aspe.*, 98(1–2): pp. 155–165, 1995.

66. Faur-Brasquet, C., et al., Modeling the adsorption of metal ions (Cu^{2+}, Ni^{2+}, Pb^{2+}) onto ACCs using surface complexation models. *Appl. Sur. Sci.*, 196: pp. 356–365, 2002.

67. Halter, W.E. and H.-R. Pfeifer, Arsenic(V) adsorption onto -Al_2O_3 between 25 and 70°C. *Applied Geochemistry*, 16: pp. 793–802, 2001.

68. Vlasova, N.N., Adsorption of Cu^{2+} ions onto silica surface from aqueous solutions containing organic substances. *Coll. Sur. A: Phys. Eng. Asp.*, 163: pp. 125–133, 2000.

69. Blesa, M.A., et al., The interaction of metal oxide surfaces with complexing agents dissolved in water. *Coord. Chem. Rev.*, 196: pp. 31–63, 2000.

70. Olsbye, U., R. Wendelbo, and D. Akporiaye, Study of Pt/alumina catalysts preparation. *Applied Catalysis A: General*, 152: pp. 127–141, 1997.

17 The Effects of Finishing and Operating Conditions on Pt Supported Catalysts during CO Oxidation

Eduardo E. Wolf

CONTENTS

17.1 INTRODUCTION

One of the least discussed catalyst preparation variables in the literature is the effect of pretreatment and the reaction environment on the resulting catalyst. Such operations, known in industry as *catalyst finishing*, are in many cases the key to activity and selectivity. This information is often not disclosed along with bulk compositions in the patent literature and is subject to secrecy agreements between catalyst manufacturers and users. In the open literature, catalyst pretreatment is often copied from existing recipes without major investigation of its effect in the final activity. For this reason, it is not uncommon to encounter widely different results on the activity of catalysts that are apparently similar. In addition, a less known effect in catalysts' performance is the effect of the reaction environment, which often modifies the surface to conditions quite different than that of the fresh catalyst. This is known in the industry as the *lining up* of the catalyst's activity. Interestingly, it is not uncommon for researchers to quote studies conducted under quite different conditions in terms of the type of material and the pressure and temperature used, to compare them with their results obtained under a completely different environment.

A case in point are results conducted under ultrahigh vacuum (UHV) and single-crystal surfaces, compared with results obtained with supported catalysts and under high pressure conditions. These vastly different conditions are referred to as the pressure gap (UHV vs. pressures of several atmospheres) and the materials gap (single crystals vs. polycrystalline particles on complex oxide supports). Although there are a handful of papers in which such studies compare favorably, these are exceptions rather than the rule. These occur when the surface is preferentially covered by one reactant (i.e., CO or carbon) and thus is rather insensitive to the environment or to the underlying surface. One of the best examples of the effect of UHV versus atmospheric pressure has been reported by none other than Somorjai[1] one of the leaders in the use of surface analysis and surface science in catalysis. Figure 17.1 displays micrographs obtained by scanning tunneling microcopy (STM) results of surface reconstruction of a few torrs of hydrogen, oxygen, and CO, showing rough and different morphologies, depending on the gas used. The same surface under UHV conditions is well ordered and shows a well-defined structure. The surfaces under high pressure are completely different because the gases restructured the well-defined surface under UHV and turned into something much more complex than the initial structure under UHV. Interestingly, this work is seldom quoted, yet it clearly demonstrates the difficulty of extrapolating from one extreme condition to another. When the issue of the heterogeneity of a supported polycrystalline particle is added to the UHV versus high pressure, then comparison between these regimes is almost untenable.

From the above, it follows that catalyst finishing often does alter the surface significantly and thus affects the resulting activity and selectivity. Although researchers are fully aware of these effects and contradictions, they are often overlooked, because researching the effect of pretreatment and the reaction is often a secondary task and adds a burden to the experimental program. Pretreatment is often not the key variable under investigation, such as a new material composition or a new structural feature of the catalyst. In addition, investigating the effect of the reaction

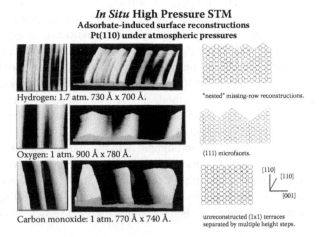

FIGURE 17.1 *In situ* STM micrographs of a Pt(110) surface under hydrogen, oxygen, and carbon monoxide at atmospheric pressure at 425 K for 5 h. (Reference 1)

conditions on the surface requires the use of methods under reaction, or *operando* conditions, as they are now known. In our laboratory, we have taken two approaches that are really not so new but have now been recast with new names that make them appear rather new. To overcome the limitations on experimental time to investigate the secondary aspects of catalysts finishing, we have readopted the use of parallel reactors for medium throughput experimentation, as opposed to high throughput experimentation. Our methodology has been summarized in the literature[2, 3] and thus, here we will only mention the essential approach rather than a detailed description.

Basically, we use infrared thermography (IRT) to detect temperature differences among an array of catalysts, which permits us to quickly ascertain catalyst composition or conditions that lead to high activity. In our studies, we have focused on exothermic oxidation reactions, wherein a temperature increase is a descriptor of catalytic activity. An array of up to 50 catalyst spots can be surveyed in a single experiment. The ones that give the highest temperature difference are then studied in a parallel flow reactor with 10 channels to directly determine activity and selectivity. Finally, the most active and selective materials are studied in a recycle reactor to determine kinetic parameters. This accelerates the discovery of new catalytic materials or new conditions that lead to higher activity and selectivity. Although high throughput experimentation has been given high press lately, one must read Mittash's article[4] on the discovery of the ammonia synthesis catalysts to realize that, in 1912, he was running 24 high pressure reactors in parallel around the clock. This permitted him to survey more than 3,000 compositions in 20,000 experiments that in 2 years led to the discovery of the commercial ammonia synthesis catalyst in 1914.

The other technique we use under operando conditions is infrared spectroscopy, which has also been used in catalysis research for more than 50 years. The new surname as an operando technique makes it sound more glamorous, but it is really an additional qualifier for *in situ* IR, which has been around for decades. In fact, the editor of this book, John Regalbuto, did his Ph.D. work using *in situ* and operando

IR during NO reduction by CO. The operando surname defines an additional qualifier to *in situ*, because the latter can occur in the absence of reaction. Another technique that we have used *in situ*, but not under operando conditions, is extended x-ray absorption fine-structure (EXAFS). Although this technique requires much more data interpretation through data fitting to models, some definitive conclusions can be drawn by direct comparison with standards obtained with well-defined samples. Figure 17.2 presents our approach schematically by likening catalyst testing to wine tasting; if it is done sequentially, is too time-consuming, whereas in parallel it accelerates significantly catalyst discovery.

Using a combination of medium throughput activity studies, operando IR, and *in situ* EXAFS, we have been able to unravel several effects of catalyst finishing, involving the nature of the precursor, crystallite size, oxidation state, and the effect of sulfur during CO oxidation. The focus in this chapter is on the main results of such studies; detailed experimental descriptions of the procedures used are presented elsewhere.[5–7]

17.2 EXPERIMENTAL SECTION

17.2.1 CATALYTIC ACTIVITY

In the work described in this chapter, no IRT results are presented because the focus is mainly on studying catalyst finishing variables rather than new catalytic materials. Thus, catalytic activities were determined using two types of flow reactors. A 10-channel

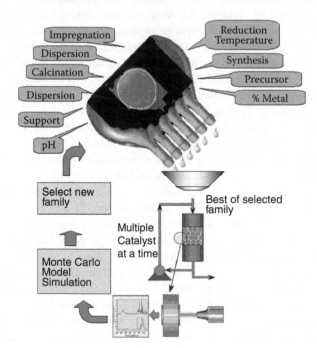

FIGURE 17.2 Schematics of parallelization approach used in the author's laboratory.

parallel microreactor was first used for high-throughput studies to obtain the main trends on the effect of pretreatment on Pt supported catalysts. Then, detailed kinetic information was obtained in a tubular flow recycle reactor. Catalyst activities were obtained under diluted reactant concentration: 1% CO in an oxidizing atmosphere with 5 or 10% O_2, balance in He. Unless specified otherwise, prior to reaction, the catalysts were reduced in hydrogen flowing at 100 cc/min at 150°C. The 10-channel microreactor is made of stainless steel and has 10 reaction wells, each of which can contain up to 200 mg of catalysts.[2,3] The reactor is interfaced on-line to a gas chromatograph (GC) through a multiple position valve, which selects the effluent from each of the 10 reaction channels for analysis. The catalyst particles were prepared by pressing the powders into pellets, crushing the resulting pellets into small agglomerates, and sieving them between 1.0 and 0.6 mm size. The reactor system is automated and computer controlled, and it has the required instrumentation to control flow rates and temperatures. The recycle reactor consists of a quartz tube (12 in. long, 1/2 in. diameter) equipped with an external diaphragm pump to provide an external recycle loop. A thermocouple is placed in a thermocouple well in the center of the catalyst bed to monitor the temperature of the bed. The effluent flow rate varied for each reaction system described between 100 and 130 cc/min, and the recycle ratio was about 20:1 to ensure operation at conditions equivalent to perfect mixing.

In all the studies reported here for CO oxidation, GC analysis of the effluent stream was carried out using a HayeSep Q column at room temperature (RT) in a GC (Varian 3300) equipped with a TC detector. The conversion data were reproducible within 5% accuracy. Reaction rates were calculated at various temperatures at less than 5% conversion to fulfill differential conversion operation in the 10-well reactor. In the recycle reactor, reaction rates were calculated in the whole conversion range due to the perfect mixing approximation obtained via the use of a large recycle ratio (> 20). Turnover frequencies were calculated from the rate and the dispersion values obtained for the freshly reduced catalysts and plotted versus 1/T to obtain activation energies.

17.2.2 EXAFS Data Collection and Analysis

Measurements using extended x-ray absorption fine-structure spectroscopy were made on the insertion-device beam line of the Materials Research Collaborative Access Team (MRCAT) at the Advanced Photon Source, Argonne National Laboratory. Details about the instrumental settings are described elsewhere.[5–7] Standard procedures based on WINXAS97 software[8] were used to extract the EXAFS data.[9] Phase shifts and back-scattering amplitudes were obtained from EXAFS data for reference compounds: $Na_2Pt(OH)_6$ for Pt-O, H_2PtCl_6 for Pt-S, and Pt foil for Pt-Pt. $Pt(NH_3)_4(NO_3)_2$, H_2PtCl_6 Pt Foil were used for the Pt^{+2}, Pt^{+4}, and Pt^0 XANES references.

In situ experiments were conducted in a controlled atmosphere EXAFS cell. The sample was pressed into a cylindrical holder with a thickness chosen to give an absorbance ($\Delta\mu x$) of about 1.0 in the Pt edge region, corresponding to approximately 100 mg of catalyst. The sample holder was centered in a continuous-flow *in situ*

EXAFS reactor tube (18 in. long, 0.75 in. diameter) fitted at both ends with polyimide windows and valves to isolate the reactor from the atmosphere. The catalysts were pretreated in the laboratory by flowing gases or reactants at temperatures similar to those used during pretreatment or reaction. After the prescribed treatment, the sample was cooled to room temperature in the gas used for pretreatment or reaction, and then the cell was isolated by closing the valves fitted at both ends of the tube and moved to the EXAFS data acquisition room.

17.2.3 OPERANDO FTIR SPECTROSCOPY

Transmission infrared spectra of pressed disks (~14 mg) of Pt/SiO_2 were collected *in situ* and under operando conditions in an IR reactor-cell placed in an FTIR spectrometer (Mattson, Galaxy 6020) at a resolution of 2 cm^{-1} and 30 scans/spectrum. The IR cell is equipped with NaCl windows, and it has connections for inlet and outlet flows, and thermocouples connected to a temperature controller to monitor and control its temperature. The spectra were obtained in absorbance mode after subtraction of the background spectrum of the catalyst's disk under He atmosphere at the corresponding temperature. The samples were pretreated at various conditions prior to study CO adsorption and reaction. During CO oxidation experiments, 10% O_2 was added to the 1% CO-He feed. In most experiments, the heating rate was 1°C/min with a total flow of 120 cc/min unless indicated otherwise.

This chapter comprises three case studies of the effect of catalyst finishing or pretreatment on the activity of the resulting catalysts for CO oxidation. These results are summarized from papers now published in collaboration with F. Gracia (who did the work as his Ph.D. thesis) and J. Miller from BP-Amoco (who did the some of the catalysts preparations and the EXAFS data collection and analysis), along with J. Kropf from Argonne National Laboratories. Thus, the focus here is to describe the overall effects of the various pretreatments and conditions used in the activity instead of detailed discussion of the observations. The reader is directed to the corresponding references for further detail. The case studies are

1. The effect of Cl-containing precursors and the pretreatment on Pt supported catalysts during CO and methane oxidation
2. The effect of oxidation, reduction, and reaction on Pt supported catalysts during CO oxidation
3. The effect of sulfur poisoning of Pt supported catalyst during CO oxidation

17.3 EFFECT OF CHLORINE ON PT-SUPPORTED CATALYSTS DURING OXIDATION REACTIONS

17.3.1 INTRODUCTION

The complete combustion of methane by Pt and Pd catalysts has been studied in relation to the control of polluting emissions from natural gas vehicles (NGVs),[10] as well as the oxidation of methane in turbines for power generation.[11] Supported

Pt catalysts are often prepared from Cl-containing precursors such as H_2PtCl_6, and it has been reported[12-24] that Cl poisons the oxidation activity of Pt. The state of the active catalyst's surface and the effect of Cl poisoning on the activity, however, have not been elucidated. Lieske et al.[25] were among the first to propose a model of the various phases that could be present in a Pt/Al_2O_3 catalyst prepared from Cl precursors as a function of pretreatment conditions. These phases, however, were not correlated with the catalyst's activity. Based on temperature-programmed reduction (TPR) of Pt catalysts oxidized at different temperatures, Hwang and Yeh[26] concluded that four types of oxide species could be formed, depending on the oxidation temperature. The formation of PtO_xCl_y complexes has also been suggested when catalysts are prepared from Cl-containing precursors.[27-29] Burch and Loader[30, 31] and Yang et al.[32] concluded that the oxidation activity of Pt and Pd catalysts is optimal for a partially oxidized and reduced surface. Simone et al.,[12] among others, reported that the presence of Cl on the catalyst reduced the methane oxidation activity of Pd/Al_2O_3, and that removal of Cl increased the catalyst's activity. Similarly, Marceau et al.[13, 14] found that elimination of Cl from Pt/Al_2O_3 catalysts at 450°C led to higher activity. Roth et al.[15] also confirmed that removal of Cl from a Pd/Al_2O_3 catalyst led to the same activity as Cl-free Pd catalysts and suggested that the active sites are PdO, which slowly deactivates to form a less active $Pd(OH)_2$. The objective of this study was to elucidate the state of the surface on Cl-free and Cl-containing Pt supported catalysts during oxidation reactions and correlate the activity with the structure of the active Pt species under oxidizing reaction conditions.

17.3.2 CATALYST PREPARATION

To investigate the effect of chlorine on the activity of Pt supported catalysts, samples were prepared from H_2PtCl_6 (with Cl) and $Pt(NH_3)_4(NO_3)_2$ (no Cl) on alumina and silica supports. Specifics on the amounts of materials used and calcination temperature used are given in Reference 5. Elemental analysis of the 1.5% Pt/Al_2O_3 (with Cl) showed a 2.1% Cl, and the hydrogen chemisorption at 25°C of the reduced catalyst corresponded to a dispersion of 82%. The hydrogen chemisorption of the 1.5% Pt/Al_2O_3 (no Cl) catalyst indicated a dispersion of 81%. In the case of the 1.5% Pt/SiO_2 (no Cl catalyst), the dispersion was 36%. After calcinations, the catalysts were reduced in flowing H_2 (200 cc/min) at 300°C for 2 h. The prereduced catalysts were pretreated in air prior to activity, IR, EXAFS, and CO chemisorption measurements.

Fractions of the Cl-free catalysts were treated with HCl using wet impregnation followed by overnight drying at 110°C to analyze the effect of Cl addition to the Cl-free catalyst from sources different from the precursor.

17.3.3 RESULTS AND DISCUSSION

17.3.3.1 Catalytic Activity

In this case, catalysts' activities were determined in the recycle reactor at atmospheric pressure. The effluent flow rate was 130 cc/min, and the recycle ratio was about 20:1 to ensure complete mixing. In these studies of the effect of Cl, prior to each run, the catalysts were pretreated in air at 300°C for 2 h to stabilize them with an oxidizing feed. In the next section, we studied specifically the effect of prereducing the catalysts. After pretreatment, the catalysts were cooled to room temperature, the feed was switched to the reaction mixture, and the temperature was increased to about 450°C. (Hydrogen chemisorption of metallic Pt particles oxidized up to 500°C showed no significant change in dispersion). Catalyst activities were measured under diluted reactant concentrations: 0.3% CO, CH_4, or C_2H_6, and an oxidizing atmosphere with 16% O_2, balanced in He.

The effect of Cl on CO, methane, and ethane oxidation reactions on Pt/Al_2O_3 catalysts is shown in Figure 17.3. The plots in the left panels correspond to the conversion at different temperatures, and those in the right, to the Arrhenius plots. The turnover frequency (TOF) values used to obtain the Arrhenius plots were obtained at low conversion (<15%) to minimize the effect of reactant concentration.

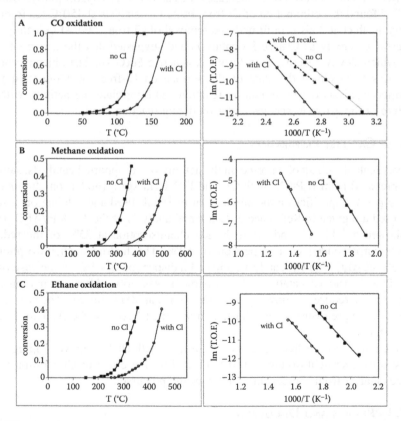

FIGURE 17.3 Oxidation reactions on Pt/alumina catalysts with and without Cl. Left panel: conversion versus temperature; right panel: Arrhenius plots. (A) CO oxidation. Data in broken line correspond to TOF recalculated using IR corrections. (B) CH_4 oxidation. (C) C_2H_6 oxidation. Feed composition 0.3% CO, CH_4, or C_2H_6; 16% O_2, balance He. Total flow rate 130 cc/min. Recycle reactor.

During CO oxidation (Figure 17.3A), even though both catalysts have nearly identical Pt loading and particle size, the catalyst prepared from H_2PtCl_6 (2.0 wt.% Cl) requires about 50°C higher temperature to obtain the same CO conversion. This corresponds to a 10-fold change in reaction rate (moles CO_2 produced per second per mole surface Pt) at 100°C. From the Arrhenius plots, activation energies of 87.1 and 59.9 kJ/mol for the Cl-containing and Cl-free catalyst, respectively, were obtained. The data in a broken line correspond to TOF values recalculated using an IR-derived correction, discussed later on.

For CH_4 oxidation, the light-off temperature (LOT, temperature at which 50% conversion is reached) is much higher than for CO oxidation. In this case, the Cl-free catalyst is also more active than the Cl-containing catalyst. As shown in Figure 17.3B, the presence of chlorine results in an increase of 100°C in the temperature, at which a CH_4 conversion of 50% is reached. At 350°C, the reaction rate of the Cl-containing catalyst is 20 times slower than that of the catalyst with no Cl. In this case, the corresponding activation energies are 107.5 kJ/mol (with Cl) and 90.9 kJ/mol (no Cl).

For the complete oxidation of ethane (Figure 17.3C), when chlorine is present, there is an 80°C shift of the conversion curve compared to that of the Cl-free catalyst. This corresponds to a 10-fold difference in the reaction rate at 330°C. Calculations from the Arrhenius plots yield similar activation energies of 68.6 kJ/mol (with Cl) and 68.2 kJ/mol (no Cl). The results presented in Figure 17.3 clearly show that Cl decreases the activity of Pt/alumina catalysts during CO, CH_4, and C_2H_6 oxidation reactions.

Addition of HCl to Cl-free catalysts. To further demonstrate the effect of Cl on the oxidation activity, HCl was added at room temperature to the Cl-free catalyst to give a Cl content of about 2 wt.%, which is similar to the Cl present on the catalyst prepared from chloroplatinic acid. After impregnation with HCl, the catalyst was dried overnight at 110°C and subjected to different pretreatments in air prior to reaction. Figure 17.4 shows the CO conversion curves as a function of the pretreatment temperature for these catalysts.

When the HCl-impregnated catalyst was reacted without any calcination treatment, there was only a minor decrease in the activity compared to the Cl-free catalyst. This suggested that Cl was adsorbed mainly on the alumina support without affecting the Pt surface. When the same catalyst was calcined at 250°C, the activity decreased significantly, indicating that, during the calcination pretreatment, Cl migrated from the support and readsorbed on the Pt surface, poisoning its activity. A further increase in the calcination temperature to 300°C of a freshly impregnated catalyst gave a catalytic activity closer to that of the Cl-containing catalyst.

The amount of Cl (added as HCl) is also a factor in decreasing the CO oxidation activity of the Cl-free catalyst. After impregnation with an HCl solution containing about 5 wt.% followed by calcination, the activity of the catalyst prepared with no Cl decreased below that of the catalyst prepared from a Cl-containing precursor. Calcination in air at 300°C lowers the LOT compared to the same catalyst pretreated at 250°C, i.e., increases the activity. This indicates that when present beyond certain saturation value, the excess chlorine is removed from the Pt particles by pretreatment in air at 300°C. For methane oxidation, however, the calcination treatment after HCl

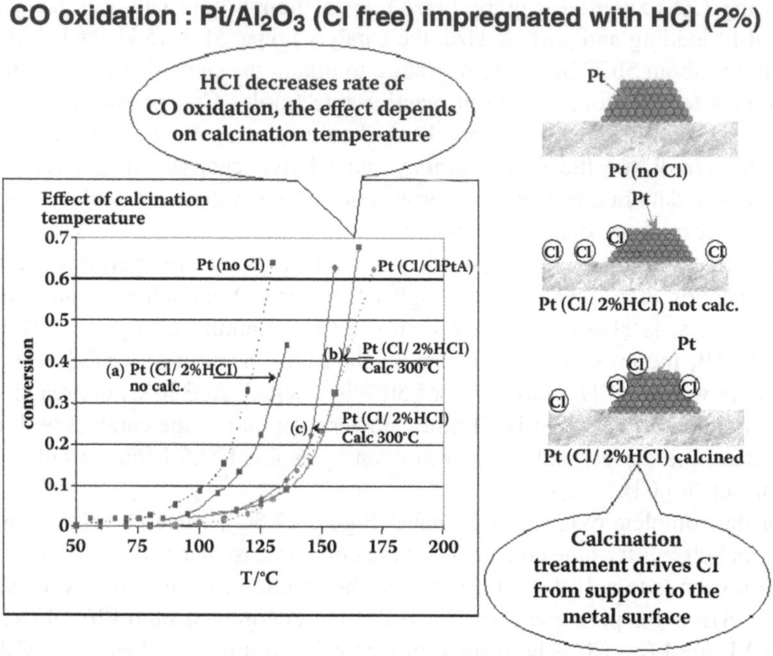

CO oxidation : Pt/Al$_2$O$_3$ (Cl free) impregnated with HCl (2%)

FIGURE 17.4 Effect of calcination temperature after impregnation with HCl during CO oxidation of Pt/alumina catalyst. (A) No calcination treatment. (B) Calcination at 250°C. (C) Calcination at 300°C. Feed composition 0.3% CO, 16% O$_2$, balance He. Total flow rate 130 cc/min. Recycle reactor.

addition had little effect on the catalyst activity. In this case, the reaction starts at a temperature higher than the calcination temperature of 300°C; therefore, heating the catalyst to the reaction temperature is equivalent to pretreating it in air. Addition of excess Cl (5 wt.% Cl) had no additional effect on the methane oxidation activity, also indicating a saturation of the Cl coverage of the Pt surface.

The poisoning effect of chlorine shown for Pt supported on alumina catalysts also occurs in the case of Pt supported on silica (Figure 17.5). After addition of Cl by impregnation with HCl, the LOT for CO oxidation on the Pt/silica catalyst is 45°C higher than with the Cl-free catalyst. At 120°C, the reaction rate of the Cl-containing catalyst is 30 times lower than that of the Cl-free one. A direct quantitative comparison of the effect of chlorine on silica and alumina cannot be made because of the different dispersion of Pt on each support. After prolonged treatments of the catalysts, the effect of Cl is different on the silica support than on alumina. As shown in Figure 17.5, after reduction in H$_2$ at 300°C for 10 h, the activity of the Cl-containing Pt/silica catalyst increases and recovers to a level similar to that of the Cl-free catalyst. This indicates that on the silica support, most of the Cl has been removed by the reduction treatment. In contrast, after treatment in H$_2$ for 40 h at 450°C, the activity for CO oxidation on the Cl-containing Pt/Al$_2$O$_3$ catalyst showed

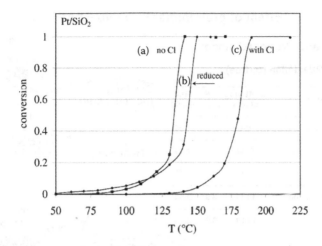

FIGURE 17.5 CO conversion versus temperature for 1.5% Pt/silica catalysts. (no Cl) 1.5% (a) no Cl. (with Cl), (b) with Cl. (reduced), (c) with Cl after reduction treatment for 10 h at 300°C. Feed composition 0.3% CO, 16% O_2, balance He. Total flow rate 130 cc/min. Recycle reactor.

no improvement. These results indicate that Cl is more strongly adsorbed on the surface of the alumina support than on silica.

17.3.3.2 *In Situ* EXAFS

A detailed discussion of the EXAFS spectrum for catalysts prepared from the Cl-containing precursor and Cl free has been presented in detail in Reference 5. Instead of repeating such analysis, here we present the end results in terms of the coordination number (CN) and scattering distances (R) that best fit the data along with a illustration representing the various mechanisms involved in the process leading to the final state of the surface. It is emphasized here that these results were obtained after an *in situ* treatment outside the x-ray room, but were not exposed to air when the cell was transferred to it under an He atmosphere. Figure 17.6 summarizes such results for the case of the 1% Pt/Al_2O_3-Cl free catalyst after various pretreatments.

As expected, when the catalyst is calcined at 250°C, the EXAFS results fit a model having Pt with six oxygen nearest neighbors with the Pt-O distance of about 2.06 Å. No Pt-Pt bonds were observed in the calcined catalysts. The XANES spectra for the calcined catalyst, $Pt(NH_3)_4(NO_3)_2$ and $Na_2Pt(OH)_6$ (see Reference Cl-WO), show that both the position and intensity of the XANES spectrum of the calcined catalyst are consistent with Pt in the +4 oxidation state. Consequently, we interpret these results as corresponding to crystallites of PtOx. After reduction of the calcined sample at 300°C, the results fit a model with a 6.6 Pt-Pt coordination number and a bond distance of 2.74 Å, corresponding to small metallic Pt particles. A fully coordinated shell in Pt metal consists of 12 nearest-neighbor Pt atoms at a distance of 2.77 Å. The EXAFS of the reduced catalyst also show a contribution from the oxygen ions of the support with a Pt-O distance of about 2.19 Å.[33] Both of these contributions are shown in the carton on the left. When the same reduced

Effect of pretreatment, *in situ* EXAFS

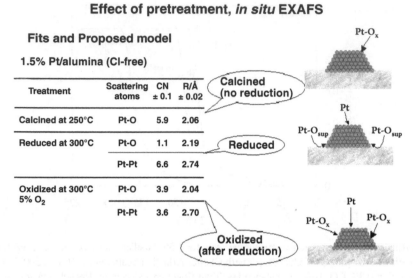

Fits and Proposed model

1.5% Pt/alumina (Cl-free)

Treatment	Scattering atoms	CN ± 0.1	R/Å ± 0.02
Calcined at 250°C	Pt-O	5.9	2.06
Reduced at 300°C	Pt-O	1.1	2.19
	Pt-Pt	6.6	2.74
Oxidized at 300°C 5% O₂	Pt-O	3.9	2.04
	Pt-Pt	3.6	2.70

FIGURE 17.6 EXAFS fits and model interpretation for the Cl free catalysts.

sample was reoxidized, the results fit a model that has a Pt-Pt coordination number decreasing to 3.6 at a distance of 2.70 Å, and a significant Pt-O contribution at a distance of 2.04 Å with a coordination number of 3.9. The XANES spectra of the calcined, reduced (with chemisorbed H_2), oxidized Pt/alumina (no Cl) catalysts, and Pt foil are also consistent with the presence of reduced and oxidized Pt atoms. The illustration depicts the surface of the reoxidized catalyst as a core of metallic Pt surrounded by oxidized Pt; the relative position of these phases is speculative, however.

The fits and model corresponding to the Cl-containing catalyst after the various pretreatments are summarized in Figure 17.7. The EXAFS and XANES spectra of the Cl-containing catalyst calcined Pt/Al_2O_3 (with Cl) fit a model where the Pt(IV) is bonded to 3.5 oxygen ions at a distance of 2.05 Å and to 2.5 chloride ions at a distance of 2.31 Å. The Pt-Cl distance is identical (within experimental error) to that in chloroplatinic acid.[34] The interpretation of this result is as a crystallite made of a Pt oxychlorinated compound PtClxOy. After reduction of this catalyst at 300°C, the data fits a model with small metallic Pt particles with a Pt-Pt coordination number of 6.9 at a distance of 2.74 Å, also with the small Pt-O coordination of 0.9 from the oxygen ions of the support at a distance of 2.21 Å. The interpretation is similar to that of the Cl-free catalysts, as shown in the corresponding illustration. It should be noted that there are no Pt-Cl bonds in the reduced catalyst, even though this catalyst originally contained 2 wt.% Cl. The coordination numbers of 6.6 and 6.9 Pt neighbors for the two reduced catalysts show that, on average, the first shell of every Pt atom is not completely coordinated to Pt, indicating small metallic Pt particles consistent with the H_2 chemisorption results, indicating that the dispersions are 0.8.

The EXAFS results of the Cl-containing catalyst reduced and then reoxidized at 300°C fit a model with a Pt-Pt distance of 2.70 Å and a coordination number

Effect of pretreatment, *in situ* EXAFS

Fits and Proposed model

1.5% Pt/alumina (Cl-containing)

Treatment	Scattering atoms	CN ± 0.1	R/Å ± 0.02
Calcined at 250°C	Pt-O	3.5	2.03
	Pt-Cl	2.5	2.31
Reduced at 300°C	Pt-O	0.9	2.21
	Pt-Pt	6.9	2.74
Oxidized at 300°C 5% O₂	Pt-O	2.0	2.05
	Pt-Cl	2.5	2.31
	Pt-Pt	0.9	2.70

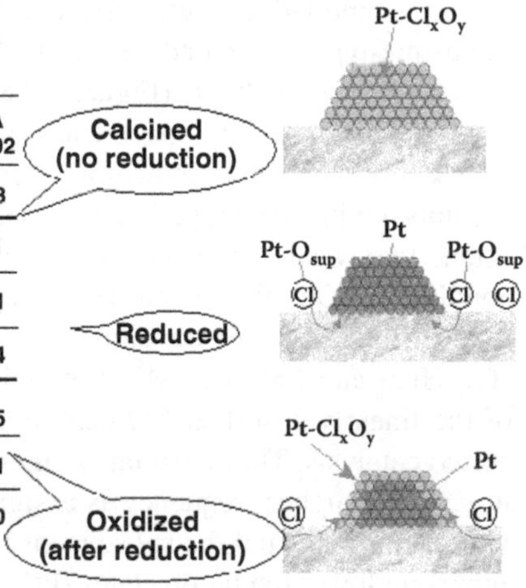

Calcined (no reduction)

Reduced

Oxidized (after reduction)

$Pt-Cl_xO_y$ Pt $Pt-O_{sup}$ $Pt-O_{sup}$ Cl

$Pt-Cl_xO_y$ Pt   Cl

FIGURE 17.7 EXAFS fits and model interpretation for Cl containing catalysts.

of 0.9, indicating the presence of small metallic Pt particles; Pt-O and Pt-Cl coordination numbers are 2.0 and 2.5 at a distance of 2.05 and 2.31 Å, respectively. Thus, although there are no Pt-Cl bonds in the reduced catalyst, upon heating in O_2 to 300°C, Pt-Cl bonds are clearly present after oxidation of the catalyst. This confirms that, as suggested by the activity results, Cl present on the support migrates to the metal surface during oxidation pretreatment. A hypothetical mechanism of Cl migration from the support and the final state of the surface after oxidation at 300°C is depicted in the lower-right corner of Figure 17.7.

17.3.3.3 CO Chemisorption and FTIR of Adsorbed CO

The amount of CO chemisorbed at room temperature was determined for several calcined and oxidized Pt/alumina catalysts. On both calcined catalysts (with and without Cl), within experimental error, there was little chemisorption of CO. Also, on the (reduced) Pt/alumina (with Cl) oxidized at 300°C, there was no chemisorbed CO. However, oxidation in air of the reduced Pt/alumina (no Cl) gave significant adsorption: 0.70 cc/g or about 40% of the Pt atoms chemisorb CO. The chemisorption results suggest that many of the metallic or low-valent Pt atoms of the Pt/alumina (no Cl) are exposed to the reacting gasses. In addition, because the metallic Pt in the oxidized Pt/alumina (with Cl) does not chemisorb CO, these atoms are not at the particle surface, thus they are likely covered by oxygen and Cl atoms.

The infrared spectra on Pt/silica oxidation catalysts were obtained in order to further investigate the effect that Cl has on the adsorption, desorption, and reaction of CO on Pt supported catalysts. SiO_2 was chosen as the support because of its higher transmissivity compared to Al_2O_3. Figure 17.8 shows the spectra for CO adsorption on Cl-free Pt/SiO_2 (Figure 17.8A) and with Cl (Figure 17.8B) in the range 2300–1800 cm^{-1}. The spectra are consistent with the presence of linearly adsorbed CO on Pt,[10, 28, 35-38] indicating that at low CO concentration (0.3% CO in He), the catalyst with no Cl rapidly adsorbs CO at the various temperatures analyzed. In addition, two weak IR bands were detected in the region 2200–2100 cm^{-1}, corresponding to CO in the gas phase. The spectra in Figure 17.8 show that the peak at 2079 cm^{-1} shifts to lower wavenumbers as the temperature increases (2065 cm^{-1} at 200°C). Herz and Shinouskis[39] observed a frequency shift (from 2120 to 2070 cm^{-1}) of the linearly adsorbed CO peak during transient reduction experiments on Pt supported catalysts. They correlated this shift with a change in the electronic state of Pt atoms. The higher frequency is assigned to completely oxidized Pt particles (Pt^{m+} with $2 \geq m > 1$), and the subsequent lower frequencies to CO adsorbed on Pt atoms represent lower oxidation states (Pt^{n+} at 2095 cm^{-1} and $Pt^{\delta+}$ at 2080 cm^{-1} with $1 \geq n > \delta$). Finally, a band at 2070 cm^{-1} is characteristic of linear CO adsorption on fully reduced Pt. Because the integrated absorbance of this peak remains almost constant and considering that the catalyst was previously calcined in air, the observed shift is attributed to a gradual reduction of partially oxidized $Pt^{\delta+}$ atoms at the higher temperatures due to the presence of CO.

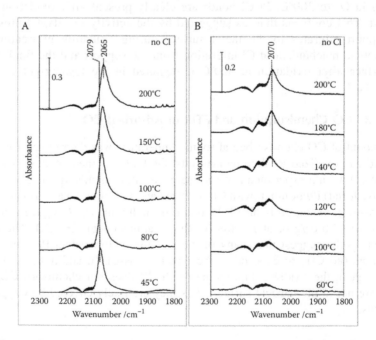

FIGURE 17.8 FTIR spectra during CO adsorption on Pt/silica. (A) 2% Pt/silica no Cl. (B) 2% Pt/silica with Cl. Feed composition 0.3% CO, balance He. Total flow rate 120 cc/min.

When Cl is present on the surface at T < 100°C (Figure 17.8B), the signal of adsorbed CO is almost below the detection level. As the temperature increases, the amount of adsorbed CO increases slightly, but it does not reach a level comparable to that observed in the Cl-free sample. These results indicate that Cl is blocking the sites for CO adsorption. To verify this conclusion, both catalysts were subjected to the same gas compositions used in the activity determinations (0.3% CO, 16% O_2, in He), and the corresponding IR spectra were collected at different temperatures (Figure 17.9).

On the Pt/SiO_2 (no Cl) catalyst, a peak at a constant wavenumber of 2077 cm^{-1} is observed at temperatures as low as 65°C. The intensity of this peak remains constant with temperature, indicating saturation of the active sites with adsorbed CO until ignition of the catalyst is reached between 140 and 145°C. At temperatures above 145°C, the surface coverage of adsorbed CO is below the detection limit, indicating complete reaction. After ignition, a peak at 2321cm^{-1} appears, corresponding to CO_2 in the gas phase. During this TPR experiment, no shift in the position of the peak at 2077 cm^{-1} is observed, indicating that in this oxidizing atmosphere, the Pt oxidation state remains unchanged. Comparison between the activity results and the IR spectra of CO oxidation on Pt/SiO_2 catalysts shows remarkable agreement regarding the temperatures at which each state of the reaction occurs (low-activity regime, ignition, and high-activity regime), allowing one to correlate the kinetic behavior reported in Section 17.3.2 with the state of the surface inferred from IR spectroscopy.

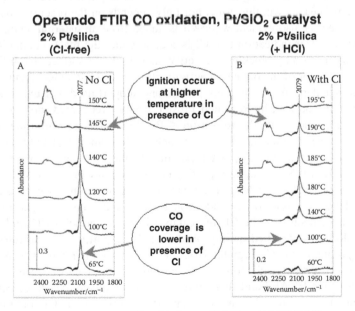

FIGURE 17.9 FTIR spectra during CO oxidation on Pt/silica. (A) 2% Pt/silica no Cl. (B) 2% Pt/silica with Cl. Feed composition 0.3% CO, 16% O_2, balance He. Total flow rate 120 cc/min.

At temperatures below 140°C, addition of chlorine leads to a significant decrease in the amount of adsorbed CO (Figure 17.9B). As the temperature increases between about 60 and 185°C, there is a slow increase in amount of adsorbed CO until initiation of the reaction. Above 185°C, the CO surface coverage decreases until, at about 190°C, there is little adsorbed CO due to consumption by the reaction. Chlorine not only decreases the number of sites available for reaction, it also slows down the ignition process. This is seen by the larger temperature difference (15°C) required for ignition with the Cl-containing catalyst compared to a 5°C temperature difference when no chlorine is present.

To determine if the rate of CO desorption was affected by Cl, after the pretreatment in air, the IR cell was purged in He at 200°C for 1 h. Then, 0.3% CO in He was introduced, and a series of spectra collected until the net absorbance was constant. At this moment, the CO was discontinued, leaving a flow of pure He. The intensity of CO adsorption band decreases with time as CO desorbs from the surface at a constant temperature for the Cl-free and Cl-containing Pt/SiO_2 catalysts. Normalizing the integrated intensity by the value at the highest coverage for both catalysts and plotting the normalized intensity versus time shows that the rate constant for CO desorption (assuming it to be a first-order process) is the same for both samples, demonstrating that CO desorption is not affected by chlorine. Furthermore, the wavenumber of the CO linear band at maximum coverage (Figure 17.8) is similar for both catalysts, indicating that the active site for adsorption is the same in both cases. These results support the conclusion that the poisoning effect of Cl is achieved by blocking the active sites for adsorption and reaction.

As it has been shown, this effect not only depends on the precursor as a source of chlorine, but is also evident when Cl is added from an external source such as HCl. This is a clear indication that Cl is mobile under the reaction conditions and leads to deactivation by blocking the active sites on the Pt surface. Even though Cl is a strong inhibitor of the oxidation activity, it is a reversible poison, and the extent of Cl coverage depends on the pretreatment and support used. For example, after a 10 h reduction, the HCl-impregnated Pt/SiO_2 catalyst recovered most of its initial activity, in agreement with the activation of Pt/Al_2O_3 catalysts, with time on-stream reported by several authors.[14, 30, 31] Nonetheless, the removal of chlorine strongly depends on the environment and temperature of the treatment as well as on the catalyst's support.[18] Most studies reporting elimination of Cl from Pt/Al_2O_3 catalysts indicate that water plays a role in the activation treatments, either when it is a product of the oxidation reaction[13] or after addition as steam under a reducing atmosphere.[40]

The kinetic results on the Cl-free catalyst impregnated with HCl indicate that, at first, chlorine is adsorbed mainly on the Al_2O_3 support,[13, 23, 29] and during calcination, a portion of the Cl migrates from the support and is readsorbed onto the Pt surface. The transport and mobility of chlorine on the catalyst's surface is also confirmed by the EXAFS and IR results. IR results indicate that no detectable CO adsorption on Pt occurs at low temperatures (60–100°C) on the Cl-containing catalyst, but as the catalyst temperature increases in the presence of oxygen and CO, the CO coverage increases. EXAFS data clearly show that there are no Pt-Cl bonds after reduction, but later in the presence of oxygen, an oxychloroplatinum phase reappears, surrounding a small core of reduced Pt atoms. Lieske et al.[25] were one of the first groups proposing the formation of oxychloroplatinum species under oxidizing conditions [$Pt^{IV}(OH)_xCl_y$] and [$Pt^{IV}O_xCl_y$], to explain the redispersion of Pt at high temperatures, 500 to 600°C. These authors found that the formation of these species requires the presence of oxygen. The interaction between platinum, oxygen, and chlorine starts at temperatures below 300°C, promoting the formation of oxychloroplatinum species. Similar species have been reported in EXAFS analysis of the preparation[41] and redispersion[42, 43] of Pt/alumina (Cl) naphtha reforming catalysts.

Thus, the kinetic and EXAFS results can only be explained if the Cl coming from HCl or evolved during reduction is reabsorbed on the support and subsequently released and readsorbed on the Pt surface during calcination. During reduction, H_2 adsorbs dissociatively on the Pt surface and reacts with the adsorbed complex to likely yield HCl, which desorbs as a gas and is transported onto the support. Recent results on chlorinated Pt catalysts supported on model flat alumina substrates[44] indicate that no Cl is observed after reduction at 300°C. Therefore, the surface-Cl interaction during reduction is affected by the porous structure of the supported catalysts.

Depending on the time, temperature, and Cl-support interaction, Cl can be eliminated or remain adsorbed on the support. In the latter case, calcination or oxidation pretreatments lead to desorption of Cl from the support and to the formation of a stable platinum oxychloride species.[46, 47] Regardless of the detailed mechanism of Cl transport from the support to the metal and vice versa, i.e., surface diffusion versus gas phase transport, all of the results show this dynamic interaction between Cl, Pt, and the support resulting in a specific catalytic activity.

On the basis of the results summarized above, and other details in Reference 5, the model shown in Figure 17.10 is proposed to explain the interaction of Cl with the surface. Upon reduction, the Pt surface is reduced, and Cl species migrate via various processes onto the support and eventually are removed from the catalysts if the proper temperature time and pretreatment conditions are used, which are

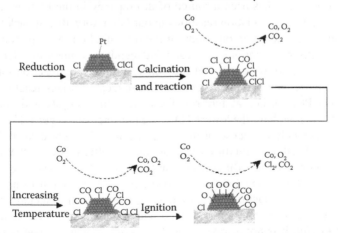

Chlorine Poisoning: Summary

FIGURE 17.10 Model depicting the Pt species after the various treatments.

critically dependent on the support. If residual chlorine is not eliminated from the catalysts, calcinations or reaction can bring it back onto the active Pt surface and reduce its activity. Interestingly, the reaction itself also removes Cl from the surface, which is the reason why, in many oxidation reactions, an increase in activity is observed. Often, bringing the catalysts to reaction conditions is enough to eliminate some of the chlorine, rendering a surface active and eliminating the blocking effect of chlorine. These interactions of pretreatment and reaction are a key factor to reach a stable activity.

17.4 *IN SITU* FTIR, EXAFS, AND ACTIVITY STUDIES OF THE EFFECT OF CRYSTALLITE SIZE ON SILICA-SUPPORTED PT OXIDATION CATALYSTS

17.4.1 INTRODUCTION

The average crystallite size and the oxidation state of the surface are two of the most important parameters determining the activity of noble metal supported catalysts. Boudart et al.[48] distinguished between catalytic reactions where the specific activity (moles/time/per active metal area) is independent of the active area, or *facile* reactions, and those in which the specific activity changes with the specific area, or *demanding* reactions. The traditional view for CO oxidation[49–58] is that this reaction is facile on noble metals at high CO partial pressures, where the reaction rate is negative order in CO concentration. In excess oxygen, however, the reaction rate is zero or positive order in CO concentration, and it has been found to be demanding. Several authors have reported particle size effects on the various elementary steps involved in CO oxidation,[59–64] as well as on the overall reaction (moles/t mass cat) at both low and high CO concentration.[65–68] Recently, the investigation of the reactivity of gold supported catalysts has indicated that the highest activity during CO oxidation, even at room temperature, is obtained on samples with highly dispersed small gold particles.[69–72] In general, the role of particle size during oxidation reactions is not yet clear, and it has been a source of discrepancy in the literature.[73, 74]

In many studies of oxidation reactions in our laboratory, dating back to the early 80s,[75–77] and more recently in our study of the role of Cl on Pt supported catalysts,[5] we have carefully measured reaction rates independent of mass transfer effects. In these studies, we have observed that the conversion and the reaction rate per unit mass were independent of Pt dispersion. Consequently, when reaction rates are calculated per Pt area, i.e., as turnover frequency, the catalysts with the larger Pt area (high dispersion) have the lowest TOF. The presence of oxygen makes it difficult to determine uniquely the cause of this result because it is not clear if the surface is metallic, oxidized, or a mixture of oxide and metallic phases.[30, 31] Whereas this is true for supported Pd and Rh catalysts, in the case of Pt catalysts, we found[5] that, as suggested by Burch and Loader,[30, 31] platinum particles seem to have a strong memory of previous reduction pretreatments, as they remain in metallic state under oxidizing conditions. Because most surface analysis techniques are conducted under vacuum or *ex situ*, it is not always clear if the same oxidation state is valid under reaction conditions at atmospheric pressure. Controlled atmosphere EXAFS and *in situ* IR are techniques that can be used under reaction conditions to probe the working surface. These techniques are used in this study, along with activity measurements, to demonstrate the relation between the particle size and the oxidation state of the working surface during CO oxidation on Pt/SiO_2 catalysts.

17.4.2 CATALYST PREPARATION

Catalysts were prepared by adding tetra-ammine platinum (+2) nitrate (PTA) to silica by two methods: wet impregnation or adsorption (Ads), and incipient wetness impregnation (IWI). Impregnation was followed by calcination in air at different temperatures to give catalysts with identical Pt loading but different dispersions. To obtain catalysts with high dispersion, PTA was adsorbed on silica at a basic pH, whereas for catalysts with lower dispersion, PTA was added by incipient wetness impregnation with no adjustment in the pH. The specifics of catalysts preparation are presented in Reference 6. Crystallite sizes were varied by changing the precursor used and the calcination temperature. The four samples calcined at temperatures up to 300°C were further reduced in hydrogen for 1 h at 300°C and evacuating for 15 min at the same temperature. Volumetric hydrogen chemisorption was determined at room temperature by the double isotherm method. The catalysts, method of preparation, and hydrogen chemisorption values (H/Pt) are given in Table 17.1.

17.4.3 RESULTS AND DISCUSSION

The effect of calcination versus reduction. In this case, catalyst activities were first obtained using the 10-port parallel microreactor to determine the main effects of particle size and pretreatment for several samples in a single run. Then, detailed kinetic information was obtained in the recycle microreactor. Because the activity was measured under oxidizing conditions, we were not sure if, under reaction conditions, the surface was oxidized or reduced, so this question was addressed first. To stabilize the state of the surface and to ascertain the role of the oxidized versus the reduced surface, prior to each run, the catalysts were pretreated either in air or

---

TABLE 17.1
Catalyst Preparation and Pt Dispersion of Reduced Catalysts

Sample number	% Pt	Method of preparation
1	2.0	Ads/calcined at 400°C. No CO chemisorption at RT.
2	2.0	Ads/calcined at 500°C. No CO chemisorption at RT.
3	2.0	Ads/calcined at 600°C. No CO chemisorption at RT.
4	2.0	IWI/calcined at 400°C. No CO chemisorption at RT.
5	2.0	IWI/calcined at 600°C. No CO chemisorption at RT.
6	2.0	Ads/calcined at 100°C/reduced at 300°C/H/Pt = 0.76
7	2.0	Ads/calcined at 300°C/reduced at 300°C/H/Pt = 0.63
8	2.0	IWI/calcined at 100°C/reduced at 300°C/H/Pt = 0.51
9	2.0	IWI/calcined at 250°C/reduced at 300°C/H/Pt = 0.29

H_2 at 200°C for 3 h. The catalyst was then cooled to room temperature and switched to the reaction mixture, and the temperature was increased to about 200°C. Catalyst activities were measured under diluted reactant concentrations: 1% CO in an oxidizing atmosphere with 10% O_2, balanced in He.

The CO conversion of samples 1–5 was measured to study the effect of the calcination temperature and preparation method in the activity. These samples were not reduced before reaction; instead, they were treated *in situ* in air at 200°C for 2 h to normalize the state of the surface, even though they had previously been calcined in air at temperatures from 400 to 600°C during preparation. One sample was calcined and reduced during preparation (sample 9, H/Pt = 0.29), and was subsequently pretreated in air, i.e., *oxidized*. The reaction rates of CO oxidation in the calcined catalysts per unit mass of catalyst, calculated as differential rates at less than 5% conversion, are listed in Table 17.2.

The calcined samples (samples 1–5) were not active at T < 125°C but did show activity at higher temperatures. These results appeared inconsistent with the fact that there was not CO chemisorption at RT in the calcined samples, so we expected no activity from these samples. It is apparent that the state of the surface had changed as the catalysts were heated up in the reaction mixture. In all cases, however, the initial activity of the calcined samples was lower than that of the prereduced/oxidized catalyst (sample 9). The light-off temperature is similar (ca. 5°C difference) for samples calcined at 400 and 500°C. For catalysts calcined at 600°C, however, the LOT is about 25 and 50°C higher than those of the 400°C calcined (samples 1 and 4) and prereduced/oxidized (sample 9) catalysts, respectively.

Portions of the calcined samples 1–5 were later reduced in hydrogen at 300°C for 1 h to measure their Pt dispersion. These values have been included in Table 17.3. After reduction, samples 2, 4, and 9 have nearly identical particle size based on their hydrogen chemisorption values. Thus, the differences in activity of these catalysts are due not only to differences in particle size, but also to the specific state of the Pt surface resulting from the different pretreatments. Clearly, the calcined

TABLE 17.2
Reaction Rate of Calcined Catalysts

Sample number	Method of preparation	Reaction rate mol CO_2 sec^{-1} gr. cat^{-1} × 10^9	
		130°C	185°C
1	Ads./calc. at 400°C/red. H/Pt = 0.47*	5.76	147.63
2	Ads./calc. at 500°C/red. H/Pt = 0.24*	3.85	139.07
3	Ads./calc. at 600°C/red. H/Pt = 0.11*	0.93	32.01
4	IWI/calc. at 400°C/red. H/Pt = 0.24*	9.13	150.61
5	IWI/calc. at 600°C/red. H/Pt = 0.09*	2.49	54.07
9	IWI./H/Pt = 0.29/oxidized	14.04	147.56

* Chemisorption measured after reduction for 1 h at 300°C.

catalysts, initially not active, most likely because of the presence of an oxide layer, became active at high temperature as the catalyst was heated in the reaction mixture. *In situ* EXAFS results, presented later on, demonstrated this point.

Particle size effects. So far, we have shown the effect of the precursor used and the pretreatment (oxidation vs. reduction) in determining the resulting activity. Another important variable, controlled by the preparation and pretreatment, is the crystallite size. Changing the crystallite size was done by selection of the precursor and the pretreatment during its decomposition, as shown for the samples listed in Table 17.3. The activity of the reduced catalysts, with dispersions ranging from 0.29 to 0.76 (samples 6–9), was measured in the 10-port reactor. The catalysts were pretreated prior to reaction in two ways: (a) oxidized at 200°C in air for 3 h, or (b) reduced in a 50% H_2-He mixture for 3 h at 200°C. After pretreatment, the CO conversion was determined from ambient to 200°C using a slow linear ramp of 0.5°C/min. Table 17.3 shows the rate per unit mass of catalysts at 100 and 130°C. Figure 17.11 (A-B) shows the TOF, or rate per surface Pt atom, at each temperature.

The results in Figure 17.11A show that, in agreement with Table 17.3, the oxidized catalysts exhibit no measurable activity below 100°C, whereas the reduced catalysts (Figure 17.11B) exhibit significant activity at these low temperatures. As the temperature increases above 100°C, the rates per gram of catalysts and TOFs of catalysts pretreated by reduction or oxidation become similar. The implications are that although the oxidized catalyst is initially inactive, during the temperature ramp in the reaction mixture, the surface undergoes a transformation at higher temperature, and active Pt sites in the preoxidized and prereduced become the same. In other words, the surface is changed by the reactants as the temperature is increased to reaction temperature in the flowing reactant mixture!

Table 17.3 also shows that the rates per unit mass on the reduced catalysts at 100°C, and reduced and calcined catalysts at 130°C, are similar even though there is more than a twofold difference in number of exposed Pt atoms between the low and high dispersion catalysts. Similar trends have been previously observed for several oxidation reactions[30, 75]; i.e., the conversion seems to be similar even though the fraction of exposed Pt is quite different. From this, it follows that when the rate per unit mass is divided by the active area to calculate the TOF, significant differences

TABLE 17.3
Reaction Rate of (Prereduced) Pt/SiO_2 Catalysts with Different Pretreatments

Sample number	H/Pt	Reaction Rate mol CO_2 sec^{-1} gr. cat^{-1} × 10^9			
		100°C		130°C	
		Calcined	Reduced	Calcined	Reduced
6	0.76	0.0	4.88	6.33	7.44
7	0.63	0.0	9.50	11.84	10.83
8	0.51	0.0	9.77	10.68	9.29
9	0.29	0.0	12.23	14.04	13.40

426 Catalyst Preparation: Science and Engineering

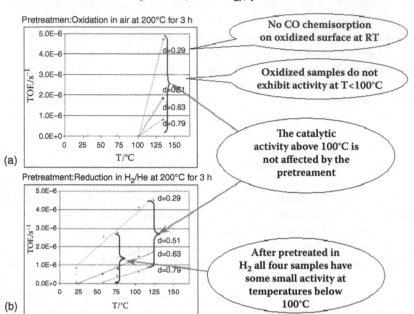

FIGURE 17.11 TOF for CO oxidation on reduced Pt/SiO₂ catalysts with different dispersion. Pretreatment: (A) Oxidation in air or (B) reduction in H₂ at 200°C. Feed composition 1% CO, 10% O₂, balance He. Total flow rate 60 cc/min. Ten-port reactor.

in TOF are obtained. Figure 17.11 shows a fivefold increase in the TOF as the dispersion decreases from 0.8 to 0.3.

The above results were obtained in the multiport parallel plug-flow reactor at low conversion and, thus, are subject to some analytic error. The characteristic autothermal behavior of a plug flow reactor further limits the conversion and the potential presence of micro hot spots can further increase the error in measuring activation energies. To obtain more accurate kinetic data devoid of heat and mass transfer effects, the rates of CO oxidation were determined with the reduced catalysts in the recycle reactor.

The fractional CO conversion in the recycle reactor, Figure 17.12A, shows again that as in the 10-port parallel reactor, the conversion is quite similar for the two catalysts having very different dispersion, consequently, the TOF of the low dispersion catalyst (low Pt area) is the highest (Figure 17.12B). The different conversion-temperature curves obtained in the 10-port reactor versus the recycle reactor are due to the better heat transfer characteristics and more isothermal conditions of the recycle reactor and higher accuracy of the GC readings at higher conversion. Because of the absence of temperature gradients, which might cause micro localized heating in the parallel reactor, the LOT is higher in the recycle reactor.

An Arrhenius plot of the TOF yields apparent activation energies (EA) decreasing catalysts dispersion. These results clearly show that CO oxidation on small Pt crystallites is a structure-sensitive reaction. It should be noted that these

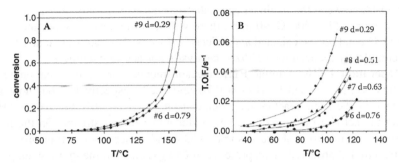

FIGURE 17.12 CO oxidation on (prereduced) Pt/SiO_2 catalysts with different dispersion. (A) Conversion versus Temperature; (B) TOF versus temperature. Recycle reactor.

differences in activation energy were not distinguishable from the conversion data obtained in the 10-port reactor. The low conversion required to obtain differential rates in the 10-port reactor does not permit accurate evaluation of E_A.

EXAFS characterization. The EXAFS and XANES results were obtained under a controlled atmosphere (not *in situ* or operando) at room temperature after the catalysts were pretreated (either in air or hydrogen) prior to data collection in the EXAFS cell. Detailed description of these results are given in Reference 6, so here we will only summarize the final end results.

The EXAFS results show that calcined catalysts are fully oxidized when calcined below 400°C, but higher calcination temperatures lead to formation of metallic Pt. The amount of metallic Pt increases with increasing calcination temperature. Hydrogen reduction leads to fully metallic particles. As expected, the dispersion decreases as the calcination temperature increases. The calcined and oxidized catalysts do not adsorb CO at room temperature, indicating no exposed metallic Pt. Nonetheless, the results shown previously indicate that both calcined and oxidized samples can oxidize CO at temperatures above 100°C. To analyze this apparent contradiction in more detail, several samples were studied using *in situ* FTIR under reaction (operando) conditions.

Temperature programmed *in situ* FTIR of CO adsorption and desorption. FTIR absorbance spectra of adsorbed CO on the reduced sample 6 (H/Pt = 0.76) and sample 9 (H/Pt = 0.29) at 100, 150, and 200°C show an intense band for the catalysts with high dispersion, and as expected, there is a smaller CO band on the catalysts with lower dispersion. For both catalysts, the band's intensity slightly decreases with increasing temperature, from 100 to 200°C, due to a small amount of CO desorption. Although the ratio of integrated CO absorbances (IA) is similar to the H/Pt ratios of both catalysts (2.7), with increasing temperature, there is a small increase in the IA ratio, e.g., from 2.5 to 2.64 to 2.7. This is due to the slightly greater percentage of CO desorption on the catalyst with lower dispersion. When similar experiments were conducted with oxidized catalysts, no CO adsorption was observed at RT, indicating no exposed metallic Pt. However, upon increasing the temperature in flowing CO to 100°C, the CO absorbance of sample number 6 increased to approximately 50% of the absorbance of the reduced catalyst, and the intensity of the band increased with time on stream. After 30 min, the difference in the CO IA between the oxidized and reduced catalyst was only about 15%. At the end of the run (T = 200°C, 90 min), there was no difference between the intensity of the CO band for the calcined and reduced catalysts. This shows that at T > 100°C, CO completely reduced the oxidized catalyst's surface to metallic Pt.

Operando FTIR. Similar FTIR measurements were conducted on reduced Pt/silica, samples 6 and 9 (H/Pt = 0.76 and 0.29, respectively), but now in the presence of the same reactant gas composition (1% CO, 10% O_2, balance He) as during the activity measurements, i.e., under an oxidizing stream. On the reduced catalysts, the intensity of the adsorbed CO band under the reaction mixture was identical to that observed in the presence of CO gas only. At the ignition temperature (around 160°C), however, there was a sudden decrease in CO absorbance reaching a value below the IR detection limit. Concurrently, there is a sudden increase in the amount of gas phase CO_2 produced. This behavior, although not shown here, exemplifies the Langmuir-Hinshelwood oxidation mechanism and is similar to that reported in previous IR studies.[5, 76] After ignition, the CO flow was discontinued while maintaining the 10% O_2 flow for 90 min at 200°C, thus the surface was expected to become oxidized again. After decreasing the temperature to 100°C, the feed was switched again to the CO/O_2 reactant gas mixture. The CO absorbance observed after the switch, however, was identical to that of the reduced catalysts, indicating that the surface was quickly reduced when the flow was resumed even under an oxidizing atmosphere (1% CO, 10% O_2, He).

The IR spectra of Pt/silica calcined at 400°C (sample 4, in Table 17.1) were also obtained after exposure of CO/O_2 at different temperatures. This sample, according to the EXAFS results, initially contains no metallic Pt. The IR results (Figure 17.13) accordingly show no CO adsorption at room temperature. At 100°C, however, there is a small, broad band at 2105 cm^{-1}. As the temperature increases, however, a band at 2077 cm^{-1}, corresponding to linearly adsorbed CO on metallic Pt, begins to increase while the band at 2105 cm^{-1} decreases.

At the same time, small amounts of gas phase CO_2 are observed. The amount of metallic Pt from the CO absorption band correlates well with the activity of this catalyst, shown in Figure 17.11. Ignition starts at about 160°C, but it is not abrupt as that for the reduced samples. Above 180°C, the surface coverage of CO is low. An illustration depicting these IR results is displayed on the right of Figure 17.13. Initially, the fully oxidized surface adsorbs small amounts of CO. As the temperature increases, in spite of the large excess of O_2, the surface starts being reduced, perhaps by an Eley-Rideal mechanism. The oxidized Pt surface continues to reduce at higher temperatures. At about 150°C, the Pt particles' surface is fully reduced with a high CO coverage. As the temperature increases further, the rate of CO oxidation increases and CO coverage decreases, as in the case of the reduced catalyst. In other words, as shown by the activity results, the state of the surface is dictated by the reaction environment. Contrary to expectations, flowing a net oxidizing mixture containing 1% CO over the Pt surface actually does not result in an oxidized surface but instead in a reduced surface at T > 100°C. Obviously, these results will vary depending on the concentration of oxygen and CO in the feed.

Herz and Shinouskis[39] reported similar IR results on Pt/Al$_2$O$_3$ catalyst. These authors observed a rapid shift of an IR band from 2120 to 2070 cm^{-1} as the reactant

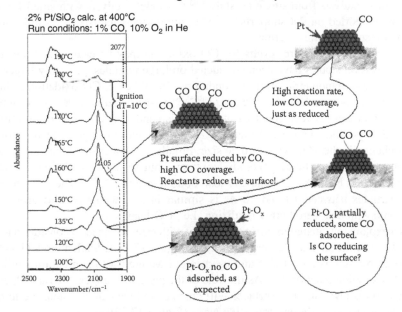

State of surface during CO oxidation, *operando FTIR*

2% Pt/SiO$_2$ calc. at 400°C
Run conditions: 1% CO, 10% O$_2$ in He

High reaction rate, low CO coverage, just as reduced

Pt surface reduced by CO, high CO coverage. Reactants reduce the surface!

Pt-O$_x$ partially reduced, some CO adsorbed. Is CO reducing the surface?

Pt-O$_x$ no CO adsorbed, as expected

FIGURE 17.13 IR spectra of 400°C calcined Pt/SiO$_2$ (sample 4) during CO oxidation. Feed composition 1% CO, 10% O$_2$, balance He. Total flow rate 120 cc/min.

mixture is switched from a 1% O$_2$/He mixture at 250°C to 1% CO/He. In another IR study on a Pt/Al$_2$O$_3$ catalyst, Anderson[78] also reported the presence of a carbonyl species band at 2125cm^{-1}. This author attributes this band to CO adsorbed at Pt^{2+} sites coexisting with metallic Pt and indicates that this species is unreactive in the presence of oxygen at temperatures between 200 and 300°C.

The model that emerges for CO oxidation is that, even in the presence of an oxygen-rich mixture, the active surface is metallic Pt. The EXAFS analyses of the catalysts calcined near 400°C are predominantly Pt^{+4} oxide before reaction. Although oxidized Pt is initially inactive at low temperature (< 100°C), the activity becomes equivalent to that of reduced Pt at T > 100°C. The FTIR results show that CO is not adsorbed on oxidized Pt at low temperature. With increasing temperature, however, oxidized Pt is reduced by CO even with a large excess of O$_2$. Thus, although metallic Pt is readily oxidized by pure O$_2$ at reaction temperatures, the presence of low CO concentrations is sufficient to reduce the oxidized Pt surface.

In addition to the effect of pretreatment on the activity, we also show that the activity per Pt surface atom also changes with crystallite size during CO oxidation; i.e., the reaction is structure sensitive. The literature results vary, depending on the conditions used and the degree of dispersion of the metal. For CO-rich conditions, where the catalytic surface is mostly covered by CO and the reaction rate is inverse order with respect to the CO partial pressure, CO oxidation has been found to be structure insensitive on Pd,[50, 53, 54] Rh,[55, 56] and Pt.[51, 58] On the other hand, under

O_2-rich conditions, CO oxidation is reported to be demanding or structure sensitive on various catalysts from single crystals[59, 66] to model catalysts with small Pd or Pt clusters supported on flat supports,[63, 66, 68] to supported catalysts,[65, 75] similar to the results presented in this section.

The previous literature results for CO oxidation may be categorized into three groups. Several studies have been conducted on large particles, generally larger than 25 Å under stoichiometric reaction conditions. In this case, CO oxidation is facile, or the TOF is independent of particle size.[49, 53, 64] Several studies have also been made on metal particles between about 10 and 60 Å, also under stoichiometric conditions,[52] and with crystallites between 30 and 75 Å.[39, 64] For these catalysts, there is only a small change in the TOF with particle size. Finally, a number of studies have shown that on small metal particles, generally less than about 20 Å, CO oxidation is structure sensitive under both CO-rich and CO-lean conditions.[60, 62, 65, 67, 68] The present study, with particles between 10 and 40 Å, is similar in size with the third group. The reaction conditions were highly oxidizing, i.e., CO-lean.

To explain the change in intrinsic activity with size, we need to consider a model of a Pt crystallite such as the one depicted in Figure 17.14A. Such a model has been proposed by Mavrikakis et al.[79] and used in our Monte Carlo simulation studies of the CO oxidation reaction.[80, 81] As the size of the particle changes, the ratio Pt atoms at the corner, edge, and flat surfaces will change, and we can calculate the number of these atoms for different crystallite sizes (Figure 17.14B).

As the crystallite size increases, the fraction of Pt atoms at low coordination sites, i.e., corners and edges, decreases dramatically, whereas the fraction of atoms

Effect of Particle Size: Proportion of Sites

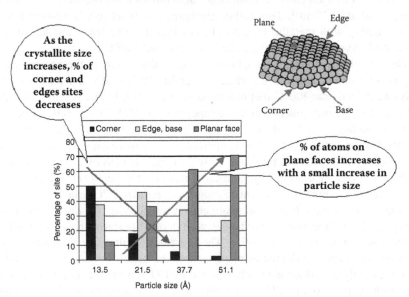

FIGURE 17.14 Crystallite size model and percentage of corner, edge, and base plane versus crystallite size.

on planar faces increases with a relatively small increase in the particle size. For an average size of 14 Å, for example, the fraction of Pt atoms in planer surfaces is only 12.5%, whereas for a particle of 38 Å, more than 60% of surface atoms are on planar faces.

A simulation of CO oxidation on the crystallite shown in Figure 17.14 has been reported by our group.[6, 80] Because a detailed description of the Monte Carlo simulations is beyond the scope of this chapter, only the main results are described next. Using the crystallite model presented in Figure 17.14, supported catalysts with dispersions 0.29 and 0.80 were considered in the simulations. Initially, the TOF of both catalysts was calculated assuming that the rates of adsorption, surface reaction, and desorption were the same in corner, edges, and plane sites. In this case, it was not possible to fit the experimental data to the theoretical results because the simulation showed that both supported catalysts have the same overall activity. We found that the experimental rates can be fit to the theory only when the rate of surface reaction on low coordination Pt, i.e., corners, edges, and so on, is at least 10 times smaller than that on the planar faces. This is reasonable if one considers that at corner and edge sites, the adsorbed reactants must overcome an additional energy barrier to diffuse and react, thus lowering the rate of reaction. Based on transient kinetic experiments on a Pt(112) single crystal, Szabo et al.[61] found that CO on Pt plane surfaces is more reactive than when adsorbed at step sites. Nieuwenhuys et al. confirmed these conclusion for the CO oxidation on Pt-Rh (111), (100), (410), and (210) single-crystal surfaces.[82] These authors report that CO oxidation occurs at a relative low temperature on flat surfaces, but at a higher temperature on stepped surfaces. As the Monte Carlo simulations indicate, for large particles, there is only a small increase in the number of more reactive planar atoms with increasing particle size, and this is offset, at least in part, by the loss in dispersion. Therefore, as observed experimentally, small changes in the TON are expected for larger particles. Our studies are consistent with the interpretation that flat surfaces have a significantly higher CO oxidation activity compared to low coordination edge and corner atoms.

17.5 THE EFFECT OF SULFUR ON Pt SUPPORTED CATALYSTS DURING CO OXIDATION

17.5.1 INTRODUCTION

The presence of adsorbed species containing sulfur on metal surfaces can block reaction sites, induce changes in surface morphology due to faceting and preferential segregation of one component in multicomponent catalysts, enhance sintering, and alter the electronic structure of the metal. Many studies have contributed to understanding these processes and their effects on catalytic activity and selectivity during hydrocarbon conversion under reducing conditions, but less is known about the effect during oxidation reactions. Although under reducing conditions, the presence of sulfur in general decreases the activity, several papers have been published reporting that during oxidation reactions, an increase in activity has been observed on Pt supported catalysts. A review of the various poisoning mechanisms by S have been presented by Gracia and coworkers.[7] Key results of that work are presented below.

According to the current literature, under reduction conditions, S compounds interact with Pt active sites, altering the state of the Pt surface. This causes short-range site blocking and long-range electronic effects that alter surface-adsorbate interactions. Under oxidation reactions, however, the local structure of the metal-sulfur active site and how it changes with differing metals, supports, and reaction conditions is not well understood.[83–85] Understanding these effects as a function of the type of metal, pretreatment, particle size, support, and reaction conditions can lead to the development of poison-resistant catalysts or catalysts with improved activity in the presence of S.

17.5.2 CATALYST PREPARATION

Similar Pt catalysts with three different dispersions as in Section 17.3, supported on silica and alumina and the homolog containing ceria, were prepared by incipient wetness impregnation of tetraammine-platinum nitrate of the support. In the case of the ceria-containing catalysts, a solution of $Ce(NO_3)_3 \cdot 6H_2O$ in 15.5 ml H_2O was added to the silica or alumina supports, followed by drying in air overnight and calcinations at 300°C in air. Specific conditions used in the preparation of these materials are reported elsewhere.[7] Table 17.4 summarizes the various catalysts studied in this case as well as their dispersion.

17.5.3 RESULTS AND DISCUSSION

Activity measurements. During oxidation reactions, it is likely that S will be present as SO_2. For gasoline with 300 ppm of sulfur, the exhaust gases entering the TWC would contain about 20 ppm of SO_2.[86] Thus, to emulate the S concentration in the exhaust gas from the engine, 20 ppm of SO_2 was added to the reactant mixture of 1% CO 10% O_2 in He and continuously fed to the catalyst. Conversion versus temperature results are presented in Figure 17.15 for Pt/silica and Pt/alumina catalysts with similar dispersion (0.63 and 0.68, respectively).

TABLE 17.4
Catalyst Preparation And Pt Dispersion of Reduced Catalysts

Sample	Catalyst	Dispersion (±0.02)
$Pt(0.10)/SiO_2$	$2.0\%Pt/SiO_2$	0.10
$Pt(0.33)/SiO_2$	$2.0\%Pt/SiO_2$	0.33
$Pt(0.63)/SiO_2$	$2.0\%Pt/SiO_2$	0.63
$Pt(0.72)-CeO_2/SiO_2$	$2.0\%Pt-5\%CeO_2/SiO_2$	0.72
$Pt(0.68)/Al_2O_3$	$2.0\%Pt/Al_2O_3$	0.68
$Pt(0.73)-CeO_2/Al_2O_3$	$2.0\%Pt-5\%CeO_2/Al_2O_3$	0.73

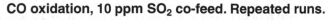

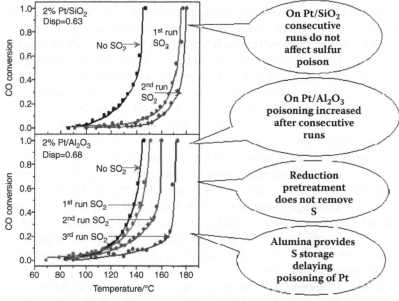

Feed: 1% CO, 10%O_2, balance He.
Flow 120 cc/min, Recycle reactor

FIGURE 17.15 Effect of SO_2 in the activity of Pt/SiO_2, Pt/Al_2O_3, and Pt-CeO_2/Al_2O_3 catalysts during CO oxidation. Reaction conditions: 1%CO, 10% O_2 in He. Recycle reactor. Temp. ramp: 1°C/min. Blue lines correspond to reaction with no SO_2. Red lines correspond to reaction with 20 ppm of SO_2.

In the absence of SO_2, Pt supported on silica and alumina show about the same activity for CO oxidation (blue squares). Upon addition of 20 ppm of SO_2 in the gas feed, the ignition temperature increases only 5°C during the first run in the alumina-supported sample, compared to 31°C in the case of the silica-supported catalyst. Because both samples have similar metal dispersion, the differences in LOT can only be attributed to the nature of the supports. In the literature, it has been proposed that the presence of SO_2 leads to the formation of sulfates on the alumina surface, which thus becomes a sulfur trap. This would allow the Pt surface to appear not to be affected as strongly as in the case of the silica support in this short time on stream (T-O-S) run. The increase in LOT is relatively small for the ceria-promoted alumina catalyst, probably for the same reason as for unpromoted alumina. Addition of cerium oxide to Pt/silica significantly decreases the LOT (by 26°C compared to Pt(0.63)/SiO_2). In the presence of S, however, the activity is the same as that of the Pt without Ce, indicating poisoning of the Ce promotional effect.

After the first reaction, the sample was cooled to room temperature in He and treated in 150 cc/min of H_2 at 300°C for 2 h. Then, the sample was again cooled to room temperature in He, the reaction mixture was started, and the reactor temperature was increased until 100% CO conversion was reached. The same procedure was repeated for the third and fourth runs. The CO conversion data for successive reactions on the Pt(0.63)/SiO$_2$ (top panel) show that the difference between the first and the second run is small on the silica-supported catalyst. A third run did not show major differences compared to the second one, showing that when silica is the support, poisoning of the Pt surface occurs almost entirely during the first run. Sulfur poisoning for the alumina-supported catalyst (bottom panel) occurs gradually between runs, with a small effect between the first and second run and considerable change after the third run. There is only a 5°C increase in the LOT between the first and second run, but the LOT increases by 18 and 33°C for the second and third runs, respectively. On both silica and alumina supported catalysts, the activity became similar after four consecutive runs. This indicates that when the extent of the poisoning has reached a pseudoequilibrium coverage, the activity becomes independent of the support. These results show that the support acts as a storage reservoir that might initially delay the effect of S reaching the Pt surface. This effect disappears when saturation coverage of the support is reached after sufficient exposure of the alumina surface to sulfur. Again, it is clear that the state of the surface is determined by the reaction. Sufficient T-O-S must be allowed to reach the saturation state, otherwise incorrect conclusions can be drawn as to the effect of the support on the activity of supported catalysts.

The turnover frequencies at various temperatures are summarized in Table 17.5. The dispersions determined by hydrogen chemisorption for the S-free catalysts were used to calculate the TOF, assuming that the number of Pt atoms exposed in the poisoned catalysts was the same as that in the S-free catalyst. The boldfaced values indicate reaction rate at 100% CO conversion, which can be calculated in the recycle reactor. For the silica-supported catalysts, the difference in the light-off temperature between the fresh and poisoned samples (ΔT_{50}) appears to depend on the average particle size (expressed as metal dispersion). For the catalyst with lowest dispersion

TABLE 17.5
Activity Results (Recycle Reactor) of the Effect of SO$_2$ During CO Oxidation

Catalyst Dispersion	Pt/SiO$_2$						Pt/Al$_2$O$_3$	
	H/Pt = 0.10		H/Pt = 0.33		H/Pt = 0.63		H/Pt = 0.68	
Rxn cond	No SO$_2$	20 ppm SO$_2$	No SO$_2$	20 ppm SO$_2$	No SO$_2$	20 ppm SO$_2$	No SO$_2$	20 ppm SO$_2$
T/°C	T.O.F./s^{-1}							
110	0	0	0.08	0.02	0.03	0.01	0.03	0.03
125	0	0	0.15	0.04	0.06	0.01	0.06	0.04
150	0.08	0	**0.57**	0.16	**0.29**	0.03	**0.28**	0.18
175	0.29	0.09	**0.57**	**0.54**	**0.29**	**0.27**	**0.28**	**0.28**
LOT/°C								
	192	213	135	160	141	172	139	144

$\Delta T_{50} = 21°C$, whereas for the catalysts with dispersion of 0.33, $\Delta T_{50} = 25°C$, and for the catalyst with highest dispersion (H/Pt = 0.63), $\Delta T_{50} = 31°C$.

In situ **and operando IR experiments during SO$_2$ cofeeding.** To further investigate the poisoning effect of SO$_2$ during CO oxidation, the reaction was carried out *in situ* under reaction conditions, i.e., under operando conditions in the IR reactor. An experiment under CO only would be *in situ* but not operando. After conducting the standard pretreatment of the wafer in 120 cc/min of H$_2$ at 200°C for 3 h, and then cooled to 100°C, the reactant mixture containing 1% CO, 10% O$_2$ and 20 ppm SO$_2$ in He was fed into the reactor. The spectra were collected at the various temperatures using a temperature ramp of 1°C/min.

The shift in the wavenumber of the L-CO band for the poisoned catalyst during the first run is only 3 cm^{-1} and remains constant with temperature until the catalyst reaches ignition. The corresponding absorbance (IA) of the adsorbed L-CO band versus temperature shows that in the presence of SO$_2$, the ignition temperature increases, as indicated by the sharp decrease in CO absorbance, consistent with the activity results in the recycle reactor. Contrary to expectations, however, before ignition, the IA of the L-CO band is similar to that during reaction without SO$_2$, and yet the ignition temperature has increased 22°C. Additional exposure to SO$_2$ in a second run leads to an additional increase in the frequency of absorption of about 5 cm^{-1}, but the ignition temperature was similar. In agreement with the activity results, the IR spectra show that SO$_2$ poisoning of Pt supported on silica is evident after the first run, showing no storage effect by the silica support.

The results of a similar IR spectra obtained on the Pt(0.68)/Al$_2$O$_3$ sample are presented in Figure 17.16, along with the spectra in wide frequency range. As in the silica supported catalyst, the integrated CO absorbance before ignition during the first run with 20 ppm of SO$_2$ was similar as in absence of SO$_2$. Unlike on Pt supported on silica, however, there was no shift in the wavenumber of the CO band, and the ignition temperature was the same as without SO$_2$ in the feed. This indicates little effect of SO$_2$ on the alumina supported Pt in the first run, which is consistent with the activity results in Figure 17.15.

The flow in the IR-cell reactor is a boundary layer-type flow; hence, it is expected that the capture of SO$_2$ in the alumina support is less in the IR reactor than in a flow reactor. Consequently, the poisoning effect during the first run with SO$_2$ is undetectable, and there is no change in the ignition temperature. When the reaction is carried out a second time with SO$_2$, however, there is a shift of 5 to 10 cm^{-1} in the wavenumber of the CO and a 16°C increase in the ignition temperature. During the third run with SO$_2$, the wavenumber has increased 15 cm^{-1} and the LOT is 35°C higher. Finally, after four consecutive runs with SO$_2$, ignition occurred at 210°C and the wavenumber of the L-CO band has shifted 25 cm^{-1} compared with that on the reduced, S-free catalyst. After the fourth run, the frequency of L-CO was similar to that of Pt/silica poisoned by SO$_2$ and did not change with temperature. It should be noted that during these runs, the catalyst was reduced before each reaction, and that the reactions were restarted at 100°C, which, as shown later, already altered the sulfur coverage and integrated CO absorbance, compared to runs started at room temperature.

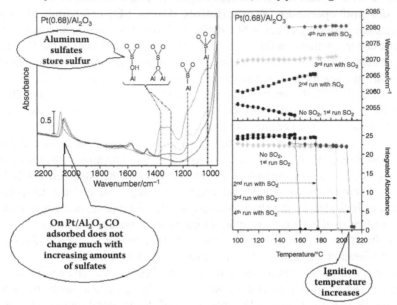

FIGURE 17.16 IR spectra and frequency shift and integrated intensity versus temperature of the L-CO band adsorbed on 2% Pt(0.68)/Al$_2$O$_3$ during CO oxidation in the presence of SO$_2$ after repeated runs. Reaction conditions: 1% CO, 10% O$_2$, 20 ppm SO$_2$ in He. Temperature ramp: 1°C/min. Upper panel = wavenumber, lower panel = integrated absorbance (IA). Black line: run without SO$_2$ and 1st run with 20 ppm SO$_2$. Green line: 2nd run with 20 ppm SO$_2$. Blue line: 3rd run with 20 ppm SO$_2$. Red line: 4th run with 20 ppm SO$_2$.

The IR results on Pt(0.68)/Al$_2$O$_3$, showing an increase in ignition temperature with increasing SO$_2$ exposure, agree very well with the increase in LOT with increasing SO$_2$ poisoning observed in the recycle reactor (Figure 17.15). Figure 17.16 also shows the entire IR spectra of the Pt(0.68)/Al$_2$O$_3$ during CO oxidation without SO$_2$ and the consecutive runs with a cofeed of 20 ppm of SO$_2$. Each IR spectrum was obtained under reaction conditions at 100°C. In every consecutive run with SO$_2$ (except for the first one), there is an increase in the absorbance of the bands at 1045, 1170, 1290, and 1360 cm^{-1}. These bands have been identified in the literature as aluminum sulfates,[12] with their suggested structure depicted in Figure 17.16.

The apparent support effect is then explained by the sulfur storage capacity of alumina, which retains sulfur on its surface, forming sulfates and thus delaying the poisoning of the Pt surface. Eventually, after the alumina surface is saturated by sulfur, both alumina and silica supported catalysts have the same LOT, i.e., the same activity. Several studies have shown that sulfur compounds (H$_2$S and SO$_2$) can form adsorbed sulfate ions on γ-Al$_2$O$_3$ or aluminum sulfate in the presence of O$_2$ and at temperatures greater than 400°C.[87–89] Chang[88] suggested that on Pt/Al$_2$O$_3$, the formation of surface sulfates occurs via spillover of oxidized SO$_3$ onto the Al$_2$O$_3$ surface.

On the other hand, the SiO_2 support, which is less reactive to the formation of surface sulfates, shows no IR bands for sulfate species, showing that no S storage occurs on the support, and thus S poisoning affects the Pt surface without the delay observed in the case of alumina. Gandhi and Shelef[86, 90] reported that in studies of SO_2 oxidation over automotive catalysts, the SO_3 formed is chemisorbed on the γ-Al_2O_3 surface in the form of aluminum sulfate. At elevated temperatures, the stored sulfur is emitted as a mixture of SO_2 and SO_3, depending on the temperature and oxygen pressure.

Accelerated poisoning. Because during the consecutive runs, the catalyst was reduced before each run, thus refreshing the surface after the previous reaction, a new sample of Pt(0.68)/Al_2O_3 was poisoned overnight with 20 ppm of SO_2 in He at 200°C (without CO or O_2) to ascertain the effect of long-term SO_2 exposure. After poisoning, the catalyst was cooled to room temperature and the reaction mixture was introduced in the reactor. In contrast with the IR results in Figure 17.16, the IR spectra in this case were obtained before reaction without reduction. The IA of the L-CO band, collected from RT and at increasing temperatures, is shown in Figure 17.17. It can be seen that the IA of the L-CO band is about 5 units at room temperature, which is low compared to the value of about 25 units for the S-free catalyst (Figure 17.16). As the temperature increases, however, the integrated absorbance also increases by a factor of about 3, demonstrating that the reaction gases are modifying the catalyst surface. Two important results are observed here: First,

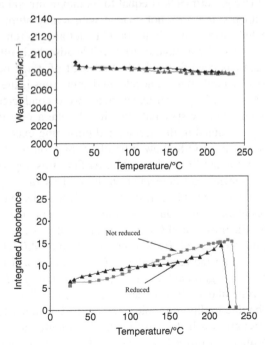

FIGURE 17.17 Frequency shift and integrated intensity versus temperature of the L-CO band adsorbed on 2% Pt(0.68)/Al_2O_3 after overnight exposure to 20 ppm of SO_2 in He at 200°C, followed by cooling to RT and exposure to reaction gases.

the long-term treatment in SO_2 in the absence of the reaction significantly poisoned the Pt surface, leading to a relatively low value of the L-CO absorbance at RT. Equally significant is the fact that, during heating in the reaction mixture to reach the reaction temperature, the reactants removed some of the adsorbed S as evident by the increase in the L-CO absorbance with temperature. The results in Figure 17. 15 also show that prereduction by H_2, before SO_2 poisoning, has little effect on the L-CO band.

When the accelerated poisoning experiment (SO_2/He, 12 h, 200°C) was conducted without cooling the catalysts to room temperature, but instead by introducing the reaction mixture at 200°C, there was no CO adsorption immediately after feeding the reactants (1% CO-10% O_2 in He without SO_2). After 1 h under reaction conditions at 200°C, however, the CO absorbance reached a level similar to that before the poisoning treatment (not shown). Upon introduction of the feed containing SO_2, the catalysts responded in a similar way as after the 4th run shown in Figure 17.16; i.e., no apparent support effect was evident because it was saturated with SO_2 and the LOT was at its highest value. This clearly shows that the state of the catalytic surface depends on the initial conditions, composition of the reactants, SO_2 concentrations, temperature, and T-O-S. Once the reaction mixture enter the IR cell in operando mode and the temperature increases, the reactants themselves are capable of removing some (but not all) of the S, until a quasiequilibrium condition is reached, and there is sufficient CO adsorbed to ignite the catalysts. The frequency of the L-CO band observed in these experiments is equal to its maximum value observed in the short T-O-S experiments and does not change with temperature. Prereduction (in H_2) of the SO_2 treated surface for 3 h at 200°C is not able to remove all the sulfur, indicating that some critical coverage of irreversible adsorbed sulfur that cannot be reduced by hydrogen has been reached, although part of it can be removed by the reactants. The LOT observed after the accelerated aging runs is the highest observed in the IR studies. No such long-term experiments were conducted in the presence of the reaction mixture, but we speculate that it will take a very long T-O-S run to reach the S coverage obtained in the accelerated poisoning experiments.

The IR results at short T-O-S show little effect of S on the IA of adsorbed L-CO, while no bridge bonded CO is detectable. At first, this appeared contradictory because adsorbed S species, not detectable in the mid IR range, were expected to occupy sites that otherwise would be available to CO, thus decreasing CO coverage as reported by Apesteguia et al.[91] Our experiments were conducted under operando conditions, and thus the presence of O_2 in the feed affected the results. The decrease in CO coverage in the presence of SO_2 was not observed on Pt/SiO_2 nor on Pt/Al_2O_3, even though there was an increase in the LOT, indicating a decrease in activity. There was, however, a significant change in the frequency of adsorbed CO in consecutive runs from about 2071 to about 2087 cm[-1]. This large decrease in frequency cannot be due to dipole-dipole interactions but rather to a weakening of the Pt-CO bond strength. Such decrease in bond strength should translate in less CO inhibition and thus in a lower LOT instead of the higher values observed. The shift in frequency can be rationalized in terms of the interaction of electron acceptor, such as adsorbed sulfur species, would decrease the back donation of the electrons from Pt to CO, thus weakening the CO-Pt bond. Similar IR results by Bartholomew et al.[83] show that at low coverage, S has little effect on the intensity of the L-CO band and that no bridge bonded CO was observed. Such experiments at low sulfur coverage are equivalent to experiments at low T-O-S as obtained in this work. A study comparing the competitive adsorption of C and CO on supported noble metal catalysts by Gandhi et al.[92] show that supported Pt coadsorbed both gases, which agrees with the explanation of the frequency shifts observed in the IR results.

EXAFS characterization of SO$_2$ poisoned catalysts. The Pt/silica with different dispersion and Pt/alumina catalysts were prereduced and tested for catalytic oxidation of CO with 20 ppm SO$_2$ and then analyzed by XAS. The XANES and EXAFS analysis (in air), Table 17.6, following the reaction shows significant amounts of oxidized Pt and the presence of Pt-Pt, and Pt-O bonds, with no Pt-S bonds. From the Fourier transform of the EXAFS, all samples have metallic Pt, thus this must be included in the XANES fit. For several samples, the height of the white line was larger than in spectra of the Pt^{+2} reference, and no fit could be obtained using only Pt^{+2} and Pt^0. However, excellent fits were obtained with a linear combination of Pt^{+4} and Pt^0. Acceptable fits were also possible using Pt^{+2}, Pt^{+4}, and Pt^0, which were about 25% less metallic Pt. Although the absolute values of the various Pt oxidation states differ with the two fits, the trends are the same.

The EXAFS were also obtained for the SO$_2$ oxidized catalysts after reduction at 250°C in H$_2$. Although a portion of the Pt-Pt EXAFS contribution overlaps with the Pt-S contribution, the fits clearly indicate that the reduced catalysts contain Pt-S as well as Pt-Pt bonds. In addition, there are no Pt-O bonds after reduction. Within the limits of accuracy, the size of the reduced particle is unchanged after SO$_2$ poisoning. Consistent with the larger metal particle size in the reduced catalysts, the Pt-Pt bond distance increases to about 2.75 Å. The presence of Pt-S in the reduced Pt/silica catalysts suggests that SO$_2$ is bonded to Pt atoms in the oxidized catalysts because silica does not form surface sulfates, as observed on alumina. In addition, because there were no Pt-S contributions to the EXAFS in the SO$_2$-oxidized sample, the SO$_2$ was bonded through an oxygen atom, i.e., Pt-O-S-O bond or similar species. For Pt(0.63)/silica, the structure of Pt is very similar after one or four reactions (and reductions) with SO$_2$ consistent with the catalytic and IR results, indicating that the Pt is rapidly poisoned. For Pt(0.68)/alumina, however, after one catalyst test,

TABLE 17.6
EXAFS Fits for *in Situ* 2% Pt Catalysts Poisoned by SO$_2$

Sample	Composition	Dispersion	Scatter	CN	R, Å	DWF (× 10^3)	E$_o$, eV
Pt1H$_2$	SiO$_2$	0.10	Pt-Pt	11.3	2.77	1.0	−1.9
Pt2H$_2$	SiO$_2$	0.33	Pt-Pt	9.0	2.76	1.5	−2.5
			Pt-S	0.5	2.30	4.0	2.2
Pt3H$_2$	SiO$_2$	0.63	Pt-Pt	7.5	2.75	1.5	−2.7
			Pt-S	0.7	2.30	4.0	1.7
Pt5H$_2$	Al$_2$O$_3$	0.68	Pt-Pt	5.5	2.74	1.5	−4.1
			Pt-S	0.3	2.30	4.0	0.9

the amount of Pt-S is significantly lower than after four reactions. As with the IR and catalytic tests, EXAFS indicates that at short times on stream, the Pt/alumina is less poisoned by SO_2 than Pt/silica.

The results presented in this case show that the effect of sulfur on Pt catalysts during CO oxidation is a dynamic phenomena that strongly depends on the catalyst's history, including: (1) how sulfur is added, (2) the pretreatment used before reaction, (3) the type of support used, (4) crystallite size, (5) the reaction conditions, and (6) the time on stream at which measurements are made. The results show how critical is the use of operando techniques to correlate the state of the surface with the catalytic activity.

It remains to explain why the activity decreases (higher LOT) if CO coverage is only slightly altered and the CO bond is actually weakened by SO_2. The explanation must reside on the other undetected adsorbate in the mid-IR range, namely oxygen, whose dissociation must be altered by sulfur. In fact, Bonzel and Ku[93] showed that on Pt(100), adsorbed sulfur considerably weakens the CO-metal bond, and that the rate of CO_2 formation on Pt(100) is completely inhibited at sulfur coverage equal or greater than 0.3, even though CO can still be adsorbed. The decrease in reaction rate was explained by the inhibition of O_2 dissociation on the Pt surface. Halachev and Ruckenstein[94] reached the same conclusion based on theoretical studies. A similar result was obtained during S adsorption on Pt and Ni single crystals,[84] and it was postulated that the effect of S on these metals is primarily electronic, reducing the availability of metal d-electrons and weakening the interaction with adsorbates. In our case, we can speculate that the S coverage was little affected by the hydrogen reduction prior to each experiment, and it did not reach a level sufficiently high to affect CO adsorption.

It is clear, however, that it was sufficient to inhibit O_2 adsorption and dissociation, thus increasing the ignition temperature. The disappearance of the bridge bonded CO band indicates that those sites were covered by S, suggesting that such sites are relevant to initiate oxygen dissociation and were responsible for the increase in the ignition temperature as S coverage of such sites increased. From the EXAFS analysis, the surface Pt forms a Pt-S bond and is Pt^{+2}, rather than metallic. O_2 likely requires metallic Pt to be dissociated. In addition, S is thought to be coordinated in threefold hollow sites on Pt. Because O_2 will require more than one Pt atom to dissociate, upon S poisoning there appear to be few regions of the surface where several S-free, metallic Pt atoms exist.

Another point evident from the activity results shown in Table 17.6 is that the TOF depends on crystallite size. This is first seen in the results obtained without SO_2, in which the TOF of the catalyst with a low dispersion (d = 0.33) has a higher TOF than the highly dispersed catalysts (d = 063). Upon addition of SO_2, the TOF decreases in all Pt/SiO_2 catalysts but remains the highest in the catalysts with d = 033. The catalysts with the lowest dispersion (d = 0.1) show the lowest activity, because in this case the low active area yields a lower activity over the increase obtained in the intermediate size crystallites versus the highly dispersed ones. In Section 17.3 we discussed these crystallite size effects on unpoisoned catalysts as interpreted via dynamic Monte Carlo simulations. The simulations agreed with experimental results only when the smallest size crystallites had corner and edge sites with 10 times lower activity than sites located on terraces. The lower activity could arise from a stronger CO-Pt bond in these low coordination sites, which in turn are less active toward CO oxidation. A similar argument can be invoked in the presence of S, which should adsorb on the sites of lower coordination, i.e., corner and edge sites. This will further make the crystallite size effect even more pronounced, as seen in the TOF at 150°C. This would be particularly important if such sites are also involved in oxygen dissociation. In the absence of S, the difference in rate in catalysts with different dispersions is only a factor about 2 times larger on the catalysts with intermediate and highest dispersion. In the presence of S, however, the difference in the TOF is a factor of about 5 with and without S.

The EXAFS results were obtained *in situ* but not under operando conditions, and thus it can only give information about the initial state of the surface. The most important result obtained from these experiments is that no Pt-S signal fit the results in the SO_2 poisoned catalysts. This suggest that, in this case, S is bonded to O species adsorbed on Pt, which agrees well with the conclusion that S is inhibiting oxygen dissociative adsorption instead of CO adsorption.

Based on the activity, operando IR, and *in situ* EXAFS studies, we proposed the model presented in Figure 17.18 to explain the results obtained during SO_2 cofeeding experiments. The relative surface coverages of CO, adsorbed S species, and adsorbed oxygen are affected by the temperature, the gas composition, and T-O-S. The latter variable determines the approach to quasi steady-state conditions. It can be then expected that this poisoning effect would be greater with higher SO_2 concentrations. Indeed, in a study of the effect of sulfur on commercially produced Pd and Pt-Rh automotive catalysts, Beck and Sommers[95] showed that increasing the sulfur concentration from 0 ppm to 10, 20, or 30 ppm of SO_2 in the reactant mixture resulted in a LOT increase during CO oxidation of 35, 40, or 50°C, respectively. After ignition, most of the CO is quickly converted to CO_2. After a 12 h treatment with

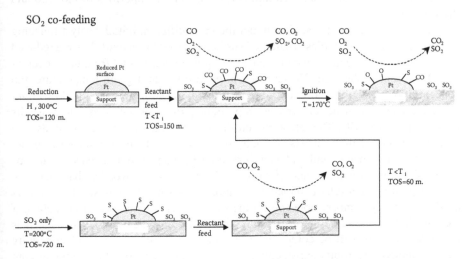

FIGURE 17.18 Schematics of the poisoning pathway depending on initial conditions and various treatments used for the Pt surface during CO oxidation and in the presence of SO_2 in the feed.

SO_2 in absence of CO or O_2 in the gas phase, the Pt surface is covered by sulfur; little CO chemisorption is observed right after resuming the reactant's flow. As the temperature is increased, CO removes part of the sulfur from the Pt surface, and CO adsorption reaches levels similar to those previously observed on the equilibrated catalysts. This effect is very similar to the effect of Cl presented in Section 17.1, suggesting a common pathway to all these catalyst finishing effects.

17.6 CONCLUSIONS

17.6.1 THE EFFECT OF CATALYST FINISHING ON THE STATE OF THE WORKING SURFACE

The results presented in this chapter clearly demonstrate that the state of the surface is a dynamic process depending on kinetics of surface transformation temperature, reactant concentration, time on stream, even on the reactor used, all of which determine the structure and composition of the Pt surface. In short T-O-S experiments, in which the surface is not equilibrated, the conditions can be different than in long-term experiments, where the surface reaches a quasiequilibrium state. Comparison of results from different research groups requires detailed examination of all the conditions used to draw valid conclusions. Although this is a common-sense conclusion, researchers often compare results at quite different conditions, e.g., single crystal versus supported catalysts, to cite a common example. Often-heated arguments about which interpretation is the correct one are just a reflection of the surfaces being under different states due to the use of different experimental parameters.

A model summarizing all the experimental results reviewed in this chapter is presented in Figure 17.19, showing the interactions between gases evolved during the precursor decomposition, pretreatment, reactants, or impurities in the feed and the surface.

The top level shows the surface under the three different initial catalyst finishing conditions used in our studies: a calcined surface covered by an oxide layer, a reduced surface, and a surface under the presence of impurities such as Cl from the precursor or SO_2 present in the feed. Under reaction conditions, at a given temperature and concentration, and given sufficient time on stream, the surface is modified by either the reactants or impurities and reaches a pseudoequilibrium state that determines its activity and selectivity. The rate at which this equilibrium is achieved depends also on the structure of the catalysts. As activity measurements are conducted in many different environments, with catalysts prepared in many different ways, and thus having different structures, and at different T-O-S, it is no wonder that there are such disparities in results obtained with apparently similar catalysts. Researchers need to carefully compare notes on all their experimental conditions to conclude if they are indeed the same before arguing about discrepancies on their results. It is unlikely that there will be coincidence on the many variables involved in activity studies, unless the results are obtained from the same group. Thus, extreme care must be observed in carefully examining the catalyst finishing step to determine what is the state of the surface under operando conditions.

Catalyst finishing effect and the state of the surface

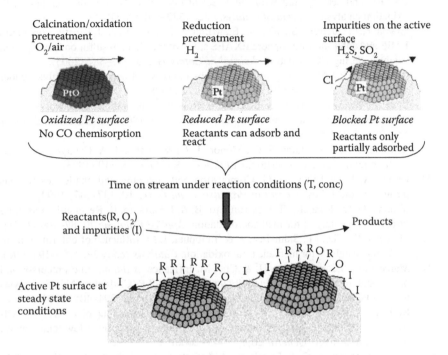

FIGURE 17.19 Model of the catalysts surface depending on pretreatment and operating conditions used.

ACKNOWLEDGMENTS

Use of the Advanced Photon Source was supported by the U.S. Department of Energy, Office of Basic Energy Sciences, Office of Science (DOE-BES-SC), under Contract No. W-31-109-Eng-38. The MR-CAT is funded by the member institutions and DOE-BES-SC under contracts DE-FG02-94ER45525 and DE-FG02-96ER45589. Funding of this work from NSF grant CTS 01 38070 is gratefully acknowledged.

REFERENCES

1. Somorjai, G. A., Hwang, K. S. & Parker, J. S. *Topics in Catalysis* 26, 87–99 (2003).
2. Li, W., Gracia, F. J. & Wolf, E. E. Selective combinatorial catalysis; challenges and opportunities: the preferential oxidation of carbon monoxide. *Catalysis Today* 81, 437–447 (2003).
3. Gracia, F., Li, W. & Wolf, E. E. *Catal. Lett.* 91, 235–242 (2003).
4. Mittash, A. *Advances in Catalysis,* Academic Press, New York II, 81 (1950).
5. Gracia, F. J., Miller, J. T., Kropf, A. J. & Wolf, E. E. Kinetics, FTIR, and controlled atmosphere EXAFS study of the effect of chlorine on Pt-supported catalysts during oxidation reactions. *Journal of Catalysis* 209, 341–354 (2002).

6.   Gracia, F. J., Bollmann, L., Wolf, E. E., Miller, J. T. & Kropf, A. J. *In situ* FTIR, EXAFS, and activity studies of the effect of crystallite size on silica-supported Pt oxidation catalysts. *Journal of Catalysis* 220, 382–391 (2003).

7.   Gracia, F. J., Guerrero, S., Wolf, E. E., Miller, J. T. & Kropf, A. J. Kinetics, operando FTIR, and controlled atmosphere EXAFS study of the effect of sulfur on Pt-supported catalysts during CO oxidation. *Journal of Catalysis* 233, 372–387 (2005).

8.   Ressler, T. WinXAS: A program for x-ray absorption spectroscopy data analysis under MS-Windows. *J. Synchrotron Rad.* 5, 118–122 (1998).

9.   Lytle, F. W., Sayers, D. E. & E.A. Stern. Report of the international workshop on standards and criteria in x-ray absorption spectroscopy. *Physica B: Condensed Matter* 158, 701–722 (1989).

10. Zwinkels, M. F. M., Jaras, S. G., Menon, P. G. & Griffin, T. A. Catalytic materials for high temperature combustion. *Catal. Rev. -Sci. Eng.* 35, 319 (1993).

11. Gupta, A. K. & Lilley, D. G. Combustion and environmental challenges for gas turbines in the 1990s. *Journal of Propulsion and Power* 10, 137–147 (1994).

12. Simone, D. O., Kennelly, T. L., Brungard, B. & J. Farrauto, R. Reversible poisoning of palladium catalysts for methane oxidation. *Applied Catalysis* 70, 87–100 (1991).

13. Marceau, E., Che, M., Saint-Just, J. & Tatibouet, J. M. Influence of chlorine ions in Pt/A12O3 catalysts for methane total oxidation. *Catalysis Today* 29, 415–419 (1996).

14. Marceau, E., Lauron-Pernot, H. & Che, M. Influence of the metallic precursor and of the catalytic reaction on the activity and evolution of Pt(Cl)/[delta]-Al2O3 catalysts in the total oxidation of methane. *Journal of Catalysis* 197, 394–405 (2001).

15. Roth, D., Gelin, P., Primet, M. & Tena, E. Catalytic behaviour of Cl-free and Cl-containing Pd/Al2O3 catalysts in the total oxidation of methane at low temperature. *Applied Catalysis A: General* 203, 37–45 (2000).

16. Malet, P., Munuera, G. & Caballero, A. Effect of chlorine in the formation of PtRe alloys in PtRe/Al2O3 catalysts. *Journal of Catalysis* 115, 567–579 (1989).

17. Pecchi, G., Reyes, P., Gomez, R., Lopez, T. & Fierro, J. L. G. Methane combustion on Rh/ZrO$_2$ catalysts. *Applied Catalysis B: Environmental* 17, L7–L13 (1998).

18. Reyes, P., Oportus, M., Pecchi, G., Fréty, R. & Moraweck, B. Influence of the nature of the platinum precursor on the surface properties and catalytic activity of alumina-supported catalysts. *Catalysis Letters* 37, 1993 (1996).

19. Baldwin, T. R. & Burch, R. Remarkable activity enhancement in the catalytic combustion of methane on supported palladium catalysts. *Catalysis Letters* 6, 131 (1990).

20. Baldwin, T. R. & Burch, R. Catalytic combustion of methane over supported palladium catalysts: I. Alumina supported catalysts. *Applied Catalysis* 66, 337–358 (1990).

21. Paulis, M., Peyrard, H. & Montes, M. Influence of chlorine on the activity and stability of Pt/Al$_2$O$_3$ catalysts in the complete oxidation of toluene. *Journal of Catalysis* 199, 30–40 (2001).

22. Peri, S. S. & Lund C. R. F. The role of chlorine in induction periods during the oxidation of methane over Pd/SiO$_{21}$. *Journal of Catalysis* 152, 410–414 (1995).

23. Cant, N. W., Angove, D. E. & J. Patterson, M. The effects of residual chlorine on the behaviour of platinum group metals for oxidation of different hydrocarbons. *Catalysis Today* 44, 93–99 (1998).

24. Marecot, P. et al. Propane and propene oxidation over platinum and palladium on alumina: Effects of chloride and water. *Applied Catalysis B: Environmental* 3, 283–294 (1994).

25. Lieske, H., Lietz, G., Spindler, H. & Volter, J. Reactions of platinum in oxygen- and hydrogen-treated Pt/[gamma]-Al$_2$O$_3$ catalysts: I. Temperature-programmed reduction, adsorption, and redispersion of platinum. *Journal of Catalysis* 81, 8–16 (1983).

26. Hwang, C.-P. & Yeh, C.-T. Platinum-oxide species formed by oxidation of platinum crystallites supported on alumina. *Journal of Molecular Catalysis A: Chemical* 112, 295–302 (1996).

27. Zhou, Y., Wood, M. C. & Winograd, N. A time-of-flight SIMS study of the chemical nature of highly dispersed Pt on alumina. *Journal of Catalysis* 146, 82–86 (1994).

28. Mordente, M. G. V. & Rochester, C. H. Infrared study of the effects of oxychlorination on Pt dispersion in Pt/Al₂O₃ catalysts. *Journal of the Chemical Society, Faraday Transactions* 85, 3495–3504 (1989).

29. Lebedeva, O. E., Chiou, W. A. & Sachtler, W. M. H. Chloride-induced migration of supported platinum and palladium across phase boundaries. *Catalysis Letters* 66, 189–195 (2000).

30. Burch, R. & Loader, P. K. Investigation of Pt/Al₂O₃ and Pd/Al₂O₃ catalysts for the combustion of methane at low concentrations. *Applied Catalysis B: Environmental* 5, 149–164 (1994).

31. Burch, R. & Loader, P. K. Investigation of methane oxidation on Pt/Al₂O₃ catalysts under transient reaction conditions. *Applied Catalysis A: General* 122, 169–190 (1995).

32. Yang, S., Maroto-Valiente, A., Benito-Gonzalez, M., Rodriguez-Ramos, I. & Guerrero-Ruiz, A. Methane combustion over supported palladium catalysts: I. Reactivity and active phase. *Applied Catalysis B: Environmental* 28, 223–233 (2000).

33. Vaarkamp, M., Modica F. S., Miller J. T. & Koningsberger D. C. Influence of hydrogen pretreatment on the structure of the metal-support interface in Pt/zeolite catalysts. *Journal of Catalysis* 144, 611–626 (1993).

34. Bazin, D., Triconnet, A. & Moureaux, P. An EXAFS characterisation of the highly dispersed bimetallic platinum-palladium catalytic system. *Nuclear Instruments and Methods in Physics Research Section B: Beam Interactions with Materials and Atoms* 97, 41–43 (1995).

35. Jackson, S. D. et al. Supported metal catalysts: Preparation, characterization, and function: II. Carbon monoxide and dioxygen adsorption on platinum catalysts. *Journal of Catalysis* 139, 207–220 (1993).

36. Hicks, R. F., Qi, H., Young, M. L. & Lee, R. G. Structure sensitivity of methane oxidation over platinum and palladium. *Journal of Catalysis* 122, 280–294 (1990).

37. Hicks, R. F., Qi, H., Young, M. L. & Lee, R. G. Effect of catalyst structure on methane oxidation over palladium on alumina. *Journal of Catalysis* 122, 295–306 (1990).

38. Bernal, S., Blanco, G., Gatica, J. M., Larese, C. & Vidal, H. Effect of mild re-oxidation treatments with CO₂ on the chemisorption capability of a Pt/CeO₂ catalyst reduced at 500°C. *Journal of Catalysis* 200, 411–415 (2001).

39. Herz, R. K. & Shinouskis, E. J. Transient oxidation and reduction of alumina-supported platinum. *Appl. Surf. Sci.* 19, 373 (1984).

40. Straguzzi, G. I., Aduriz, H. R. & Gigola, C. E. Redispersion of platinum on alumina support. *Journal of Catalysis* 66, 171–183 (1980).

41. Lagarde, P. et al. *J. Catal.* 84, 333 (1983).

42. Normand, F. L., Borgna, A., Garetto, T. F., Apestegiua, C. R. & Moraweck, B. *J. Phys. Che.* 100, 9068 (1996).

43. Borgna, A., Garetto, T. F., Apestegiua, C. R., Normand, F. L. & Moraweck, B. *J. Catal.* 186, 433 (1999).

44. Borg, H. J., Oetelaar, L. C. A. V. D. & Niemantsverdriet, J. W. Preparation of a rhodium catalyst from rhodium trichloride on a flat, conducting alumina support studied with static secondary ion mass spectrometry and monochromatic XPS. *Catalysis Letters* 17, 81–95 (1993).

45. Castro, A. A., Scelza, O. A., Benvenuto, E. R., Baronetti, G. T. & Parera, J. M. Regulation of the chlorine content on Pt/Al_2O_3 catalyst. *Journal of Catalysis* 69, 222–226 (1981).
46. Barbier, J., Bahloul, D. & Marecot, P. Effect of chloride on sintering of Pt/Al_2O_3 catalysts. *Catalysis Letters* 8, 327 (1991).
47. Contescu, C., Macovei, D., Craiu, C., Teodorescu, C. & Schwarz, J. A. Thermal-induced evolution of chlorine-containing precursors in impregnated Pd/Al_2O_3 catalysts. *Langmuir* 11, 2031–2040 (1995).
48. Boudart, M., Aldag, A., Benson, J. E., Dougharty, N. A. & Girvin Harkins, C. On the specific activity of platinum catalysts. *Journal of Catalysis* 6, 92–99 (1966).
49. McCarthy, E., Zahradnik, J., Kuczynski, G. C. & Carberry, J. J. Some unique aspects of CO oxidation on supported Pt. *Journal of Catalysis* 39, 29–35 (1975).
50. Engel, T. & Ertl, G. *Advances in Catalysis,* Academic Press, New York, (1979).
51. Cant, N. W. Metal crystallite size effects and low-temperature deactivation in carbon monoxide oxidation over platinum. *Journal of Catalysis* 62, 173–175 (1980).
52. Cant, N. W. & Donaldson, R. A. Infrared spectral studies of reactions of carbon monoxide and oxygen on Pt/SiO_2. *Journal of Catalysis* 71, 320–330 (1981).
53. Ladas, S., Poppa, H. & Boudart, M. The adsorption and catalytic oxidation of carbon monoxide on evaporated palladium particles. *Surface Science* 102, 151–171 (1981).
54. Kieken, L. & Boudar, M. In *New frontiers in catalysis: Proceedings of the 10th international congress on catalysis,* Budapest, July 19–24, 1992, *Studies in surface science and catalysis,* Guczi, L., Solymosi, F. & Tetenyi, P. (Eds.), Elsevier, Amsterdam, (1993).
55. Oh, S. H., Fisher, G. B., Carpenter, J. E. & Goodman, D. W. Comparative kinetic studies of CO---O_2 and CO---NO reactions over single crystal and supported rhodium catalysts. *Journal of Catalysis* 100, 360–376 (1986).
56. Oh, S. H. & Eickel, C. C. Influence of metal particle size and support on the catalytic properties of supported rhodium: CO---O_2 and CO---NO reactions. *Journal of Catalysis* 128, 526–536 (1991).
57. Wong, C. & McCabe, R. W. Effects of high-temperature oxidation and reduction on the structure and activity of Rh/Al_2O_3andRh/SiO_2 catalysts. *Journal of Catalysis* 119, 47–64 (1989).
58. Rainer, D. R., Koranne, M., Vesecky, S. M. & Goodman, D. W. CO + O_2 and CO + NO Reactions over Pd/Al_2O_3 Catalysts. *J. Phys. Chem. B* 101, 10769–10774 (1997).
59. Hopster, H., Ibach, H. & Comsa, G. Catalytic oxidation of carbon monoxide on stepped platinum(111) surfaces. *Journal of Catalysis* 46, 37–48 (1977).
60. Belton, D. N. & Schmieg, S. J. Effect of Rh particle size on CO desorption from Rh/alumina model catalysts. *Surface Science* 202, 238–254 (1988).
61. Szabo, A., Henderson, M. A. & Yates, J., John T. Oxidation of CO by oxygen on a stepped platinum surface: Identification of the reaction site. *The Journal of Chemical Physics* 96, 6191–6202 (1992).
62. Zafiris, G. S. & Gorte R. J. CO oxidation on $Pt/[\alpha]$-Al_2O_3(0001): Evidence for structure sensitivity. *Journal of Catalysis* 140, 418–423 (1993).
63. Meusel, I., Hoffmann, J., Hartmann, J., Libuda, J. & Freund, H.-J. Size-dependent reaction kinetics on supported model catalysts: A molecular beam/IRAS study of the CO oxidation on alumina-supported Pd particles. *J. Phys. Chem. B* 105, 3567–3576 (2001).
64. Santra, A. K. & Goodman, D. W. Catalytic oxidation of CO by platinum group metals: From ultrahigh vacuum to elevated pressures. *Electrochimica Acta* 47, 3595–3609 (2002).

65. Akubuiro, E. C., Verykios, X. E. & Lesnick, L. Dispersion and support effects in carbon monoxide oxidation over platinum. *Applied Catalysis* 14, 215–227 (1985).

66. Stara, I., Nehasil, V. & Matolin, V. The influence of particle size on CO oxidation on Pd/alumina model catalyst. *Surface Science* 331–333, 173–177 (1995).

67. Nehasil, V., Stara, I. & Matolin, V. Size effect study of carbon monoxide oxidation by Rh surfaces. *Surface Science* 352–354, 305–309 (1996).

68. Heiz, U., Sanchez, A., Abbet, S. & Schneider, W.-D. Catalytic oxidation of carbon monoxide on monodispersed platinum clusters: Each atom counts. *J. Am. Chem. Soc.* 121, 3214–3217 (1999).

69. Haruta, M. Size- and support-dependency in the catalysis of gold. *Catalysis Today* 36, 153–166 (1997).

70. Haruta, M. Catalysis of gold nanoparticles deposited on metal oxides. *CATTECH* 6, 102–115 (2002).

71. Grunwaldt, J.-D., Kiener, C., Wogerbauer, C. & Baiker, A. Preparation of supported gold catalysts for low-temperature CO oxidation via "size-controlled" gold colloids. *Journal of Catalysis* 181, 223–232 (1999).

72. Wolf, A. & Schuth, F. A systematic study of the synthesis conditions for the preparation of highly active gold catalysts. *Applied Catalysis A: General* 226, 1–13 (2002).

73. Goodman, D. W., Peden, C. H. F., Fisher, G. B. & Oh, S. H. Comment on "Structure sensitivity in CO oxidation over rhodium," by M. Bowker, Q. Guo, Y. Li, and R.W. Joyner. *Catalysis Letters* 22, 271–274 (1993).

74. Bowker, M., Guo, Q., Li, Y. & Joyner, R. W. Structure sensitivity of CO oxidation over rhodium. *Catalysis Letters* 22, 275–276 (1993).

75. Carballo, L. M. & Wolf, E. E. Crystallite size effects during the catalytic oxidation of propylene on Pt/[γ]-Al$_2$O$_3$. *Journal of Catalysis* 53, 366–373 (1978).

76. Kaul, D. J., Sant, R. & Wolf, E. E. Integrated kinetic modelling and transient FTIR studies of CO oxidation on Pt/SiO$_2$. *Chemical Engineering Science* 42, 1399–1411 (1987).

77. Lane, G. S. & Wolf, E. E. Characterization and Fourier transform infrared studies of the effects of TiO$_2$ crystal phases during CO oxidation on Pt/TiO$_2$ catalysts. *Journal of Catalysis* 105, 386–404 (1987).

78. Anderson, J. A. *Journal of Chemical Society, Faraday Transactions* 88, 1197–1201 (1992).

79. Mavrikakis, M., Stoltze, P. & Norskov, J. K. *Catal. Lett.* 64, 101–106 (2000).

80. Gracia, F. & Wolf, E. E. *Chem. Eng. J.* 82, 291–301 (2001).

81. Gracia, F. J. & Wolf, E. E. Non-isothermal dynamic Monte Carlo simulations of CO oxidation on Pt supported catalysts. *Chemical Engineering Science* 59, 4723–4729 (2004).

82. Siera, J., Silfhout, R. V., Rutten, F. & Nieuwenhuys, B. E. *Stud. Surf. Sci.* 71, 395–407 (1991).

83. Bartholomew, C. H., Agrawal, P. K. & Katzer, J. R. *Sulfur Poisoning of Metals* in New York: Academic Press, (1982).

84. Oudar, J. & Wise, H. *Deactivation and Poisoning of Catalysts,* Marcel Dekker, New York (1985).

85. Somorjai, G. A. On the mechanism of sulfur poisoning of platinum catalysts. *Journal of Catalysis* 27, 453–456 (1972).

86. Shelef, M. & Graham, G. W. Why rhodium in automotive three-way catalysts? *Catal. Rev.-Sci. Eng.* 36, 433–457 (1994).

87. Deo, A. V., Lana, I. G. D. & Habgood, H. W. Infrared studies of the adsorption and surface reactions of hydrogen sulfide and sulfur dioxide on some aluminas and zeolites. *Journal of Catalysis* 21, 270–281 (1971).

88. Chang, C. C. Infrared studies of SO_2 on [γ]-alumina. *Journal of Catalysis* 53, 374–385 (1978).

89. Okamoto, Y., Ohhara, M., Maezawa, A., Imanaka, T. & Teranishi, S. Hydrogen sulfide adsorption on alumina, modified alumina, and molybdenum trioxide/alumina. *J. Phys. Chem.* 90, 2396–2407 (1986).

90. Gandhi, H. S. & Shelef, M. Effects of sulphur on noble metal automotive catalysts. *Applied Catalysis* 77, 175–186 (1991).

91. Apesteguia, C. R., Brema, C. E., Garetto, T. F., Borgna, A. & Parera, J. M. Sulfurization of Pt/Al_2O_3---Cl catalysts: VI. Sulfur-platinum interaction studied by infrared spectroscopy. *Journal of Catalysis* 89, 52–59 (1984).

92. Gandhi, H. S., Yao, H. C., Stepien, H. K. & Shelef, M. Evaluation of three-way catalysts—Part III, Formation of RhO_3, its suppression by SO_2 and re-oxidation. *SAE paper 780606* (1978).

93. Bonzel, H. P. & Ku, R. Adsorbate interactions on a Pt(110) surface. I. Sulfur and carbon monoxide. *The Journal of Chemical Physics* 58, 4617–4624 (1973).

94. Halachev, T. & Ruckenstein, E. Poisoning and promoting effects of additives on the catalytic behavior of metal clusters. *Journal of Catalysis* 73, 171–186 (1982).

95. Beck, D. D. & Sommers, J. W. "Impact of sulfur on three-way catalysts: Comparison of commercially produced Pd and Pt-Rh monoliths" *in Catalysis and Automotive Pollution Control III, Proceedings of the Third International Symposium (CAPoC3)*, Frennet, A. & Bastin, J.-M. (Eds.), Elsevier, Amsterdam (1995).

18 Catalysts for New Uses: Needed Preparation Advances

Bernard Delmon

CONTENTS

ABSTRACT

Different reasons stimulate advances in catalyst preparation. A personal view will be proposed, based on the longstanding interest of the author in catalyst preparation, but perhaps also with some bias due to another strong interest, actually in selective and complete oxidation reactions. Some aspects will be examined in some detail, considering scientific possibilities as well as the chemical engineering aspect.

1. The developments of ultrafast reactions for energy generation (catalytic combustion), and in the near future, the production of useful molecules both demand that the catalyst texture be carefully adapted to the control of reactive radicals population.
2. Coprecipitation does not, by far, achieve chemical homogeneity at the atomic level in solids containing different catalytically active elements

(e.g., silica-alumina, pore walls of mesoporous solids, multicomponent oxidation catalysts). New approaches are needed to achieve this goal.
3. Discoveries are needed for creating hybrid catalysts combining several effects for enhancing selectivity.
4. Methods must be developed to support heteropolyacids in order to take full advantage of their ability to achieve difficult reactions.

Additional demands are due to the fact that new types of reactors appear as very promising: reverse-flow and ultrafast reactors, microreactors, reactors for photocatalysis, and so on. Suitable catalysts not only need new chemical formulations and textures, but also preparation processes adapted to the corresponding shaping techniques. Not all cases can be examined in this chapter. However, much emphasis will be laid on the role of a comprehensive approach in the scientific and chemical engineering research for catalyst fabrication. A few examples will illustrate the discussion. Suggestions may come from other fields of science and technology.

18.1 INTRODUCTION

This contribution constitutes an attempt to meet the wish of the organizer of the special symposium, The Science and Engineering of Catalyst Preparation. His desire was to have a contribution speculating on perspectives for the future of catalyst preparation. The philosophy he proposed for the oral communication was summarized as follows: Promising Scientific Avenues, an impressively broad topic indeed. The present contribution is an attempt to correspond to this objective, although it can just present personal views concerning selected problems.

It was emphasized in the past that the development of catalysis was the result of a sort of push-pull interaction. Catalysis is crucial for modern chemical industry, and contacts between laboratories and industry are generally intense. The problems industry is faced with can stimulate (or should stimulate more) research in Academia (the *pull* effect). On the other hand, new ideas and discoveries in Academia are one (modest) factor that contributes to *push* innovations in industry. Stronger incentives to innovations result from the considerable influence that energy and raw materials cost, environmental protection, sustainability in general, and profitability exert on industrial developments. Catalyst preparation is one aspect of catalysis where this push-pull effect has, or should have, a particularly strong influence. The present contribution, inspired by the Promising Scientific Avenues challenge, will therefore attempt to be close to the spirit of the symposium title.

Combinatorial chemistry can perhaps help discover new catalyst formulations for reactions presently of particular interest, such as oxidations or ammoxidation, and generally all reactions of alkanes. Reactions traditionally made in different kinds of processes are frequently shown to be also activated by heterogeneous catalysts (e.g., epoxidations). Reactors of unexpected design allow surprisingly selective reactions (e.g., monoliths for the oxidative dehydrogenation of light alkanes). However, the distance often remains long between these discoveries and the manufacture of active and selective catalysts adequately structured for particular use in an industrial reactor inserted in an industrial plant.

The objective of the present chapter is to outline typical problems of heterogeneous catalysis encountered in the development of catalyst and processes. Examples will be selected of difficulties encountered in innovative development, and when possible, some ways to circumvent these difficulties will be mentioned or just suggested. These show up:

(1) In preparing catalysts corresponding exactly to the demands of catalytic science concerning highest activity and selectivity
(2) In responding to the demands of processes with respect to catalyst shape and structure
(3) In developing a comprehensive approach (a fully fledged chemical engineering approach) in the simultaneous development of catalysts and reactors

This certainly can suggest avenues if this is understood as wide general directions for work, but a more precise analysis of the problem shows that there are actually only very specific ways or approaches to reach goals, a multitude of narrow passages or tracks, rather than avenues.

18.2 CATALYSTS FOR ULTRAFAST REACTIONS

Ultrafast catalytic reactions have been known for a long time and have been exceptionally successful in industry. These are the oxidation of ammonia to nitrogen oxides (with platinum gauze) for the manufacture of nitric acid and the Andrussov process for hydrocyanic acid. In the field of fine chemistry, the BASF process for the oxidation of isoprenol (2-methylbut-1-ene4-ol) to isoprenal for the synthesis of citral, with a residence time of 1 ms (1), constituted a landmark in selective oxidative dehydrogenation catalysis. Last but not least, auto exhaust purification using monoliths constitutes the most conspicuous development of ultrafast reactions. More recently, the removal of volatile organic compounds has been developed, and high-temperature catalytic combustion integrated in energy production processes is the object of intense development programs.

Ultrafast reactions offer an excellent example of the need for a chemical engineering approach in catalyst fabrication, namely the necessity to look with equal attention at the reaction and the corresponding demands, as well as to the catalyst and the corresponding demands. Auto exhaust catalysts constitute by far the major industrial product of the catalyst manufacturing industry. They can be considered as mature products. However, two processes in development are emerging as offspring of the monolith technology. These are catalytic combustion (2) and the production of olefins by oxidative dehydrogenation along with the pioneers' work of Huff and Schmidt (3). Research in this field took some time to progress, but it finally developed and gained fame with metal gauzes or, more systematically, with monoliths (see, for example, References 4–8). This kind of system has been modelled in detail (progress of reaction, heat profile, etc.) (see, in Reference 2, contributions of P. Forzatti, pp. 19–31, and also of M. Lyubovsky et al., and Reference 9). The intervention of radicals has been proven and elucidated at least partially; mass and heat transfer calculated at several scales, especially in the washcoat; and much information is already available. But the developments came principally from the chemical engineering side, and it can be wondered whether sufficient investigations have been conducted concerning other aspects. There is certainly a need for specially adapted methods for deposition of catalytic material on monoliths or other supports. Those developed for auto exhaust devices are a starting point but cannot constitute a solution for the new processes, because the demands are different.

The reason fundamentally is that organic radicals play an essential role in these processes. More precisely, the reaction takes place at temperatures where radicals spontaneously form in the gas phase. Without control, the reaction will go to complete oxidation or deposition of carbon, according to the reaction conditions. If olefins or selective oxidation of products is desired (as also in the case of methane selective oxidation or oxidative coupling), radicals must be trapped or quenched. This depends not only on the nature of the catalyst surface, but also crucially on catalyst porosity and catalyst shape. A very stimulating article that perhaps did not attract sufficient attention is the one of Quah and Li (10). The authors show that very high flow rates of gases passing through a gauze, or meshed catalyst, strip away methyl radicals and prevent them from adsorbing on the catalyst surface and being fully oxidized. These radicals can dimerise (oxidative coupling) or also react to other useful products, although the yield in oxygenated products mentioned in literature is not yet as attractive as the yield in olefins. There is, therefore, a challenge to produce a catalyst with a surface texture allowing this removal of radicals, a typical problem where gas-flow studies and search for innovative preparations in cooperation need to be developed. In this respect, it would perhaps be useful to have a look at the studies that USSR conducted in the 50s and 60s on what was then called *homogeneous-heterogeneous reactions*.

There is, therefore, much work to do on the chemical engineering side. But the task is still more important for the chemistry of catalytic materials and deposition techniques. A personal feeling is that the activity of the catalytic substance should essentially be due to its external surface or that it should possess no porosity or only very wide pores. Nanoclusters (e.g., of monodispersed TiO_2 on mesoporous silica, Reference 11), micelles (e.g., Reference 12), nanometer crystallites of zeolites (see, for example, Reference 13), and mesoporous solids could be candidates. An ample choice is possible in a period where nanomaterials are the object of intense research (see, for example, Reference 14).

The deposition and strong attachment of such highly dispersed catalysts on supports are not easy. New problems will arise with the introduction of new supports permitting flow with practically no pressure drop, such as cloths (15, 16) or knitted gauze (17). Besides the field of catalysis, however, different sectors of industry already use deposition technologies able to achieve similar tasks, and many laboratories are scientifically supporting the corresponding innovations or extensions. Surface treatments and industrial applications of coating constitute well-established technologies. Catalysis literature contains reports on the use of chemical vapor deposition (CVD) or organometallic CVD (for example, Reference 18), plasma deposition (for example, the review of Reference 19), sputtering, ion implantation, and use of colloidal interactions (many authors; see, for example, Reference 20). This indicates that some roads or avenues are already available for the development of catalyst adapted for ultrafast reactions where organic radicals are involved. There is little doubt that research in the directions suggested above will develop in the near future.

18.3 PRODUCTION OF HIGH SURFACE AREA CATALYTIC MATERIAL BY COPRECIPITATION TECHNIQUES

In contrast to the preceding situation, the challenge for improved coprecipitation presently concerns more chemical engineering than chemistry and is situated at the other end of the dimension scale of phenomena, in which chemical engineering has to find solutions, namely nanometers rather than decimeters or more. This concerns mainly the preparation of supports. In addition to classical silica-alumina, many publications and several industrial developments have dealt with silica-titania or silica-zirconia, alumina-titania, and ceria-zirconia. The hope was either to modify the properties of one component by introducing a doping agent in the structure or to create new compounds. This has been particularly the case with silica-alumina when this association was discovered to possess very specific acid sites due to Al-O-Si bonds. Nearly all these supports have been prepared by coprecipitation, and often almost exclusively using this technique. Considerable progress has been achieved for controlling surface area and porosity (see, for example, References 21 and 22, and the review paper in Reference 23). But besides the considerable knowledge acquired scientifically and industrially with zeolites, the atomic scale structure of amorphous supports and that of the wall of mesoporous solids has not progressed very much.

Taking silica-alumina, we have shown that the composition is incredibly far from homogeneity (24). We investigated by analytical electron microscope a series of 32 samples of different compositions prepared industrially or using the more elaborate technique called the sol-gel method (16 samples of each category). Between 50 and 75 points were analyzed for each sample. No one single sample was homogeneous at a 40 nm scale. In samples prepared by coprecipitation, the lack of homogeneity appeared even at the micrometer scale, with bulky zones composed exclusively of silica. For the best supports, fluctuations of the Si/Al ratio were of 10–20%. Two typical examples are given in Figure 18.1 and Figure 18.2. In view of these results, it is not surprising that acidity strength measurements only give a very blurred picture, to the point that correlation between selectivity (and also activity) and nature of acid sites very often fails to provide useful conclusions.

Several programs have been conducted for improving homogeneity in coprecipitation (e.g., Reference 25 and references therein, and Reference 26). The approach was typically a chemical engineering one, but data showing important progress in improving homogeneity have not been published. It seems that the attention was finally focused on a better control of pore size uniformity.

The method using the cohydrolysis of alkoxides was apparently studied first in the 60s for making high-temperature-resistant ceramics for space programs. It became known as the sol-gel method, although coprecipitation also produces a gel,

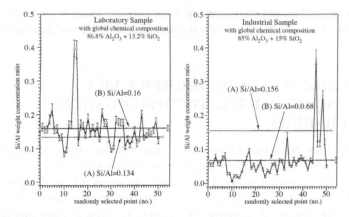

FIGURE 18.1 Microanalysis of silica-alumina (around 13–15% SiO$_2$) with the electron microscope. The resolution is about 40 nm, supposing that the effect of lateral diffusion of electrons in the very thin layer of particles in the sample used is negligible; (A) and (B) correspond to average compositions using different statistical models (24).

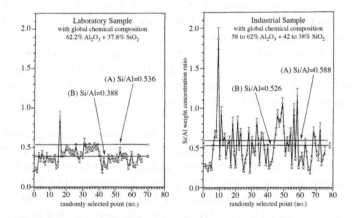

FIGURE 18.2 Microanalysis of silica-alumina (around 38% SiO$_2$) with the electron microscope. Other comments are identical to those of Figure 18.1 (24).

and the name tends presently to be used for classical coprecipitation as well as for other methods. The common obstacle to perfect homogeneity in these methods is the fact that the reaction leading to precipitation does not occur at the same rate for the two or several components supposed to coprecipitate. In the case of the alkoxide method, mildly chelating agents, or alternatively the choice of different alkoxy groups for each cation, can make the reactivity of one component closer to that of the other. But success is limited. The crucial step is the introduction in the starting liquid mixture of the reactant triggering precipitation (for example, sodium hydroxide in classical coprecipitation or acidified water in the sol-gel technique). In spite of very energetic mixing, the compositions at the scale of several nanometers are different from place to place, and reaction occurs at different rates, producing gel particles of different composition and often of different size, as well.

Chemical engineers know that mixing is still a unit process resisting spectacular improvements in spite of the invention of many devices. A molecular-scale homogeneity is perhaps possible, using a concept (Figure 18.3) borrowed from ideas developed for microdevices. The speculative idea is to make thin films of the two components to mix (nr 1 and nr 2) or of imperfect mixed material, to intercalate sheets of type 1 between sheets of type 2, as a sort of millefeuille, and to repeat the operation several times. The problem is that a perfect process based on this principle would need about 22 or 23 operations. (The calculation has nothing to do with the Avogadro number.) Other approaches are the relatively frequent use of ultrasound (see, for example, Reference 27) or that of microwaves (e.g., Reference 28, concerning the preparation of VPO catalysts). These techniques can be used alone or in addition to those already mentioned. But, to our knowledge, no comparative studies have been published. With respect to microwaves, it should be known that microwaves, in addition to supplying heat to a reacting medium and to making molecules move, can selectively activate chemical bonds (29). Tuning the wavelength could perhaps allow us to activate selectively reactions between two different alkoxides in the sol-gel technique. As a partial conclusion, the above section outlines the perspectives that many different possible contributions of chemical engineering can open.

Concerning the chemistry counterpart, the challenge is also enormous. The simplest idea would be to flocculate colloids constituted of species M, N, P, and so on, carrying opposite charges in order to obtain $-M^+-N^--M^+-P-M^+$ clusters. There are very few publications dealing with this possible method. It seems that current ideas correspond to very complicated approaches, like the synthesis of organometallic molecules containing all the desired metals in the desired proportions that would be calcined or transformed in other ways in very mild conditions to a solid solution or a compound containing these two or several elements.

A relatively practical approach could be derived from the sol-gel process. If the alkoxides contained several elements in the right proportion, uniformity of composition could be achieved for these elements. The difference with conventional sol-gel method would be to start from building blocks, namely complex alkoxides containing two or several elements, instead of individual bricks. Conceptually, this is the way zeolites are constructed. Zeolites of different compositions and structures can be made because the building blocks are different. Figure 18.4 schematically

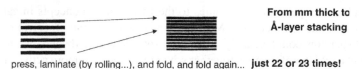

From mm thick to
Å-layer stacking

press, laminate (by rolling...), and fold, and fold again... just 22 or 23 times!

FIGURE 18.3 The concept is to start by arranging sheets of the different substances to associate, to stack them alternatively on top of the pile, to press or laminate, and to fold the resulting ribbon, and to repeat the process again and again. In view of perfect mixing, the worst situation is that where no diffusion from sheet to sheet takes place. Supposing that the starting material is constituted of a piece of one material placed on top of a similar piece of the second material, both of 1 cm thickness, the operation should be repeated 22 or 23 times for achieving homogenization at the nanometer scale.

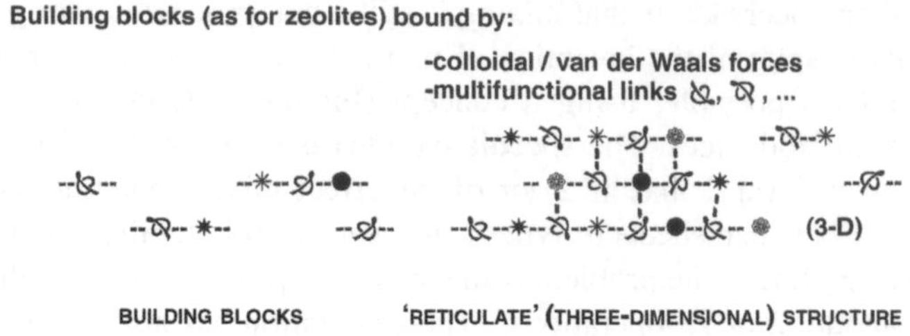

Building blocks (as for zeolites) bound by:

-colloidal / van der Waals forces
-multifunctional links

BUILDING BLOCKS   'RETICULATE' (THREE-DIMENSIONAL) STRUCTURE

FIGURE 18.4 Preparation of highly homogeneous, multielement precursor of catalysts or supports: creation of a bi- or tridimensional structure from molecular building blocks containing several elements, by using multifunctional links (like the acid-alcohols used in the citrate process).

suggests building blocks possibly containing several elements and the bidimensional or tridimensional linking of these building blocks for making a multielement material. In zeolite synthesis, colloidal and van der Waals forces attach the building blocks to each other. Formally, chemical links could do the same. This is the case in the citrate process (30–32). It is still used successfully, and its potential to solve problems is well known. The use of acid-alcohols and citric acid in the citrate process is an example that could be generalized.

As the problem in the sol-gel process is the lack of selectivity in the formation of bonds between different alkoxides, an idea advanced by some investigators is to attach different functions to the parts containing the different elements to associate. These functions could be selected in such a way that they react only with a single matching kind of function. This is schematically represented in Figure 18.5. The arrows and respective selectively matching clefts symbolize molecular recognition or similar interactions. A sort of elaborate Lego structure could be constructed in principle.

In conclusion, making catalysts containing several elements homogeneously distributed in the solid at the subnanometer scale remains a major challenge and deserves more research than has presently been done. Considered independently, neither the chemical engineering nor the synthetic chemistry approaches alone seem to be able to provide simple solutions. Breakthrough may reasonably be expected from a cooperative work between both fields of scientific activity. Even imperfect homogeneity could be a valuable objective for lack of real molecular dispersion, because phase cooperation now comes to the forefront of concepts in catalysis, in particular in oxidation reactions. Most methods used with the objective of ultimate atomic dispersion could actually achieve a very fine interdispersion of different phases and could therefore give, together with other methods, more active or selective catalysts because contacts between phases would be more numerous.

18.4 HYBRID CATALYSTS

In the development of catalysis, the necessity appeared to increase the dispersion and diminish the consumption of active material. This led to the preparation of catalysts composed of several phases intimately associated together, and specifically

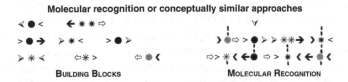

FIGURE 18.5 Speculation: possible use of molecular recognition for making precursors by association of different elements in a specific way.

to supported catalysts. Different sorts of cooperation between phases were also discovered, in particular bifunctional catalysis, and this led to multiphase catalysts, especially the multicomponent selective oxidation catalysts. Presently, other reasons appear for associating substances of very different natures. This constitutes a need in materials science in general, but also in catalysis. This is leading to the very rapid development of hybrid materials, a field that is becoming very fashionable. Quite logically for materials science, much attention is given to the potential of associating organic substances, principally polymers, with inorganic compounds. In catalysis, a trend in the same direction was detected when fluorinated substances or silicone-covered supports for catalysts were mentioned in literature. Recently, in particular, interesting results have even been found in the case of supercritical reactions on supported micellar catalysts (33, 34). The motivation was to increase selectivity by acting on differences in hydrophilicity between reactants, intermediates, or final products.

More broadly, a recent review paper presents many typical examples of hybrid porous substances that could perhaps become useful as mesoporous adsorbents or catalysts (35). Fundamental studies on the mechanism of catalytic reactions, especially when complicated molecules or complicated mixtures are reacted, will certainly suggest nanometer-scale structures that can now be constructed in laboratories. Just to suggest how diverse the applications could be, let us cite a hybrid composed of cellulose and silica (36). This catalyst is biodegradable, only leaving silica, a normal component of soil. It could, therefore, be dumped in principle. Hybrid catalysts typically constitute a broad new field, an avenue or, more exactly, several boulevards. The essential problem will rest in fabrication at a large scale. The difficulty is considerable in all fields of nanotechnology. The example of fullerene nanotubes, and still more of catalysts based on these nanotubes, shows that the boulevards will not immediately allow easy progress.

In addition to these general comments, it may be adequate to mention here an example of potential practical interest, but possibly corresponding of a very particular case, actually a narrow alley. Oxynitrides were first studied as catalysts by Grange. One member of the family studied in his laboratory is extremely active and selective in the ammoxidation of propane to acrylonitrile (37–39). Even with the nonmodified, nondoped catalysts, a selectivity of 60% is easily achieved at 500°C with a productivity (e.g., per kilogram per hour) over 10 times higher than with the catalysts mentioned in literature. The catalyst is obtained by the action of ammonia on amorphous $AlV_{1-x}O_y$. This mixed oxide remains amorphous below 400°C, allowing easy initiation of the reaction. The structural units of this starting material seem to be identical to those existing in $AlVO_4$, with covalent links between vanadium and aluminum. But a strange situation appears when the structure of the oxynitride is studied (Figure 18.6). Vanadium forms various bonds with different nitrogen species, including $N = N$ moities. But aluminum remains in a pure oxygen coordination. It seems to constitute a sort of alumina-type sublattice, with essentially AlO_6 octahedral coordination and some AlO_5 pentahedral arrangements, similar to some structural features found in $AlVO_4$. The structure is that of an open alumina-like lattice serving as host for VN_x species (40). This structure collapses only around 600°C, to alumina and vanadium nitride VN.

The synthesis of these materials through a sort of topotactic (or chimie douce) process suggests that nanotechnology applied to hybrids will also develop for inorganic-inorganic systems. This concerns purely inorganic chemistry aspects. But new problems linked to manufacture arise also for organic-inorganic hybrids. For borides, carbides, nitrides, and phosphides, problems to solve are the obtention of high porosity precursors, transformation at temperatures low enough to maintain a high surface area, and adequate mechanical properties. In the specific case of vanadium oxynitrides described in some detail above, the coprecipitation of the oxide precursor is crucial (41)

18.5 SUPPORTING HETEROPOLY COMPOUNDS

In the fields of acid-base and oxidation catalysis, heteropolyacids (HPAs) and their derivatives have demonstrated a remarkable capacity to make difficult reactions with

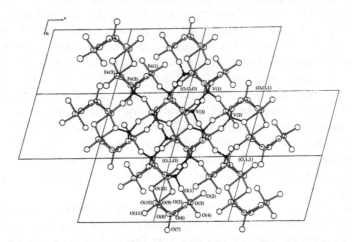

FIGURE 18.6 A preferential breaking of Al-O-V bonds during the partial nitridation is possible in solids possessing the essential structural features of $AlVO_4$ (40), as represented in the figure in the case of $FeVO_4$. This justifies the findings of ^{27}Al and ^{51}V NMR on oxynitrides of different compositions (39) and strongly suggests, together with other data, that the material is a sort of composite, essentially formed by an open three-dimensional network of aluminum oxide containing vanadium nitride clusters of different compositions as guests.

high selectivity. A recent special issue is dedicated to them (42). They are used at the laboratory scale, but few industrial processes use them in practice. Notable exceptions concern principally Japanese processes for selective oxidation. HPAs are relatively costly and very fragile thermally and mechanically. The structural fragility has been demonstrated in several studies. Among recent ones, Reference 43 and other articles by the same authors could be cited, as well as more recent contributions (44–46). There is, therefore, an incentive to find suitable materials for supporting and hopefully stabilizing them.

Several kinds of substances have been examined as potential supports. Some hope has been laid on inorganic solids containing the central ion of the HPA structure, especially silica. It is difficult to assess the degree of success of the technique. The answer is probably in the hands of industry. One publication considered the possible regeneration of silica supported HPA (47). Many other attempts dealt with carbon and organic polymers. The use of porous carbon as support led to rather pessimistic conclusions (48). Polymers are probably more promising. For example, the Keggin structure of tungstophosphoric acid on a polyvinyl alcohol hydrogel is stable, and the catalyst permits the esterification of acetic acid with isoamyl alcohol (49). Polyaniline was used to support the mixed HPA $H_4PMo_{11}VO_{40}$ (50). The acid centers of the HPA were strongly inhibited, but a high redox activity in the hexanol conversion to hexanal rather than hexene was observed. Both examples correspond to reactions that take place at low temperatures, and organic polymers can act as support. But reactions for which HPAs and derivatives exhibit outstanding activities are oxidation by molecular oxygen or oxidative dehydrogenation at temperatures exceeding 200°C. These are the reactions where robust HPAs could be particularly useful. The lack of success until now with totally inorganic compounds constitutes an incentive to explore hybrid catalysts involving HPA compounds.

A conclusion to this section is that the objective of immobilizing heteropolyacids and their derivatives looks like a formidable task. There is not yet any avenue or even a narrow alley for using polyhetero compounds in heterogeneous catalysis in a practical way. The problem is to break open (to blast open!) completely new ways.

18.6 BRIDGING THE GAP BETWEEN HOMOGENEOUS AND HETEROGENEOUS CATALYSIS

An objective often mentioned in catalysis, actually still a dream, is to make heterogeneous catalysts possessing more of the remarkable properties of homogeneous ones. The tethering (or anchoring) operation merely immobilizes homogeneous catalysts, without modifying much their properties, and essentially constitutes a way to facilitate separations. One could thus consider that this does not really bridge the gap, in the sense that the essence of homogeneous catalysis is not changed, and that the corresponding studies do not shed much light on the intimate mechanisms with heterogeneous catalysts. In a simplified way, let us just consider two fundamental aspects of catalysis that are particularly useful for selectivity. The hope would be to take simultaneously advantage of the powerful and specific effect of organic ligands and of the relatively rigid coordination of transition metal atoms in a solid matrix.

Supported organometallic catalysts could in principle make this dream reality. Metal crystallites modified by ligands, on the one hand, or by attachment to a surface, on the other hand, partially correspond to this ideal picture. But considering the first possibility, it is clear that the magnificent success of supported organometallic catalysis is due more to a grafting, or more precisely a tethering, of organometallic molecules by one of their side chain, namely immobilization, rather than to electronic interaction with the catalytically active metal. It is striking in this respect that supported organometallic catalysts in their majority are attached to silica (51). This led to remarkable scientific developments and useful applications. But silica is not likely to permit any substantial electronic interaction or to provide an adequate ligand field or a rigid, electronically active coordination. Other supports either make tethering impossible or destroy complexes or just add the specific, unmodified function of the support, to make a bifunctional catalyst. The hope would be that both organic structures and atoms belonging to the solid should be involved to cooperate in a new type of activity. The approach probably pertains to surface science, with the difficulties inherent to the difference of techniques used for metals and oxides, and the enormous difficulty to attach elaborate structures on surfaces lending themselves to surface studies.

18.7 A FEW WORDS CONCERNING OTHER SCIENTIFIC AVENUES AND THEIR PLACE IN PRACTICAL CATALYST FABRICATION

The preparation of supported catalysts constitutes a frequent activity in laboratories. A very large proportion of catalyst sales is represented by supported catalysts. No section in the present contribution deals with the preparation of supported catalysts. This is a personal choice. The reason is that fundamental information as well as practical indications are generally available in literature, and there is a natural trend in this field to continuously improve knowledge. The barriers for making a supported catalyst precisely corresponding to design are much less hindering than difficulties in the other fields examined above. This opinion is perhaps that of the organizer of the special symposium The Science and Engineering of Catalyst Preparation. He can very rightly speak on perfectly justified grounds of an engineering of catalyst preparation. What is perhaps lacking is mainly a well-ordered data bank, where the methods used for supporting active compounds could be mentioned with Internet links to the best corresponding publications. Colloidal interactions, chromatographic effects, diffusion processes, remobilization and redeposition of ions in pores, and phenomena during drying and calcination are the object of very good publications. Incidentally, access to these data would avoid many wrong interpretations in publications.

Other steps in catalyst preparation could deserve similar comments: shaping, activation, and so on. Ceramic science and the principles of chemistry of solids are often ignored in spite of their potential contribution to catalyst shaping or aging prevention.

We leave to specialists the task to describe the desired new approaches in the preparation of zeolites and mesoporous materials. Much work remains to be done for the scale-up of fabrication of MCM and similar materials. The manufacture of specially structured catalysts for very compact devices, like on-board fuel cells and automobile reforming units, is certainly able to bring much information valuable for the fabrication of many specific catalysts in environmental protection, especially for reverse-flow reactors. The demands of photocatalysis also impose constraints.

Other topics are more or less relevant to the general context of the present contribution. For example, this would be the manufacture of molecularly imprinted catalysts (footprint catalysts) possessing an activity much higher than those described in literature until now. The manufacture of already activated catalysts (already in use in hydrotreatment or, like Raney catalysts, in hydrogenation) could also be the object of stimulating remarks. But only the corresponding specialists could express valuable opinions.

18.8 PERSPECTIVES

Outside observers in the world often express some skepticism and disappointment concerning the innovation potential of heterogeneous catalysis. Many think that large programs based on topics made popular for different reasons did not bring sufficient practical applications, in consideration of their cost. These programs in the past corresponded to sorts of worldwide fashions such as SMSI, amorphous alloys, materials exhibiting high-temperature superconductivity, sol-gel techniques, and oxidative coupling of methane. There is no doubt that these programs had some stimulating effects. But their impact on fundamental knowledge in catalysis was often limited. Conversely, innovative processes have been the result of comprehensive approaches conducted by industry, like the riser bed reactor of DuPont with its special catalyst for maleic anhydride (and furan) manufacture, titanium silicalite (ENI), the direct hydroxylation of phenol with air and hydrogen in acetic acid as solvent (Tosoh Co.), or the direct synthesis of citral by BASF (already mentioned in Reference 1).

Presently, 90% of the large and middle-size production processes in the chemical industry are based on catalysis. In the comprehensive effort aiming at the development of new catalytic processes, catalyst preparation constitutes just one part amid other contributions. But this part led to conspicuous innovation, and this because the chemical engineering aspect of catalyst industrial development was considered simultaneously with the basic science aspects and the development of the whole process. This remark fully justifies the ambition of the special symposium The Science and Engineering of Catalyst Preparation. It also justifies the fact that the selection of topics in the present contribution was essentially not influenced by fashion. In our opinion, the word *engineering*, when catalyst preparation is considered, should certainly apply to the unit operations of catalyst manufacture. But it should also point to the comprehensive approach that includes the first steps in the discovery of an interesting catalytic reaction as well as the industrial process development. With that in mind, one must conclude that the most promising avenue is to give full consideration to both the science and the engineering aspects of catalyst preparation.

REFERENCES

1. W.F. Hölderich, in *New frontiers in catalysis,* L. Guczi, F. Solymosi, P. Tétényi (Eds.), Elsevier, Amsterdam, p. 127, 1993.
2. 5ᵗʰ Workshop on catalytic combustion, S.K. Kang, S.I. Woo (Guest eds.), Catal. Today, 83, pp. 1–297, 2003.
3. M. Huff, L.D. Schmidt, J. Catal., 149, 127, 1994.
4. L.D. Schmidt, 12ᵗʰ Intern. Congr. Catalysis, A. Corma, F.V. Melo, S. Mendioroz, J.L.G. Fierro (Eds.), (Series Surface Science and Catalysis No. 130 A), Elsevier, Amsterdam, Vol. 1, p. 661, 2000.
5. A.S. Bodke, D. Henning, L.D. Schmidt, Catal. Today, 61, 65, 2000.
6. A.S. Bodke, D. Henning, L.D. Schmidt, S.S. Bharadwaj, S.J. Maj, J. Siddal, J. Catal., 191, 62, 2000.
7. R.P. O'Connor, L.D. Schmidt, J. Catal., 191, 245, 2000.
8. J.J. Krummenacher, K.N. West, L.D. Schmidt, J. Catal., 215, 332, 2003.
9. A. Beretta, P. Forzatti, J. Catal., 200, 43, 2001.
10. E.B.H. Quah, C.-Z. Li, Appl. Catal. A, 250, 83, 2003.
11. A. Tuel, L.G. Hubert-Pfalzgraf, J. Catal., 217, 343, 2003.
12. S.C. Tsang, K.M.K. Yu, A.M. Stele, J. Zhu, Q.J. Fu, Catal. Today, 81, 573, 2003.
13. E.W. Beers, T.A. Nijhuis, N. Aalders, F. Kapteijn, J.A. Moulijn, Appl. Catal. A, 243, 237, 2003.
14. S.E. Park, R. Ryoo, W.-S. Ahn, C. W. Lee, J.-S. Chang, *Nanotechnology in mesos- trucutured materials* (Series Surface Science and Catalysis No. 146), Elsevier, Amsterdam, 2003.
15. Y. Matatov-Meytal, M. Sheintuch, Appl. Catal., 231, 1, 2002.
16. Y. Matatov-Meytal, Y. Shindler, M. Sheintuch, Appl. Catal B: Environmental, 45, 127, 2003.
17. J. Pérez-Ramírez, F. Kapteijn, K. Schöffel, J.A. Moulijn, Appl. Catal. B: Environ- mental, 44, 117, 2003.
18. A.E. Aksoylu, J.L. Faria, M.F.R. Pereira, J.L. Figueiredo, P. Serp, J.-C. Hierso, R. Feurer, Y. Kihn, P. Kalck, Appl. Catal. A, 243, 357, 2003.
19. Ch.-J. Liu, G.P. Vissokov, B.W.-L. Jang, Catal Today, 72, 173, 2002.
20. D. Didillon, E. Merlen, T. Pagès, D. Uzio, in *Scientific bases for the preparation of heterogeneous catalysts,* B. Delmon, P.A. Jacobs, R. Maggi, J.A. Martens, P. Grange, G. Poncelet (Eds.), (Series Surface Science and Catalysis No. 118), Elsevier, Amsterdam, p. 41, 1998.
21. R. Takahasi, L. Sato, T. Sodesawa, M. Yabuki, J. Catal., 200, 197, 2001.
22. N. Yao, G. Xiong, S. Sheng, M. He, K.L. Yeung, in *Scientific bases for the preparation of heterogeneous catalysts,* E. Gaigneaux, D.E. De Vos, P. Grange, P.A. Jacobs, J.A. Martens, P. Ruiz, G. Poncelet (Eds.), (Series Surface Science and Catalysis No. 143), Elsevier, Amsterdam, p. 715, 2002.
23. J. Cejka, Appl. Catal. A, 254, 327, 2003.
24. C. Sârbu, B. Delmon, Appl. Catal. A, 185, 85, 1999.

25. J.P. Reymond, G. Dessalces, F. Kolenda, in *Scientific bases for the preparation of heterogeneous catalysts*, B. Delmon, P.A. Jacobs, R. Maggi, J.A. Martens, P. Grange, G. Poncelet (Eds.), (Series Surface Science and Catalysis No.. 118), Elsevier, Amsterdam, p. 118, 1998.
26. B. Brizzo, G. Bellussi, in *Oxide-based systems at the crossroads of chemistry*, A. Gamba, C. Colella, S. Coluccia (Eds.), (Series Surface Science and Catalysis No. 140), Elsevier, Amsterdam, p. 401, 2002.
27. S.C. Emerson, C.F. Coote, H. Boote III, J.C. Tufts, R. LaRoque, W.R. Moser, in *Scientific bases for the preparation of heterogeneous catalysts*, B. Delmon, P.A. Jacobs, R. Maggi, J.A. Martens, P. Grange, G. Poncelet (Eds.), (Series Surface Science and Catalysis No. 118), Elsevier, Amsterdam, p. 773, 1998.
28. U.R. Pillai, E. Sahle-Demessie, R.S. Varma, Appl. Catal A, 252, 1, 2003.
29. A. Loupy (Ed.), *Microwaves in organic synthesis*, Wiley-VCH, Weinheim, 2002.
30. C. Marcilly, Ph. Courty, B. Delmon, J. Amer. Ceram. Soc., 53, 56, 1970.
31. C. Marcilly, H. Ajot, Ph. Courty, B. Delmon, Powder Tech., 7, 21, 1973.
32. P.G. Menon, B. Delmon, in *Handbook of heterogeneous catalysis*, G. Ertl, H. Knözinger, J Weitkamp (Eds.), Wiley-VCH, Vol. 1, section 2.1.6 C, pp. 112–114, 1997.
33. S.C. Tsang, K.M.K. Yu, A.M. Steele, J. Zhu, Q.J. Fu, Catal. Today, 81, 573, 2003.
34. J. Zhu, S.C. Tsang, Catal. Today, 81, 673, 2003.
35. S. Inagaki, in *Nanotechnology in mesostructured materials*, S.E. Park, R. Ryoo, W.-S. Ahn, C. W. Lee, J.-S Chang (Eds.), (Series Surface Science and Catalysis No. 146), Elsevier, Amsterdam, p. 1, 2003.
36. I. Portugal, S. Magina, M.A. Batista, D. Evtiouguine, 4th European Congress of Chemical Engineering, Granada, Sept. 21–25, 2003, Session 10, comm. O-10.
37. M. Florea, R. Prada Silvy, P. Grange, Appl. Catal. A, 255, 289, 2003.
38. H.M. Wiame, M.A. Centeno, L. Legendre, P. Grange, *submitted*.
39. H. M. Wiame, L. Bois, P. L'Haridon, P. Lentz, J. B-Nagy, P. Grange, *submitted*.
40. B. Robertson, E. Kostiner, J. Solid State Chem., 4, 29, 1972.
41. N. Blangenois, M. Florea, P. Grange, R. Prada Silvy, S.P. Chenakin, J.-M. Bastin, N. Kruse, B.P. Barbero, L. Cadús, Appl. Catal. A, 263, 163, 2004.
42. I. Kiricsi, A. Molnar (Guest eds.), Appl. Catal. A, 256, pp. 1–137, 2003.
43. G. Mestl, H. Knözinger, R. Schlögl, Appl. Catal. A, 210, 13, 2001.
44. N. Dimitriatos, J.C. Vedrine, Appl. Catal. A, 256, 251, 2003.
45. P.A. Jalil, M. Faiz, N. Tabet, N.M. Hamdam, Z. Hussain, J. Catal., 217, 292, 2003.
46. P.A. Jalil, N. Tabet, M. Faiz, N.M. Hamdam, Z. Hussain, Appl. Catal. A, 257, 1, 2004.
47. G. Baronetti, H. Thomas , C.A. Querini, Appl. Catal. A, 217, 131, 2001.
48. A. Lapkin, Catal. Today, 81, 611, 2003.
49. L.R. Pizzio, C.V. Cáceres, M.N. Blanco, in *Scientific bases for the preparation of heterogeneous catalyst*, E. Gaigneaux, D.E. De Vos, P. Grange, P.A. Jacobs, J.A. Martens, P. Ruiz, G. Poncelet (Eds.), (Series Surface Science and Catalysis No. 143), Elsevier, Amsterdam, p. 731, 2002.
50. G. Zhou, T. Cheng, X. Guo, W. Li, Y. Bi, K. Zhen, Chinese J. Catal., 25, 189, 2004.
51. C. Copéret, M. Chabanas, R. Petroff Saint-Arroman, J.M. Basset, Angew. Chem. Int. Ed., 42, 156, 2003.

Index